Engineering and Health in Compressed Air Work

RELATED BOOKS FROM E & FN SPON

Buried Structures: Static and Dynamic Strength
P.S. Bulson

Engineering the Channel Tunnel
Edited by C. Kirkland

High Altitude Medicine and Physiology
M.P. Ward, J.S. Milledge and J.B. West

The Human Side of Safety
E. McKenna and I. Glendon

Pile Design and Construction Practice
M.R. Tomlinson

Pipejacking and Microtunnelling
J.C. Thomson

Risk Analysis for Process Plant, Pipelines and Transport
Edited by J.R. Taylor

Safety Analysis: Principles and Practice in Occupational Safety
L. Harms-Ringdahl

Soil-Structure Interaction: Numerical Analysis and Modelling
Edited by J.W. Bull

The Stability of Slopes
E.N. Bromhead

Tunnelling '94
The Institution of Mining and Metallurgy

Tunnelling: Design, Stability and Construction
B.N. Whittaker and R.C. Firth

Underground Excavations in Rock
E. Hoek and E.T. Brown

Underwater Inspection
M.R. Bayliss, D. Short and M. Bax

Engineering and Health in Compressed Air Work

Proceedings of the International Conference, Oxford, September 1992

Edited by
F.M. Jardine and R.I. McCallum

CRC Press
Taylor & Francis Group
Boca Raton London New York

CRC Press is an imprint of the
Taylor & Francis Group, an **informa** business
A SPON PRESS BOOK

CRC Press
Taylor & Francis Group
6000 Broken Sound Parkway NW, Suite 300
Boca Raton, FL 33487-2742

First issued in paperback 2019

ISBN-13: 978-0-419-18460-7 (hbk)
ISBN-13: 978-0-367-40195-5 (pbk)

A catalogue record for this book is available from the British Library

Visit the Taylor & Francis Web site at
http://www.taylorandfrancis.com

and the CRC Press Web site at
http://www.crcpress.com

Contents

Preface

Engineers will continue to adopt and adapt compressed air techniques for tunnelling and caisson projects for many years to come. In doing so they will seek and rely upon the knowledge and advice of medical experts and health and safety professionals. This, the main message from the conference of which this volume is the proceedings, re-emphasises the collaboration between civil engineering, medicine and occupational health that started to develop a century ago and which has strengthened ever since.

For more than thirty years, CIRIA's purpose has been to carry out collaborative projects that bring together different interests and disciplines and lead to better, more efficient and safer construction. Over most of those years, CIRIA provided a collaborative mechanism for improvements in the UK in relation to working in compressed air. When, therefore, the Compressed Air Working Group of the Medical Research Council's Decompression Sickness Panel put forward the proposal for an international conference, it was appropriate that CIRIA should take on its organisation, not alone, but with strong support from the Health and Safety Executive, the British Tunnelling Society and the MRC Decompression Sickness Panel. That support was given in many ways, particularly in the time and advice from individuals in those organisations. The conference itself received further endorsement as one of the events in the European Year of Safety, Hygiene and Health Protection at Work.

The outcome of the conference was a reaffirmation of the value of compressed air methods in civil engineering construction and of how they can be extended to new applications. As such, the conference provided a benchmark for the revision of regulations. Specific matters highlighted by the papers and discussions are:

Engineering in compressed air

- The flexibility, economy and good-health record of work at below 1 bar
- The versatility of the method for shaft sinking and in support of TBMs
- Positive control when dewatering and groundwater exclusion are not viable options
- Guidance on predicting air losses
- Scope for use of oxygen in decompression and therapy
- Decanting probably needing special permission
- Moves towards European standardisation of air locks and TBM chambers
- Scope for adaption of hyperbaric methods for deeper, higher pressure civil engineering work.

Health in compressed air

- Importance of careful medical selection of workers
- Improved control and integration of the medical and engineering management systems
- Beneficial use of oxygen for decompression and therapy
- Improved, descriptive classification of decompression illnesses
- Insights into the causation of bone necrosis
- Improved methods for comparing decompression sickness incidence
- Medical research findings on physiological effects
- The need for more reliable reporting of decompression illnesses.

Compressed air regulations

- Urgent need for revision in UK and elsewhere
- Flexibility needed in their framing to permit adjustment to new developments.

The Proceedings of the conference are presented in the following order: following the opening address by Viscount Ullswater, the three invited keynote lectures are given in Part One; the papers to the individual technical sessions followed by the reports of the relevant discussion sessions are contained in Parts Two to Nine. Part Ten is the report of the open discussion with which the conference concluded. At the end of the volume, are a glossary of terms, with abbreviations and units of pressure used in the conference and the list of delegates who attended.

Note about the papers

All of the papers given in the technical sessions result from rigorous peer review of drafts by two, and often three, referees from disciplines appropriate to the subjects of the paper. Authors responded to the refereeing comments in submitting their papers. The papers in a pre-print volume were available to the delegates before or, for a few late submissions, at the start of the conference. Except for those few who were unable to attend, lead authors presented main points of their papers in the sessions prior to discussion by the delegates. Subsequent to the conference, all the papers have been edited for typographical errors and to format.

Note about the reports of the discussions

The reports of the discussions are based on:

1. Written records made by the contributors to the discussion following the session.
2. Written replies of authors in response to questions.
3. Notes taken during the discussions by persons (usually two) asked to record the subjects covered and who took part.
4. In the case of the final session, which was entirely discussion, the report is based on notes taken by four persons.

While the purpose of these reports of the discussions is to show the coverage of the subjects discussed and who took part, readers are asked to recognise, first, that the discussions were informal and, secondly, that the reports are inevitably subjective and abbreviated. In what were often vigorous, rapid exchanges not subsequently recast by the participants into formal expression in writing, any record of the discussion will be imperfect, and would remain so even after several iterations of consultation with the contributors. If the questions, answers and comments attributed to individuals misrepresent what they actually said – whether by omission, mishearing or misunderstanding – the editors ask for their forbearance.

Acknowledgements

CIRIA gratefully acknowledges the help and support of the following individuals and organisations:

- Health and Safety Executive
- Committee and Members of the British Tunnelling Society
- MRC Decompression Sickness Panel and the Members of the Compressed Air Working Group
- Organising Committee of the Commission of the European Communities for the European Year of Safety, Hygiene and Health Protection at Work.

Throughout the planning and preparatory work for the conference, the organisers were guided by the advice of the Conference Advisory Committee whose members were:

Mr I H Elliott (Chairman)	Mott MacDonald Group
Mr J Buchanan	Consultant
Dr H V Hempleman	Compressed Air Working Group
Mr F M Jardine	CIRIA
Mr R F Jennion	Kilroe Civil Engineering Ltd
Dr J D King	London Hyperbaric Medical Service
Mr D R Lamont	Health and Safety Executive
Professor R I McCallum	Institute of Occupational Medicine, Edinburgh

Some forty persons acted as referees for the papers; many of the authors expressed to the organisers their thanks to the referees for the comments on their papers, and CIRIA and the editors express our gratitude to them too.

Within CIRIA, while several people contributed to the conference, two in particular deserve special mention. Louise Harrison took over the administrative arrangements, gently but effectively creating order from masses of detail, and made delegates to the conference feel welcome: it is due to Louise's work that the event was so well run. To Sarah Macpherson we and the authors owe special thanks for her typing or retyping, word-processing and formatting these proceedings. Each page reflects the way the care and quality of her work adds value to the work of others. From its inception to this final document, Sarah's contribution to the conference has been continuous – as, indeed, it is on many other CIRIA projects. It is good to be able to record here our thanks and the regard and affection in which she is held.

As much as anyone, the conference owed its success to Jim Buchanan's enthusiasm and drive. His energy kept us to programme and his attention to detail kept us to the mark on quality. Jim put in a great deal of work behind the scene.

Finally, we were well looked after at St Catherine's College, Oxford. The service we received and facilities of the college helped to create the right environment for what was an enjoyable as well as technically successful conference.

The Conference sponsors

CIRIA, the Construction Industry Research and Information Association, is an independent, private sector, non-profit distributing association of organisations concerned with construction which exists to carry out research and to provide information of benefit to members and the industry at large. Founded more than 30 years ago by industry and Government, CIRIA focuses on providing best practice guidance to professionals that is authoritative, convenient to use and relevant to their needs. CIRIA was the medium for the preparation and publication of *Medical Code of Practice for Work in Compressed Air* in 1973, i.e. Report 44, now in its third edition.

The British Tunnelling Society (BTS) is the UK's professional forum for engineers involved with tunnels and underground construction, and is the British section of the International Tunnelling Association. As well as its regular meetings at the Institution of Civil Engineers, of which it is an associated society, BTS also sponsors professional development courses on tunnelling and conferences which promote and improve tunnelling practice and safety.

The Decompression Sickness Panel of the Medical Research Council (MRC) was set up in 1957 to study the occupational health problems of workers exposed to imposed changes of pressure in their working environment. The Panel led research on the decompression illnesses, bone necrosis and on the need for improved decompression procedures in compressed air work. A registry of medical records, radiographs and hyperbaric exposure details of UK compressed air workers (and, later, divers) was set up in 1964 in Newcastle upon Tyne under the auspices of the Panel. Since 1986, the Panel's Compressed Air Working Group has continued to review new knowledge about the effects of compressed air work on health, to stimulate research, and to be a source of advice on healthier working practices. The initial proposal for this conference was made by the Group.

The Health and Safety Executive (HSE) is the UK governmental agency set up through the Health and Safety Commission to enforce and administer the *Health and Safety at Work, etc. Act 1974* and the regulations made under the Act. HSE also provides and publishes advice on legal health and safety standards, technical topics and working practices. Commissioning and undertaking research on health and safety matters, HSE supported the Decompression Sickness Central Registry and analysis of the exposure records.

Opening address

Viscount Ullswater
Parliamentary Under Secretary of State, Department of Employment

Ladies and Gentlemen, first of all may I say how delighted I am to be here this morning in my role as an Employment Minister to open this international conference on work in compressed air. I am reliably informed that this is the first time since 1965 that such a conference has been convened, to provide the opportunity for engineers, scientists, doctors, health and safety practitioners and lawyers with a common interest in the health and safety of compressed air workers to discuss common problems and common experiences and I would like to welcome delegates to the conference and, in particular, I extend a warm welcome to those from other countries.

This morning, inevitably, I will be talking about developments that are centred in Europe, but I am sure that much is happening beyond our European borders. Those of you from outside Europe will have much to share with your European colleagues, and I hope, will learn something of the developments here.

The timing of this conference is particularly opportune, coinciding as it does with the European Year of Health and Safety and with the UK Presidency of the Council of Ministers. Our objectives for both are to demonstrate that we take health and safety very seriously indeed. In particular we aim to concentrate on the effective implementation and enforcement of both existing and new legislation.

With 25 new European Directives to be implemented in 1992/93, the requirements of legislation will be a major theme of your discussions, and one I shall focus on this morning. The events organised to mark the Year offer important opportunities to make practical sense of the substantial body of community measures in the field of health and safety.

More than this, the Year should act as a catalyst to put the subject of health and safety more firmly on everyone's agenda. We need to get the message across at every level about the vital importance of good standards of health and safety at work.

Health and safety cannot be divorced from the effective management of any operation. And therefore we have a three-pronged message:

1. Safe management is efficient management

Engineering and Health in Compressed Air Work Edited by F.M. Jardine and R.I. McCallum. Published in 1994 by E & FN Spon. ISBN 0 419 18460 0

2. Safety pays – and finally
3. Good health is good business.

All of this brings me to the new body of legislation framed to bring this message home. The *Management of Health and Safety at Work Regulations 1992* which implement the Framework Directive on Health and Safety at Work in the UK, are central to the development of good safety management. These regulations come into force on 1 January 1993 and will make more explicit the duties employers already have under the general requirements of the *Health and Safety at Work Act*. Similar legislation will be coming into force in the other member states.

Although these regulations, and the EC legislation they implement, apply to all workplaces, they have a particular relevance to working in hazardous areas or with hazardous processes, such as compressed air.

The main duties imposed on employers, and in some cases the self-employed, will be:

1. To assess the risks to employees and anyone else who might be affected by their work activities.

2. To introduce arrangements for the planning, organisation, control, monitoring and review of any preventive and protective measures adopted as a result of the findings of the risk assessment.

A suitable and sufficient risk assessment will reflect what it is reasonable to expect an employer to know about the hazards in the workplace. The purpose of risk assessment is to help the employer or self-employed person to determine what measures they should take to comply with their legal duties to secure effective health and safety protection.

During the course of their work, many employers already carry out risk assessments on a day-to-day basis without, perhaps, realising this is what they are doing. They will note changes in working practice, recognise faults as they develop, and take whatever steps are needed to correct them. The requirement in the new regulations to assess risks systematises this process.

Other Directives extend duties to the design and planning stage of an operation, as well as its execution. The *Temporary or Mobile Construction Sites Directive* will be of particular relevance to work in compressed air. The Directive has to be implemented in member states by 31 December 1993, and its object is to reduce accidents and ill health in construction by placing appropriate new duties and responsibilities on clients, designers and contractors during design, planning and site operation.

Changes in legislative requirements are not only at European level. The Health and Safety Executive is currently revising the *Work in Compressed Air Special Regulations*. Whilst the revision has been planned for some time, HSE will take the opportunity to bring the new Regulations into line with both the latest scientific and medical knowledge available and the new European Directives.

All of these legislative changes will act to bring beneficial change to one of Europe's most hazardous industries. As many of you will know, the construction industry has been identified as one of the high risk sectors for special attention during the European Year. On the basis of accidents alone it is no surprise that the EC has shown particular concern for this sector. Construction provides only about 8% of total employment in the EC and yet accounts for 16% of notified accidents. In the UK, France, Germany and Spain combined there are around three deaths in this sector every day.

Whilst these figures are disturbing enough, it must be borne in mind that there is considerable under-reporting of non-fatal injuries. Research shows that in the UK, only about one third of reportable non-fatal injuries are actually notified to Health and Safety Executive. None of this begins to take account of the longer term impact that various forms of construction work may have on worker's health.

Tunnelling provides us with a perfect example of the risks to health and safety. It is, of course, vital to the infrastructure of a country and it plays an essential part in any country's construction industry. Compressed air tunnelling techniques present particular health and safety problems, but in recent years, advances in technology have reduced the need for workers to be exposed to compressed air. For example, modern sophisticated tunnel boring machines have reduced the need for workers to be exposed to compressed air except for carrying out maintenance to the cutting head. But concern about the health effects which can arise from work in compressed air remain and I am sure you are all aware of these and the research which is continuing in order to find ways to overcome them. The Medical Research Council Decompression Sickness Panel and its Compressed Air Working Group have played a significant role alongside others internationally in the development of our understanding of these problems.

In addition to health hazards, compressed air tunnelling poses particular safety problems. Rapid emergency escape from the workplace is not permitted without special provision, because of the time needed to decompress. This particularly affects the removal of workers who are injured and require medical attention. It also makes it more difficult to escape from fire, the risk of which is increased by the pressurised atmosphere.

Of course, the use of compressed air is not confined to the construction industry. Many of you here today are from the diving industry, an essential adjunct to the offshore sector. This too is an area where legislative change – at both EC and UK level – is underway to improve safety management. We expect the EC to agree a Directive on diving and a review of the UK Diving Regulations is currently underway with a view to further improving standards, particularly with regard to onshore diving operations, where there have been deaths in recent years.

No matter the sector in which compressed air is used, the need for assessment and re-appraisal is constant. This conference allows you the opportunity to do both, precisely because it brings together specialists from such a wide range of differing backgrounds to share experience and knowledge of an important aspect of health and safety.

In closing Ladies and Gentlemen, we in the UK are aiming for the European Year of Health and Safety to have a practical and lasting effect on health and safety at work. I therefore urge you, in the spirit of the European Year, not to lose sight of how the complex and technical issues which you will be discussing can help you to find and manage practical solutions to engineering problems: solutions which are safe, healthy and efficient.

I would like to wish you an enjoyable and thought-providing conference.

Thank you for listening.

Part One
Keynote Lectures

1 An engineer's perspective on legislative changes

J.M. Anderson
Health and Safety Executive

1 Introduction

I was pleased to receive an invitation some time ago from the Chairman of the organising committee of this Conference to deliver the first Keynote lecture at this important event. I had never been asked before to give any sort of talk with the general heading of 'keynote' but I was comforted by the thought that no title was required to me there and then. Another plus point was that whatever had to be in writing from me could wait until after the conference which, at that time, seemed a comforting distance away.

In the intervening period, I made up my mind to share some thoughts and opinions with you about the nature of legislation and how it was changed over the years – and in particular how it is about to change again in the European context. A keynote address on the subject of occupational health and safety law in this the European Year of Health and Safety seems particularly appropriate.

Everyone knows that dangerous human activities are best subject to some forms of rules – however loosely they may be defined – and formal statutory controls are only one part of the 'system' that society puts in place. There is a whole technical 'infrastructure' that is also in place in developed countries – by this I mean non-statutory codes of practice, national standards, rules prepared by contractors' associations, company rules prepared by individual employers, or even just site specific rules for the very local situation. Nevertheless the existence of statutory controls – when and how they come into being and how they change and overlap – is of the first importance for all practitioners in any particularly dangerous activity. At the present time, more changes are underway, and it now seems opportune to spend a little time reviewing the past, considering where we are at present, and looking towards the immediate future.

I quite deliberately put a personal dimension into the title. My background is as a civil engineer with employment as an engineering trainee under agreement and employment spells with a national contractor, two firms of consulting/design engineers, and a major client organisation (the Central Electricity Generating Board). One of my first jobs was on a hard rock tunnel in the highlands of Scotland as a

Engineering and Health in Compressed Air Work Edited by F.M. Jardine and R.I. McCallum. Published in 1994 by E & FN Spon. ISBN 0 419 18460 0

student. Pairs of students used to work at one end of the tunnel while drilling and blasting proceeded at the other end. I vividly remember getting severe headaches by the end of most days, headaches which I never had before, and thankfully have never had since. It was not until many years later (after I had joined the then Factory Inspectorate) that I discovered that I was, in fact, being poisoned at work.

Nowadays, I know quite a lot more about occupational health and safety at work in the construction industry.

What I am to say on legislation is my perspective over these 22 years as a civil engineer, as a civil servant in the central government service contributing to policy matters, and as a primary enforcement inspector for compressed air tunnelling matters in the United Kingdom for over 10 years.

2 Legislation

Reasons for the initiation or amendment of safety legislation in the United Kingdom (and I suspect that the same is true of other countries) come under four main headings:

1. Actions by Members of Parliament concerned with the social conditions of persons at work and a desire to improve these conditions by means of regulation and enforcement.
2. As a result of disasters which have demonstrated the need for more detailed legislation to control more specific health and safety issues. Here I would quote two examples. The dreadful Aberfan disaster, where there was a slide of a tip of colliery waste into a school in South Wales, was followed by the present Mines and Quarries Tips Regulations. Secondly, there was the Piper Alpha disaster in the North Sea and Lord Cullen's subsequent report. This has led to new regulations on the presentation of formal structured safety cases for all UK off-shore structures.
3. New and revised legislation proposed by the government department itself, through the mechanism of the *Health and Safety at Work Act etc.* 1974.
4. More recently, to put in place, by means of UK legislation, the Directives of the European Commission.

All four of these processes are open to public input and scrutiny. Making new legislation, and I can only speak of the health and safety variety, is a complicated and tricky business, and clearly, at the end of the day the very last thing everybody wants is 'bad' law.

2.1 UK Legislation

The very first attempt to control working conditions in factories – came about with the *Health and Morals of Apprentices Act* of 1802 which was precipitated by revelations of the working conditions of apprentices shipped from the south to work in the mills of the north. Two Visitors were appointed to enforce the Act – a magistrate and a clergyman. It is said that 'where these visits were paid at all they seem to have been largely worthless owing to the ignorance or complacency of the Visitors' (HMSO, 1980). It is an interesting thought that as long ago as 1802 two criteria had emerged as essential qualities for enforcement persons!

The true ancestry of the present-day legislation and enforcement department was the 1833 Act to 'Regulate the Labour of Children and Young Persons in the Mills and Factories of the United Kingdom'. With this Act there were to be appointed 'by warrant under his sign manual during His Majesty's pleasure four persons to be Inspectors of Factories'. Each was paid £1,000 pa and provided with a horse.

£1,000 pa was a lot of money in 1833, and these persons were clearly to be seen to be of considerable importance.

At the beginning, each of the four was of equal status and had a geographical territory. Minutes of their half-yearly meetings are still extant and, I am told, a great deal of their time was spent on the interpretation of the Act, the difficulties of administration, and their travel and subsistence expenses. The title of Chief Inspector of Factories was created in 1878 and exists today. The first Medical Inspector of Factories was appointed in 1898 but he was preceded by a Dr Robert Baker who was on the staff of the Inspectorate in 1834, one year after it all started.

Other health and safety legislation followed and eventually we find ourselves between the wars with the *Factories Act 1937.* This is the Act which permitted the introduction of the *Work in Compressed Air Special Regulations 1958*. In the preparation of notes for this address I discovered from another researcher that a Departmental file from the 1950s still existed relating to the preparation of these first regulations on compressed air (Buchanan, 1992). I have recovered this file from the Public Record Office at Kew. (As an aside, these special regulations, the detailed consideration of which dates back to 1953, were originally to apply to 'Diving Operations and Work in Compressed Air'.)

The regulations were based on known medical and engineering accidents, risks and hazards. There was a 1936 Report prepared by the Institution of Civil Engineers in London containing guidance for engineers and contractors engaged in compressed air caisson and tunnelling work (ICE, 1936). They are in the nature of prescriptive legislation, laying down minimum standards in the provision of plant, etc.

The story of how the decompression tables in the 1958 Regulations became unacceptable to industry practitioners and how they were in effect replaced by the Blackpool Tables from the CIRIA Report 44 (CIRIA, 1973) is well known to most of you present. Safety provisions for tunnelling work in general are covered by the *Construction (General Provisions) Regulations, 1961*.

What I would like to do is move on to the new legislative situation with the coming of the *Health and Safety at Work Act* 1974. Lord Robens had been appointed by the then Minister to look into aspects of health and safety at work, which although it encompassed five government departments, with many codes and regulations, still left at least five million employees without statutory protection.

Several basic changes were made:

1. A change from the definition of 'premises' and what you do at specific places of work, to the concept of 'persons at work'.
2. A statutory obligation for 'the provision and maintenance of plant and systems of work that are, so far as reasonably practicable, safe and without risks to health'.

The key words here, that I got very excited about at the time, were **safe systems of work, so far as reasonably practicable, and without risks to health**. So we now had the 1958 Regulations and a new layer of more general conceptual safety legislation. This new Act was a great help to specialist enforcement inspectors like myself. Quite often one had to be concerned not just with individual nuts and bolts as specified in the 1958 Regulations, but also with the wider **system** of work.

The 1958 Regulations and the 1974 Act co-exist together today.

As part of the considerations of both the framework and the content of the UK revision to the 1958 Regulations Donald Lamont and I undertook some research into the form and detail of legislation in the field of compressed air work that existed elsewhere in the world and, especially, in Europe. The potential hazards associated with compressed air work are well understood, and the detail of the legislation would logically be similar wherever one went in the world. We made as direct a comparison as we thought possible and presented a paper to the Tunnelling '91 Conference in London (Anderson and Lamont, 1991). Clearly we cannot claim that the information contained in this paper is still up-to-date, but some of you may find the information we gathered to be of some interest and value. We would welcome comments and updates from delegates here for this Conference. A keynote address of this kind is not the place to look into the detail of this paper, but I cannot help but mentioning in passing my own particular favourite set of regulations, and that is the set from Canada. At the same time I am filled with admiration for the detail of the regulations from the former USSR.

2.2 EC Directives and new UK legislation

In 1989 the European Commission passed a Directive known as the 'Framework Directive' on occupational health and safety. This is but one of a series of Directives binding on the UK government, the provisions of which (if they are not already part of our national legislation), have to be made into UK law by means of a new set of UK Regulations. One of the central requirements of this Directive is for the employer to make a structured assessment of the risks to which his employees are exposed while they are at work. In principle, making 'risk assessments', is nothing new – we all do it all the time – instinctively or otherwise. What is new is that it is now part of our legislative framework (both here in the UK and in Europe) from 1 January 1993: these new regulations in the UK are the *Management of Health and Safety at Work Regulations 1992*. This 'hazard identification and risk assessment' approach is part of the level playing field throughout the EC on workplace safety legislation. You have in your proceedings of this Conference the paper by Wright and Marks (1994) which covers a great deal of the background.

This new legislation really does go back to basics, and it is worth looking at one of the main regulations in detail, i.e.

> 'Every employer shall make a suitable and sufficient assessment of:
> (a) the risks to the health and safety of his employees to which they are exposed whilst they are at work; and
> (b) the risks to the health and safety of persons not in his employment arising out of or in connection with the conduct by him of his undertaking,
>
> for the purposes of identifying the measures he needs to take to comply with the requirements and prohibitions imposed upon him or under the relevant statutory provisions'.

As a reminder of the definitions:

Hazard is something with the potential to cause harm.
Risk expresses the likelihood that the harm from a particular hazard is realised, and so reflects both the likelihood that harm will occur and its severity.

The new legislative approach calls for a systematic appraisal of the fundamental nature of hazards. In the context of compressed air tunnelling this includes ground collapse, noise, fire, as well as the pressurised atmosphere. If fire is a hazard, what is the *risk* of fire? Is the risk high or low? What can burn? What can cause ignition? What could be the likely consequences? How often is it likely to happen? How can we take specific hazards out of the system? The last step is to deal with the residual significant hazards by 'health and safety arrangements', i.e. not just by, say, a method statement but in a much wider context of 'planning, organisation, control, monitoring and review', i.e.

HEALTH AND SAFETY ARRANGEMENTS (Regulation 4(i))

Every employer shall make and give effect to such arrangements as are appropriate, having regard to the nature of his activities and the size of his undertaking for the effective planning, organisation, control, monitoring and review of the preventative and protective measures.

I suggest that this new formal, structured and rigorous risk assessment and management approach could be the right approach to the question of the possible introduction of oxygen into the routine decompression of tunnelling workers in the UK. In his paper to this conference, Dr Kindwall (1994) from Milwaukee states:

'The Health and Safety Executive in the UK has been against change in the present tunnelling decompression procedures until there is an assurance that "no new risk is added".'

I certainly agree with Dr Kindwall; the HSE is in the business of reducing risks to employees and is not very interested in adding new ones!

I now look to the immediate future.

Council Directive 92/57/EEC, on the implementation of minimum safety and health requirements at temporary or mobile construction sites, was passed by the EC in June 1992. It is a 'daughter' Directive of the Framework Directive, and must be read in conjunction with it. This will give rise to new UK legislation, i.e. *Construction (Design and Management) Regulations 199_*, which have to be in place by 31 December 1993. To this end, the consultative document was issued in October 1992, with the period of consultation expected to be to the end of January 1993. The thrust of the new regulations is about the management of the construction process from client's inception through the design and construction stages to the use stage of the project.

The European Commission last year published proposals for the creation of a European Agency for safety and health at work. I believe that this is a very interesting and important development. Among stated aims are:

1. To set up and coordinate a network for the purpose of exchanging information and experience.
2. To organise training courses for experts and exchanges between experts in different Member States.
3. To promote co-operation in monitoring the application of measures relating to health and safety.

It is no secret that both the Health and Safety Commission and Executive would like to see this agency located in the United Kingdom.

3 Concluding remarks

It is salutary to note that the incidence rates of accidents in UK construction are not decreasing (Figure 1).

In regard to risk and hazard issues, the following summary, expressed in general terms, should be applied.

- If possible avoid the risk altogether
- Combat risks at source
- If possible adapt work to the individual
- Take advantage of technological progress
- Measures need to be part of a coherent policy/approach
- Where possible adopt measures to protect the whole workplace

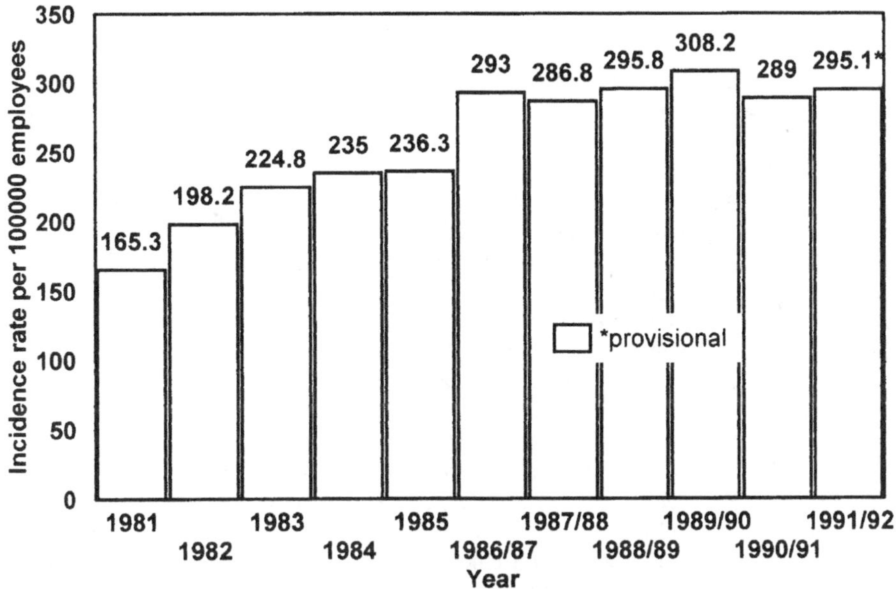

Fig.1. Fatal and major injury incidence rates for employees in construction in UK

- Make sure everyone understands what they need to do
- Promote strong organisational health and safety culture.

Finally, I was a little stung by one of the authors in one of the papers who wrote:

> 'As the 21st century approaches, it is clear that the construction industry and its regulators must abandon trying to solve high-tech problems using 19th century tradition'.

I am certainly not speaking for the industry, but I do know quite a lot about what happens within the major regulator of health and safety matters in the UK; and I think that this is unfair. I hope I have demonstrated that although legislation has indeed its historical roots in the 19th century, legislation does move with the times, and I firmly believe that the concepts being reinforced in our statute at this time are appropriate for the addressing any high-tech problem the 21st century may throw at us.

References

Work in Compressed Air Special Regulations 1958 (SI 1958 No 61)
Construction (General Provisions) Regulations 1961 Part IV: Excavations shafts and tunnels
Health and Safety at Work Act etc. 1974
Management of Health and Safety at Work Regulations 1992 (SI 1992 No 2051) and *Approved Code of Practice*, HMSO, London, 1992
Medical Code of Practice for Work in Compressed Air. CIRIA Report 44, first published 1973, 3rd Edition 1982
A brief history of the Factory Inspectorate, HMSO, London, 1980
Anderson, J.M. and Lamont, D.R. (1991) A comparison of international legislation concerned with tunnelling in compressed air, Proc. *Tunnelling '91*, Institution of Mining and Metallurgy, London
Buchanan, J. (1992) personal communication
European Commission (1991) EC Agency document COM(90)564 on proposals for establishing a European Agency for Safety and Health at Work, Brussels
European Commission (1992) Council Directive 92/57/EEC - on the implementation of minimum safety and health requirements at temporary or mobile construction sites, Brussels, June, 1992
Health and Safety Commission (1992) *Consultative document on proposals for Construction (Design and Management) Regulations*, Health and Safety Commission, London
Institution of Civil Engineers (1936) *Report of the Committee on regulations for work carried out under compressed air.* January 1936, Institution of Civil Engineers, London
Kindwall, E.P. (1994) Optimum schedules for caisson decompression. In: *Engineering and Health in Compressed Air Work* (eds. Jardine, F.M. and McCallum, R.I.) E&FN Spon, London
Wright, F.B. and Marks, L. (1994) Some medico-legal aspects of modern work in compressed air. In: *Engineering and health in compressed air work*, (Jardine, F.M. and McCallum, R.I. eds.), E&FN Spon, London

Discussion following Mr Anderson's lecture

The Conference Chairman, Mr I H Elliott, thanked Mr Anderson for so relevant and stimulating a lecture. He noted the contribution that CIRIA Report 44 had given to enforcement of good health and safety practice, suggesting that its success owed much to its having been produced by industry. Mr Elliott asked how regulations are formulated elsewhere in Europe and what approach is being used in UK.
Mr Anderson. Each country is having to revise its current regulations to fit with the Framework Directive. Compressed air working is of such a special nature, the end products of the revision have to be defined carefully. When drafted by HSE there will be a period for industry consultation. It is likely that they will be quite

different in shape from the 1958 Special Regulations. The regulations, in themselves, will be kept to a minimum, but with an approved code of practice and possibly guidance notes.

Professor D H Elliott. The term 'likelihood' was used in the context of risk assessment. For the benefit of those who might associate this term with a mathematical approach, could you please confirm that this word, in contrast to 'probability', has been used to imply a qualitative, and not necessarily quantitative, process of risk assessment?

Mr Anderson. Yes, generally it will be qualitative. There will be an approved code of practice to set out what will be required.

Dr G C Voigt. To what extent are or will governmental regulatory agencies in the UK and other EC countries be able to provide information and consultation services to the industries they regulate? I ask this question because it is my experience that those agencies viewed as accessible and helpful by the private sector are less likely to be regarded as threatening in their role as enforcers.

Mr Anderson. A whole range of supplemental guidance is being prepared, although it is recognised that it cannot be prescriptive. This approach was taken successfully for controlling risks for the Channel Tunnel works, for example.

Mr J C Le Péchon. I found extremely interesting the way HSE is taking into account directives from EEC. Being involved in drafting French legislation for hyperbaric situations at work I can appreciate that both countries are working in similar directions. Enlarging the topic from premises to persons has been a step. The French rules have enlarged the topic from professions (e.g. divers or compressed air workers) to the protection against hazards which may induce risks associated with a hyperbaric work situation whatever the reason for being under pressure. This new approach is enforced by *décret* 1990 and has been offered (September 1992) to EEC as draft for a future directive.

Dr A Evans. What happens to the health-and-safety plan and file at the end of the construction and commissioning periods?

Mr Anderson. The plan, no longer being needed, is scrapped. The file is handed to the client for use in relation to maintenance and possible subsequent repair or renovation of the works).

Dr R W Hamilton. You used the magic word 'oxygen'. Will the new HSE rules permit the use of oxygen decompression, or is it not necessary for them to do so?

Mr Anderson. It is not necessary for them to allow it specifically.

Dr E Kindwall. First, I want to say how impressed I have been with Mr Anderson's thoughtful presentation. I realise that in the past I have been impatient and critical of what I have perceived as a reluctance to change in both the American and British regulatory bodies, and perhaps if the remarks in my paper have been slightly acid, that is a reflection of my previous impatience! However, today I owe apologies to Mr Anderson, as I find nothing at all wrong with his reasoned approach to appraisal of selective risk as outline in his lecture, and I feel that none of us should have a problem working within that framework.

(Reporters: J. Buchanan, F.M. Jardine)

2 Bone necrosis

D.N. Walder
University of Newcastle upon Tyne, UK

1 Introduction

Bone necrosis is the non-specific term used to describe the radiographic appearance of a special sort of damage to bone. It has many causes one of which is related to changing the pressure of the environment, i.e. (a) exposure to compressed air, (b) diving and possibly (c) ascending in an aeroplane to altitude. The Americans call it dysbaric osteonecrosis. In the compressed air industry it was originally known as caisson disease of bone but more recently by the men as 'bone rot'.

It is important, not the least from a medical-legal point of view (at present the going rate for successful claims against the employer is around £100,000 per joint), to be aware of the numerous other conditions with which it is associated, that is the differential diagnoses. Fortunately it is not difficult for doctors to exclude the non-pressure related causes of bone necrosis by physical examination and tests.

Bone necrosis is a serious hazard; at best it leads to changes in the bone structure which are detectable on a radiograph (in the USA this is accepted as evidence of an industrial injury which can attract compensation) and at worst it leads to collapse of weight-bearing bone surfaces leading to severe disability in the form of pain and limited movement affecting one or more major joints.

Cases were first described in compressed air workers in 1910, but these were all late cases in which a great deal of secondary osteoarthritis was superimposed on the original lesions and the men had been forced to attend hospital because of the severity of their symptoms.

No attempt was made to assess the prevalence of the condition until 1957 during the construction of the 1st Dartford Tunnel when we were able, with the assistance of the Medical Research Council, to start a study of bone lesions in symptom-free compressed air workers. In 1964 a Registry was established in Newcastle upon Tyne to collect from many contracts radiographs and compressed air records not only of men who had suffered from bends but also from those who had not.

Engineering and Health in Compressed Air Work Edited by F.M. Jardine and R.I. McCallum. Published in 1994 by E & FN Spon. ISBN 0 419 18460 0

2 Identification

2.1 Radiology

Because the onset of bone necrosis is insidious and does not give rise to symptoms in the early stages, it is usually first detected by radiological surveys of the skeleton.

Interestingly only certain sites of the skeleton are usually affected: the head of the humerus at the shoulder joint and occasionally its neck, the head of the femur at the hip joint and occasionally its neck, the lower end of the shaft of the femur, and the upper end of the shaft of the tibia. Neither of the latter two sites usually spreads to involve the knee joint.

To ensure that all radiologists use a universally understood description of the various types of bone lesions which are seen in compressed air workers we have devised a classification (Table 1).

Table 1. Radiological classification of bone necrosis in compressed air workers

(A)	Juxta-Articular Lesions		
	1.	Dense areas, with intact articular cortex	
	2.	Spherical segmental opacities	
	3.	Linear opacity	
	4.	Structural failures:	
		(i)	Translucent subcortical band
		(ii)	Collapse of articular cortex
		(iii)	Sequestration of cortex
	5.	Osteoarthritis	
(B)	Head, Neck and Shaft Lesions		
	1.	Dense areas	
	2.	Irregular calcified areas	
	3.	Translucent areas and cysts	

You will note that there are 'juxta-articular' lesions which are situated immediately beneath a joint surface of a bone, and 'head, neck and shaft' lesions which are well clear of any joint surface.

Figure 1 shows a juxta-articular lesion affecting the right hip joint. As this shows, in some cases the dead bone breaks up or splits off so that the joint surface becomes distorted. When the normally smooth surface of the joint is disturbed, pain will occur and there will be limitation of the range of movement at this joint. Consequently bone necrosis can be a seriously disabling condition.

Figure 2 shows a shaft lesion, which affects the centre of the bone. It does not weaken the bone nor does it give rise to pain. In fact it does not seem to matter very much, but it does indicate that, as with a juxta-articular lesion, something has gone wrong with the safe management of the compressed air worker.

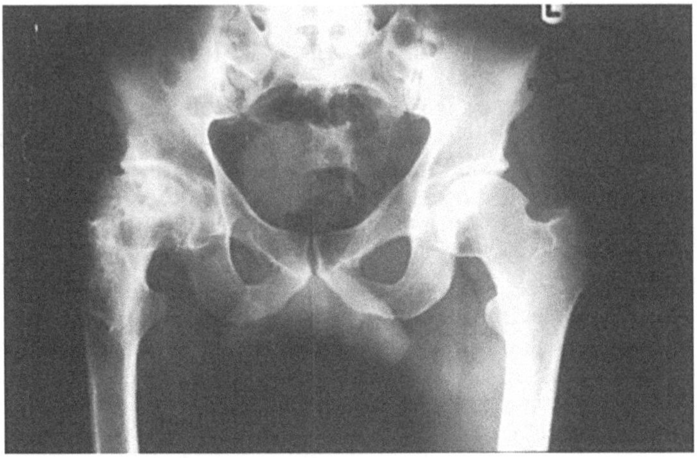

Fig.1. Radiograph showing a juxta-articular lesion of right hip joint

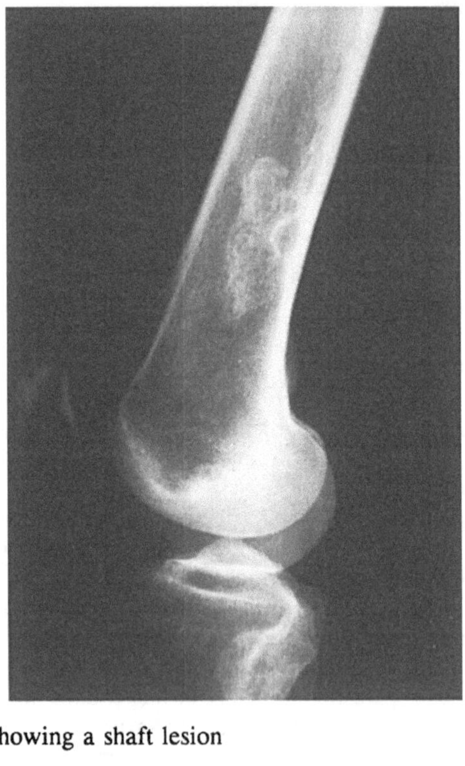

Fig.2. Radiograph showing a shaft lesion

Unfortunately, radiology is not an ideal way of studying bone necrosis lesions in compressed air workers because:

1. The changes in the radiological picture take many months to develop, making it difficult to say precisely on which exposure the damage occurred.
2. It gives no indication of when a juxta-articular lesion will go on to collapse with the consequent onset of pain and loss of movement. About 26% of juxta-articular lesions go on to collapse.
3. It gives no indication of the full extent of the damage. Figure 3 shows a lesion which on radiological examination appears only to be affecting a small volume of the head of the bone, but when examined under the microscope it was seen that the whole of the head had been affected. The body has been able to naturally repair the bone up to the dense line seen on the X-ray. Beyond that line no repair has occurred and the dead bone is liable to collapse at any time.

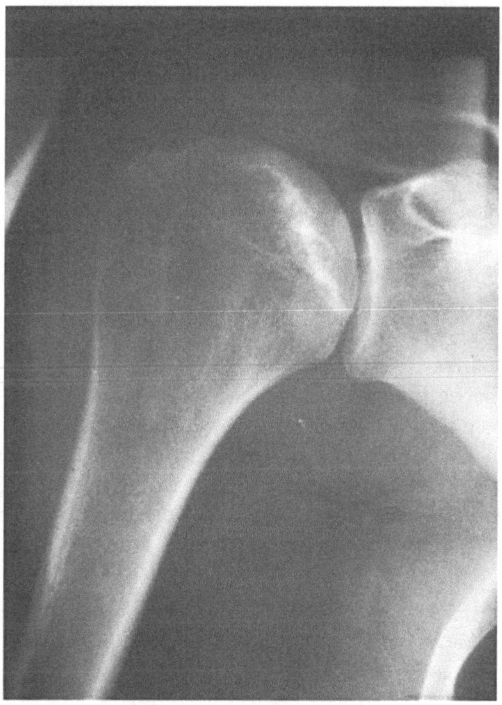

Fig.3. Radiograph showing a lesion at head of bone

For these reasons we have tried to find other tests which would be more satisfactory than radiography for:

- detecting bone damage at an early stage (that is within days or weeks rather than months of its occurrence)
- estimating accurately the likelihood of collapse of a juxta-articular lesion
- giving an accurate assessment of the full extent of a lesion.

2.2 Biochemical tests

When part of a bone and its marrow are damaged, many substances are released into the circulation; and we have looked to see if detecting a rise in the level of any of these in the blood or urine would be helpful.

Unfortunately we are forced to conclude that, at present, there is no satisfactory biochemical test which can be applied to man for the early detection of bone necrosis.

Even if a satisfactory biochemical marker could be found it would still not indicate the site or extent of the damaged bone and so we turned to look at other methods of detecting bone damage.

2.3 Radio-isotope scanning

By the use of injected radio-active bone-scanning agents such as Technecium Tc 99m-labelled diphosphonate it has proved possible to detect damaged areas of bone within a few days of their occurrence.

It is such a sensitive indicator, however, that it will show up areas of damaged bone which are so small that they heal spontaneously and never reach the stage of showing up on a radiograph or, for that matter, would ever give rise to symptoms or disability. Moreover it still involves radiation, and in any case, before it can be used routinely, much more information about the correct interpretation of bone scans is required.

2.4 Nuclear magnetic investigation (NMI)

This is an even more recently introduced technique which apparently has no ill effects. The subject is placed in a magnetic field which produces an internal resonance of hydrogen nuclei resulting in the emission of radio frequency waves which are then translated into an image of the part of the body being scanned. This technique is very sensitive in showing up bone and particularly fat changes, but here again it is too sensitive and shows up changes due to minor trauma as well as bone necrosis.

The upshot of all this is that we still depend on radiography for the diagnosis of bone necrosis.

3 How prevalent is bone necrosis?

The prevalence of bone necrosis in compressed air workers has been variously reported as being from zero to 75%. The radiographic surveys carried out by the MRC Decompression Sickness Panel in the UK on several large contracts showed that about 18% of the men had definite bone necrosis and that in 12% the lesion was next to a joint surface (juxta-articular) and therefore likely to give rise to symptoms and disability in due course.

Not all of the classically involved sites are equally affected. As can be seen in Table 2 the lower end of the femur is most frequently damaged and the shoulder joint is much more frequently affected than the hip joint.

Table 2 Distribution of definite bone lesions (977 lesions in 383 men)

Site	Number of lesions	Percent total lesions
Humeral head	230	24
Femoral head	118	12
Humeral neck	65	7
Femoral neck	32	3
Lower femur	373	38
Upper tibia	159	16
Total	977	100

Many men have more than one site affected. I know of 97 men each of whom has four sites or more involved. The condition is bilateral in as many as one third of the men affected.

4 Why does bone necrosis occur?

It is generally assumed that both bone necrosis and decompression sickness occur solely because bubbles arise in the body as a result of inadequate decompression.

Table 3 Decompression sickness and aseptic bone necrosis

Decompression sickness	Number of men with bone necrosis		
	No	Yes	Total
No	2978	52	3030
Yes	1228	147	1375
Total	4206	199	4405

$X^2 = 176.1$ $p < 0.0001$. Excluding 757 men without sufficient data.

Because bone necrosis affects the same regions of the skeleton as the bends, i.e. the limbs and not the central skeleton, and because there is a statistically significant relationship ($p < 0.0001$) between those who suffer from bone necrosis and those who have had decompression sickness it has naturally been tempting to assume that the one condition follows the other (see Table 3). However, not all those who suffer from decompression sickness go on to develop bone necrosis – only 10% go on to bone necrosis – and not all those with bone necrosis have ever suffered from the bends (25% of those with bone necrosis have never had decompression sickness). Although the sites affected by the bends correspond closely with those affected by bone necrosis no clear relationship between the specific site of an attack of the bends and the subsequent appearance of a lesion of bone necrosis has ever been established.

There are other features common to decompression sickness and bone necrosis which suggest that they may be related. For example, there is evidence that age and obesity, both traditionally associated with the susceptibility to decompression sickness, may also be important in the susceptibility to bone necrosis (see the paper by Dr Evans to this conference).

Bone necrosis, like decompression sickness, is confined to the limbs and does not affect the central skeleton. This distinction suggests that, in order to understand the mechanisms involved, attention must be focused on the differences between the two types of bones involved rather than on the tissues outside the bones, since these latter are the same at both sites, (e.g. muscles, ligaments and periosteum). All could give rise to pain when stretched by bubbles.

4.1 Blood circulation of bones

The circulatory arrangements of the bones of the central skeleton and those of the limbs are the same.

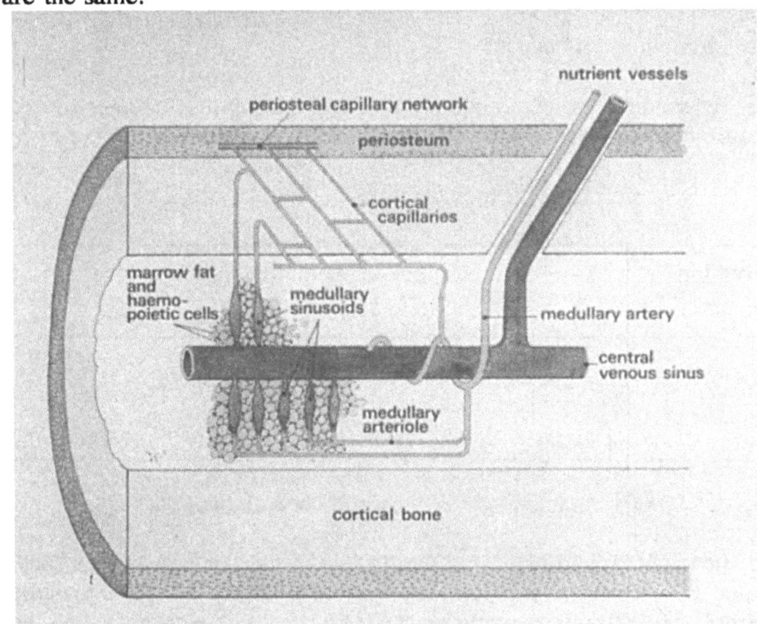

Fig.4. Scheme of blood circulation in bone

As represented in Figure 4 a nutrient artery enters the bone *via* a nutrient canal. Once inside the marrow cavity the artery gives rise to two systems essentially in parallel.

1. One system goes to supply cortical bone. Most of the vessels drain centrifugally into the periosteal capillary bed, but some turn back to traverse the marrow and enter thin-walled sinusoids which drain into a central vein.
2. The other system traverses the marrow and then enters thin-walled sinusoids to join a central sinus. The sinusoids are arranged around the central sinus like the branches of a fir tree. The central sinus is drained by veins which pass out of the bone through the nutrient canal and into the veins of the general circulation.

The bone blood vessels are supplied with pain nerve fibres and it is the distension of sinusoids which causes a dull aching pain. Such distension is the explanation for the pain of osteoarthritis which orthopaedic surgeons relieve by osteotomy which allows the sinusoids to deflate and results in the immediate abolition of the pain. It therefore seems reasonable to suggest that the pain of bends results from the distension of sinusoids by intravascular bubbles. The pain would be ill defined and ill located, i.e. it could be reported as hip, thigh or knee in the lower limb and shoulder, arm or elbow in the upper limb.

4.2 Bone marrow

But why should bubbles grow in the limb bone circulation rather than in the central skeletal bone circulation? Well, herein lies the important difference between a limb bone and a central skeletal bone. In the adult (i.e. those over 18 years old) the limb bone marrow cavity is packed with fat cells (yellow marrow). Nitrogen is very soluble in fat so that the fat cells of the yellow marrow hold a great deal of gas, whereas the central skeleton bone marrow cavity is packed with blood-forming cells (red marrow) in which nitrogen is not particularly soluble.

4.3 Bubbles

In whatever manner bubbles arise, a bubble entering a venous sinusoid of a limb bone will find itself in a richly fatty region supersaturated with gas and will consequently rapidly grow to distend the sinusoid and give rise to the pain of bends.

Bubbles similarly finding their way into the sinusoid of a central skeletal bone are unlikely to enlarge as the red marrow does not hold any quantity of nitrogen. Furthermore some developing red cells will leave the sinusoid as is their nature. No distension of the sinusoid will result and consequently no pain will be felt.

Bubbles in venous sinusoids of limbs could also account for the shower of bubbles which are detected by those using ultrasonic (Doppler) devices on compressed air workers who have exercised after being decompressed, since exercise will promote bone blood flow through the venous sinusoids and their vascular connections with the overlying attached muscles.

The suggestion therefore is that bends pain results from the distension of sinusoids by intravascular bubbles. Whole-body compression will reduce the size of the bubbles and relieve the pain of bends. Incidentally another important conclusion is that the bubbles causing decompression sickness are intravascular.

Clearly this is not the whole story. We know that not every attack of decompression sickness results in bone necrosis. Bone necrosis is related most strongly to long exposures at high pressure. There must be some other factor associated with limb bones and their fat-rich marrow.

4.4 Fat cells

Dr Lent Johnson, an American bone pathologist, also recognised that the sites where bone necrosis occurs in compressed air workers are those in which fatty marrow exists. He believed that fat cells are particularly likely to develop intracellular bubbles during decompression. Such bubbles would expand the cells which, being surrounded by rigid bone, could only expand at the expense of the blood vessels that lie between them. The blood vessels would be squashed and the blood supply to the cells in the surrounding bone cut off. This would result in the development of bone necrosis.

It is a good idea, but I have never been able to produce bubbles in fat cells as a result of decompression after an exposure to pressure. However, we wondered whether fat cells might change their volume during an exposure to pressure or on decompression due, for example, to the osmotic effects of dissolved gases, as suggested by Dr Brian Hills some years ago. We therefore carried out experiments in which we measured the size distribution of suspensions of isolated living fat cells before and after exposure to raised air pressures of more than 28 psig for periods of 4 h or more.

We found that all the fat cells increased in volume. Further investigations enabled us to show that this was due to the raised partial pressure of the oxygen in the air at pressure. Nitrogen by itself had no affect on the size of fat cells.

4.5 Hypothesis

As a consequence of these findings we have developed the following hypothesis which would account for the location of bone necrosis in specific parts of the skeleton and for the fact that the risk of developing bone necrosis is highest after long duration exposures (4 h or more) to high pressure (30 psig or more). See Table 4.

From Figure 4, which is a longitudinal section through a long bone such as the humerus or femur, it can be seen that the blood vessels in the marrow cavity of the bone are closely packed around by the fatty marrow.

To explain how bone necrosis comes about we need to realise that the marrow cavity of bones is really a bony honeycomb in which the fat cells are packed (Figure 5).

Table 4. Aseptic bone necrosis and the Blackpool Tables for 891 men in relation to their maximum working pressure and maximum working time

↑	+2 59 men -57 Prevalence 3.5%	+65 416 men -351 Prevalence 15.6%
30 psig		
↑	+0 69 men -69 Prevalence 0.0%	+6 347 men -341 Prevalence 1.7%
	→4h→	

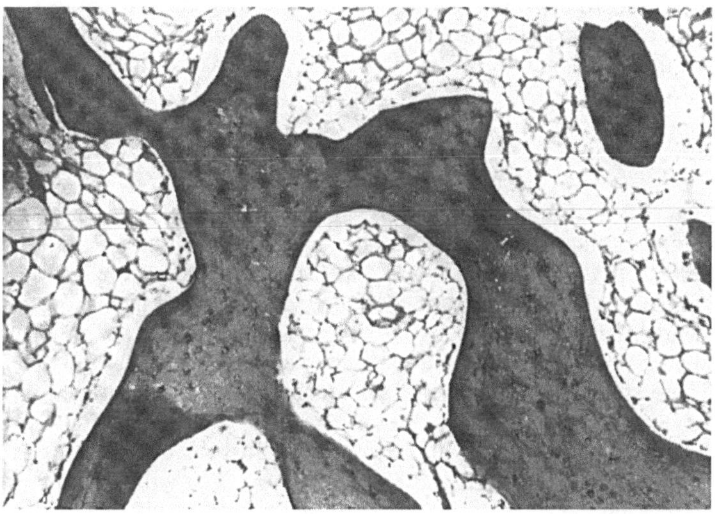

Fig.5. Marrow cavity of bone

When the marrow fat cells confined in the rigid compartments of the trabecular bone to be found in the head, neck and lining the walls of the shafts of long bones, swell as the result of a long hyperbaric exposure (i.e. a long exposure to a raised partial pressure of oxygen), they can only do so at the expense of the space occupied by the neighbouring blood vessels. These consequently become squashed and narrowed. This results not only in a reduction of the blood flow to the fat cells which consequently, on decompression, desaturate more slowly than expected, but it also results in the likelihood of any bubbles in the circulation becoming trapped in the narrowed vessels. Such bubbles, finding themselves in a high supersaturated environment, will not only persist but even grow in size. The outcome of all this is a prolonged stoppage of the circulation to the adjacent marrow and bone cells which then die.

Dead fat cells release breakdown products which stimulate new bone formation and damage the surrounding bone trabeculae thus giving rise, for instance, to one of the typical appearances seen on a radiograph of bone necrosis, the thick dense line. It is stressed that such a mechanism could not occur in the bones of the central skeleton because there the trabecular compartments are filled with blood-forming tissue (red marrow) which is not subject to swelling as a result of hyperbaric exposure.

To produce bone necrosis then, not only are bubbles from a bad decompression required, but also a predisposing factor, namely a prolonged exposure to a raised partial pressure of oxygen, to create the circumstances for the entrapment and persistence of the bubbles in the vulnerable areas of the skeleton.

5 In summary

The suggestion is therefore that (1) bends pain results from the distension of venous sinusoids by intravascular bubbles which have enlarged when they reached the gas-rich environment of the fatty marrow and that (2) bone necrosis results from the swelling of fat cells which narrows the blood vessels supplying them in their trabecular compartments. This results in the trapping of bubbles which in turn causes the necrosis of the fat cells and the subsequent release of breakdown products. These stimulate the bone changes recognised on a radiograph as caisson disease of bone.

Figure 6 shows where the bone marrow in an adult is yellow and consists of fat and where the bone marrow is red and blood-forming. You will remember that the characteristic distribution of bone necrosis is in the fat areas only.

Thus it is suggested that both bone necrosis and decompression sickness occur as a result of bubbles but the point of action is different; distension of sinusoids gives bends, trapping of bubbles gives fat necrosis which results in bone necrosis. The two conditions do not necessarily happen at the same time.

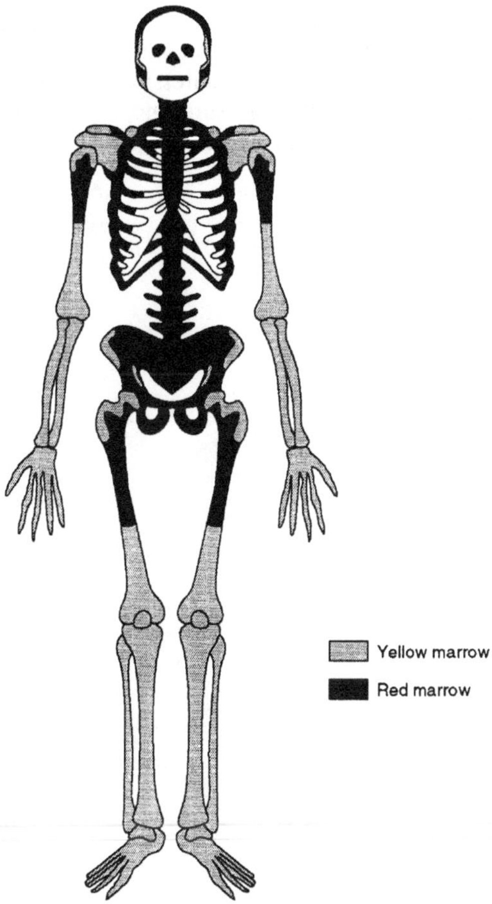

Fig.6. Yellow and red marrow in adults

The essential ingredients for decompression sickness are:

1. Bubbles
2. Fatty marrow
3. Venous sinusoids.

The essential ingredients for bone necrosis are:

1. Bubbles
2. Fatty marrow
3. Honeycomb structure of bone.

In my opinion the conclusion, therefore, is that to minimise the risk of suffering from bone necrosis, not only must the compressed air worker avoid bubble formation by submitting to an adequate decompression procedure, but he must also avoid breathing oxygen at raised partial pressure (more than 0.6 bar) for prolonged periods (more than 4 h) as occurs when working in pressures of >30 psig.

3 Decompression illness
Keynote Lecture

D.H. Elliott
Chairman, MRC Decompression Sickness Panel, UK

1 Introduction

The title 'Decompression Illness' is broad in its scope and thus represents the potential of what could be covered, but space requires that I am selective. Within this vast subject I wish to concentrate on three particular aspects. First, the definitions of the various decompression disorders need to be questioned. Consideration will then be given to clinical aspects and, as there is not time for a detailed scientific review, I will concentrate on research priorities in decompression illness. Finally, it is timely at this meeting to review the history and accomplishments of the Decompression Sickness Panel, a co-sponsor of this meeting and to look at its future.

2 The nature of decompression illness

Many medical conditions have been identified as being associated with decompression ever since man learned how to make excursions from the surface of this planet, either in flight or to depth. The first descriptions of associated illnesses were clinical and were without any understanding of the underlying pathogenesis. One of the first descriptions of decompression illness was made by Triger (1845):

> '.... Two workers, after passing 7 consecutive hours in compressed air, experienced rather keen pains in the articulations, half one hour after leaving the shaft. .... rubbing with spirits of wine soon relieved this pain in both men; they kept on working the following days.'

Generally, over the next decades, accumulated experience led to several attempts to categorise the various dysbaric conditions. Bert in 1877 quoted a classification by Dr Alphonse Gal, who described the first neurological cases among Greek sponge

Engineering and Health in Compressed Air Work Edited by F.M. Jardine and R.I. McCallum. Published in 1994 by E & FN Spon. ISBN 0 419 18460 0

divers. These were the 'diving diseases which begin suddenly:

- the fleas
- muscular pains
- pains and inflammations of the ears
- serious symptoms and paraplegia.'

From such early beginnings, which were based solely upon observed manifestations, the more recent classifications of the decompression illnesses have evolved. These reflect an improved understanding of the underlying mechanisms, and tend to be dependent on perceptions of pathogenesis, e.g.:

- middle ear barotrauma
- pulmonary barotrauma and arterial gas embolism
- decompression sickness
- dysbaric osteonecrosis.

For the purpose of case-management, particularly by personnel in the absence of a medical officer, a distinction was made by the U.S. Navy between 'mild' and 'serious' decompression sickness, so that different treatments could be applied as appropriate.

The later introduction of Type 1 and Type 2 decompression sickness was based on a similar distinction between those cases with 'limb-bends' or other minor manifestations only (Type 1), and those, with or without concurrent 'bends', that had a more serious condition (Type 2). This was used not only for guidance in treatment but was also an effective tool for the retrospective evaluation of the relative safety of the decompression tables.

The recognition of 'arterial gas embolism' (AGE) as a distinct decompression disorder was associated with submarine escape training in which an overt or concealed 'pulmonary barotrauma' is a recognised hazard. This led to a separate treatment algorithm which has been effective operationally. Based upon an understanding of the hazards of rapid ascent in the absence of a significant gas-load in the tissues, and supported by the findings at autopsy and from animal studies, the 'burst lung' terminology was introduced. This was easily correlated with the mechanistic diagnosis of arterial gas embolism which, with a single term, displaced the various descriptive diagnoses among submarine escape trainees such as hemiplegia and disorientation.

However there is now a growing realisation that there are other, equally valid, pathogenic pathways for clinical manifestations after decompression. The mechanistic labels may now be misleading and, on some occasions, actually wrong.

'Dysbarism' is certainly a term to be considered in this context. Correctly the term excludes isobaric bubble illness due to counter-diffusion, even though this illness may be used in the laboratory to simulate aspects of bubble pathology. 'Dysbarism' applies to the effects of any change of environmental pressure, both an increase and a decrease. It is a term correctly applied to compression effects and thus includes not only middle ear barotrauma but, in the broadest sense of the word, also the high-pressure neurological syndrome (HPNS). For this reason dysbarism is too broad for use in the decompression disorders though one must recognise that it is occasionally still used in this context.

The decompression disorders include traditional 'decompression sickness' (DCS) which may be considered to be that illness which arises from the presence of bubbles formed on decompression from gases that have become dissolved in the tissues of the body during a sojourn at a greater environmental pressure. There are a number of clinical syndromes which are considered to be typical of this condition, but a difficulty in diagnosis is that there can be cases which present with manifestations which are similar to those of arterial gas embolism which is assumed to be secondary to 'pulmonary barotrauma'. In the relatively uncomplicated environment of submarine escape training the clinical manifestations are predominantly neurological and do not include limb-bends and other features of 'classic decompression sickness'. Although some neurological presentations among divers with omitted decompression and submariners with no significant gas-load may be indistinguishable, an ascending paraparesis (so-called "spinal bends") is not found among the submarine escape trainees.

Current concerns about the true nature of the decompression syndrome have been highlighted by recent papers such as those concerned with:

- cerebro-vascular abnormalities with HMPAO (Adkisson *et al.*, 1989)
- retinal pigment and vascular changes (Polkinghorne *et al.*, 1988)
- the presence of neurological signs and symptoms following deep diving (Todnem *et al.*, 1989)
- unsuspected neuropathology at autopsy (Palmer *et al.*, 1987; Palmer *et al.*, 1990)
- the role of the inter-atrial septal defect (Moon *et al.*, 1989; Wilmshurst *et al.*, 1989)
- biphasic decompression illness (Neuman and Bove, 1987).

Another recently described phenomenon is that of 'yo-yo' diving: multiple shallow dives which cumulatively fall within the no-stop limit but occasionally leading to paraparetic manifestations (Douglas and Milne, 1991).

None of these fit with the traditional and tidy clinical concepts of simple 'decompression sickness' and 'arterial gas embolism'. The manifestations, where present, do not fall exclusively into either category and some alternative pathogenetic pathways seem to be invoked.

Concurrent with this realisation that there are probably more than two pathways to the diverse outward manifestations seen in clinical practice, there is also a realisation that even experienced diving doctors make these traditional diagnoses inconsistently (Smith *et al.*, 1992). This supports a proposal that diagnosis should return to being descriptive and should not make guesses about the causative underlying pathology which cannot be validated (Francis and Smith, 1991).

Other phenomena exist which also demonstrate the complexity of the decompression illness.

A degree of adaptation may arise from regular decompressions by compressed-air workers and divers seemingly permitting them to shorten their predicted decompression times with impunity. Such 'acclimatisation' is quickly lost during a break from regular daily work after which resumption of the same pattern of diving with minimal decompression may no longer be associated with immunity from decompression problems (Paton and Walder, 1954).

In contrast, there seems to be a tendency, particularly among those sports divers making several dives daily during a week-long vacation, to run into trouble around the fifth day. This may be because dissolved gas loads in the slowest tissues may not have been entirely eliminated before the next day's diving, even though the table or computer has not suggested the presence of any residual gas.

Another disconcerting observation is that of the occasional untoward effect of a double dive. A mere breath-hold dive to retrieve an anchor after completing an appropriate decompression for a compressed-air dive was described by Melamed (1984) to lead to sudden death on more than one occasion. An explanation, that the second dive reduced the diameter of venous gas emboli sufficient for them to slip through the pulmonary filter, may be speculative but can be applied to other unexplained cases of neurological decompression illness.

Laboratory studies suggest that a high level of C5a complement may be associated with a greater degree of susceptibility (Ward *et al.*, 1987), though this like so many research findings is not necessarily confirmed in other studies (Hjelde *et al.*, 1992).

These examples cannot be reviewed in detail here but are offered as evidence that decompression illness cannot be considered as two distinct illnesses, DCS and AGE, but that there are many outward presentations which possibly are inter-related. Though we know more today about decompression pathology than ever before, the one conclusion is that there is so much more to be researched. Even so, the true pathology of acute decompression sickness in man may never be fully known simply because it is so elusive. The condition is usually transient, it is not accurately reproducible in animals, and many bubbles found in man at autopsy are post-mortem artifacts.

A return to a system of descriptive reporting in place of misleading labels needs to be encouraged wherever this can be recorded accurately and the data made available for analysis. For instance, although pain is so difficult to describe, a recent workshop of the Undersea and Hyperbaric Medical Society (Nashimoto and Lanphier, 1991) suggests that it may be possible to distinguish between cases with sharp, localised and superficial pain in the region of a joint, and those with a more dull, diffuse and deeper pain. There are hypothetical reasons why each might be associated with a different pathogenetic mechanism and why only one of these two might be associated with the possible development of osteonecrosis. Only meticulous clinical detective work in many centres and over a number of years could ever answer such questions.

The patterns of observed decompression illnesses from many treatment centres are indeed different; for example contrast the symptoms of the Caribbean 6-day sports diver with those of the UK 50-metre sports diver and compare both with those of the compressed air worker. A potential success for descriptive reporting would be a comparative study of the occurrence of manifestations leading to the potential identification of a number of decompression illness syndromes, not just the two classic diagnoses. With clarification of such clinical categories, possibly related to the nature of the precipitating dive, perhaps related to intrinsic biochemical factors, may come distinct therapeutic approaches aimed towards the specific clinical condition of a case.

3 The treatment of decompression illness

Perhaps because the pathogenesis and natural time-course of the various presentations of decompression illness are largely unknown, there is also uncertainty about the optimum treatment. Indeed, given the wide variety of pressure profiles and gas mixes that precede decompression illness, the response of the individual to a given decompression insult, and the delay that may occur before treatment can begin, it may be too simplistic to believe that all therapeutic decisions can be solved by adherence to a universal algorithm.

Within the debate about treatment it must never be forgotten that the early return of the stricken diver or compressed air worker to raised environmental pressure often provides an immediate and complete resolution of the illness. The subsequent therapeutic decompression procedures are designed to be sufficiently slow to prevent a relapse.

In submarine escape training accidents the quadriplegic may be restored to full function within seconds during the recompression, but controversy exists about the appropriate depth of recompression to be selected. The depth of total relief is not practical because in some cases relief might not be achieved. For instance, pressure may reduce only the length of a longitudinal intravascular bubble but not its diameter. In all types of decompression illness, an arbitrary depth, or arbitrary increment of depth, is usually selected.

Recent studies suggest that, with concurrent oxygen-breathing, a shallow depth such as 18 m (60 fsw) may be more effective than a deeper recompression for cases of decompression illnesses arising at sea-level after exposure to compressed air. For cases of 'blow-up' or for cases arising during decompression from mixed-gas 'bounce' diving, immediate recompression to much deeper depths can be dramatically effective. In contrast, for symptoms arising during saturation decompression, the depth of relief may sometimes be only 2 m or so (6 fsw) deeper than the depth of their onset.

Another therapeutic variable, which can be controlled independently of ambient pressure, is the partial pressure of oxygen. Though commonly 100% oxygen is administered at 18 m (282 kPa) there is a view that such a high tension may cause excessive vaso-constriction and that a lesser oxygen concentration may be more effective.

The need for further research in this area is emphasised by the continuing controversy about the nature of the inert gas, nitrogen or helium, to be used as the oxygen-diluent in the treatment of compressed air decompression illness. The use of ancillary drugs may have a rational basis within the currently known mechanisms, but the effectiveness of drug treatments in decompression illness has yet to go much beyond the relief of symptoms. The blocking of specific pathological processes is still at the research phase. Beyond oxygen and pressure, only the restoration of fluid balance can be said to have a place in therapy that is beyond doubt but, even here, one can debate which intravenous fluid is the optimum choice.

Given the variability between the different cases, the geographical scatter between different treatment centres, the relatively small number of cases that do occur in each centre, and the difficulty of setting up a full clinical trial, the only practical solution appears to be the pooling of knowledge. A central decompression accident data bank with descriptive information for analysis is needed. For instance, to what extent do different therapies affect outcome? What is the relationship of immediate post-recompression outcome to 6-week, 6-month and even 6-year outcome? In therapy there are so many questions and so few answers.

4 Long-term decompression sequelae

Osteonecrosis provides a useful basis for the consideration of other possible long-term consequences of decompression. The recognition of the problem of bone necrosis was based upon a number of early clinical cases and then, using standardised radiological procedures and agreed diagnostic criteria, the prevalence of asymptomatic necrosis was assessed within the populations at risk. The establishment of a central registry led to a better understanding of the causative factors and the natural progression of the condition. There are still a number of unanswered questions but at present further progress has been limited by underfunding.

The longer-term sequelae in the central nervous system of overt decompression illness and of asymptomatic decompression bubbles are a particular cause for concern. Clinical examinations and special imaging, electrophysiological and neuro-psychometric investigations have so far failed to provide the necessary reassurances.

Among those exposed to raised environmental pressure and the subsequent decompression are a number of other conditions such as decrements in pulmonary function, which have been described but which, until the contribution of oxygen exposure and other contributory factors is known, should not be considered as a variety of decompression illness.

There is a solution but it may not come to fruition without much concerted effort by us all. It is again the pooling of data into a central registry for research. The occupational illnesses of diving and tunnelling are academically challenging but, because of the worldwide geographical scatter of divers and compressed-air workers who numerically form a relatively small occupational group among other workers, their illnesses are not easy to study in any other way.

From this outline emerges an idea of the broad scope for needed research, from the papers of this meeting come some examples of detail within their scope and from both comes a perception of the amount of research that needs to be done. In addition to specific research-related work, such as the study of bone necrosis, the identification of priorities in such research, the assessment of their feasibilities, recommendations for funding and the monitoring of progress, are tasks that have been undertaken by the MRC Decompression Sickness Panel.

5 The Decompression Sickness Panel

The Decompression Sickness Panel was set up in 1957, under the Chairmanship of Professor Sir William Paton FRS, to study the occupational health problems of workers exposed to unnatural changes of environmental pressure. In particular, it focused on the need to provide safe decompression for compressed air workers and on the clinical problems of decompression illness and dysbaric osteonecrosis. The Panel became the only comprehensive body in the UK dealing with man's exposure to hyperbaric environments and its meetings continue to be regularly attended by international visitors.

The early reported cases of bone necrosis led to the Panel initiating a radiological survey of compressed air workers to detect possible pre-symptomatic lesions in apparently healthy workers. Under the Chairmanship of Professor D N Walder the Panel used its experience to define radiological diagnostic criteria and develop a classification that it used worldwide. As part of this survey an MRC Decompression Sickness Registry of compressed air workers (and, later, divers) was set up in Newcastle upon Tyne in 1964 comprising medical records, bone x-rays and exposure records.

Dissatisfaction with the British Decompression Tables (1958) for compressed air working led to a review of decompression procedures in other countries. In 1965 a CIBA symposium 'The Decompression of Compressed Air Workers' was initiated by Panel members. The meeting summarised current knowledge and raised other issues such as the possible long-term neurological sequelae of compressed air work. In 1966 new decompression tables ('the Blackpool tables') were produced by Dr H V Hempleman, a member of the Panel. A few years later the Panel's *Medical Code of Practice for Work in Compressed Air* was published by the Construction Industry Research and Information Association (CIRIA).

During the 1970s the membership of the Panel was enlarged to cover health problems among professional divers, focusing upon the prevalence of osteonecrosis.

By 1981 some 283,000 manned-decompressions had been coded in the Central Registry and the funding for this data-base was transferred from the MRC to the relatively new Health & Safety Executive. Under the Panel Chairmanship of Professor R I McCallum, the occupational health records of some 12,500 compressed air workers and 7,000 divers were transferred to the Registry's magnetic tape archives.

In 1984 HSE funding ceased, a preference for short-term problem-orientated research projects seeming to be more attractive to authority than a long-term longitudinal study of occupational health. Their decision was contested vigorously and now, with the continuing questions of long-term health even higher in priority on the public agenda, the lack of any data input on health or exposure to hazards of compressed air workers and divers during the last 8 years, seems contrary to HSE's own objectives.

In 1987 Professor D H Elliott took over Chairmanship of the Panel. The current terms of reference of the Panel are:

- to provide a forum for medical and physiological research associated with diving and compressed air work
- to exchange relevant scientific information
- to promote and co-ordinate appropriate research
- to advise, as appropriate, on the formulation of guidelines, codes or regulations and on the implications of their introduction.

To enhance the output of the Panel which by now had a very broad remit, three working groups were created to reflect its principal areas of research interest.

The Compressed Air Working Group was set up in 1986 under the Chairmanship of Dr H V Hempleman, (formerly Superintendent, of the Admiralty Research Establishment's Physiological Laboratory at Alverstoke).

Contracts for work in compressed air are intermittent and geographically widespread and, for 30 years, the Panel has provided continuity of information and advice in this field. The Group is currently providing the HSE with scientific advice in relation to the revision of the *Medical Code of Practice for Work in Compressed Air*.

The Long-Term Health Effects Working Group was set up in 1986 under the Chairmanship of Dr M J Halsey (Division of Anaesthesia, MRC Clinical Research Centre, Harrow). Awareness of the permanent damage that diving and compressed air working can cause to cells, tissues and organ systems, even in the absence of acute symptoms, is growing. The diversity in the long-term changes caused in the body due to high pressure and decompression is reflected in the broad range of current research. Permanent changes to specific areas of the central nervous system, as well as effects in bones, retina, lungs, the vascular system, kidneys, liver and chromosomes are now being investigated. The Working Group has been concerned with the evaluation of current research in the field, with a view to identifying both the problems and the most promising approaches towards their solution. To this end the proceedings of a Workshop 'Diagnostic Techniques in Diving Neurology' organised by the Panel, were published in 1987. The Working Group's standing in this field was emphasised by the large attendance at their open meeting on neurological sequelae of diving at the Second International Symposium on Man in the Sea, held in the USA in November 1988. The objective of developing liaison with fellow researchers in North America and in the Pacific was met and should facilitate standardisation of investigative procedures internationally. The Long-Term Health Effects Working Group is also responsible for a research project which was funded by the HSE. This is a cross-sectional study of divers and matched controls to test the hypothesis that detectable, but not necessarily clinically significant, neurological damage may be an occupational consequence of diving, even in those who have no known episode of decompression illness. Subsequent to a preliminary finding of objective neurological abnormalities among experienced professional divers, the responsibility for this project is to be transferred to the Offshore Safety Division of the HSE.

The Diving Accidents Working Group was established in 1989. The role of the Group is to consider the causes, pathogenesis, prevention and treatment of the acute decompression illnesses. The setting up of a diving accident data-bank was considered to be of the highest priority in order to make progress in defining the natural progression of different varieties of decompression illness and for the purpose of assessing the effectiveness of different treatment protocols.

The Decompression Sickness Panel is accountable to the Medical Research Council who, as part of its regular review of cost-effectiveness in research, assessed in detail in 1991 the Panel's current and proposed work. After much debate it has now been confirmed that the Panel has a unique role to play and with some administrative changes, will continue its work. It is acknowledged to be unlike the main Sub-Committees of the MRC in that the Panel acts as a forum for the recognition and assessment of problems. For this task, the membership of researchers and clinicians is strengthened by engineers and by senior representatives of industry, the Navy, the Police, recreational and other diving organisations, as well as by several divisions of the Health & Safety Executive.

The Panel believes that it has a valuable future and will continue to adapt its objectives and structure according to new developments and changing circumstances. Research on the effects of exposure to raised environmental pressure will continue to be of importance so long as man continues to work at pressure on tunnelling and similar engineering projects and to be exposed to the undersea environment, whether for work or recreation.

The importance of the Decompression Sickness Panel is that it is the only body available to review and co-ordinate medical research related to a unique environment in which the interests of many different authorities meet, but in which none has exclusive responsibility.

References

Adkisson, G.H., McLeod, M.A., Hodgson, M., Sykes, J.J.W., Smith, F., Stack, C., Torok, Z. and Pearson, R.R. Cerebral perfusion deficits in dysbaric illness. *The Lancet*, **ii**, 119-122, 1989

Bert, P. 1877: *Barometric Pressure*. Hitchcock, M.A. & F.A., translators. Columbus, Ohio: College Book Co., 1943

Douglas, J.D.M. and Milne, A.H. Decompression sickness in fish farm workers: a new occupational hazard. *BMJ*, 1991, **302**: 1244-5

Francis, T.J.R. and Smith, D.T. *Describing decompression illness*. Forty-second Undersea and Hyperbaric Medical Society Workshop. Bethesda: UHMS, 1991

Hjelde, A., Brubakk, A.O. and James, P.B. Complement activation in divers following repeated air/heliox dives. *Undersea Biomed. Res.*, **19 Suppl.**, 50, 1992

Moon, R.E., Camporesi, E.M. and Kisslo, J.A. Patent foramen ovale and decompression sickness in divers. *The Lancet*, **i**, 101-102, 1989

Melamed, Y. Personal communication, 1984

Nashimoto, I. and Lanphier, E.H. *What is bends?* Forty-third Undersea and Hyperbaric Medical Society Workshop. Bethesda: UHMS, 1991

Neuman, T.S. and Bove, A.A. Severe refactory decompression sickness resulting from no-decompression dives and pulmonary barotrauma: Type III decompression sickness. In: *Underwater and Hyperbaric Physiology IX*. Bove, A.A., Bachrach, A.J. and Greenbaum, L.J. eds. Undersea and Hyperbaric Medical Society: Bethesda, Maryland, pp 985-991, 1987

Palmer, A.C., Calder, I.M. and Hughes, J.T. Spinal cord degeneration in divers. *The Lancet*, **ii**, 1365-1366, 1987

Palmer, A.C., Calder, I.M. and Yates, P.O. Cerebral vasculopathy in divers. pp 137-144 in EUBS 1990. *Proceedings from the Joint Meeting on Diving and Hyperbaric Medicine*. Amsterdam: Foundation for Hyperbaric Medicine, 1990

Paton, W.D.M. and Walder, D.N. *Compressed Air Illness. Special Report Series, Medical Research Council, No. 281*. London: HMSO, 1954

Polkinghorne, P.J., Sehmi, K., Cross, M.R., Minassian, D. and Bird, A.C. Occular fundus lesions in divers. *The Lancet*, **ii**, 1381-1383, 1988

Smith, D.J., Francis, T.J.R., Pethybridge, R.J., Wright, J.M. and Sykes, J.J.W. Concordance: a problem with the current classification of diving disorders. *Undersea Biomed. Res.*, **19 Suppl.**, 40, 1992

Todnem, K., Nyland, H., Dick, A.P.K., Lind, O., Svihus, R., Molvaer, O.I. and Aarli, J.A. Immediate neurological effects of diving to a depth of 360 metres. *Acta Neurol. Scand.* **80**, 330-340, 1989

Triger, 1845 quoted by Bert, P., 1877: *Barometric Pressure*. Hitchcock, M.A. & F.A., translators. Columbus, Ohio: College Book Co., 1943

Ward, C.A., Weathersby, P.K., McCullough, D. and Fraser, W.D. Identification of individuals susceptible to decompression sickness. In: Bove, A.A., Bachrach, A.J. and Greenbaum, L.J. eds. *9th International Symposium on Underwater and Hyperbaric Physiology*. Bethesda: Undersea and Hyperbaric Medical Society, 1987

Wilmshurst, P.T., Byrne, J.C. and Webb-Peblow, M.M. Relation between interatrial shunts and decompression illness in divers. *The Lancet*, **ii**, 1302-1306, 1989

Part Two
Construction Projects

4 Controlled caisson sinking – design and performance on Contract 4 of the Greater Cairo Wastewater Project

J. L. Vincett and F.N. Rowley
Tony Gee and Partners, Cobham, UK

Abstract
This paper describes the design and operation of sinking techniques for concrete caissons under compressed air to form access shafts to the sewer tunnels of Contract 4 of the Greater Cairo Wastewater Project. The contractor's rationale for the control of risk, and the benefits obtained by the development of shaft access equipment to improve both safety and productivity are described. The contractor's experience with the technique is outlined and some details of the safety record achieved in compressed air working are given. The paper is presented from the viewpoint of design consultants appointed by the contractor, Lilley-Misr Group, to develop and detail their conceptual proposals for the construction techniques adopted.
Keywords: Caisson design, Air locks, Compressed air, Caisson sinking control.

1 Introduction

The Greater Cairo Wastewater Project has been undertaken in order to alleviate severe overloading of the original sewer system for the city of Cairo built in 1907 and extended in the 1920s (Roberts and Flaxman, 1985).

Lilley Construction of the UK, in joint venture with local partner Misr Engineering, were in 1984 awarded Contract 4 for the construction of sewer tunnels from Souk el Samak to Abdeen on the East Bank. Tunnel construction using compressed air techniques was chosen because of the combination of sandy ground conditions and high water table. The works included eight access shafts and three vortex drop-shafts. The shafts to be constructed had been designed by the Engineer as reinforced concrete cylinders of between 6 and 10 m in internal diameter and from 20 to 24 m in depth. As well as forming permanent access shafts to the tunnel system, they would at some locations provide assembly areas for the tunnelling machines in readiness for the tunnel drive. Figure 1 illustrates the contract layout.

Engineering and Health in Compressed Air Work Edited by F.M. Jardine and R.I. McCallum. Published in 1994 by E & FN Spon. ISBN 0 419 18460 0

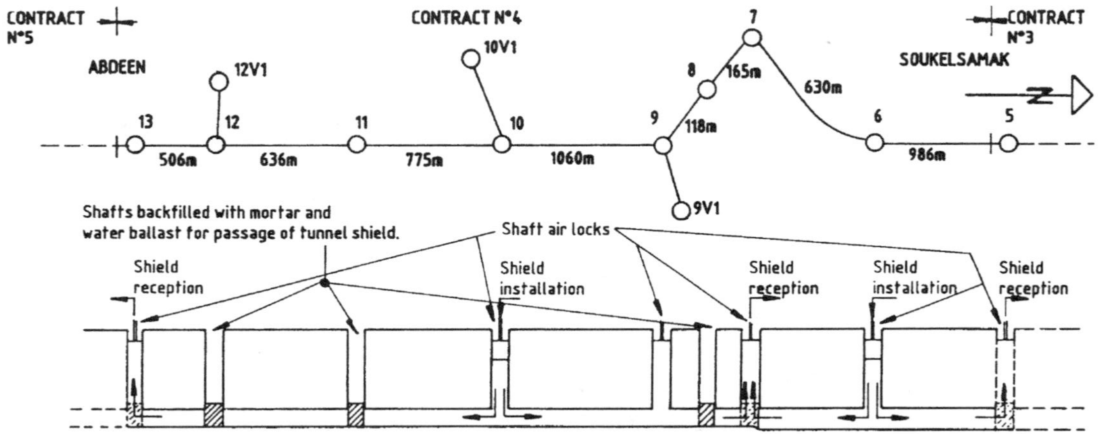

13	12	12V1	11	10V1	10	9V1	9	8	7	6	Shaft Number
11.7	10.3	7.7	11.7	7.7	11.7	7.7	11.7	12.9	12.9	11.7	Outside diameter (m)
9.0	8.0	6.0	9.0	6.0	9.0	6.0	9.0	10.0	10.0	9.0	Inside diameter (m)
20.52	22.78	19.99	23.30	22.15	24.16	21.47	23.43	22.78	23.90	24.06	Depth (m)
1.3	2.9		3.1		3.6		3.3	3.0	2.9	2.6	Depth to water (m)
1.92	1.99	1.71	2.02	1.86	2.05	1.84	2.01	1.98	2.1	2.16	Maximum air pressure (bar)
3.55	4.95	3.1	3.55	3.1	3.55	3.1	3.55	3.95	3.95/4.95	4.15	Plug below tunnel invert (m)
9.97	11.68	11.54	12.75	13.70	13.59	13.02	12.86	11.83	11.95	11.91	Depth to caisson toe (m)
7.07	8.83	8.94	9.85	11.10	10.89	10.42	9.98	8.68	8.95	9.01	Height of caisson (m)
8	5	5	8	5	8	5	8	8	8	8	Dia. of air deck hole (m)

Fig.1. Caisson sites on Contract 4

The background to the Engineer's design is fully described by Kell *et al* (1985), Darling and Drake (1985), Hall and Gray (1985) and Elliott and Spolton (1985).

The caissons would provide the access to the tunnelling works making them a crucial element in the sequence of handling plant and the flow of men, muck and materials which would govern the rate of progress of the contract. Shafts to be used for tunnelling machine assembly were on the critical path for construction operations and a delay in caisson construction would affect the entire contract.

Lilley-Misr took fundamental decisions to provide separate air locks for men and materials in order to avoid decanting during the decompression of workers and to maximise the throughput capacity of both locks. They appointed Tony Gee and Partners to develop and prepare detailed designs to implement their proposed methods, including air decks, the air locks to be used for the passage of men and materials, and safety interlocks intended to prevent accidental misuse of the locks.

The methodology for sinking the caissons was to be developed into a scheme involving hydraulic equipment capable of maintaining positive control over the caissons at all times, and ballasting sequences indicating safe limits for use of compressed air.

2 Design requirements

2.1 The site

The shafts are in very restricted sites within the existing street system of central Cairo. The sewer tunnels follow street alignments wherever possible in order to minimise the effects of any construction settlements on property and buildings.

All the shafts are in highly permeable soils with groundwater within a few metres of the surface and at times higher, leading to flooding of the streets by groundwater and raw sewage which the construction of the new sewer system is intended to relieve.

Typically, shaft construction would require building the caisson shell on soft silt or fill deposits, sinking the caisson through fill and silts into sands, and excavation to formation in sand or gravel layers. Considerable variability was expected in the upper horizons, particularly as records indicated possible hidden foundations and disused canal retaining walls which could cross a shaft site.

Buildings were close to the construction sites, and some were of historical and religious significance including an important mosque (which can be seen in Figure 2) sensitive to any loss of round support during excavation. Basements in neighbouring buildings are common, their waterproofing easily disturbed by variations in groundwater level, or by untoward migrations of pressurised bentonite or air from caisson sinking or tunnelling operations. In recognition of these potential problems the Engineer had specified a substantial programme of ground improvement by grout injection at critical locations such as tunnel/shaft connections, which was extended during the course of the contract – see 4.2 below.

Fig.2. General view of caisson during sinking

2.2 The access shafts

Contract 4 required the construction of eleven shafts which function either as access to the tunnel system, or as vortex drop-shafts to control sewage flows. Use of the shafts for installation and reception of tunnelling shields during construction had also been considered by the Engineer. Figure 1 summarises the dimensions of the shafts.

The Engineer had designed the shafts for construction in two stages as illustrated in Figure 3. The upper shaft was to be constructed above ground and sunk as a caisson until the toe was 2.5 m above the level of the tunnel crown. The lower portion of the shaft contained the tunnel connection eyes, and formed the lining for the bottom plug which provides the bottom seal and ballast against flotation. This section was designed to be constructed by an underpinning technique using pre-cast concrete shaft rigs against which the *in-situ* reinforced concrete lining could be cast, and which would then be undercut by excavation for the next layer of rings. The tunnel eyes were unreinforced to facilitate break-out or reception of a tunnelling shield. The completed caisson had a reinforced concrete roof, with access manholes to street level suitable for operational and maintenance requirements.

Excavation was to be under compressed air in order to control water pressures as excavation progressed and prevent blowing of the excavation floor. Once the plug was cast it was possible to revert to free air, remove the air deck, and assemble the

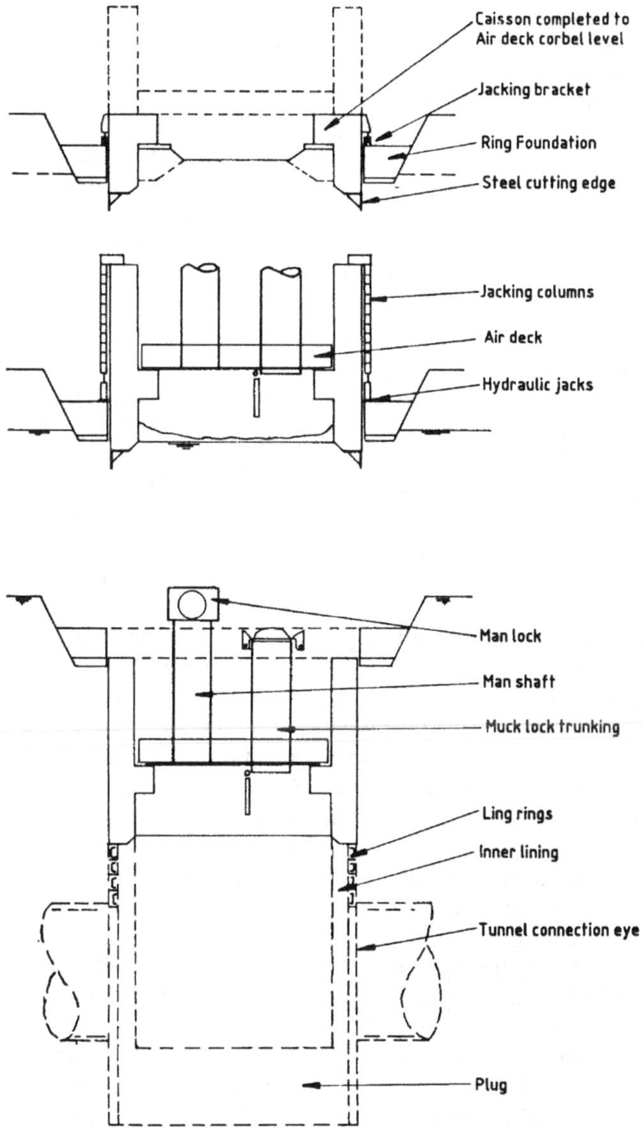

1. Start of caisson construction

2. Start of caisson sinking

3. Underpinning to complete caisson

Fig.3. Construction sequence

tunnelling machines. The air deck was to be replaced and compressed air reapplied for break-out operations at the commencement of tunnel driving.

2.3 The contractor's requirements

The locking arrangement giving access to compressed air workings used by British contractors for many years consists of an integrated man and muck lock known in various stages of its development as the Davis or Gowring pattern. All operations for transfer to men and materials are carried out manually.

Its principal drawback in terms of operative safety is the nature and difficulty of access and egress. The decompression chambers for men are in the form of blisters on the side of the top chambers, with exit to free air through a side door. The chamber is of very restricted proportions with standing room only. If staged decompression or recompression because of medical problems is needed the men must be decanted to a separate lock. A common shaft is used for muck, skips and man access, using vertical rung ladders which are exposed to any muck falling from the skip. A typical arrangement is illustrated in Figure 4.

For the Cairo contract, Lilley wished to improve this system in the combined interests of production, working conditions and safety. The layout would change to use separate shafts and air locks for the passage of men and materials. Spiral stairs would be used in preference to open-rung ladders, and men using the ladders would not be exposed to falling muck. A multi-chamber man lock would be used, in a horizontal configuration mounted on a tee-head, allowing staged decompression without decanting, with a separate small chamber allowing alternative entry or exit when the other was in use.

The muck-lock chamber would be substantially larger than the Gowring type, allowing use of a 2.5 m^3 capacity bucket instead of the traditional 0.8 m^3 bucket. Doors and clamps would be powered rather than manually operated, and a full system of interlocks would be used to link the operation of bottom door, top door and exhaust valve to prevent any possibility of incorrect operation of the doors.

All air lock components were manufactured on a modular basis to provide interchangeability on site, and allow flexibility in set-up, e.g. to permit larger decompression lock chambers to be used at locations requiring more men on a working shift. Once tunnel construction was sufficiently advanced the air lock could be transferred from its mounting on the air deck to a pressure bulkhead within the tunnel, exploiting the horizontal configuration of the lock.

The pre-formed concrete section of the caisson shaft was to be constructed to full height prior to sinking. A solid foundation would be required for this to avoid unacceptable settlements of the soft surface silts, weakened by high water tables, or the possibility of tilting because of hidden obstructions. The contractor required a temporary works system which would then control the critical early stages of sinking the caissons prior to obtaining a 'bite' in the more uniform sands below. Positive control of level was to be provided to maintain the caisson alignment in all possible variations of ground conditions and to allow positive steering to control the plan position of the shaft axis. The capacity provided would allow control even if a solid obstruction were encountered on a segment of the caisson perimeter.

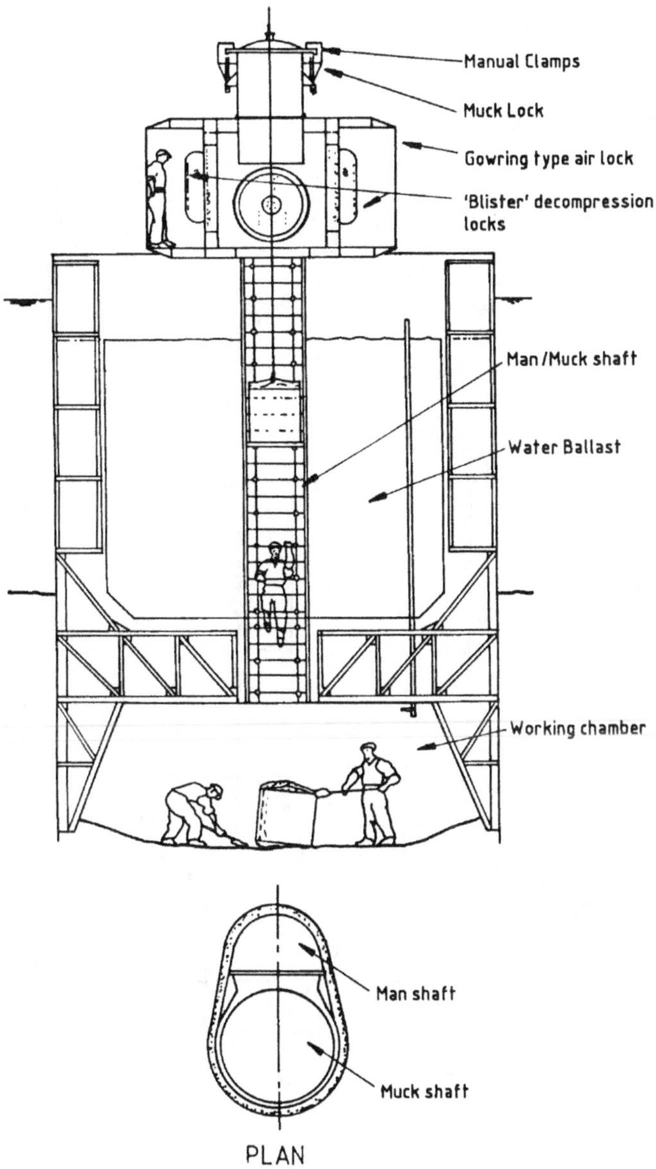

Fig.4. Steel caisson with Gowring-type lock: New Khedive Ismail Bridge Cairo (after Wilson and Sully, 1949)

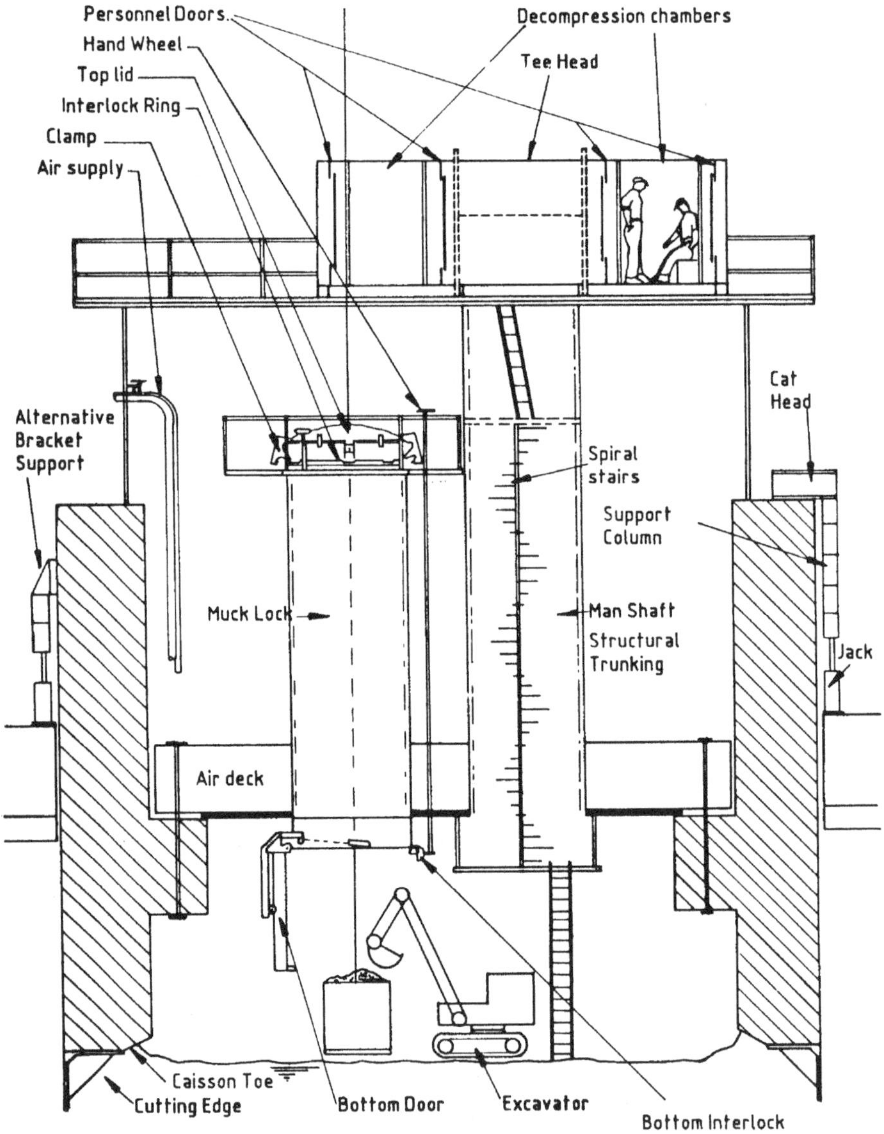

Fig.5. Caisson air and lowering equipment

It is usual to control the positional alignment of the axis of a caisson by controlled differential excavation around the caisson toe, but the indications were that the variability of the ground, and the possibility of a potential runaway if a soft silt layer were encountered would be particular risks at the Cairo sites. The use of bentonite injection was planned to reduce skin friction on the sides of the shaft; this, together with the provision of a positive support and control system, would give the ability to cope with these potential situations.

Mechanical excavation, rather than manual labour, was to be used within the working chamber, and provision would be required in the air decks for the passage of services including air supply, electricity, bentonite, water and hydraulic power for the Smalley mini-excavator which would work within the chamber. Another constraint on the design of the air decks was the need for easy dismantling for transfer in the congested city streets between shaft locations. Hence an all-steel design would be used, capable of easy re-assembly and re-use. Plate junctions would incorporate bolted seals rather than welded joints.

These requirements, together with some preliminary sketches of proposed site and component layouts provided the basis of the brief on which the air decks, man locks, muck locks, caisson support and lowering equipment were to be designed, and detailed air pressure and ballasting charts prepared for the excavation sequence at each site. Figure 5 illustrates the arrangement adopted, drawn at the same scale as Figure 4 for comparison.

3 Construction equipment

3.1 Air decks and structural trunking

Two sizes of air deck were required to suit clear openings between permanent concrete corbels of 8.0 and 5.0 m dia. in 10 (or 9) and 8 (or 6) m dia. shafts respectively which were designed to support the air deck. The design working pressure was to be 2.76 bar (40 psi). The air deck also had to support the ballast required to counteract buoyancy and overcome friction during the later stages of sinking operations, and design had to allow for support of this ballast without the benefit of air pressure within the working chamber. At the contractor's option, the ballast could take the form of kentledge weights placed on the main beams of the deck, or water or sand, which would act directly on the steel lining plates. To minimise the design requirements for the deck and hold-down system, limitations on air pressures were set at 2.07 bar (30 psi) unballasted or 2.76 bar (40 psi) with 51 kN/m^2 uniformly distributed ballast, and a maximum of 189 kN/m^2 ballast in free air, with corresponding limitations for use of the smaller deck.

In order to seal plates at bolted splices, overhanging cover plates with mastic caulking were used. At the junction between plate and concrete corbel a hollow rubber tube in a formed channel made an 'O' ring-type seal.

The hold-down clamps utilised high yield reinforcing bar with rolled threads and 'Macalloy' type nuts each end, passing through holes formed in the concrete corbel to steel yokes over the main beams of the air deck.

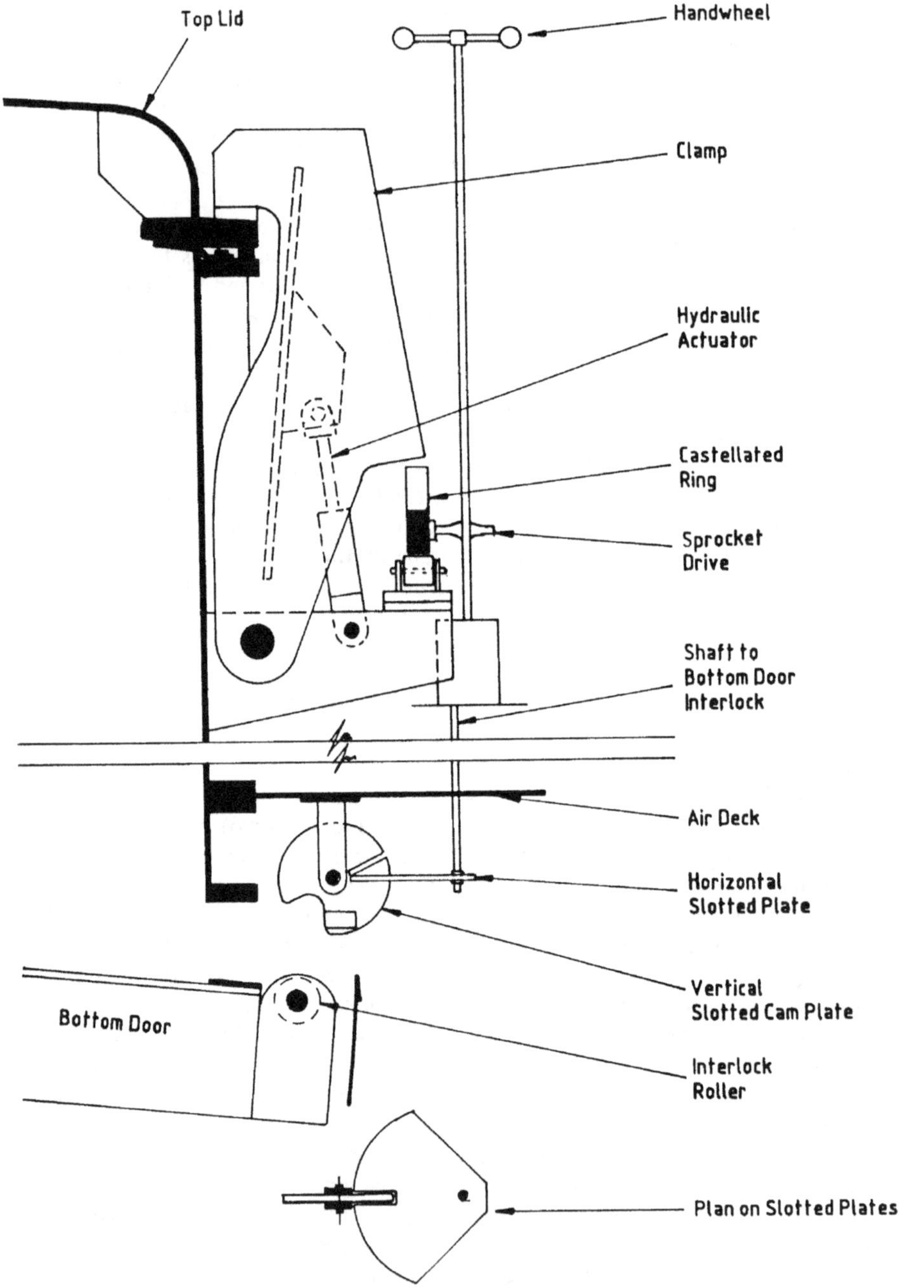

Fig.6. Schematic layout of muck-lock mechanical interlock

Twelve flanged service connections and sealed-through fixings for lifting-hoist attachments were also provided.

The structural trunking for both muck and man locks comprised a 12mm thick plate with egg-crate type stiffening, attached to a hollow section support frame. Since hydraulic rams were used to close the door, a swinging counterweight was not required, and use of hollow sections kept the weight of moving parts to a minimum. The closure rams shut the door against a simple rubber seal, and release of air from the upper chamber removed any further need for support.

The top lid, being captive on the crane rope, was made of an unstiffened semi-elliptical plate to minimise weight, with a solid mating flange. Three guide plates located the lid as the bucket was lowered into the shaft, and an upstanding rim forced a fabric reinforced neoprene flap against the flange to make a seal. Ten hydraulically actuated clamps swung over the flange to clamp it securely, before the air pressure was applied, in a similar manner to the manual clamps used on a conventional Gowring lock. The angles of the clamping plates ensure that friction held the clamps in place under load, and backup from the interlock ring (described below) ensured that the clamps were fail safe. The muck-locking system required 24 h operation passing 20 skips per 8 h shift to remove all excavated material from the shafts. Fail-safe operation, proof against operator error, was considered paramount in its design, in view of the very large number of operating cycles to which it would be subjected.

A mechanical interlock system (Figure 6) was devised linking pressure in the muck-lock trunk with the bottom and top doors. A guiding principle for the design of this system was that it should be fail safe, and the principles of its mechanical operation be readily seen by the operator.

The key to the system was a castellated steel ring, mounted on rollers and passing under the rear of the top lid clamping plates around the circumference of the top lid assembly. In the shut position, the ring blocked the clamp plates from opening under the full air pressure without other assistance. The ring was locked in this position by a pneumatic actuator piston linked to the air pressure within the lock chamber. It was linked *via* a hand-wheel and chain drive with a vertical shaft to a slotted cam plate mounted alongside the bottom door. With this door closed a second slotted and weighted plate could swing down through the slot, releasing a roller follower which would allow the bottom door to be opened when the pressures in lock and working chambers were equalised.

To reverse the procedure, the bottom door would be closed by hydraulic rams, thus picking up the swinging follower and bringing the slots in the cam plates into alignment. The exhaust valve from the lock chamber would then be opened. When fully discharged, pressure on the pneumatic actuator would be released so that the ring could be rotated by the operator, using the hand wheel to rotate the bottom cam slot to lock the bottom door and to bring the ring castellation into line with the top clamps so that they could be retracted to release the top lid, and also lock the exhaust valve open to prevent inadvertent repressurisation. The cam slot arrangement at the bottom door would prevent rotation of the ring unless the bottom door were fully closed.

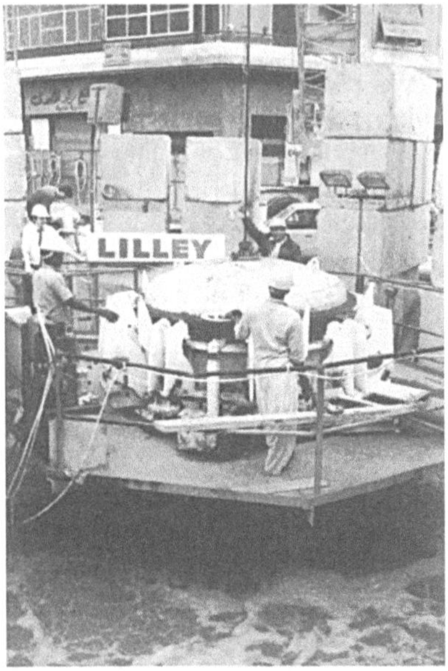

Fig.7. Muck lock top lid showing interlock in use

Fig.8. Caisson-lowering jacks

The system was supplemented by an electric traffic light arrangement to confirm the bottom door position to the surface operators.

Apart from a minor problem with welding to the bottom door support frame, the system proved suitably robust to provide reliable service throughout the contract. The arrangement of the interlock is shown in Figure 6. The top lid of the muck lock is shown in Figure 7.

3.4 Caisson lowering equipment

3.4.1 Foundations
The caisson, weighing up to 1200 tonnes, had to be completed at ground level before sinking, requiring a substantial foundation to distribute its weight to the soft and variable silty clays and fill at street level, but without obstructing the sinking of the caissons. A reinforced concrete ring foundation was used outside the periphery of the caisson shell, strong enough to support the entire caisson when fully suspended. This provided a solid base for constructing the works and gave a positive initial alignment for the commencement of sinking operations.

3.4.2 External support
The caisson cutting edge, and the lower part of the wall including the air deck corbel was cast on a slab foundation set as low as water levels at the site would permit.

The ring foundation was then cast around the cutting shoe, leaving pockets which would be used for bentonite injection to lubricate the later stages of caisson sinking. A ring of column packs was set up on the foundation bearing against brackets fixed to the side of the caisson wall. These were of sufficient capacity to support the weight of the completed caisson. The concrete caisson was then cast to its full height, supported on the packs and on timber cribbing set up beneath the inner corbel. The facility for relief of load to start lowering operations was provided by means of sand jacks. Steel catheads were then stressed down to the head of the wall, and a stack of modular packs inserted under the catheads, so that the caisson could be supported, in effect suspended, from hydraulic jacks at the base of the stack.

The packs were fabricated from column sections with end plates, and were maintained in correct alignment by clamps to screw fixings cast in the caisson wall. The pack length was somewhat shorter than the throw of a jack, allowing lowering of the caisson through its full height, if necessary by cyclic lowering of jacks and removal of packs. It was anticipated that once the soft fill and silt had been penetrated the full weight of the caisson would be taken by the cutting toe at its base and the jacking system would then become redundant. This in fact proved to be the case, and on later caissons the jacking arrangement was only provided up part of the caisson wall height. At some of the firmer sites the fixings were provided but the ground proved adequate to support the caisson from the start of sinking operations. Figures 8 and 9 illustrate the jacking arrangement with the caisson supported for construction of the upper walls.

Fig.9. Caisson supported on jacking system

3.4.3 Jacking system

The jacking system was required to fulfil two main functions. During caisson-lowering cycles it would be necessary to steer the descent. This was done by arranging the jacks in three banks on separate hydraulic circuits which could be independently controlled. Pressure-compensated flow valves allowed control of the operation.

During excavation the caisson load would be shared between the bearing capacity of the cutting edge toe, any air pressure applied to the working chamber, friction on the caisson outer walls, and the jacking circuits. Progressive excavation would remove the support from the caisson toe, but extra air pressure might be applied to control water levels. The jacking system would be required to maintain the position of the caisson during this phase for the security of operators within the chamber. Monitoring of jack gauge pressures would allow control of excavation to be maintained to assist even penetration. If an obstruction to the toe were encountered this would be indicated by a relief of jack pressure on the appropriate bank, and the system provided the capability to support the caisson in true alignment, while the obstruction was removed.

By this means the contractor was provided with a construction system capable of providing positive control to caisson sinking at all times, whatever conditions of soft variable ground or local obstructions were encountered. As experience of the ground was gained it proved possible to balance excavation and sinking without assistance from the jacks, but the system was necessary for initial caisson construction and was always available to be brought into use if a subsequent problem were encountered.

3.4.4 Ballasting

As described above, the sinking of a caisson under compressed air is a continuing balancing operation between weight of caisson, bearing capacity at the caisson toe, skin friction, groundwater pressure, applied air pressure, and buoyancy. As the caisson is sunk it becomes necessary to add ballast to counteract buoyancy and maintain sufficient pressure on the cutting edge to continue downward movement. It was intended that the problem of skin friction would be largely removed by the use of bentonite injected into the annulus between caisson and ground. Weak soil and high water tables constrain the range of air pressures and net caisson weights which are acceptable, although the jacking system greatly increases the window of safe options open to the contractor.

In the second phase of construction, while underpinning to formation level, increased air pressure with depth would require further increases in ballasting to ensure an appropriate factor of safety against caisson uplift.

A series of charts was prepared for the contractor indicating safe limits for ballasting at any stage of excavation. Ballast, self-weight, buoyancy and air pressures were balanced with appropriate factors against bearing capacity at the caisson toe, skin friction on the sides, and available jacking restraint capacity to construct charts showing the range of acceptable ballast at any stage of sinking, and a possible sequences of ballasting. Adjustment of parameters would permit a ready

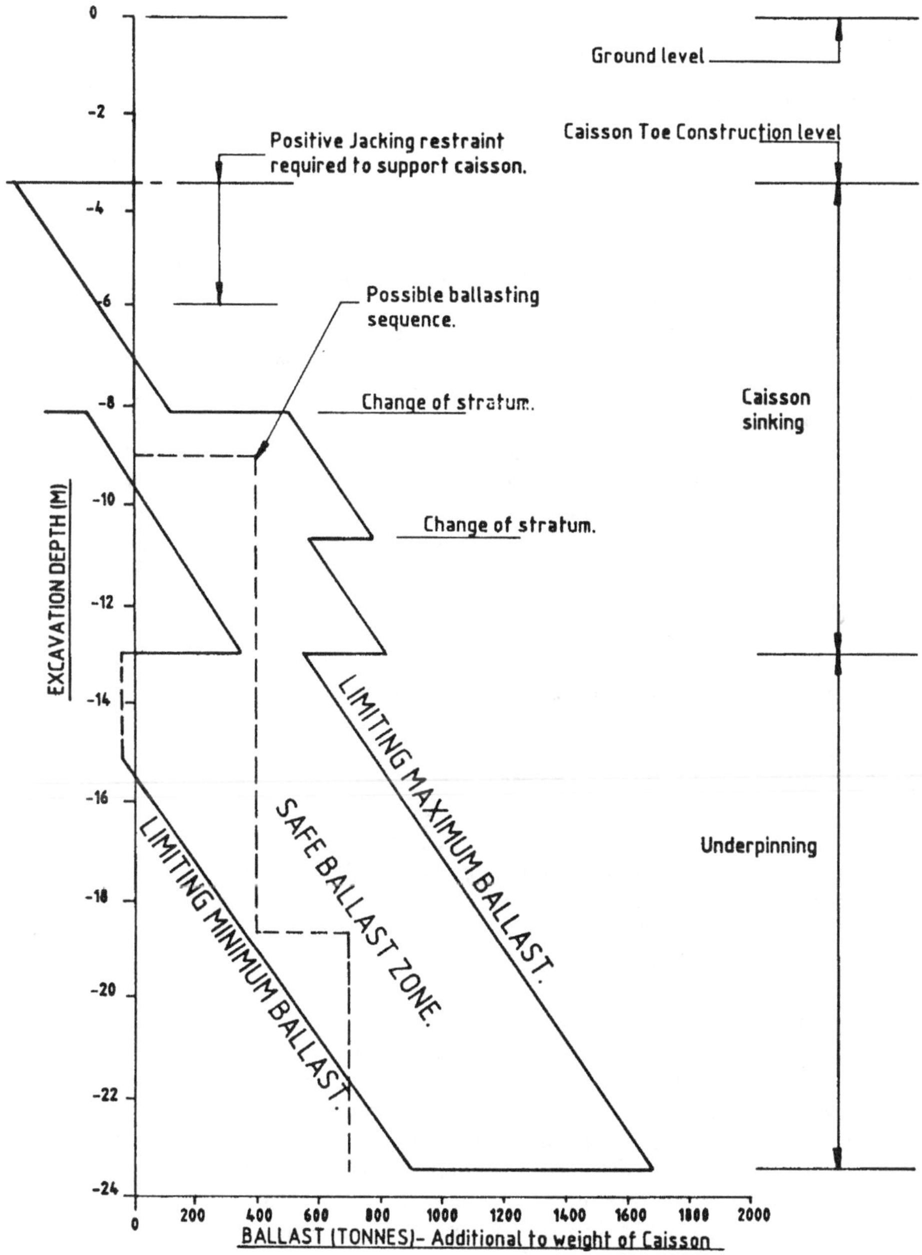

Fig. 10. Caisson-ballasting diagram for Shaft 9

reassessment of ballasting requirements for changes in site circumstances. A typical chart is illustrated on Figure 10. The calculations involved would readily lend themselves to a spreadsheet computer programme with graphical output, to increase further the flexibility of site control over the balancing of excavation progress, compressed air pressures and ballast provided.

4 Performance

4.1 Caisson sinking

No particular problems were reported with caisson sinking on the Lilley-Misr contract. Apart from the muck lock and man lock door units, and specialist hydraulic equipment, the hardware was all fabricated out of imported British steel by a local fabricator under expatriate supervision. The construction hardware performed as designed, although some difficulties were experienced balancing loads during commissioning of the hydraulic lowering system. This was resolved when it was found that amendments made to the design of the hydraulic circuits by the equipment suppliers produced a control system too sensitive for the piston pumps used. Reversion to the original design proved satisfactory. The mechanical excavators were modified to use remote hydraulic power, and were successful in considerably reducing the size of excavation gang traditionally reckoned as ten men per air lock per 110 m^2 of open excavation.

The bentonite lubrication functioned well although, because of the high groundwater table, the results on the surface could be messy if a temporary imbalance of air pressure were applied to the excavation.

Excellent accuracy in alignment of the caissons was achieved, and minimal blowing down of the caissons was used, showing the efficacy of the bentonite.

Ground conditions were found to be more straightforward than prudence had dictated in design of the lowering system, and it was possible to progress the excavation of later caissons by traditional control techniques. However, fittings associated with the system were cast into the caisson walls at all times, so that the control system could be immediately attached if necessary. At one location a brick-lined canal was encountered during the early stages of caisson sinking. This was broken through without undue difficulty.

4.2 Underpinning

This operation proved to be more difficult than had been anticipated, for a number of reasons. At times, there were substantial losses of air from the working chamber because of the nature of the ground exposed in the chamber floor, and when underpinning, and also through joints between the pre-cast segments used for the initial lining. Hydrophilic gaskets were used between the segments, and although these performed well on completion, the air pressure during construction frequently prevented water reaching the gaskets so that their design swelling did not occur. Grout caulking was used to seal leaks. Even so, air-supply requirements were extremely high at some locations, peak demand exceeding 2.4 m^3/s (5000 cfm) at six shafts with a maximum of 4.7 m^3/s (10000 cfm) giving rise to safety concerns as back-up supply capacity was exhausted and the limiting capacity of the available equipment approached. A programme increasing the scope of curtain grouting was instructed by the Engineer to alleviate this problem.

Preferential flow paths existed in the ground, and on occasions surplus air, mixed with groundwater and bentonite from the lubrication sleeve migrated considerable distances, causing unexpected flooding of streets and building basements remote from the working sites.

Installation of the secondary reinforced concrete lining initially proved slower than planned, because of the heavy reinforcement and multiple lifts of curved shutters to be erected under the air decks. Foreman carpenters and steelfixers with previous compressed air experience were brought in to expedite this portion of the works.

Table 1 illustrates the relative rates of progress achieved for caisson wall construction on the surface, followed by caisson sinking, in comparison with underpin excavation and erection of lining rings followed by construction of the *in-situ* lining under compressed air conditions. The benefits of maximising work elements constructed in free air by advance construction of the caisson walls are clearly apparent.

Table 1 Comparison of caisson and underpinning techniques

Lilley Misr – Contract 4 Shaft 10	9 m dia., 108 m^2 projected area
Construction	
Caisson-wall construction	10.69 m in 96 x 12 h shifts = 0.22 m/day
Underpin *in-situ* lining construction	10.54 m in 132 x 12 h shifts = 0.16 m/day
Excavation	
Caisson sinking	10.0 m in 27 x 8 h shifts = 1.11 m day
Underpin excavation	10.54 m in 68 x 8 h shifts = 0.47 m day

4.3 Safety

A good record was achieved for the compressed air working. Two centres containing medical locks, emergency breathing and fire-fighting apparatus were established, which were continuously manned throughout the compressed air working.

All expatriates were thoroughly examined in the UK for their suitability for compressed air work before travelling to Cairo, and a similar procedure was adopted for the local employees.

The application and observation of the guidelines governing working in compressed air were strictly enforced at all times achieving the decompression sickness rate of 0.74% noted in Table 1 (Fawcett and Reed, 1988).

No accident occurred in working the air locks, although one incident did occur on dismantling the lock, when a local labourer removed with a burning torch the support to the lid door on which he was standing. This serves to emphasise the need for constant vigilance, and fail-safe equipment with a readily observable method of operation wherever possible on civil engineering works.

Table 2 Comparison of Contract 4 and the New Khedive Ismail Bridge works

Item	**Unit**	**Lilley-Misr Contract 4**	**Dorman Long New Khedive Ismail Bridge**
Number of caissons	no.	11	7
Shaft depth	m	24.0	17.0
Caisson penetration	m	13.6	13.0
Area of excavation floor (min. to max. size caisson)	m^2	45–131	95–214
Excavation statistics			
Skip size	m^3	2.5	0.61
Skips per 8 h shift	no.	20	60
Excavation rate (bulked)	m^3/h	6.3	4.6
Excavation (net)	m^3/h	5.0	2.73
Compressed air statistics			
Maximum air pressure	psig (bar)	34 (2.34)	33 (2.28)
Air consumption – max. per shaft	cfm	1500 to 10000	1500
Sickness rates			
All compressed air operations	%	0.56	0.46
Decompressions	no.	54811	14400
Pressure over 14 psig (1 bar)	%	0.74	
Decompressions	no.	41764	-

5 Historical comparisons

Cairo has presented engineers with similar problems in the construction of groundwork throughout its history, and under European influence compressed air played an early role for contractors. Between 1869 and 1872 the Kasr el Nil bridge was constructed connecting the west and east bank regions of the city; a French contractor built the main pier foundations using wrought iron caissons, sunk under compressed air to a depth of 19m and filled with limestone masonry in a lime mortar.

The New Khedive Ismail bridge was built to replace this structure between 1931 and 1933 by the Dorman Long Company, using steel caissons to construct the foundations under compressed air (Stephen, 1934/5). A comparison of the progress achieved with that on the recent sewer contracts is of interest.

G.B. Gowring is believed to have prepared the designs of the air locks for this contract, and as a footnote, the site agent was T. Biggart, father of A. Biggart, Technical Director for Lilley-Misr.

Figures 4 and 5 show the relative proportions of the two caisson systems, and Table 2 presents some comparable statistics.

The faster excavation rate obtained by the modern mechanised equipment is illustrated, part of the gain arising by the reduced bulking of material when placed by mechanical excavator in a larger skip.

The paper on New Khedive reports frequent use of 'blowing down' to sink the caisson – a method to overcome skin friction on the shaft wall by controlled reduction in air pressure, which can often result in sudden movement with possible tilting of the caisson and substantial inflows of ground and consequent surrounding settlement. A skirt was fabricated around the cutting edge to allow bleeding of air between the external wall and the ground to release friction without sudden blowing and some success was reported. The modern technique of forming a bentonite skin has largely removed the necessity for such undesirable operations.

Stephen's paper also discusses the difficulties arising from the adjacent previous bridge foundations, which caused tilting of the new caissons and in one instance prevented it from reaching its design depth. The sickness rates reported on the two contracts are remarkably similar. However the reference level at which discomfort was reported as sickness is not known, and the values quoted may not therefore be directly comparable.

6 Conclusion

The access shafts for Contract 4 of The Greater Cairo Wastewater Project were successfully and safely constructed by contractor Lilley-Misr using the methods described above. Techniques employed illustrate the development of equipment since earlier use of compressed air caissons in the city.

References

BS 5500 *The Design of unfired fusion welded pressure vessels*

Darling, R.S. and Drake, J. Greater Cairo Wastewater Project. Studies and master plan. *Proc. Instn. Civ. Engrs.*, Paper 8923. Vol. 78, Aug. 1985

Elliott, I.H. and Spolton, R.L. Greater Cairo Wastewater Project. Design of west bank stage 1 works tunnels. *Proc. Instn. Civ. Engrs.*, Paper 8926. Vol. 78, Aug. 1985

Fawcett, D. and Reed, R.J. Facing the Facts in Cairo. *Tunnels and Tunnelling*, June 1988

Hall, T.D. and Gray, M.W. Greater Cairo Wastewater Project. Rehabilitation works. *Proc. Instn. Civ. Engrs.*, Paper 8924. Vol. 78. Aug. 1985

Kell, A.D., Bosker, C.M. and Moir, I.W.M. Greater Cairo Wastewater Project. Organisation, financial arrangements and contract strategy. *Proc. Instn. Civ. Engrs.*, Paper 8922. Vol. 78. Aug. 1985

Roberts, D.G.M. and Flaxman, E.W. Greater Cairo Wastewater Project. History, development and management philosophy. *Proc. Instn. Civ. Engrs.*, Paper 8921, Vol. 78. Aug. 1985

Stephen, F.W. (1934/5) The New Khedive Ismail Bridge, Cairo, Egypt. *Proc. Instn. Civ. Engrs.*, Vol. 240. 1934/1935. Part 2

Wilson, W.S. and Sully, F.W. Compressed Air Caisson Foundations. *Instn. Civ. Engrs., Engineering Division Papers*, 1948/1949, Vol. 7

Acknowledgements

This paper is published with the permission of Lilley Construction and the authors acknowledge the support of Mr R J Reed and Mr A R Biggart, formerly Construction Director and Technical Director respectively of Lilley Construction, in its preparation.

5 Use of compressed air at Lake Lothing, Lowestoft

J.G. Campbell, D.R. Gutteridge and A.P. Smith*
Charles Haswell and Partners Limited, London, UK
**Costain Engineering and Construction, Maidenhead, UK*

Abstract
This paper describes the compressed air tunnelling carried out for a new cable crossing of Lake Lothing in Lowestoft. A description of the project is given together with a summary of the construction methods. Details of the compressed air plant are provided as well as pressures used. The number of recorded decompression sickness events are also discussed.
Keywords: Tunnelling, Compressed Air, Decompression Sickness, Decant.

1 Introduction

Lake Lothing is a tidal waterway approximately 106 m wide in Lowestoft, East Anglia. Prior to the mid 1980s a single electric cable crossing existed with the cable being buried beneath the bed of the waterway. The Eastern Electricity Board wished to increase the capacity of the crossing and chose to install new cables within a tunnel to be built beneath the waterway.

Charles Haswell and Partners were appointed in 1985 to design and supervise construction. Costain Civil Engineering Ltd were appointed to build the tunnel, shafts and associated works. The tunnel was driven under compressed air.

This Paper briefly describes the construction of the tunnel and gives an appreciation of the medical aspects of the compressed air working.

2 Description of project

The project comprised one 3.05 m and one 2.74 m internal diameter shaft both approximately 19 m deep connected by a 1.96 m internal diameter tunnel, about 126 m long (Figures 1 and 2). The 3.05 m diameter south shaft was slightly deeper than the north shaft to allow for drainage. The shafts and tunnel were built in precast concrete bolted segments.

Engineering and Health in Compressed Air Work Edited by F.M. Jardine and R.I. McCallum. Published in 1994 by E & FN Spon. ISBN 0 419 18460 0

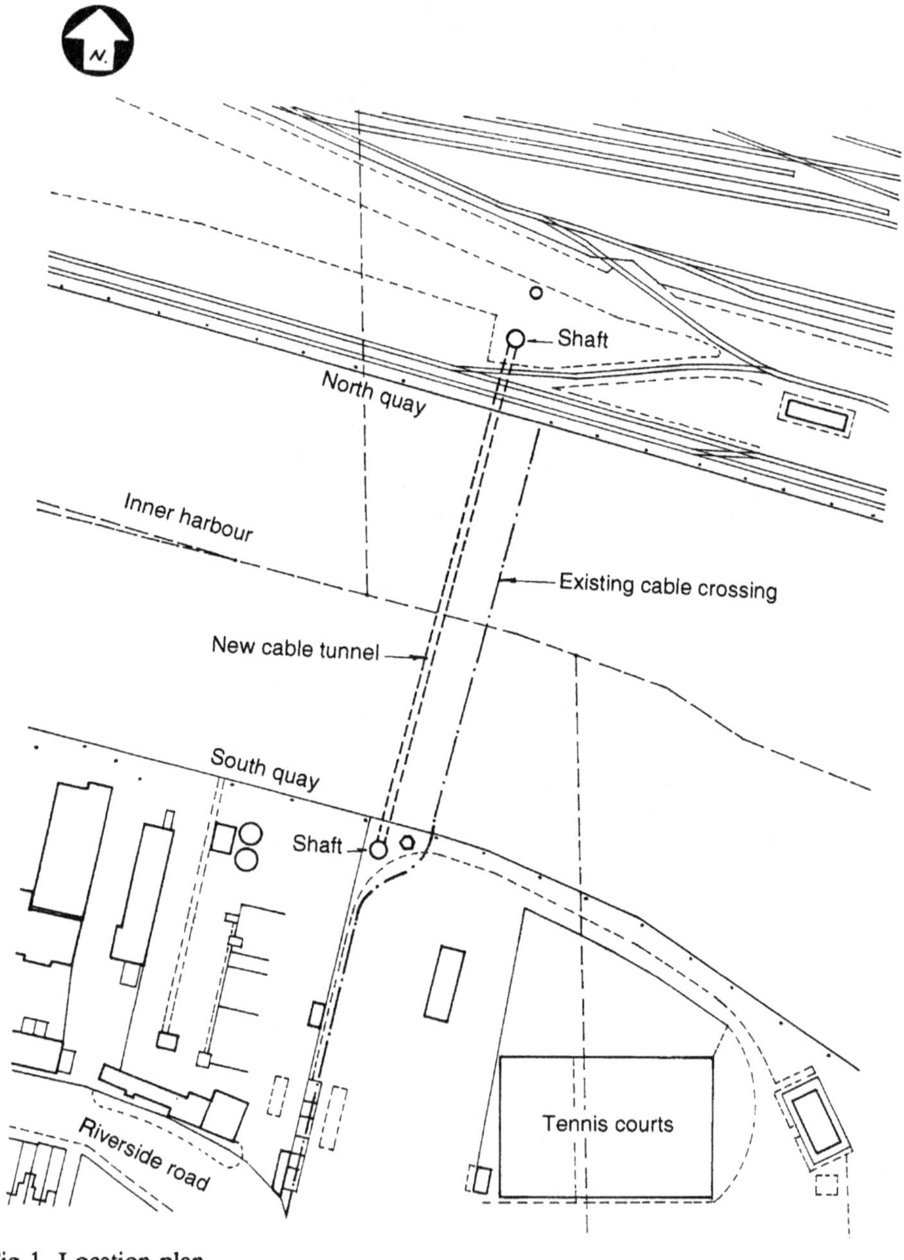

Fig. 1. Location plan

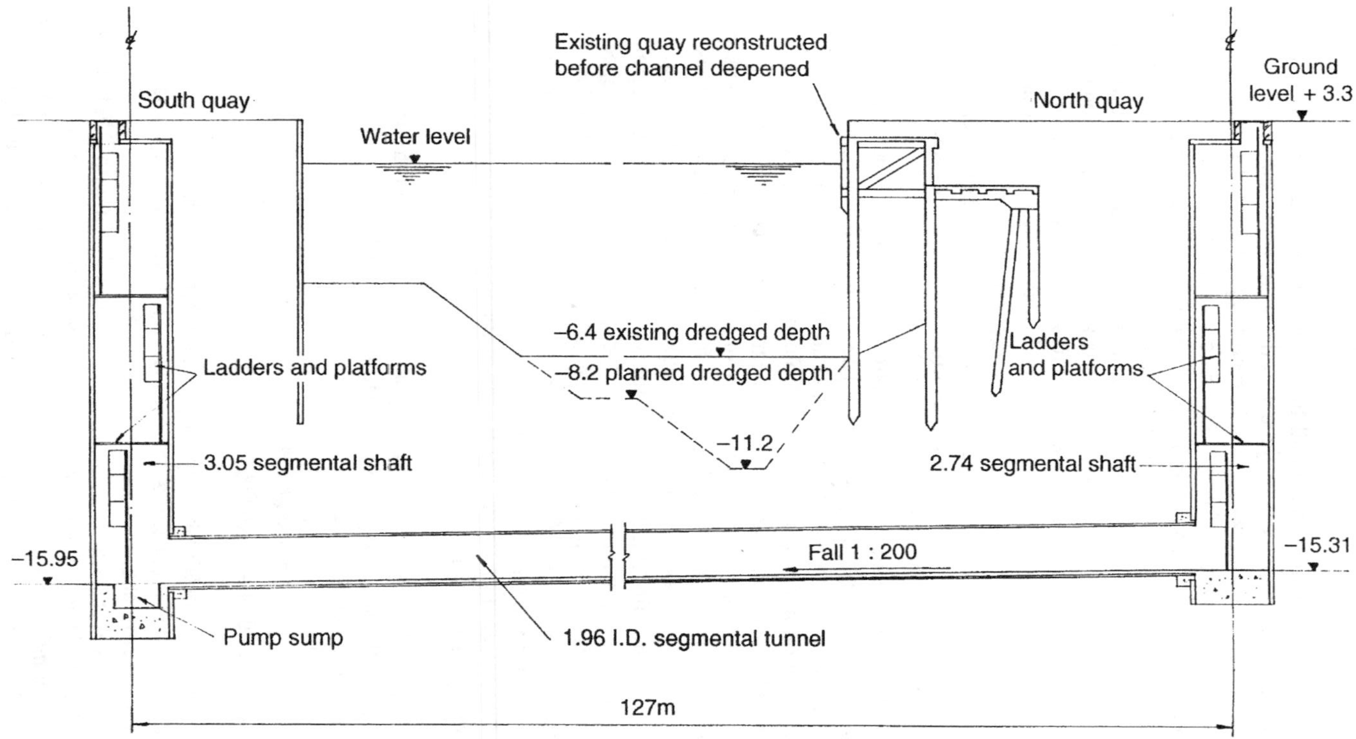

Fig.2. Longitudinal section through tunnel and shafts

Boreholes were sunk for the project and these were examined in conjunction with boreholes carried out for other works adjacent to the waterway. No boreholes were carried out from within the waterway. The borehole logs indicated made ground overlying sand with some layers of clay. The sand sometimes contained silt and/or gravel. The total tidal range was about 3 m, and the ground water table responded tidally over a smaller range. Quay level and ground level were approximately 2 m above high water level. The tunnel invert depth adopted was about 18 m below quay level at the banks of the waterway. This was a compromise between obtaining the required cover to minimise the risk of blowouts and tunnelling in minimum hydrostatic pressure. The cover to the existing bed of the waterway was about 6.5 m although there were plans to dredge the waterway, removing about 2 m of material with a further 3 m being excavated close to the north shore. The toe of existing piling was about 4 m above the tunnel crown. The mean water level was about 13 m above the tunnel crown.

The shafts were sunk as caissons with the base plugs being tremie concreted prior to pumping out and casting of the base slabs. The south shaft was strengthened with a reinforced concrete lining to withstand internal air pressure. Each shaft took about seven weeks to build on a single shift basis. Problems were mostly confined to difficulties with old timber piles and tie rods to earlier quay construction.

The tunnel was constructed by hand within a shield and under compressed air. Driving was uphill from the south shaft on a 3 x 8 h shift 5 days a week basis. The rate of progress achieved was 30 m/wk on average after driving started. The ground excavated was estuarine silts and sands. The tunnel was mucked using a cylindrical lock skip (designed for use with the vertical air lock). It was landed on a specially adapted bogie so that it lay at an angle of 30° for filling at the face. The bogie was pushed by hand on a single 24" (605mm) gauge track, the tunnel not being long enough and the shaft not being big enough for a locomotive.

The shield was lowered down the south shaft prior to fitting the air deck. Work carried out in compressed air included:

1. Fitting of the tail skin which was hung above the shield on the shaft wall, lowered and attached once the shield had been buried.
2. Breaking into reception shaft.
3. Removal of shield internals.
4. Bench concreting in both shafts.
5. Cleaning and pointing the tunnel.

Tunnel alignment was established by the use of a gyro-theodolite as the air deck could not be removed once tunnelling had started. This work was done as a sub-contract by British Aerospace.

The segment joints were sealed using Hydrotite and pointed with 'Waterplug'.

To receive the shield at the north shaft the base was filled with 5 N/mm^2 concrete. The remainder of the shaft was filled with sand. The shield hood was buried in the weak concrete. The shield was gutted, the skin left in and lined with concrete. The remainder of the weak concrete was removed and the junction grouted.

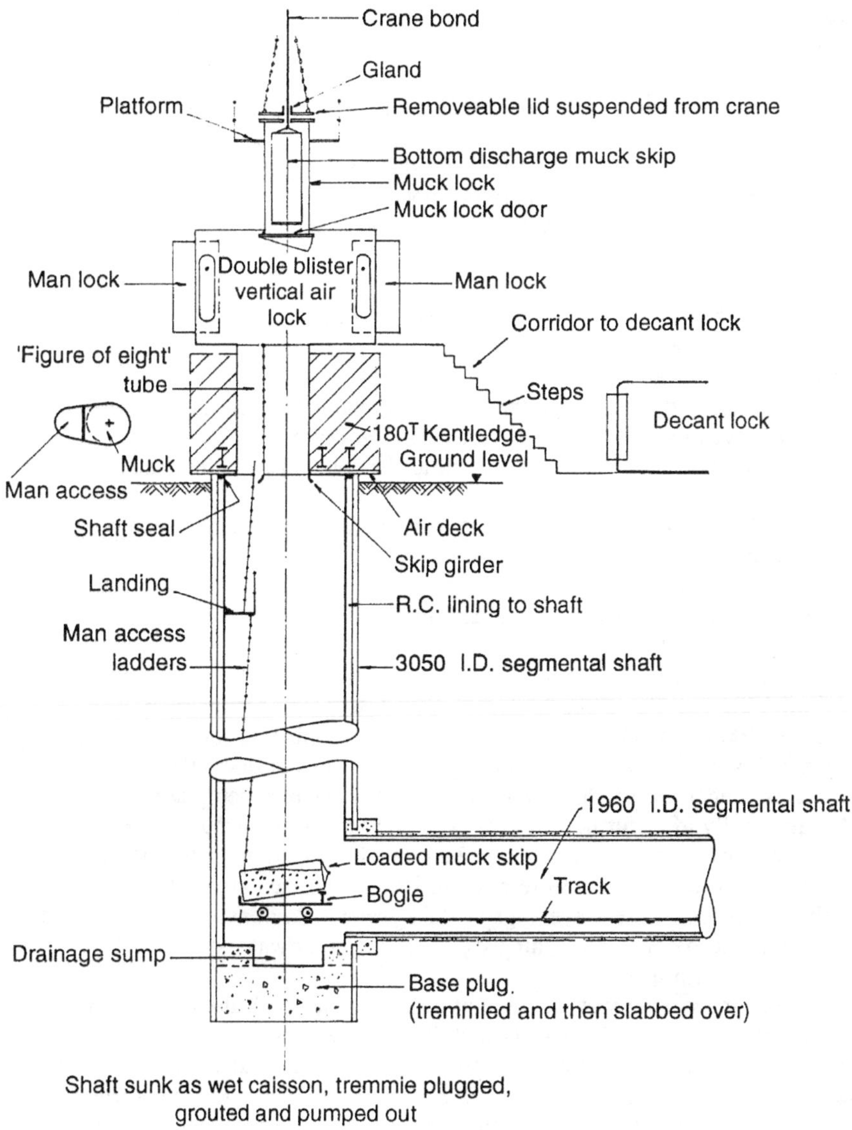

Fig.3. Section through lock and shaft

3 Use of compressed air

The short length of tunnel and consideration of ground conditions together with hydrostatic pressure meant that the most economical construction method was compressed air. It was felt that the use of compressed air was the only practical solution because:

1. The set up costs for a TBM for such a short drive were disproportionally high.
2. There was very limited space for shafts due to the quay wall tie rods. Large shafts for pipe jacking were therefore not feasible.
3. Ground treatment was difficult because of the presence of silts and the tunnel being under a waterway.

The limited length of tunnel also dictated the use of a decant lock, access to and from the tunnel being *via* blister locks.

The air deck was located on top of the south shaft at ground level because of the small shaft diameter and to facilitate the loading of about 180 tonnes of kentledge. Sealing was achieved by the use of cork gaskets (See Figure 3).

The compressed air plant installed adjacent to the south shaft comprised:

- 4 no. 1000 cfm (28.37 m^3/min) compressors (2 no. in use at any one time).
- 1 no. decant lock.
- 1 no. medical lock.
- 2 no. 350 kVA standby diesel generators.

The maximum amount of compressed air used was about 2500 cfm (56.74 m^3/min) but only for a short while at launch to get under the quay wall. Thereafter one 1000 cfm (28.37 m^3/min) compressor running at about half capacity was adequate, i.e. 12% of available capacity. There was considerable loss of air when tunnelling under the south quay wall with many bubbles being seen. Air loss reduced dramatically after further progress and comprehensive grouting. Bubbles in the harbour were ever present throughout the tunnelling operations despite pointing as tunnelling progressed. These were not confined to the face area. They were particularly noticeable on clear evenings at slack water. Pointing of the tunnel did not appear to affect the quantity of air used. No difficulty was experienced in passing under the north quay wall.

Compressed air working was carried out in accordance with the recommendations in the third edition of CIRIA Report 44 and the lock keepers were trained accordingly. All medical examinations and X-Rays were carried out at the North Sea Medical Centre at Great Yarmouth. The Appointed Doctor for the contract was Costain's own Occupational Health Medical Officer, Dr. David Shashall, who also attended some of the examinations. He was closely involved throughout the time the project was 'under air.' The medical aspects of the supervision were very carefully planned and included a visit to a similar installation in Bideford. The medical examinations were very thorough and included long-bone x-rays.

The lock keepers reported to Dr Shashall and all statistics were sent to him. He also regularly kept in contact with the medical lock attendants and the site agent. The medical lock attendants were carefully selected and they were in seeing that the requirements of CIRIA Report 44 were met. The average air pressure was 25 psig (1.72 bar) although there were some minor variations (see Figure 4). The tidal range did not appear to affect the air pressure required. Pressure was increased occasionally when the face became more damp. There was no obvious point of balance of air pressure and water pressure, the whole face usually being damp. The pressure was not raised to obtain a completely dry face because of the risk of a 'blow-out'.

Compressed air working commenced on 18 February and was completed on 19 June 1986, a total of 79 working days. Three-shift working was generally adopted throughout the tunnel drive. The total number of exposures was 1135 of which 523 were for less than four hours. Fifteen cases of Type 1 decompression sickness were reported and treated by recompression therapy, there being no Type II cases and no retreatments. Some of the workmen stated they had previous symptoms but had not reported these. The fifteen cases of decompression sickness amount to 1.32% of the total compressions.

Table 1 gives details of each case of Type 1 decompression sickness and Table 2 gives details of previous compressions for each of the patients. The relationship of each of the incidents to the working pressure is also shown on Figure 4.

The fifteen incidents of decompression sickness related to ten men as shown below by trade.

Resident Engineer	1
Labourer/miner	3 cases (1 man)
Miner	5 cases (4 men)
Pit-bottom man	4 cases (2 men, 1 man had 3)
Carpenter	1
Engineer	1

Of these cases, five men had previously worked in compressed air, four had not and there are no available records of the remaining person. Of the two personnel treated on three separate occasions, the labourer/miner had previously worked in compressed air, the pit bottom man had not. Both were engaged in heavy manual labour and worked long shifts as shown on Table 3.

Six cases of decompression sickness occurred after work on the day shift, seven after the back shift and two after the night shift. The ages of workers suffering decompression sickness are shown on Table 1.

A total of forty-six medical examinations were carried out for the project. Fifteen people examined did not work in compressed air. The reasons for not working were:

Table 1 Record of incidence of decompression sickness

NO	DATE	WORKER (Age)	PREVIOUS SHIFT (AND PRESSURE) psig	MAXIMUM TREATMENT PRESSURE psig	TIME AFTER DECOMPRESSION OF TREATMENT h	LOCATION OF DISCOMFORT	REMARKS
1	25/3/86	MINER C (25)	DAY (25)	32	5.67	BOTH KNEES	UNSATISFACTORY ACCLIMATISATION
2	25/3/86	MINER E * (34)	BACK (25)	28	2.5	RIGHT KNEE	
3	3/4/86	MINER A * (34)	NIGHT (25)	28	3.0	RIGHT KNEE	HEAD COLD, STOPPED FROM CA WORK UNTREATED EVENT 24/3/86
4	3/4/86	PIT BOTTOM A * (35)	BACK (25)	28	6.0	LEFT KNEE	
5	4/4/86	ENGINEER (30)	DAY (25)	30	7.5	BOTH ANKLES	STOPPED FROM CA WORK
6	8/4/86	MINER B (32)	DAY (25)	28	1.25	RIGHT KNEE	PREVIOUS UNREPORTED CASES POOR ACCLIMATISATION NOT SUITED TO HIGH PRESSURE
7	9/4/86	RESIDENT ENGINEER (37)	BACK (25)	27	-	LEFT KNEE AND ANKLE	MULTIPLE EXPOSURES TO 2.75 HRS. POOR ACCLIMATISATION
8	12/4/86	MINER A * (34)	BACK (25)	28	3.5	BOTH KNEES	HEAD COLD
9	15/4/86	MINER E * (34)	BACK (25)	27	5.25	RIGHT KNEE	POOR ACCLIMATISATION
10	19/4/86	PIT BOTTOM A * (35)	DAY (25)	28	3.5	RIGHT KNEE	ONE UNTREATED PREVIOUSLY UNREPORTED EVENT 11/4/86
11	25/4/86	PIT BOTTOM B (23)	BACK (24)	33	10.4	LEFT KNEE	HEATER OFF IN LOCK. COMPLAINTS OF BEING CHILLY
12	25/4/86	MINER A * (34)	NIGHT (24)	28	5.5	LEFT KNEE	
13	25/4/86	MINER D (29)	BACK (24)	29	7.7	BOTH ANKLES	HEATER OFF IN LOCK. COMPLAINTS OF BEING CHILLY
14	5/5/86	PIT BOTTOM A * (35)	DAY (25)	28	1.25	RIGHT KNEE	
15	5/6/86	CARPENTER (40)	DAY (25.5)	34	2.5	RIGHT KNEE	ONE UNTREATED PREVIOUSLY UNREPORTED EVENT 4/6/86

Table 2 Previous (eight) exposures prior to Type 1 event

1	DATE	19/02/86	19/03/86	20/03/86	21/03/86	24/03/86			
	PRESSURE psig	14.5	15	25	25	25			
	DURATION hrs	1.48	0.47	4.56	6.03	8.04			
2	DATE	18/02/86	19/02/86	20/03/86	21/03/86	24/03/86			
	PRESSURE psig	15	18	25	25	25			
	DURATION hrs	2.02	1.45	4.10	6.08	7.52			
3	DATE	20/03/86	21/03/86	24/03/86	25/03/86	26/03/86	27/03/86	01/04/86	02/04/86
	PRESSURE psig	25	25	25	25	25	25	25	25
	DURATION hrs	4.56	6.03	8.04	8.20	8.29	7.50	6.20	8.11
4	DATE	31/03/86	01/04/86	02/04/86	03/04/86	04/04/86	07/04/86	08/04/86	09/04/86
	PRESSURE psig	22	24	25	25	25	25	25	25
	DURATION hrs	0.08	6.20	8.11	8.12	7.35	8.11	7.49	8.05
5	DATE	20/03	01/04/86	26/03/86	27/03/86	01/04/86	02/04/86		
	PRESSURE psig	25	24	25	25	22	24		
	DURATION hrs	0.21	6.20	3.05	0.28	3.05	0.25		
6	DATE	23/03/86	24/03/86	25/03/86	26/03/86	01/04	02/04/86	03/04/86	04/04/86
	PRESSURE psig	25	25	25	25	22	25	25	25
	DURATION hrs	0.24	1.00	7.56	7.50	4.56	7.54	7.49	7.43
7	DATE	21/03/86	24/03/86	25/03/86	26/03/86	27/03/86	01/04/86	02/04/86	03/04/86
	PRESSURE psig	25	25	25	25	25	22	25	25
	DURATION hrs	0.13	7.52	0.14	0.24	0.12	3.05	0.21	2.43
8	DATE	01/04/86	24/03/86	04/04/86	07/04/86	08/04/86	09/04/86	10/04/86	11/04/86
	PRESSURE psig	25	25	25	25	25	25	25	25
	DURATION hrs	6.20	0.30	7.35	8.11	7.49	8.05	8.10	8.08
9	DATE	02/04/86	02/04/86	07/04/86	07/04/86	08/04/86	09/04/86	10/04/86	11/04/86
	PRESSURE psig	25	25	25	25	25	25	25	25
	DURATION hrs	7.57	8.11	7.43	7.59	8.10	7.54	7.56	6.14
10	DATE	07/04/86	03/04/86	09/04/86	10/04/86	14/04/86	15/04/86	16/04/86	17/04/86
	PRESSURE psig	25	25	25	25	25	25	25	25
	DURATION hrs	8.11	7.59	8.05	8.10	7.53	7.58	8.17	8.04
11	DATE	15/04/86	08/04/86	17/04/86	18/04/86	21/04/86	22/04/86	23/04/86	24/04/86
	PRESSURE psig	25	25	25	25	25	25	24	24
	DURATION hrs	8.06	7.49	8.15	4.36	7.24	7.47	7.56	7.43
12	DATE	15/04/86	16/04/86	17/04/86	18/04/86	21/04/86	22/04/86	23/04/86	24/04/86
	PRESSURE psig	25	25	25	25	25	24	24	24
	DURATION hrs	7.58	7.59	8.04	7.59	8.04	8.11	8.12	7.22
13	DATE	15/04/86	16/04/86	17/04/86	18/04/86	21/04/86	22/04/86	23/04/86	24/04/86
	PRESSURE psig	25	25	25	25	25	25	24	24
	DURATION hrs	8.06	8.17	8.15	4.36	7.24	7.47	7.56	7.43
14	DATE	23/04/86	16/04/86	25/04/86	28/04/86	29/04/86	30/04/86	01/05/86	02/05/86
	PRESSURE psig	24	25	25	25	25	25	25	25
	DURATION hrs	8.12	7.59	6.47	8.01	7.46	7.32	7.11	5.43
15	DATE	28/05/86	24/04/86	01/08/86	02/06/86	03/06/86			
	PRESSURE psig	24	24	25.5	2.5	25.5			
	DURATION hrs	8.12	7.22	5.53	6.22	8.11			

Unfit at examination	3
Failed lock test	1
Medical lock attendant	3
Left site	1
Failed first shift (claustrophobia)	1
Electricity Board engineer	1
Site Safety Officer did not enter tunnel	1
Did not work on project	4

Table 3 Exposures of men suffering multiple decompression sickness

	Number of exposures of:		
Exposure times	<4 h	>4-8 h	>8 h
Pit-bottom man A	7	50	14
Labour/miner A	2	28	19

Of those who were unfit, one had previously worked in compressed air in the Far East with no reported problems. However, he had a collapsed lung in his youth which bars him from such work in this country. Of the 31 who were passed fit, eighteen had previously worked in compressed air. Details of hearing loss noticed at the time of the medical examination i.e. prior to employment, are given in Table 4. No known further hearing tests were taken at the end of the work except for one person who had had partial deafness for some time. No deterioration was noted.

The decompression procedures were strictly controlled by experienced lock attendants. The size of the shaft meant that decanting was unavoidable. However, the decant lock was placed very close to the shaft and a 'dedicated' corridor provided to it so that all the men were quickly transferred without any absconders and pressurised within the permissible time limits. There was only one known incidence of the time limit been exceeded when a seal folded back on the door. Depressurisation was required to resolve the difficulty.

The decant lock was supplied by Tickhill Plant. It could hold 6 or 7 men comfortably and was covered with plywood to shelter it from the sun. Blankets, thermal suits and electric heaters were provided within.

Decompressions were generally carried out in accordance with Tables 18 and 19 of CIRIA Report 44, i.e. total times of 95 and 106 minutes for men who were in the tunnel for a full shift. Hot drinks were provided and a constant watch was kept on the occupants by the lock-keeper. No one was allowed to go to sleep. Reading material in keeping with the interests of the occupants was provided.

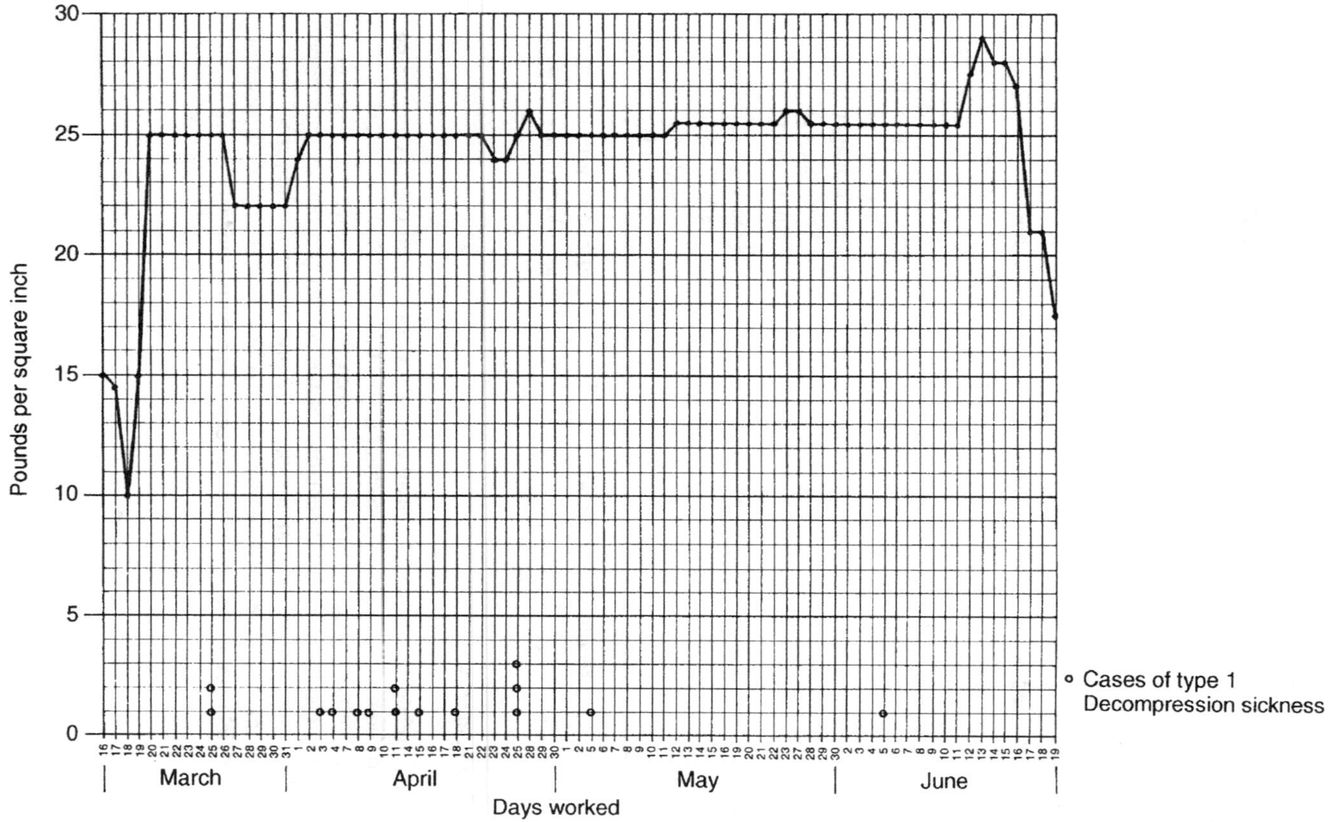

Fig.4. Working pressure and DCS incidents

Table 4 Results of hearing assessment

Total examined[1]	46
Number with hearing loss[2]	25 (54.3%)
Those with hearing loss and previous work in 'air'	20 (80%)
Trades (nos.) of those with hearing loss who had worked previously in 'air'	Miner (8) Labourer/miner Pit boss (2) Medical lock attendant (2) Carpenter (1) Project director (1) Tunnel inspector (1) Project manager (1) Engineer (1) Unknown(1)
Their ages	Over 50 (1) 40-50 (6) 30-40 (8) 25-29 (5)

Notes: 1. One record not available. 2. Defined as a loss of more than 20dB at any frequency

4 Concluding remarks

The data above relate to a relatively small tunnel project carried out in compressed air. The use of compressed air with decanting was the most practical means available for the project in question. It can be seen that much care was taken in the planning of the project and in implementation of the correct procedures with respect to the use of pressure in excess of atmospheric. Notwithstanding the precautions taken, it is seen that improvements can be achieved in the reporting of all cases of decompression sickness so that an accurate record can be obtained.

Discussion on Session 1: Construction projects – I

The session opened with a showing of a video of the works carried out by the Lilley-Misr Consortium on Contract 4 of the Cairo Wastewater Project, Egypt, which was presented by **Mr R J Reed**.

Mr J L Vincett presented his paper about the caisson sinking on that contract. **Mr P T Simpson** presented the paper on the Lake Lothing tunnel, on behalf of the authors, his colleagues, who could not be present. Following these presentations, the session chairman, **Mr J Buchanan** invited questions.

Mr R T Slocombe. On the two projects, had different kinds of side lock been considered to avoid decanting? Could we have more information about the interlocks on the doors.

Mr R S Hoad. At Lowestoft there was only one borehole on each side of the water crossing. At the time my firm advised the Contractor to undertake special bathymetric and sparker surveys to try to prove the uniformity of sediments and the absence of any buried gravel channel. Would you agree that this sort of site investigation should be carried out by the Engineer as part of the routine design of sub-aqueous tunnels?

Mr Buchanan. What were the air losses from the Cairo caissons in relation to the air pressures balancing the hydrostatic head?

Dr E P Kindwall. Was artificial ground freezing, e.g. by liquid nitrogen, considered to stabilise the ground for the Cairo caisson sinking?

Mr A Smith. In answer to Mr Slocombe, as I was Contract Manager for the Lake Lothing project I can explain that it was not practically possible to use a T-lock for direct decompressions for the following reasons.

The shaft diameter of 3.03 m i.d. meant that the air deck had to go at ground level, thus the kentledge (180 t) had to go on top. This meant that the kentledge obstructed any access to the air shaft below the blister lock to connect a T-lock.

I also confirm that a pneumatic interlock prevented unintentional removal of the muck lock lid when the lock was pressurised.

Mr Vincett. In response to the questions from Mr Slocombe, Mr Buchanan and Dr Kindwall:

1. Use of blister locks for decompression without decanting. Although it is possible to use a blister lock for decompression, it is restricted in size – a tee-head lock can be extended to any size required to accommodate the size of gang working on a shift.
2. Interlock to air exhaust prior to removing muck lock lid. It was confirmed that an interlock plate required the exhaust valve to be open, prior to releasing the clamps on the lid to the chamber. Further, a pneumatically operated pin prevented the interlock ring from moving to the 'top lid open' position until the air in the lock was fully discharged.

3. Air losses from the working chamber. It was necessary to apply sufficient air pressure to hold water levels down to the working chamber floor. During the underpinning phase, the large height to the air deck results in an excess of air pressure over external water head at the underside of the air deck, resulting in aggravated conditions for air loss. Use was made of ground freezing and/or chemical consolidation for commencing tunnel drives, but not for caisson sinking. Bentonite was used as a lubricant for caisson sinking, and to assist with sealing air losses around the caisson perimeter.

Mr Reed. Ground freezing was used later on Contract 1 at Cairo, using the brine system.
Mr Simpson. What Mr Hoad suggests would be beneficial.

(Reporters: A. Evans, F.M. Jardine)

6 Compressed air tunnelling for the Hong Kong Mass Transit Railway

J.G. Campbell and D.R. Gutteridge
Charles Haswell and Partners Limited, London, UK

Abstract
This paper describes the compressed air tunnelling that was carried out for construction of the three main stages of the Hong Kong Mass Transit Railway. Also covered is a description of the trial tunnels built prior to the main construction and which included research into the use of compressed air in Hong Kong. Data on the number of exposures to compressed air are given for various contracts. The number of incidences of decompression sickness events are described with a breakdown into Chinese and European workers.
Keywords: Hong Kong, Tunnelling, Compressed Air, Decompression Sickness, Exposures.

1 Introduction

Studies for the Hong Kong Mass Transit Railway (MTR) continued over many years culminating in a proposed system which was to be built in stages. Construction of the first stage, the Modified Initial System, commenced in 1975. Additions to the railway have continued until relatively recently, the last part being the Eastern Harbour Crossing.

Much of the tunnelling was carried out using compressed air in varied ground conditions. Air pressures up to 3.45 bar (50 psig) were employed.

This paper describes the development of compressed air tunnelling in Hong Kong and relates to specific contracts. Medical data are presented where they are available.

2 Geology

The area in and around Hong Kong is formed from marine and deltaic deposits of the Palaeozoic and Lower Jurassic periods and acid volcanic rocks of the Upper Jurassic. They were intruded by a large acid granite batholith during the Cretaceous

Engineering and Health in Compressed Air Work Edited by F.M. Jardine and R.I. McCallum. Published in 1994 by E & FN Spon. ISBN 0 419 18460 0

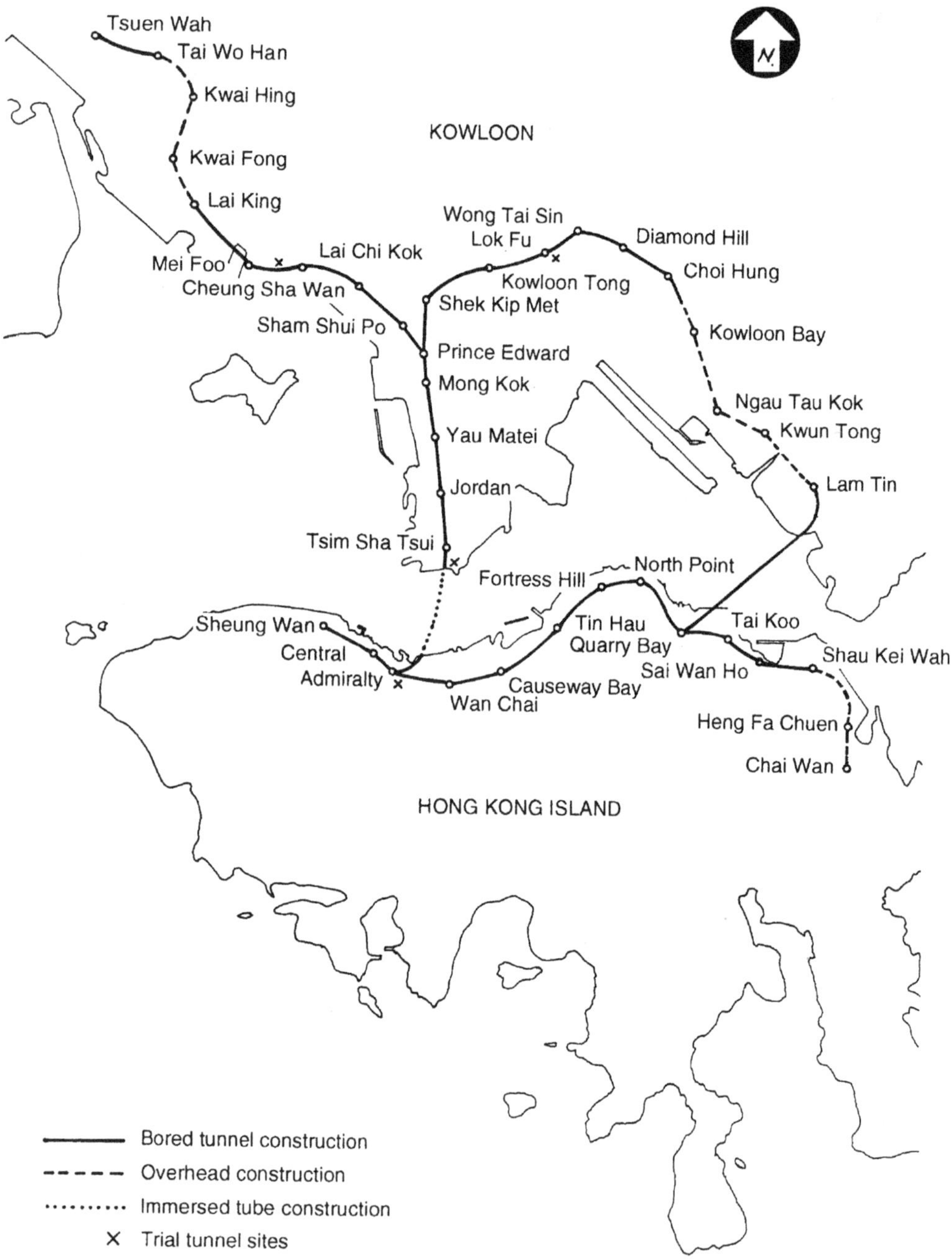

Fig. 1. Underground network of Hong Kong

period.

The fresh granite weathers to depths of 60 m or more through stages of slightly, moderately, highly, to completely weathered and finally to residual soil. It becomes progressively softer with weathering but varies widely from being dense and almost impermeable to being loose and permeable. A troublesome feature in the soft weathered granite is the random and frequent presence of relatively unweathered boulders (corestones).

When planning the MTR, it was considered that over 60% of the tunnelling would be in soft ground below the water table. The soft ground would comprise weathered granite, marine clays and silts, colluvium, alluvium and fill.

3 Trial tunnels

Soft ground tunnelling was virtually unknown in Hong Kong prior to the commencement of construction of the MTR. It was therefore decided to design and construct some trial tunnels and shafts to gain further knowledge of the ground conditions and test various tunnelling techniques.

The work involved was experimental in nature and was carried out on a cost reimbursement basis with a fee element. The works envisaged comprised a shaft and short length of tunnel at each of four sites selected so as to encounter the most varied and difficult conditions likely to be experienced on the MTR project (Figure 1). The contract proposed the use of compressed air in three of the tunnels and two of the shafts. However, because of delays in obtaining plant and other problems, compressed air was only used at two of the sites, these being at Admiralty and Lai Chi Kok.

In the absence of legislation for compressed air working, the recommendations set out in CIRIA Report 44 were adopted for this work. These requirements dictated that special measures had to be taken with air temperatures and humidity.

The climatic conditions in Hong Kong result in a mean maximum temperature in July of 30.5 °C but with the temperature often exceeding 32 °C. The mean relative humidity generally exceeds 80% from early March until August.

Information on rock temperatures indicated that this was constant at a depth of 10 m at about 27 °C. It was considered that higher temperatures might be encountered although seasonal variations would be small. In the event, ground water temperatures of between 26 and 29 °C were recorded.

Both these temperature conditions exceed the CIRIA Report 44 prescribed working requirements. These are that the temperature of the air supplied should not exceed 21 °C with a maximum wet bulb temperature in the working chamber of 27 °C. A ventilation and cooling system including recycling of air from the working chamber was therefore designed with the objective of maintaining temperature in the working chamber at 24 °C at 100% humidity. In the event the refrigerant plant in conjunction with the cooling tower did successfully deliver to the working chamber air at 19 °C with a humidity reading of approximately 75%, although recycling of air was not achieved due to difficulty with valves and air losses.

3.1 Admiralty trial

Shaft excavation was carried through fill, soft marine deposits and completely weathered granite down to fresh granite. The total depth of the shaft was about 19.4 m, the groundwater level being about 2.85 m below ground level. The upper section of the shaft was supported by Franki piles forming a contiguous pile wall with a pitch circle diameter of 6.55 m.

Compressed air was required below about 15 m in depth as the ground became live. Vertical airlocks were installed in the shaft and excavation recommenced with an air pressure of 0.69 bar, which progressively increased to 1.72 bar equating to the full hydrostatic head.

A 2.59 m internal diameter tunnel was constructed from the base of the shaft in compressed air, the initial pressure being 1.24 bar which was subsequently raised to 1.65 bar. The compressed air was effective in keeping the tunnel dry and allowed the tunnel face of completely weathered granite to be maintained without additional support. On removal of the air, a seepage of 0.57 l/s was evident.

The average air consumption during excavation was approximately 2.41 m^3/min or 0.34 $m^3/min/m^2$ of face area. The work was carried out in March and April. The relative humidity at the surface and the tunnel face were 65-70% and 85-90% respectively. The temperature at the tunnel face was around 28°C.

3.2 Lai Chi Kok trial

The purpose of this trial tunnel was to obtain information on tunnelling in soft marine silts and sands and to determine whether normal soft ground tunnelling techniques would be appropriate.

Shaft excavation was through fill, clayey silt and sand with boulders and then bedrock. The initial part of the tunnel was excavated in rock followed by completely weathered granite and then mixed soft ground consisting of black marine silty sand in the upper third, a variety of lenses of sand and silt at the axis and coarse sands with cobbles and small boulders in the invert.

With tunnel advance the black marine deposits continued in the upper part of the face, but the clay and silt content increased and in some areas it took on a soft cohesive consistency. The lenses of mixed material at tunnel axis were later replaced by a larger depth of coarse sand with cobbles and boulders.

A horizontal air lock was installed in the tunnel while in rock and compressed air at a pressure of 1.38 bar was applied before the last blast. The shield was then quickly brought up to the face.

The completely weathered granite again proved excellent for tunnelling under compressed air and air losses were extremely low.

The air pressure was increased to a maximum of 1.65 bar to balance the full hydrostatic head when a layer of coarse sand with cobbles was present in the invert. This corresponded with an increase in air consumption and evidence of air escaping was seen at the surface of a drainage nullah about 100 m from the tunnel face.

The work was carried out in May and June; the relative humidity readings were 75-80% and 85-95% at the surface and in the tunnel respectively. The temperature at the tunnel face averaged 30°C.

3.3 Decompression sickness in trials

At Admiralty there were four cases of decompression sickness and at Lai Chi Kok only one. All were Type I and not serious.

The cases at Admiralty were all Chinese workers and three of these occurred during the initial work in the shaft. The case at Lai Chi Kok was a European non-shift worker. Table 1 sets out data with respect to decompression sickness.

Table 1 Incidence of decompression sickness in trial tunnels

Class of worker	Number of Decompressions	Number of Recompressions	R/D %
Chinese shift	1306	3	0.229
Chinese non-shift	1157	1	0.086
All Chinese	2463	4	0.162
European shift	395	Nil	Nil
European non-shift	354	1	0.282
All European	749	1	0.134
TOTAL	3212	5	0.156

4 Modified Initial System

The Modified Initial System (MIS) was built between 1975 and 1979 by a number of international and local contractors. The contracts were let on a design-and-construct basis, although tenderers were given site investigation data and details of the assumptions made by the Engineer in preparing his outline design. These assumptions included the provision of compressed air in places.

Construction methods envisaged by the Engineer conformed directly to the expected ground conditions. All drives were relatively short and ground conditions variable, hence full-face mechanical excavation was not considered viable. In soft ground, hand methods or manual shields were indicated using compressed air where required.

Compressed air tunnelling was adopted on five contracts:

Contract 201 - Lok Fu to Wong Tai Sin, running tunnels
Contract 203 - Kowloon Tong to Lok Fu, running tunnels
Contract 107 - North Nathan Road, running tunnels
Contract 108 - Jordan to Waterloo (now called Yau Ma Tei), running tunnels
Contract 109 - Tsim Sha Tsui – Admiralty – Central, running tunnels excluding the immersed tube across the harbour.

Table 2 Incidence of decompression sickness on Modified Initial System

Contract No.	Length of drive in CA (m)	Ground conditions[1]	Pressure (bar (psig))		No. of exposures	Decompression Sickness (DCS) events			DCS rate (%)	Permanent lining type (i.d.)[2]	Progress (m/wk)		Remarks
			Min.	Max.		I	II	Total			Min.	Max.	
201	1,225	A+B+C	0.3 (4.4)	1.8 (26.1)	28,255	21	-	21	0.074	PCC (5.1m)	19.5	61.0	Low DCS incidence
203	378	E+D+G	0.6 (8.7)	1.1 (15.95)	36,500	-	-	-		*In-situ* concrete (4.9m)	7.8	25.0	-
107	(16 faces worked)	C+F+G	0.7 (10.15)	2.45 (35.5)	226,554	1,538	4	1,54 2	0.68	SGI and PCC (Varied from 11.5m to 4.9m)	See Table 4	-	See Table 3
108	(6 faces worked)	C+B	-	1.0 (14.5)	-	-	-	-	-	*In-situ* concrete (4.9m and 11.5m)	6.0	24.0	No reported case of bends
109	2,500 (6 faces worked)	G+B+A+C	1.0 (14.5)	2.3 (33.4)	79,676	317	-	317	0.398	PCC (4.9m) SGI (5.0m)	3.6	28.0	-

(1) A = Fill B = Colluvium C = CWG D = HWG E = Alluvium F = Fresh Granite G = Marine deposits
(2) PCC = Precast concrete segmental lining SGI = Spheroidal graphite iron segmental lining

4.1 Contract 201 – Lok Fu to Wong Tai Sin

Twin tunnels in the section from Lok Fu to Wong Tai Sin were driven consecutively by a Bade shield. This was an entirely soft ground section through fill material, colluvium and completely weathered granite. The initial drive of 86 m was in free air aided by ground water lowering using deep wells. Air locks were then installed in the tunnels and the remainder of the 615 m drive was completed under compressed air with a maximum pressure of 1.8 bar.

It can be seen from Table 2 that there was a low incidence of decompression sickness, only 21 cases out of a total of 28,255 exposures being reported (0.074%).

4.2 Contract 203 – Kowloon Tong to Lok Fu

An access shaft was sunk approximately midway between stations and the 450 m long tunnels to Kowloon Tong driven under compressed air. Temporary support of steel ribs and lagging was used prior to casting an *in-situ* concrete permanent lining. This led to a larger area of exposed ground compared with the Engineer's assumption that shields and segmental lining would be used. Fairly high air consumption resulted.

During the initial stages of the Up-track drive it was not possible to raise the air pressure above 0.65 bar because of supply limitations. At this pressure hydrostatic balance was achieved at 1 m below axis level resulting in slurrying in the tunnel invert and settlement of the arch ribs. An additional air supply pipe was installed which resulted in air consumption rising to 300 m^3/min. The *in-situ* lining was brought close to the tunnel face which allowed a pressure of 1 bar to be maintained and the air consumption eventually to drop to 180 m^3/min.

A comparison of the air consumption related to air pressure and exposed tunnel length is given in Figure 2.

4.3 Contract 107 – North Nathan Road

Contract 107 comprised all the tunnelling work between Waterloo, Argyle (now called Mong Kok) and Prince Edward Stations. Between Waterloo and Argyle, there were four running tunnels and two crossover tunnels while from Argyle to Prince Edward there were four running tunnels, reducing to three north of the high-level crossover. The total tunnel length comprised about 2580 m of 4.9/5.0 m internal diameter and 240 m of 10.0/11.6 m internal diameter. All the tunnels were driven in compressed air except for a short length at the northern end of the contract. Some statistics relating to the compressed air works are given in Tables 2, 3 and 4.

The ground encountered varied considerably, with the predominant material being completely weathered granite containing a varying and occasionally high proportion of fresh granite corestones. Other materials were marine deposits and from moderately weathered to fresh granite. The water table was generally 2 m below ground level. Cover varied from 8 to 22 m.

Coarse marine sand of uniform particle size and up to 2 m thick was encountered in the crown of one tunnel drive. This resulted in a doubling of air consumption.

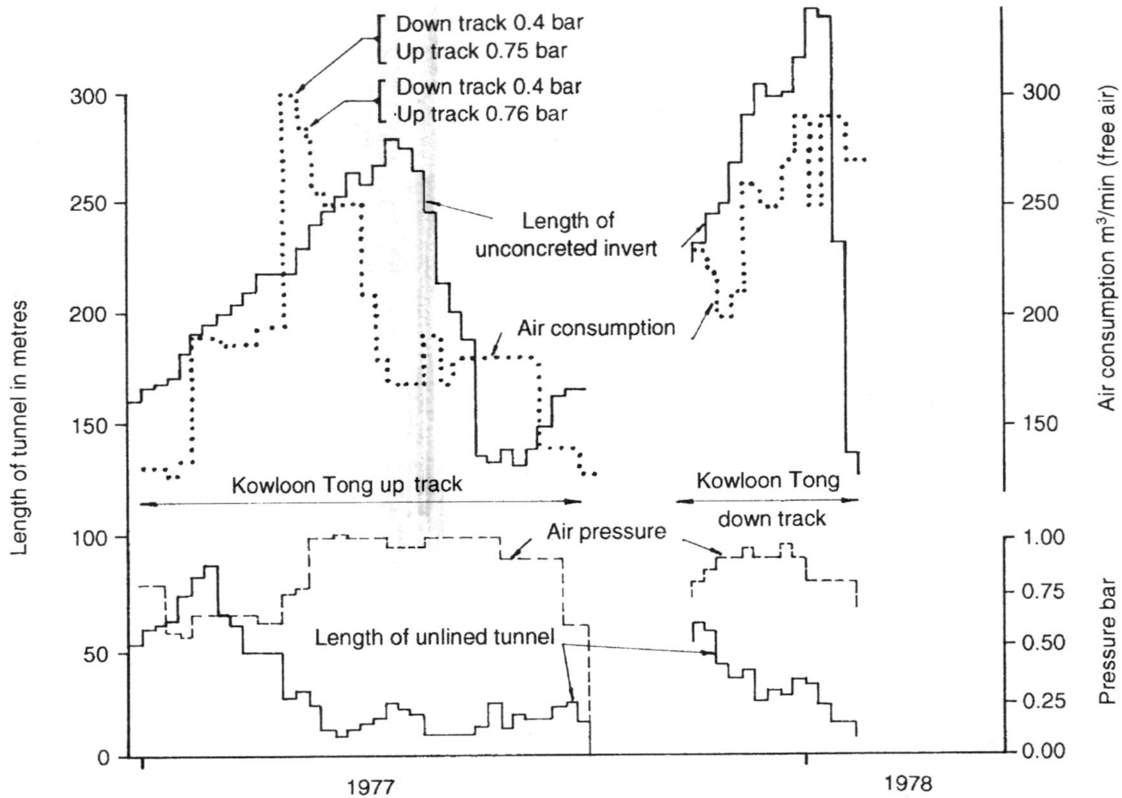

Fig.2. Comparison of air consumption with air pressure and exposed tunnel lengths: Kowloon Tong to Lok Fu

Table 3: Contract 107: Incidence of decompression sickness (DCS)

Pressure* (lbf/sq.in) (bar)	European			Chinese			Total		
	Exposures	DCS events	DCS rate (%)	Exposures	DCS events	DCS rate (%)	Exposures	DCS events	DCS rate (%)
14-16 0.95-1.09	3 244	1	0.031	20 697	3	0.014	23 941	4	0.017
16-18 1.09-1.22	2 859	1	0.035	15 081	2	0.013	17 940	3	0.017
18-20 1.22-1.36	812	5	0.616	6 209	6	0.097	7 021	11	0.157
20-22 1.36-1.50	1 121	4	0.357	7 722	21	0.272	8 843	15	0.283
22-24 1.50-1.63	1 546	6	0.388	9 933	24	0.242	11 479	30	0.261
24-26 1.63-1.76	1 470	12	0.816	6 999	27	0.386	8 469	39	0.460
26-28 1.76-1.90	1 275	15	1.176	9 856	74	0.751	11 131	89	0.800
28-30 1.90-2.04	3 695	31	0.839	25 327	211	0.833	29 022	242	0.834
30-32 2.04-2.18	10 097	174	1.723	79 289	673	0.849	89 386	847	0.948
32-34 2.18-2.31	1 516	24	1.583	13 888	178	1.282	15 404	202	1.311
34-36 2.31-2.45	279	-	-	1 121	19	1.695	1 400	19	1.357
36-38 2.45-2.58	390	1	0.256	2 128	26	1.222	2 518	27	1.072
14-38 0.95-2.58	28 304	274	0.968	198 250	1264	0.638	226 554	1538	0.679

* Tables in the Hong Kong regulations are in imperial units. Contract 107 used gauges in imperial units. All other contracts used metric pressure gauges.

Table 4 Contract 107: Compressed air statistics and construction dates

	Waterloo to Argyle	Argyle to Prince Edward
Installed capacity, m^3/min	842	365
Installed horsepower	4290	2180
Standby diesel generator capacity, kVA	1875	625
Working pressure, bar	0.5-2.5	0.5-2.5
Tunnel temperatures, °C:		
average	27	27
maximum	32	32
Construction dates:		
Commenced sinking access shafts	August 1976	December 1976
Last tunnel ring built	July 1979	November 1978
Substantial completion	December 1979	July 1979

Ground treatment was used for the construction of a high-level crossover. It was so successful that the pilot tunnel was constructed in free air. Free air would also have been satisfactory for the enlargement, but the running tunnels beneath were pressurised to over 2 bar at the time and it was considered prudent to use a nominal pressure of 0.7 bar.

Two compressed air installations were required with provision being made for additional compressors or interchanges of machines if required. At Hamilton Street, the installed capacity amounted to 29,800 cfm (843.8 m^3/min of low pressure air, 21,000 cfm (594.7 m^3/min) of chiller/dryer and 13,000 cfm (368.1 m^3/min) of standby using diesel generators. The installed capacity at Playing Field Road was 12,900 cfm (339.8 m^3/min). Both installations were of sophisticated design in that it was necessary to provide air conditioning, water cooling, chiller/dryers for the low pressure air together with standby diesel generators and in general a 'fail-safe' system. As more faces opened up, and with continuous grouting operations, the temperatures in some faces rose to 33 °C and water-cooled spot coolers were introduced into the tunnels with successful results.

Complete medical facilities were provided at each of the sites with three medical locks at Hamilton Street and one at Playing Field Road. A qualified medical doctor was employed full time supported by an experienced expatriate medical lock attendant.

At one time 16 faces were being worked concurrently with more than 700 men working in compressed air. The shifts had to be staggered so that three full decompressions could be run in the man locks in each 8 hour shift.

Some modifications were made to the Engineer's assumed tunnel lining for the crossover sections as a result of concern over the high excess air pressure at the crown of a large-diameter tunnel.

As a result flat-bottomed and elliptical linings were used.

4.4 Contract 108 – Jordan to Waterloo

As half of the tunnelling to be constructed between Tsim Sha Tsui and Waterloo Stations would be in rock the Contractor elected to adopt rock tunnelling techniques throughout. Compressed air tunnelling was carried out adjacent to the stations where soft ground occurred. In all there were six compressed air working locations with two faces working at any one time.

4.5 Contract 109 – Tsim Sha Tsui to Central (Hong Kong Island)

Contract 109 included all the tunnels between Tsim Sha Tsui and Central Stations with the exception of the immersed tube crossing the harbour.

The tunnels north of the immersed tube were driven for much of their length in a mixed face of soft ground and rock. Colluvium and marine deposits were evident for a short length. Fill and marine deposits were more frequent south of the harbour and a large section of the tunnelling was through weathered granite with corestones.

All tunnels were shield driven in compressed air and/or with the aid of ground treatment and were started from three on-line shafts. Sixteen items of compressed air plant were installed.

The tunnels either side of the harbour were driven under compressed air up to and into a specially grouted fill prior to socketing into the immersed tube. Some data on the incidence of decompression sickness are included in Table 5.

Table 5 Contract 109: Incidence of decompression sickness (DCS): pressure exceeding 1 bar but not exceeding 2 bar

	Exposures	DCS Events	DCS rate, %
European	3 300	3	0.090
Chinese	24 422	164	0.671
Japanese	9 956	82	0.823
Korean	41 970	68	0.162
Total	**79 674**	**317**	**0.398**

5 Tsuen Wan Extension

The Tsuen Wan Extension (TWE) was built between 1977 and 1982. The works included 22 km of rail track of which 10.41 km are in tunnel.

Unlike the MIS, the TWE was let on the basis of the permanent works being designed by the Employer's consulting engineer. Compressed air was required for ground stabilisation on three of the four contracts. These were:

Contract 301 - Prince Edward to Cheung Sha Wan running tunnels

Contract 302 -	Cheung Sha Wan to Lai Wan (now called Mei Foo) running tunnels
Contract 305 -	Kwai Hing to Tsuen Wan running tunnels

As the contractors for the three contracts had also been employed on MIS compressed air tunnelling contracts, the same designs and equipment were used for low-pressure installations.

5.1 Contract 301 – Prince Edward to Cheung Sha Wan

Compressed air was used for sinking shafts by the pneumatic caisson method for the tunnels between Prince Edward and Sham Shui Po. On completion of sinking, an airdeck was installed in the shaft to allow construction of the access adits to the tunnels together with an initial length of the running tunnels. Horizontal airlocks were installed in the access adits to allow completion of the work.

The plant arrangement on this section of the Contract produced 414 m^3 of low-pressure compressed air per minute. Standby capacity was provided by diesel generator producing 720 kW.

Boiler-type airlocks were used, sized to accommodate 34 persons (43 m^3) for the Up-track tunnel and 42 persons (54 m^3) for the Down-track tunnel.

Table 6 details the incidence of compressed air sickness.

5.2 Contract 302 – Cheung Sha Wan to Lai Wan

Mid-point shafts were sunk for both sections of tunnel, the shafts being on-line for the sections between Cheung Sha Wan and Lai Chi Kok and off-line in a section of rock tunnel between Lai Chi Kok and Lai Wan.

The tunnels between Cheung Sha Wan and Lai Chi Kok were generally in completely weathered granite with some corestones. Air decks were installed in the shafts with three-compartment man locks at the shaft top to allow tunnel driving to commence. The air locks were installed within the tunnel after a sufficient length had been driven.

Between Lai Chi Kok and Lai Wan the initial tunnel drives were all in rock. The tunnel drives to Lai Wan Station were completed by rock tunnelling techniques in completely weathered granite in compressed air and using *in-situ* reinforced concrete airlocks. The tunnels to Lai Chi Kok were driven in mixed rock/soft ground, marine deposits and variable weathered granite. Boiler locks were used in these drives. This section of tunnels was driven from the same site as one of the trial tunnels.

Compressed air pressures of up to 2.4 bar were employed. The installed capacity of the compressed air plant was 264 m^3/min. Standby capacity was provided by diesel generators with a total output of 1620 kW.

5.3 Contract 305 – Kwai Hing to Tsuen Wan

Tunnelling conditions on this contract were generally a mixed face of soft ground and rock with a very short length completely in soft ground. Compressed air pressures of less than 1 bar were used.

Table 6: Incidence of decompression sickness on Tsuen Wan Extension

Contract No.	Length drive in CA (m)	Ground conditions[1]	Pressure (bar (psig))		No. of exposures	Decompression Sickness (DCS) events			DCS rate (%)	Permanent lining type (i.d.)[2]	Progress m/wk		Remarks
			Min	Max		I	II	Total			Min	Max	
301	1,200	C+A+E+G	1.4 (20.3)	3.45 (50.0)	122,992	883	4	837	0.68	PCC (5.5m)	1	42	-
302	300	G+C	0.95 (13.8)	2.4 (34.8)	124,258	454	0	454	0.37	PCC (5.3m) SGI (5.5m)	1	29	-
305	300	V	-	0.9 (13.1)	-	-	-	-	-	*In-situ* concrete (5.0m)	8	9	No records registered

(1) A = Fill B = Colluvium C = CWG D = HWG E = Alluvium F = Fresh Granite G = Marine deposit V = Mixed face

(2) PCC = Precast concrete segmental lining SGI = Spheroidal graphite iron segmental lining

6 Island Line

Compressed air was used extensively during construction of the Island Line including the following contracts:

Contract 402 - Running tunnels Sheung Wan to Central, the over-run tunnel to the west of Sheung Wan and the station tunnels at Sheung Wan.

Contract 403 - Tunnels from Central to Admiralty and Central to Sheung Wan.

Contract 404 - Running tunnels Admiralty to Wan Chai, the Wan Chai station platform tunnels, passenger adits and part of the length of running tunnels from Wan Chai to Causeway Bay.

Contract 405 - Tin Hau Station platform tunnels, running tunnels Causeway Bay to Tin Hau, sundry cross passages and passenger adits and upper level of Causeway Bay Station.

6.1 Contract 403 – Central to Admiralty

The extension to the station at Central was constructed in bored tunnel, using compressed air in conjunction with ground treatment without a shield. The 8 m diameter platform tunnels were constructed as an enlargement of the running tunnels, in a mixed face of marine deposits overlying colluvium (one tunnel) and fill overlying marine deposits and weathered granite in the other tunnel. The air pressure was maintained at under 1 bar reducing seepage at the face to an insignificant amount.

6.2 Contract 404 – Admiralty to Wan Chai

The upper running tunnel from Admiralty to Wan Chai Station was constructed initially in an area of ground treatment using a form of New Austrian Tunnelling Method (NATM) in free air. The ground varied from completely weathered to slightly weathered granite and blasting techniques were employed as required. As the tunnel approached the intersection with Hennessy Road, the old water front, it was intended to construct a shield chamber for the erection of a tunnel shield to be used in conjunction with compressed air. Whilst excavating for the shield chamber adjacent to the rock/soft interface a cave in occurred and the tunnel was filled with completely weathered granite. On the surface a large hole appeared, this was filled with rock from Tai Koo Station excavation and cement bentonite ground treatment. Tunnelling resumed six months later and the drive was excavated with no further mishaps. Air pressure of 3.3 bar was required.

6.3 Contract 405 – Causeway Bay to Tin Hau

The upper platform tunnel of Causeway Bay Station was excavated using ground treatment and compressed air. An annulus of ground treatment was placed above the top of the tunnel to reduce air losses. The tunnel was excavated in NATM using a combination of fibreglass rock bolts, shotcrete and ribs and mesh, and the 7.6 m diameter cast iron tunnel lining was erected at a distance back from the tunnel face.

7 Contract documents

At the time of the trial tunnels there was no existing legislation in Hong Kong which covered compressed air working. Following discussions with the Labour Department during the trial tunnels and later during the preparation of the Contract Documents for the Modified Initial System, the Hong Kong legislation under the Factories and Industrial Undertakings (Work in Compressed Air) Regulations 1975 were approved. These were essentially similar to the CIRIA Report recommendations other than the provision of hot drinks not being obligatory except for persons undergoing treatment in a medical lock, and the deletion in the Hong Kong regulations of reference to air inlet temperature. Reference to the working chamber temperature alone is considered sufficient, since the air inlet temperature must be low enough to keep the ambient temperature in the working chamber at the regulation level.

The Specification for low-pressure compressed air working called for the following minimum requirements in respect of the capacity of air supply and emergency standby:

1. Capacity to be provided at the rate of 5 m^3/min of free air/m^2 of excavated face.

2. Additional standby capacity of 50% of the above.

3. Standby electrical supply to be installed sufficient to provide the 50% standby air supply capacity referred to above.

The volume of 5 $m^3/min/m^2$ was based on the figure of $12D^2$ quoted by Hewett and Johannesson (1922) and was supported by data obtained from driving the trial tunnel at Admiralty. Greater air losses were experienced at Lai Chi Kok.

8 Concluding remarks

It can be seen from the foregoing that compressed air was successfully used to construct driven tunnels in soft ground for the Hong Kong Mass Transit Railway. Use of compressed air was found to be economic particularly when considering the very varied nature of the soils encountered.

The local labour adapted well to working in the environment and cases of decompression sickness were relatively low. The majority of the work was carried out with due regard to health and safety, the Hong Kong legislation being based on CIRIA Report 44. However, there were several incidences where mistakes were made and illegal incidents occurred.

9 References

Cater, R.W., Shirlaw, J.N., Sullivan, C.A. and Chan, W.T. (1984) Tunnels constructed for the Hong Kong Mass Transit Railway, *Hong Kong Engineer*, Oct., pp 37-49

Haswell, C.K. and Umney, A.R. (1978) Trial tunnels for the Hong Kong Mass Transit Railway, *Hong Kong Engineer*, Feb., pp 15-23

Haswell, C.K., Umney, A.R., Hall, P., Hansen, F.J., Storey, F.G., Archer, G.O. and Langfield, R.A. (1980) Hong Kong Mass Transit Railway Modified Initial System : design and construction of the driven tunnels and the immersed tube, *Proc. Instn Civ. Engrs.* Part 1, 1980, 68, Nov., pp 627-655

Haswell, C.K. and Campbell, J.G. (1985) Compressed air tunnelling with special reference to the Hong Kong Mass Transit Railway, *Proc. Conf. Eurotunnel '83*

Haswell, C.K. (1987) Tunnels for the new Hong Kong Metro, *Tunnels and Tunnelling*, November 1987, pp 31-35

Hewett, B.H.M. and Johnannesson, M. (1922) *Shield and compressed air tunnelling*, McGraw Hill, New York, pp 234-5

Howat, M.D. (1987) Tunnelling through Hong Kong marine deposits, *Hong Kong Engineer*, 1987, April, pp 15-21

Howat, M.D. and Cater, R.W. (1983) Ground settlements due to tunnelling in weathered granite, *Proc. Int. Symp. on Engg. Geol. and Underground Constrn.*, Int. Assoc. Engg. Geol., Lisbon, Portugal

McFeat-Smith, I., Turner, V.D. and Bracegirdle, D.R. (1985) Tunnelling conditions in Hong Kong, *J. Hong Kong Institution of Engineers*, 1985, pp 13-24

Acknowledgements

The authors wish to express their thanks to the Hong Kong Mass Transit Railway Corporation for permission to publish this paper.

7 Use of compressed air for shaft sinking and tunnelling on Contracts 3, 12 and 12/5A of the Greater Cairo Wastewater Project, Egypt

D.J. Court
Edmund Nuttall Ltd., Camberley, UK

Abstract
This paper deals with the installation, operation and supervision of compressed air works for the construction of shafts and tunnels in the water-bearing sands of Cairo. The paper covers the plant installations, health and safety, training and general operations and provides information on the numbers of decompressions and details of decompression sickness, etc. The paper shows the different ways in which compressed air was used to carry out the construction work and the effect of the compressed air on the surrounding areas: it also shows how the contractor changed methods to limit the use of compressed air on the latest contract.
Keywords: CIRIA Report 44, Decompression Sickness, Compressed Air, Caissons, Tunnelling.

1 Introduction

Since the beginning of 1985, the Anglo-Egyptian Cairo Wastewater Consortium (CWC) has been engaged in the construction of shafts and sewer tunnels for the Greater Cairo Wastewater Project. CWC is a consortium of UK companies comprising Balfour Beatty Construction Limited, Cementation International Ltd., Edmund Nuttall Ltd. and Tarmac Overseas Ltd. together with Egypt's largest civil engineering contractor, The Arab Contractors (Osman Ahmed Osman & Company).

The Greater Cairo Wastewater Project is one of the largest ever undertaken and is designed to upgrade the existing overloaded and outdated sewerage system with a new gravity-fed system capable of coping with all the sewerage demands of a city with a population in excess of 12 million. The project is split geographically into the East and West Banks of the Nile. The West Bank is funded by the United States (US) and the East Bank has to date been funded by an initial grant of £50 million from the United Kingdom (UK) together with UK-backed loans totalling £185 million. The total East Bank Project when completed will consist of 33 km of branch tunnels feeding into a 12 km long main spine tunnel running the full length of the city (Figure 1).

Engineering and Health in Compressed Air Work Edited by F.M. Jardine and R.I. McCallum. Published in 1994 by E & FN Spon. ISBN 0 419 18460 0

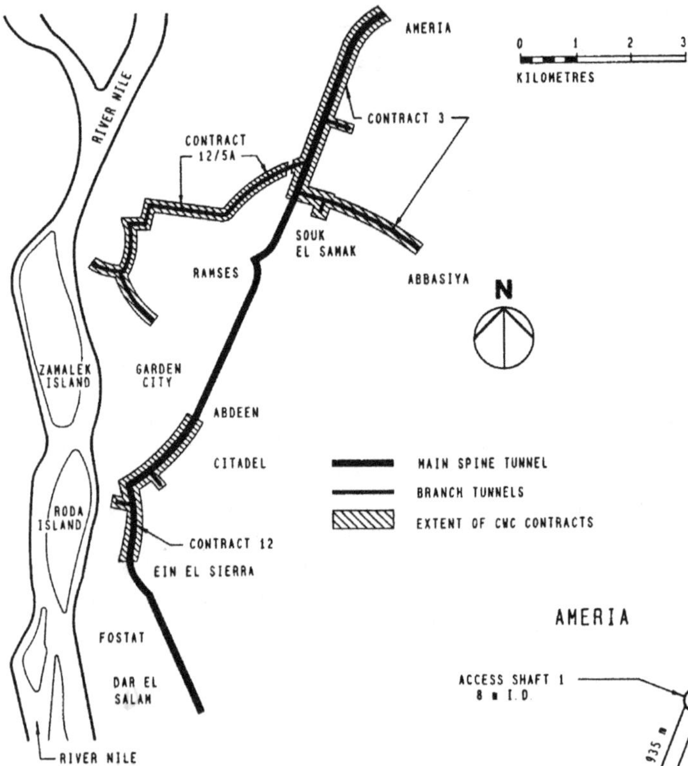

Fig. 1. East Bank Tunnels

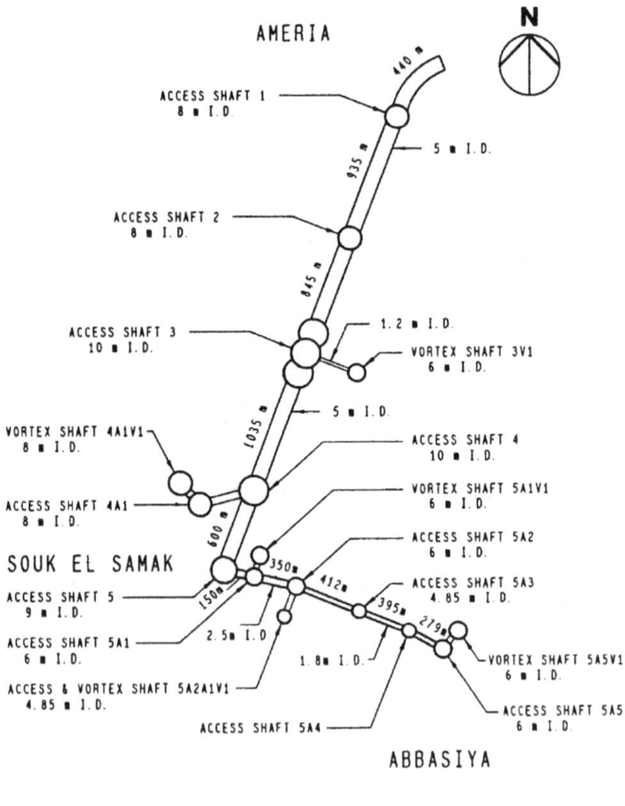

Fig. 2 Contract 3: Ameria to Souk el Samak

The main spine tunnel finishes at Ameria in the north east of Cairo where the new Ameria Pumping Station lifts the sewage up into a series of culverts through which the sewage will flow out to a new treatment plant at Gabal el Asfar.

To date four out of the six tunnel contracts have been constructed such that the main spine tunnel together with 10 km of branch tunnel is now complete. CWC have constructed three of these contracts, i.e. Contracts 3 and 12 which comprised 7.3 km of main spine tunnel together with 2.8 km of branch tunnel and Contract 12/5A which comprised 6.5 km of branch tunnel.

With the water table in the East Bank area of Cairo being no more than 2 m below ground level all the tunnels and shafts have had to be constructed in saturated sands, gravels and silts under heads of water as great as 26 m and as such extensive use was made of compressed air to a maximum working pressure of 38 psig (2.6 bar).

2 Contract 3: Ameria to Souk el Samak

Contract 3 (Figure 2) was the first contract to be awarded to CWC and commenced in February 1985. The contract consisted of 4 km of 5.0 m finished diameter segmentally lined tunnel from Ameria Pumping Station to Souk El Samak with a maximum depth to invert of 23 m together with 16 no. shafts, ranging from 4.85 to 10 m internal diameter and 2 km of segmentally lined branch tunnel ranging in finished diameter from 2.8 to 1.2 m at an average depth of 15 m.

The contract documents required the main tunnel to be driven with a slurry tunnelling machine and envisaged that all other tunnels would be driven using shields under compressed air. The shafts were designed by the Engineer to be sunk as caissons for the first 10 m and then completed with a segmental lining by underpinning in compressed air.

At the outset, CWC proposed changing the method of shaft construction to full-depth caissons sunk under compressed air. This method was felt by CWC to have numerous advantages not least of which was the fact that by caisson sinking full depth there would be a considerable reduction in the requirement for compressed air. Subsequently the proposal, which was accepted by the client and the Engineer, has been proved truly successful, and CWC have now sunk 48 no. shafts in this manner in Cairo.

In a similar vein CWC also proposed modifying Shaft 3 which was intended to be the main tunnel drive shaft for both slurry machines. The original concept was of a 10 m I.D. shaft from which CWC would have had to have constructed two 7 m diameter launch chambers. As the ground was expected to be a coarse-grained single-sized sand the consumption of compressed air would have been significant and a large volume of ground treatment would have been required. Instead CWC proposed and constructed a triple-cell caisson (Figure 3) with the outer cells forming the launch chambers, thus removing the need for large volumes of compressed air and ground treatment.

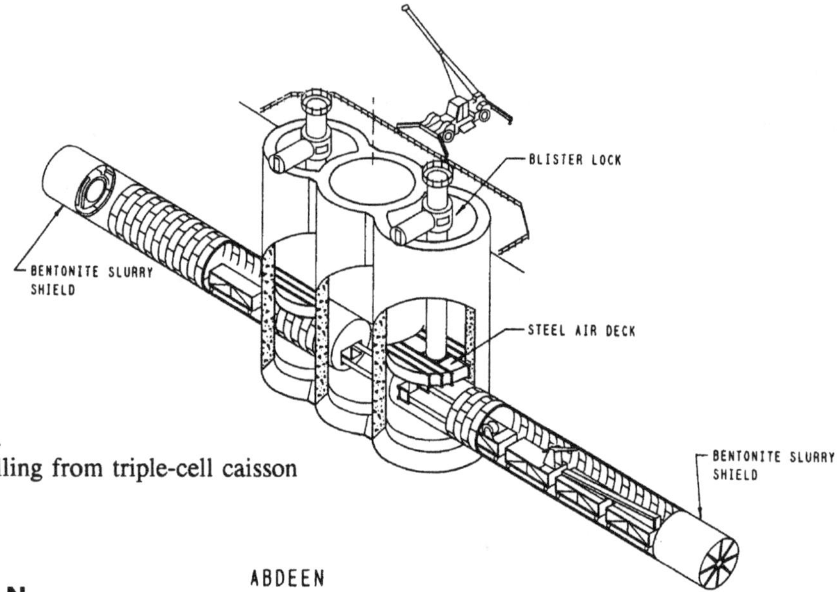

Fig. 3.
Tunnelling from triple-cell caisson

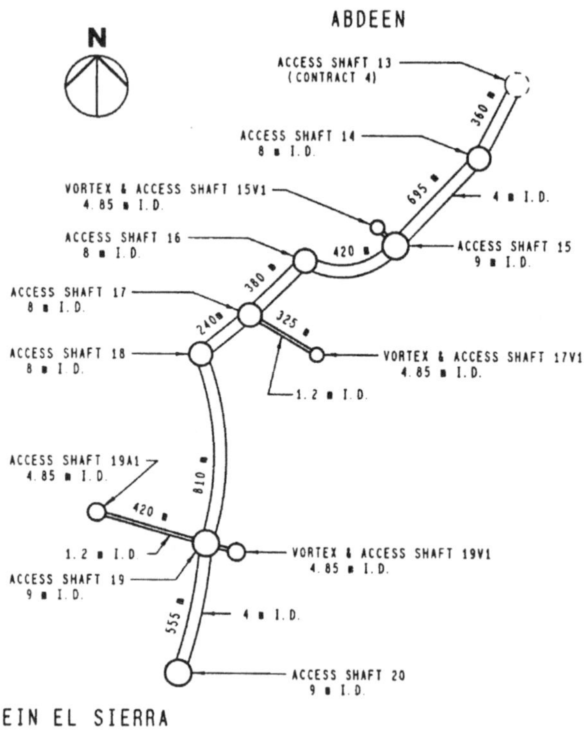

Fig. 4. Contract 12:
Abdeen to Ein el Sierra

The small-diameter branch tunnels were constructed using hand shields under compressed air; and the short transition chambers required for future contracts were constructed by hand using compressed air in ground that had been chemically consolidated prior to the start of the tunnelling.

At each entry and exit at a shaft, a plug of ground was consolidated using chemical ground treatment.

3 Contract 12: Abdeen to Ein el Sierra

Contract 12 (Figure 4) was awarded in 1986 and comprised 3.3 km of 4.0 m finished diameter segmentally lined main spine tunnel at an average depth to invert of 17 m below ground level, together with 0.8 km of branch tunnel of 1.2 m finished diameter at an average depth of 12 m below ground level and 11 no. shafts ranging from 4.85 to 9 m internal diameter.

Unlike the Contract 3 tunnels located within coarse sands, the Contract 12 tunnels were expected to lie within silty sands to sandy silts except for a short section at Shaft 20 which would be excavated through the Mokattam limestone.

For this contract there was no contractual requirement for slurry machines and all the tunnels were planned to be constructed with shields under compressed air. As on Contract 3, the shafts were redesigned and sunk as full depth caissons using compressed air other than Shaft 20 which, being located in the rock outcrop, was sunk by conventional underpinning in free air.

The main tunnel was driven using three semi-mechanised shields except for drive 13-15 (Figure 4) where a slurry machine was used. The two small branch tunnels were driven using a conventional hand shield. At the exit and entries to all shafts a plug of ground was treated using the jet grouting system. This technique was used on Contract 12 because of the silty nature of the ground for which the type of chemical treatment used on Contract 3 would have been inappropriate.

4 Contract 12/5A: Boulac branch tunnels

Contract 12/5A (Figure 5) was started in March 1990; it comprised 2.8 km of 2.5 m and 3.7 km of 1.8 m finished diameter segmentally lined tunnel at an average depth of 15 m below ground level together with 20 no. 6 m and 1 no. 8 m internal diameter shafts.

On this contract CWC took the opportunity to redesign the primary lining and, by using earth pressure balance tunnelling machines, reduced the requirements for compressed air considerably. As on both Contracts 3 and 12, the shafts were sunk as full depth compressed air caissons, but in this instance ground treatment was not used.

Fig. 5.
Contract 12/5A:
Boulac branch
tunnels

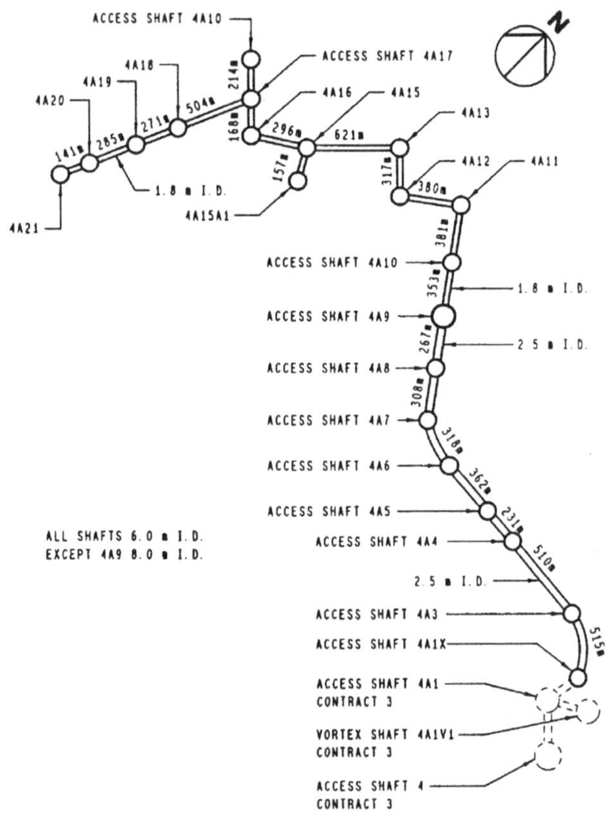

Fig. 6.
Compressed air plant:
Contracts 3,12&12/5A

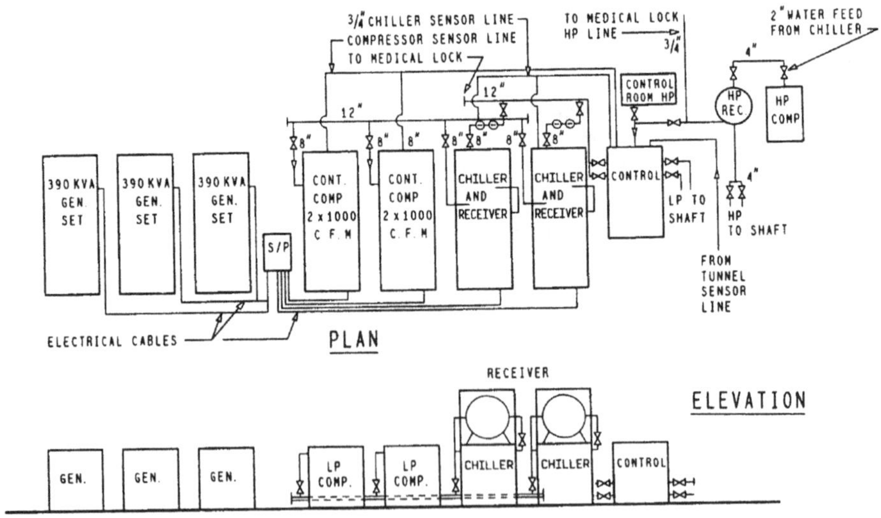

5 Construction methods: Contract 3

5.1 Compressed air plant

At the outset it was decided that as the contract had a large number of shafts and short drives with relatively short construction periods it would be best to devise a simple mobile system of compressed air plant that would lend itself to rapid movement around the sites. Thus all plant was containerised and a basic 1000 cubic feet per minute (cfm) (1700 m^3/h) compressed air set comprised:

1 no. 20 ft (6m) container, housing 2 no. 1000 cfm LP compressors (1 running, 1 standby).
2 no. 20 ft containers each housing 1 no. 390 kVA diesel generator.
1 no. containerised cooling plant with two separate chillers for compressor cooling and air chilling.
1 no. containerised LP air receiver.
1 no. containerised control cabin.

This system proved very flexible and was used for all compressed air work on Contract 3. In most cases more than one 1000 cfm set-up was used at each location as shown in Figure 6.

5.2 Shaft sinking

All shaft sinking was carried out under compressed air using vertical blister locks for men and materials access though a steel air deck into the working chamber (Figure 7). The use of blister locks meant that all decompressions had to be carried out by decanting and, for this reason, a separate decant lock was situated at each shaft-sinking site.

On Shaft 3, where the caisson consisted of three interconnected cells, three blister locks were used, thus enabling all three working chambers to be mucked out at the same time. Man access, however, was limited to two locks: one for access and one for egress. This allowed the larger gangs to be decompressed together.

5.3 Main spine tunnel

On the main slurry drives on Contract 3 (apart from early work under compressed air for which access was gained via Shaft 3), the tunnels were driven in free air. The two slurry machines, of a German design, relied upon a bubble of compressed air in the excavating chamber to maintain the face pressure balance (Figure 8).

In order to allow for access to the face of the machine, each machine was fitted with a small two-compartment lock. These locks had to be used both for access and for decompression and therefore, whenever possible, working times were limited to reduce the decompression times.

For the slurry machine to be able to complete a drive, the receiving shaft and the tunnel immediately behind the machine had to be pressurised with compressed air. In these situations it was necessary to construct a 3 m thick sand/cement mortar bulkhead across the tunnel into which was built a horizontal material lock at rail

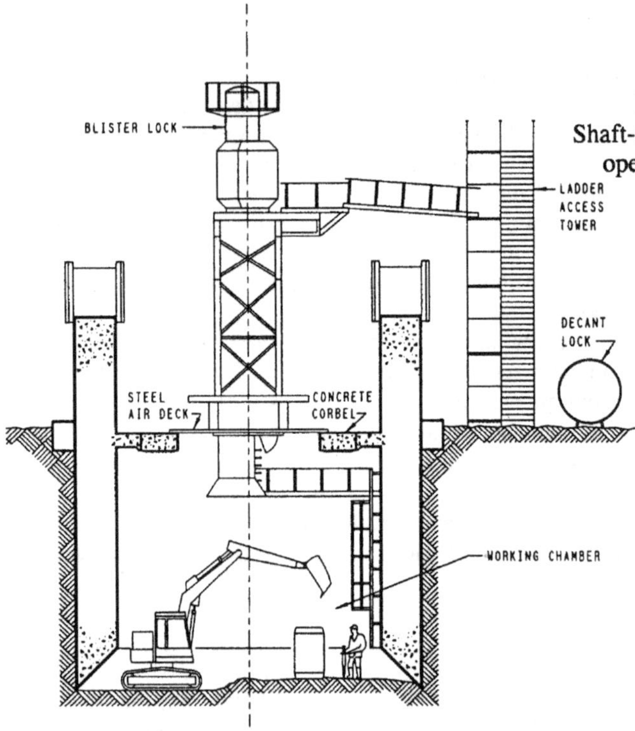

Fig. 7. Shaft-sinking operations

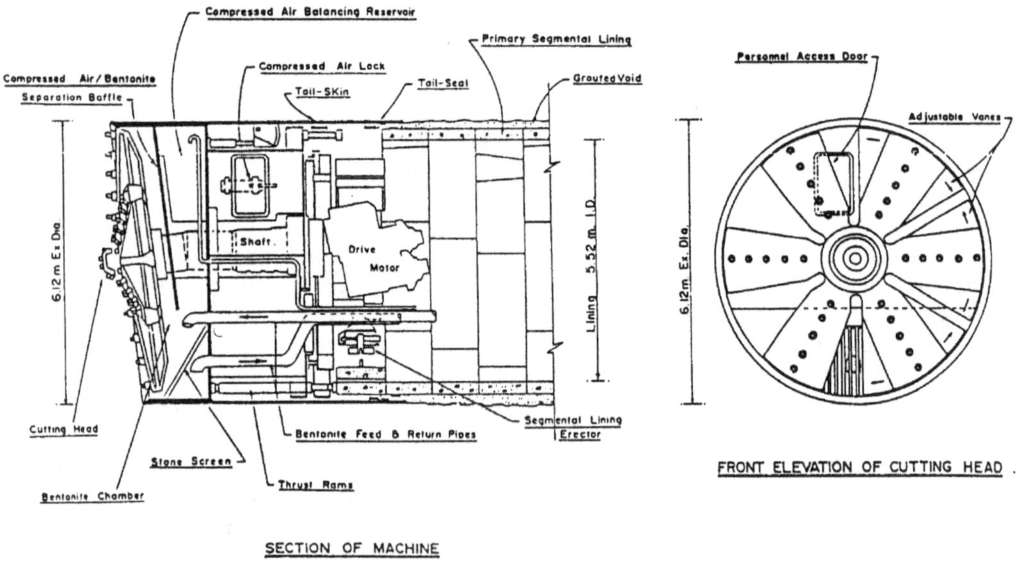

Fig. 8. Main features of Hydroshield used on Contract 3

level with a three-compartment man-lock above. These locks were formed using 2 m long boiler lock tubes bolted together with door units. Man access into the tunnel was *via* the muck-lock with egress through the decomposition lock.

All the compressed air for the main tunnel was supplied from containerised units on the surface at Shaft 3.

5.4 Branch tunnels and chambers

For the branch tunnels and transition chambers on Contract 3 two stems were used. Where a stub tunnel, transition chamber or short length of tunnel had to be built, all work was carried out through the blister lock and air deck in the shaft with decompression by decanting.

For the two main branch drives, 5A2 – 5A5 and 5A2 – 5 (see Figure 2), the shields were initially launched and operated *via* the blister lock. Once sufficient tunnel had been constructed a materials lock was installed using prefabricated sectionalised bulkhead doors fitted to sacrificial rings built in the circle joint between two rings of tunnel lining. After the lock had been installed the air deck was removed from the shaft and a decant lock installed in a stub tunnel at pit bottom. Man access and egress to the tunnel was then *via* the materials locks with decompression being carried out by decanting.

In one section of the tunnel, where it was not possible to install a decant lock at pit bottom, man access was gained from a side entry shaft through the blister lock with just muck and segments going through the horizontal lock in the tunnel.

6 Construction methods – Contract 12

6.1 Shaft sinking

Following the success of the mobile containerised compressed air set-ups on Contract 3, Contract 12 shafts were sunk using identical compressed air plant (Figure 6). Again, as on Contract 3, the access on most shafts was by the standard blister lock but on Shafts 19 and 16, from where it had been planned to drive the main tunnels, a purpose-built system was used. This system comprised a separate man-lock and muck-lock system as described below.

6.2 Main spine tunnel

At the start of the contract, it was envisaged that three semi-mechanised shields (Figure 9) would construct the main spine tunnel under compressed air except for the section of drive in the rock where the tunnel would be built in free air. Because of the drive lengths it was decided to drive all the initial lengths using access through the shaft air deck and for this purpose a specially designed system was used.

The materials lock consisted of vertical 2 m diameter boiler lock tubes with two pneumatically operated door sections forming the muck lock. This system enabled a 3 m^3 shaft skip to be used for muck removal. For man access there was a separate 1 m diameter vertical tube connected from the air deck up to a four-compartment horizontal man-lock on the surface. Three compartments of the lock were used as a

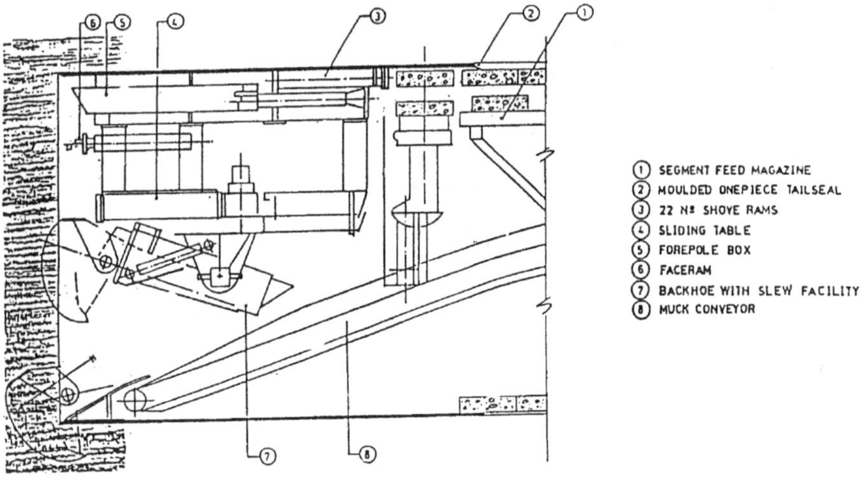

Fig. 9. Semi-mechanised shield – Contract 12

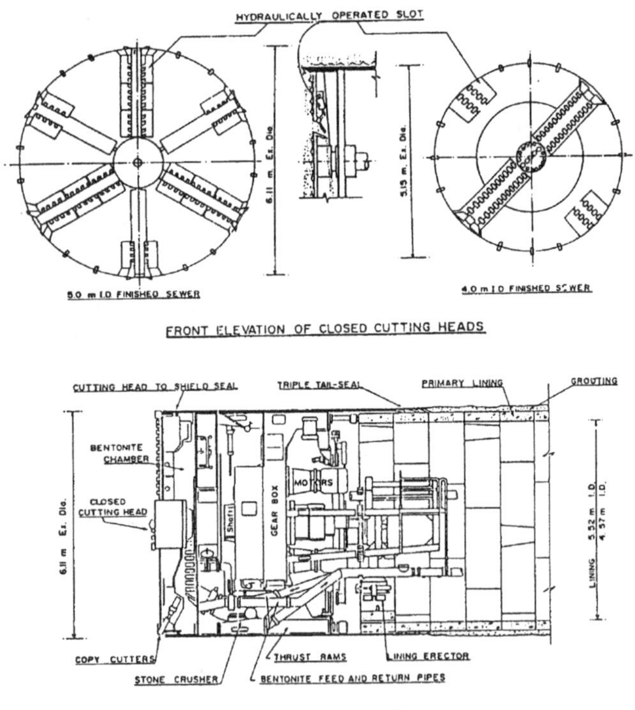

Fig. 10.
Main features of Markham/Okumura machine used on Contracts 4 & 12

traditional decompression lock with one compartment on the opposite side of the access shaft for entry only.

For drives 20 – 19 and 19 – 18 a containerised compressed air set up was used identical to those used for Contract 3 (Figure 6). At Shaft 16 where there were to be drives from 16 – 15 – 14 – 13 and 16 – 17 – 18, a static compressor and generator house was established with five 2000 cfm and two 1000 cfm compressors available to supply air to the tunnel. The generated power was supplied by 390 kVA generators.

As it turned out this static system never fulfilled its function. During the drive 16 – 15 the ground conditions deteriorated until the face was having to be fully boxed for every ring and the total capacity of the compressor house was running at all times. It was obvious that tunnelling could not continue in this manner and therefore CWC had to look at either moving the compressed air set up to Shaft 15 or constructing the tunnels using a different method.

Lilley-Misr were at that time tunnelling towards Shaft 13 with their Markham/Okumura slurry machine (Figure 10) and an agreement was reached whereby CWC would hire from Lilley-Misr and take over the operation of the machine at Shaft 13 and drive it to Shaft 15 (Figure 4).

As at Shaft 5 on Contract 3, a horizontal lock was installed in the tunnel between Shaft 13 and 14 in preparation for the final break-through at Shaft 15. This proved fortuitous because after some 400 m of the drive from 14 to 15 the tail seals failed completely and the slurry drive had to be completed under compressed air to keep the tunnel dry.

Further problems occurred on the rock drive from Shaft 20 when the inflow of water through the limestone was so large that early installation of compressed air was required to control it. In this instance a mortar bulkhead was formed approximately 60 m into the drive into which was built a horizontal muck lock and man lock. This drive was then carried out using a semi-mechanised shield with explosives all under compressed air.

6.3 Branch tunnels and chambers

On Contract 12 there were two 1.2 m finished diameter branch tunnels to construct. Both of these drives were constructed using hand shields in compressed air through silty sand and sandy silt. On the drive from 17 – 17V1, a distance of 325 m, man and material access was *via* a blister lock in the shaft. On the drive 19 – 19A1, a distance of 420 m, the access was by the main tunnel lock system. This enabled the mucking to be carried out faster because of the ability to store muck in a large muck skip at pit bottom. Compressed air was supplied by the mobile containerised plant.

7 Construction methods – Contract 12/5A

Having proved the caisson-sinking techniques on Contracts 3 and 12, CWC carried on with the same technique on Contract 12/5A. To improve sinking times overall, construction techniques were modified and all shafts were generally sunk in 7 – 10 days. Compressed air usage was of the order of 500 – 1000 cfm. For the short launch chambers and reception chambers traditional compressed air hand work was carried out but thereafter the tunnels were all built in free air. For this contract CWC elected to use Lovat Earth Pressure Balance (EPB) machines (Figure 11) together with a specially designed 'CWC' ring comprising six identical trapezoidal shaped segments 1 m wide. Sealing was by hydrophilic rubber seals and the rings were jointed with curved bolts on the seat joints and dowels on the circle joints. The EPB machines were launched in free air through a rubber launch seal. Tunnelling proceeded in free air with the annulus between the tunnel machine skin and the back of the segments sealed with three rows of grease impregnated wirebrush seals.

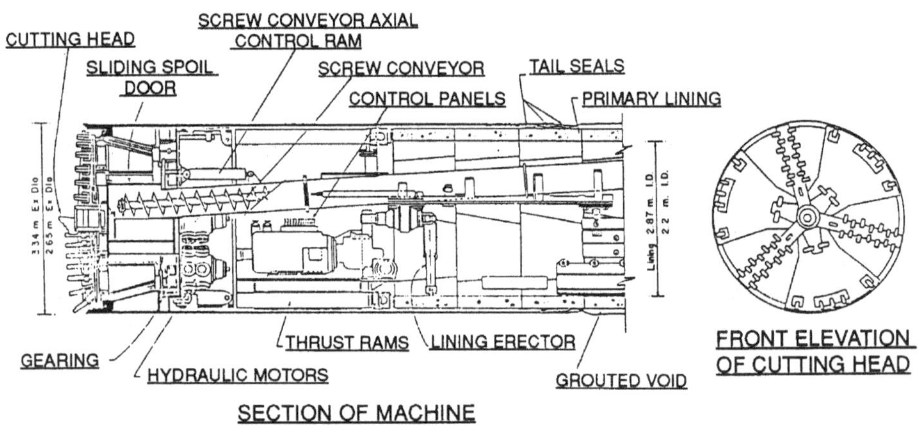

Fig. 11. Earth pressure balance machine: Contract 12/5A

8 Effects of compressed air working

On both Contracts 3 and 12 there were many instances when the demand for compressed air on a tunnel drive far exceeded the worst case that could be predicted from the empirical formula $Q=CD^2$. In fact on one drive on Contract 3 where the diameter of the face was 3.2 m the peak usage of compressed air was 12,500 cfm which equates to a C factor of 113 (in imperial units). In these situations the containerised compressed air plant was really used to its best advantage enabling additional capacity to be hooked into an existing plant both quickly and efficiently (Figure 6). Overall, CWC had plant capable of supplying a total of 42,000 cfm at any one time and for long periods; when both contracts were at their peak as many as nine sites would be under compressed air at any one time. Across most of Cairo the sands are overlain by clays and silts, which gave rise to air leaking to the surface often as much as 400 m from the actual face. In addition, the compressed air constantly lifted the water table, temporarily flooding basements and creating bubbling pools of water in streets that were flooded. In one particular case a local resident drilling a borehole at the front of his house penetrated the sand layer at the same time as a caisson some 300 m away. The result was quite spectacular with the borehole becoming a geyser of mud, slurry, compressed air and sand!

Nevertheless, in spite of the millions of cubic feet of compressed air pumped into the strata underlying Cairo, no permanent damage was caused to any structures and only minor damage was caused to existing sewer lines. Over the whole of the three contracts only one major loss of air was experienced when on the 16 – 15 drive on Contract 12 the tunnel rapidly decompressed as a result of a 'blow out' at the face. No major losses of air were experienced as a result of plant failure, proving the choice of compressed air plant and its operating systems.

9 Health and safety

9.1 General

Within the Contract Documents it was specified that the Contractor was to use the 3rd Edition of CIRIA Report 44, *Medical Code of Practice for Work in Compressed Air*, for the medical aspects of all compressed air working. With this in mind, knowing that the work was to be carried out in Egypt where tunnelling under compressed air was virtually unknown, CWC proceeded to set up systems that ensured the safe and efficient operation of the works without the requirement for outside help, such as medical decompression facilities.

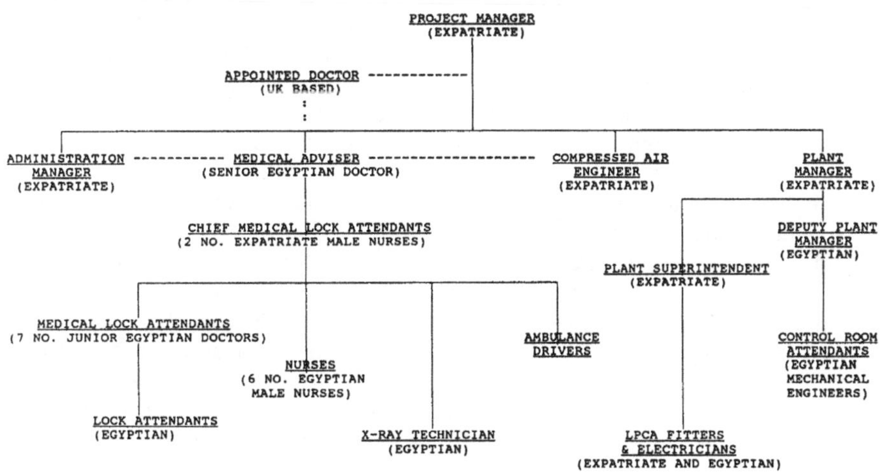

Fig. 12. Compressed-air management: Contracts 3 and 12

9.2 Staffing and responsibilities (see Figure 12)

9.2.1 Plant supervision

In order to ensure the correct operation of the compressed air plant, the plant department was staffed with senior expatriates with compressed air experience. Once plant had arrived on site and the first compressed air set-up installed a period of intensive training was instigated to train the expatriate and local fitters, electricians, lock attendants and control room attendants in the operation and maintenance of the compressed air system and how to deal with breakdowns and emergencies. During the initial months of compressed air work each set-up had an expatriate in attendance to ensure the safe working of the system. As the local staff became familiar with the operations and proved themselves competent in the running of the plant and dealing with breakdowns or emergencies, the expatriate cover was reduced to just shift cover rather than continuous individual cover.

9.2.2 Medical supervision

The Appointed Doctor, as required under CIRIA Report 44, was Dr K.P.Walker O.B.E., T.D. of the Executive Medical Centre, Croydon. Dr Walker who is UK based was responsible for the initial medicals for all expatriates to determine their fitness to work in compressed air and to review the report of their end-of-contract medical which for practical purposes was done in Cairo. The Appointed Doctor was also responsible through the Project Management for the implementation of proper procedures for the medical aspects of compressed air working. In Egypt, CWC employed as their Medical Adviser a senior Egyptian doctor, Dr Victor Ikladius. This doctor was resident on site and had attended courses on compressed air in the UK. The Medical Adviser was responsible for conducting all intermediate and end-of-contract medical examinations, the records of which were sent to the Appointed Doctor, together with the other responsibilities as defined in CIRIA Report 44. Ideally CWC would have preferred the end-of-contract medical examination to be carried out by the Appointed Doctor, but it was realised early on that on returning to the UK expatriates were loath to give up time to attend for a medical. Hence they were handed their air tickets only after their medical examination in Cairo had been completed. During the initial period of Contracts 3 and 12, CWC employed two experienced expatriate male nurses as senior medical lock attendants. Their dual role was both to train and supervise the Egyptian lock and medical lock attendants and to establish good working procedures.

In addition to the senior medical lock attendants, the Medical Adviser had under his control medical lock attendants who were all junior Egyptian doctors, lock attendants all of whom were Egyptian, an Egyptian x-ray technician, Egyptian male nurses and the ambulance drivers. With Contracts 3 and 12 spread out over an area of 40 sq. km. within the congested central part of the city of Cairo, it was necessary for CWC to have one central medical centre and two satellite centres strategically placed within the area to deal with all problems that might arise. Each centre was manned 24 hours/day by medical lock attendants and nurses on a shift system and was fully equipped to handle cases of decompression sickness and accidents. In addition, at the central medical centre there was an x-ray room and surgery to cope with general sickness, minor accidents, monthly medicals and annual x-rays.

9.2.3 Safety supervision

In order to ensure the smooth operation of the compressed air operations, CWC further employed one senior expatriate engineer with compressed air experience as the compressed air/safety engineer. It was this engineer's duty to monitor all aspects of the compressed air work to ensure that all procedures and documentation were being properly carried out. To assist the expatriate engineer in his duties CWC employed an Egyptian safety officer who was responsible for general site safety and for regular gas testing in the compressed air workings checking for carbon monoxide, carbon dioxide, hydrogen sulphide, oxygen and nitrous gases.

9.3 Documentation

At the start of the contract CWC compiled a Manual of Procedures and Documentation for working in compressed air which was formally issued to all senior expatriates and local staff. In addition a small booklet written in English and Arabic and entitled *Advice and instructions to men working in compressed air* was formally issued to all the expatriate and local workforce prior to their commencing work in compressed air. Further formal documents were issued to man-lock keepers and medical lock attendants defining exactly what they were required to do. At every site where compressed air was in use a full set of the above documents were displayed on a notice board adjacent to the lock entrance. All compressed air workers were issued with a compressed air card which had their photograph affixed to it together with instructions in English and Arabic on where the medical locks were situated. Prior to compressed air work starting, the Medical Adviser visited all the local hospitals in the area to advise them of our work and to explain the symptoms of decompression sickness. All compressed air workers were given a monthly medical at the Medical Centre together with a set of joint x-rays annually and at the end of an individual's contract.

Any worker suffering from Type II decompression sickness was immediately excluded from working in compressed air again and, if necessary, was repatriated.

A full set of documents was maintained for every compressed air worker which included:

- Medical examination reports
- Compressed air health register
- Decompression sickness case sheet
- Lock test report
- Individual compressed air record.

In addition, medical lock attendants completed a weekly report of all compressed air set-ups together with a daily return and weekly return of all decompressions carried out in the period both non-medical and medical. Lock attendants completed

Table 1 Exposures and decompression sickness: Contract 3

Working pressure psig	No. of exposures for various working periods			Total	No. of cases of decompression sickness following various exposures						Incidence rate %	
					Type I			Type II			Type	
	<4 hrs	4-8 hrs	>8 hrs		<4 hrs	4-8 hrs	Total	<4 hrs	4-8 hrs	Total	I	II
<14	5853	4120	5009	14982								
14-15.5	60	167		227		1	1				0.44	
16-17.5	5153	7619	93	12865	1	2	3				0.02	
18-19.5	2527	4172		6699		8	8				0.12	
20-21.5	1716	2990		4706		26	26		1	1	0.55	0.021
22-23.5	6598	12370		18968	1	32	33		2	2	0.17	0.011
24-25.5	1456	3087		4543		46	46				1.00	
26-27.5	1700	2873		4573	7	35	42				0.92	
28-29.5	794	1697		2491	2	19	21	1		1	0.84	0.040
30-31.5	537	690		1227	3	40	43		2	2	3.50	0.163
32-33.5	44	140		184		7	7		1	1	3.80	0.543
Total	26438	39925	5102	71465	14	216	230	1	6	7	0.32	0.010

Table 2 Exposures and decompression sickness: Contract 12

Working pressure psig	No. of exposures for various working periods			Total	No. of cases of decompression sickness following various exposures						Incidence rate %	
					Type I			Type II			Type	
	<4 hrs	4-8 hrs	>8 hrs		<4 hrs	4-8 hrs	Total	<4 hrs	4-8 hrs	Total	I	II
<14	4058	2365	3100	9533								
14-15.5	318	350	331	999								
16-17.5	3097	7018	603	10718								
18-19.5	2798	9895		12693		3	3		1	1	0.02	0.008
20-21.5	8591	29107		37698		33	33		3	3	0.09	0.008
22-23.5	5850	30223		36073		18	18		3	3	0.05	0.008
24-25.5	134	262		396		5	5				1.30	
26-27.5	107	226		333								
28-29.5	116	204		320		5	5				1.60	
30-31.5												
32-33.5												
Total	25079	79650	4034	108763		64	65		7	7	0.06	0.006

Table 3 Decompression sickness rates in different races (exposures over 1 bar): Contracts 3 and 12

Race	Staff (No.)	Exposures (No.)	Incidence of decompression sickness	
			Type I (%)	Type II (%)
British	229	41901	84 (0.20)	9 (0.021)
Etyptians	1673	106085	190 (0.18)	5 (0.005)
Sudanese	118	32242	20 (0.06)	0 (0)

Table 4 Decompression sickness rates in different races (all exposures): Contracts 3 and 12

Race	Staff (No.)	Exposures (No.)	Incidence of decompression sickness	
			Type I (%)	Type II (%)
British	229	35813	84 (0.23)	9 (0.025)
Etyptians	1673	91872	190 (0.21)	5 (0.005)
Sudanese	118	28028	20 (0.07)	0 (0)

Table 5 Decompression sickness rates for decanting and normal decompressions: Contracts 3 and 12

Type of decompression	Decompressions (No.)	Medical decompressions for Type I (%)	for Type II (%)
Normal	64,201	138 (0.21)	6 (0.009)
Decanting	91,512	158 (0.17)	8 (0.009)

a standard Lock Attendant's Register and control-room attendants completed a compressed air shift log that was kept with the relevant barograph records. All changes of the pressure in the working chamber were only carried out by authorised staff members who had to sign the control-room attendant's sheets for each pressure change.

10 Statistical analysis of compressed air records

10.1 General

With three contracts having been completed over a period of six years all utilising compressed air, CWC have a large volume of information from which general statistics have been extracted for this paper.

10.2 Contracts 3 and 12

Over the period of Contracts 3 and 12 i.e. between June 1985 and September 1988, a total of 180,228 exposures were recorded of which 155,713 were at pressures of 14 psig (1 bar) or above and therefore required stage decompression in accordance with CIRIA Report 44.

From this total number of exposures there were 14 occurrences of Type II and 294 occurrences of Type I decompression sickness (Tables 1 and 2). Of the Type II decompression sickness nine were Expatriates and five were Egyptians, whereas of the Type I decompression sickness eighty four were Expatriates, one hundred and ninety were Egyptian and twenty were Sudanese (Tables 3 and 4). Overall this represents a percentage incidence rate against exposures above 14 psig (1 bar) of 0.009% for Type II and 0.19% for Type I. These figures can be translated into incidence rates of

Type I - 1 case / 530 decompressions
Type II - 1 case / 11122 decompressions

From Figure 13 where is plotted the overall monthly incidence of Type I against the number of exposures per month, it can be seen that following the initial high incidence rate at the end of 1985 the general monthly incidence rate remained well below 0.5%. The early incidence rate is almost certainly due to lack of acclimatisation. In August 1985 two small shafts were sunk with final pressures only just over 14 psig (1 bar) followed by a month of no compressed air work at all. This was then followed by a period of intense activity with large shafts such as Shaft 3 being sunk. In addition to physical acclimatisation there was also the problem of ignorance on the part of the local workforce who initially were wary of compressed air working and it is quite possible that a few of the initial Type I cases were in fact not Type I at all. At all times CWC's response with regard to the local workforce has been to assume Type I and treat accordingly thus ensuring all genuine cases were not missed. In addition to the above it can be seen from the individual contract details that Contract 3 had a far greater incidence of bends than Contract 12 and this is likely to be explained by reasons such as the following :

1. Contract 3 was the first contract so that many miners who worked on Contract 12 had already acclimatised on Contract 3 before being transferred.
2. Contract 3 had almost 20% of all its exposures at pressures in excess of 25 psig (1.75 bar), pressures which were usually worked for relatively short periods at times such as when bottoming-up a shaft, hence there were no

steady periods in which the men could acclimatise.

The exact opposite was the case on Contract 12 where the majority of exposures were in tunnels working regular shift work at steady pressures for months. Another comparison carried out was that of comparing the time of onset of the symptoms of decompression sickness after decompression was completed. This showed that in the first hour when men are still on site 34% of Type I and 50% of Type II cases would have occurred and in the first two hours 61% and 71% respectively (Figures 14 and 15). This suggests that it would be better to have a longer period of two hours on site after decompression rather than the present one hour. Further comparison would seem to indicate that Egyptians are less susceptible to the bends than expatriates, but this may be more attributable to the expatriates' life style rather than anything else. Comparison between decanting and normal decompression (Table 5) with respect to the incidence of decompression sickness tends to suggest decanting is better. However, this is more likely to be explained by the fact that more decanting exposures will have been at low pressures and for shorter periods, as all the shaft sinking was carried out using decanting.

A further record was made as shown in Tables 6 and 7 of the various symptoms exhibited by those personnel suffering decompression sickness.

10.3 Contract 12/5A

On Contract 12/5A the use of compressed air was limited to work of a short duration and was generally carried out at lower pressures, i.e. up to 27.5 psig (2.0 bar). Following completion of all shaft sinking and major compressed air work there had been a total of 8354 exposures below and 14071 exposures above 14 psi (1 bar) (Table 8). From these 14,071 exposures there were 32 cases of Type I and no cases of Type II (Figure 16). The incidence rate for Type I was therefore 0.23%. This figure can be translated into an incidence rate of 1 case/440 decompressions which is slightly worse than the overall rate for Contracts 3 and 12. This is again probably explained by the fact that all decompressions were by decanting and there was no steady continuity of compressed air work.

The onset of the symptoms of decompression sickness after decompression was completed was similar to Contracts 3 and 12 with 44% and 63% of Type I cases occuring within one and two hours respectively (Figure 17). The symptoms exhibited by those personnel suffering decompression sickness is shown in Table 9.

11 Comparisons of working methods

Following completion of the three contracts 3, 12 and 12/5A, it is clear that the lessons learnt on the first two contracts were put to very good use on the third. On all three contracts CWC's change to full-depth caissons proved to be an extremely wise decision, one that reduced the compressed air requirements for shaft construction considerably. In addition because of the experience gained on Contracts 3 and 12, CWC were able to construct all the shafts and hand tunnels on contract

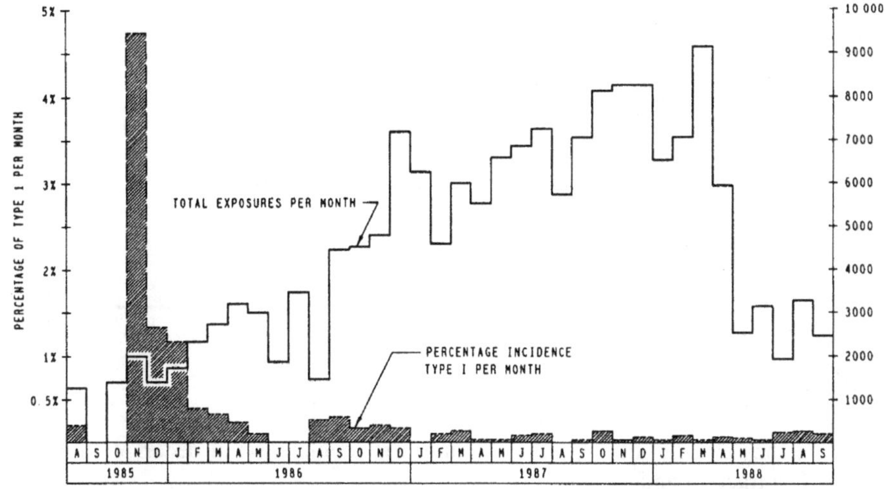

Fig. 13. Exposures and decompression sickness rates for Contracts 3 and 12

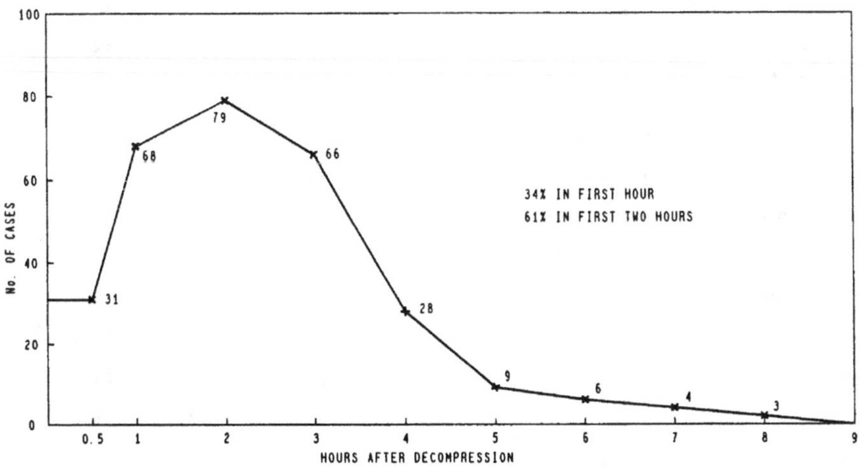

Fig. 14. Times of onset of Type I symptoms after completion of decompression: Contracts 3 and 12

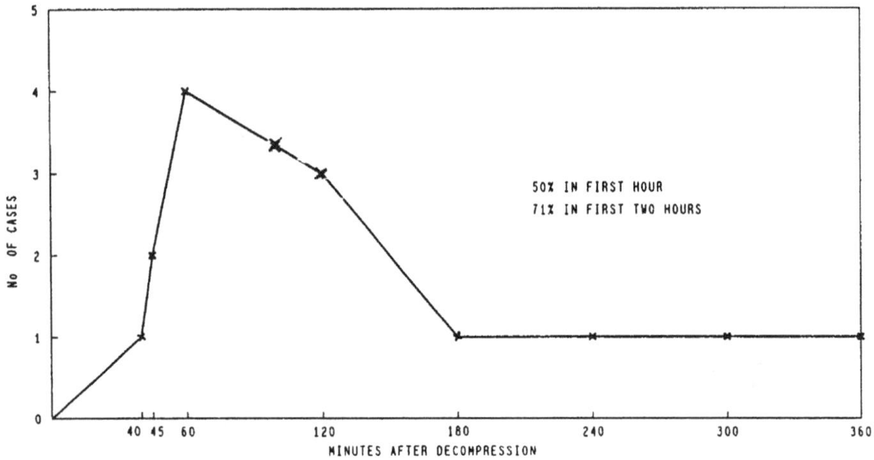

Fig. 15. Times of onset of Type II symptoms after completion of decompression: Contracts 3 and 12

Table 6 Symptoms manifested in 14 cases of Type II decompression sickness: Contracts 3 and 12

Symptomatology		Number	Percentage of cases exhibiting symptoms
Dizziness		13	92.8
Confusion		2	14.3
Disorientation		5	35.7
Eye Signs	Nystagmus	-	
	Diplopia	8	85.7
	Tunnel vision	4	
Sensory impairment	Hyperaesthesia	2	14.3
	Paraesthesia	3	21.4
Motor weakness or loss		2	14.3
Collapse		4	28.6
Nausea and vomiting		9	64.3
Joint pains		5	35.7
Respiratory signs	Dyspnoea	2	
	Coughing	2	42.8
	Chest pain	2	

Note: Cases may exhibit more than one symptom.

Table 7 Symptoms manifested in 294 cases of Type I decompression sickness: Contracts 3 and 12

Site	Pain	Cutaneous	Cases exhibiting symptoms (%)
Shoulder	127	-	43.0
Axilla (armpit)	5	-	1.7
Arm	4	2	2.0
Elbow	8	-	8.2
Hip joint	24	-	2.7
Thigh	44	-	14.9
Knee	170	-	57.9
Ankle	28	1	9.9
General arthralgia	8	-	2.7

Note: Cases may exhibit more than one symptom

Table 8 Exposures and decompression sickness occurrence: Contract 12/5A

Working pressure psig	No. of exposures for various working periods			Total	Type I	
	<4 h	4-8 h	>8h		No.	%
<14	3475	2493	2386	8354		
14-15.5	918	1290	775	2983	1	0.03
16-17.5	777	1252	445	2474	2	0.08
18-19.5	1336	2034	459	3829	2	0.05
20-21.5	880	1323	294	2497	7	0.28
22-23.5	371	946	318	1635	6	0.37
24-25.5	131	288	45	464	7	1.51
26-27.5	92	59	33	184	7	3.80
28-29.5						
30-31.5						
32-33.5						
Total	7985	9685	4755	22425	32	0.14

1. All DCS related to exposures of between 4 and 8 h
2. There were no Type II cases.

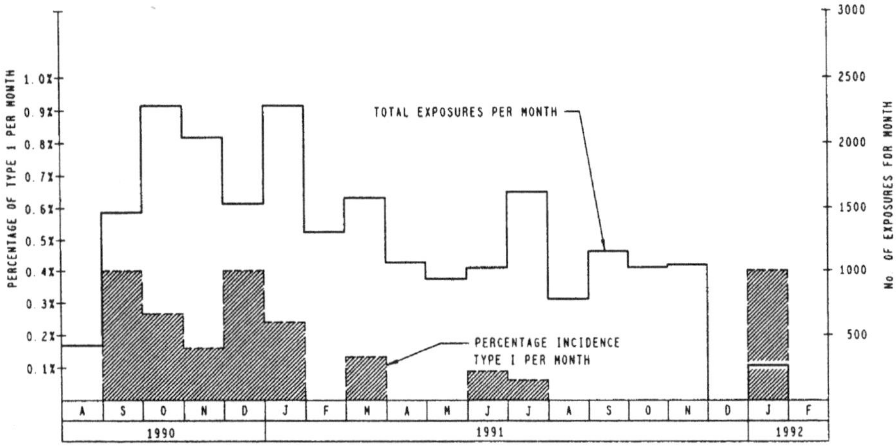

Fig. 16. Exposures and decompression sickness rates: Contract 12/5A

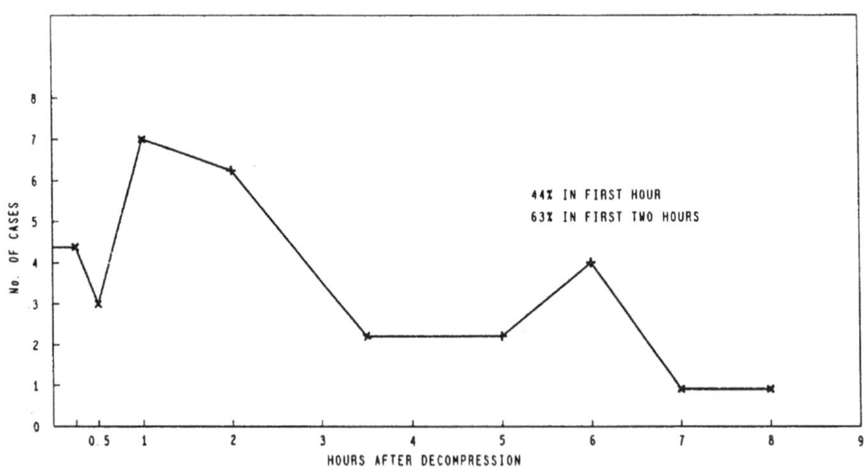

Fig. 17. Times of onset of Type I symptoms after completion of decompression: Contract 12/5A

12/5A using Egyptian personnel only with no expatriate supervision. On the tunnelling side, the use of earth pressure balance tunnelling machines with the much wider segmental lining and hydrophilic seals on Contract 12/5A produced tunnelling speeds of up to 42 m/day and 210 m/week. Furthermore the tunnelling was carried out in free air and the completed tunnel was dry and did not need caulking, thereby ensuring that the tunnel was available for secondary lining within 4 weeks of the TBM breaking through. When this is compared with the compressed air drives of the same size on Contract 3 it can be seen that tunnelling was of the order of 7 times faster on 12/5A not allowing for the additional time needed on the compressed air drives for caulking and general cleaning. Overall on Contract 12/5A the improvements in construction methods enabled 21 no. shafts and 6.5 km of tunnel to be completed in just 18 months with minimal compressed air usage.

Table 9 Symptoms manifested in 32 cases of Type I decompression sickness: Contract 12/5A

Site	Pain	Cutaneous	Percentage of cases exhibiting symptoms
Shoulder	7	-	21.8
Arm	1	-	3.1
Elbow	1	-	3.1
Hip joint	-	-	-
Thigh	1	-	3.1
Knee	25	-	78.1
Ankle	3	-	9.4
General Arthralgia	1	-	3.1

Note: Cases may exhibit more than one symptom

12 Concluding remarks

The operating systems and plant used to carry out Contracts 3, 12 and 12/5A (Table 10) have proved to be a successful choice and enabled considerable flexibility to respond to varying conditions. At the peak of construction there were upward of

Table 10 Major plant used in Contracts 3, 12 and 12/5A

6.11 m O.D	Bentonite slurry machines (German design)	- 2 No.
5.2 m O.D	Bentonite slurry machine (Japanese design)	- 1 No.
4.1 m O.D	Hand shield	- 3 No.
3.35 m O.D	Earth Pressure Balance TBM	- 1 No.
2.67 m O.D	Earth Pressure Balance TBM	- 1 No.
3.35 m O.D	Hand shield	- 1 No.
2.66 m O.D	Hand shield	- 1 No.
2.05 m O.D	Hand shield	- 1 No.
Containerised Generators – total generator power		- 17 MW
Compressors and associated equipment – total capacity		- 42000 cfm
Rough terrain mobile cranes (20 – 80 t)		- 17 No.
Crawler cranes (40 – 60 t)		- 5 No.
Gantry cranes (5 – 10 t)		- 5 No.
Excavators and loading shovels		- 14 No.
Items of heavy transport (Lorries, Hiabs, Artics)		- 27 No.
Items of light transport (Cars, Pick-up, Mini-buses)		- 114 No.
Piling hammers		- 3 No.
Concrete batching plants		- 2 No.
Truck mixers		- 8 No.
Concrete pumps		- 5 No.
Grouting equipment		- 53 No.
2'-0" Gauge tunnel locomotives		- 25 No.
2'-0" Gauge rolling stock (muck skips, segment bogies)		- 174 No.

200 expatriates and 2000 Egyptians (including resident engineer staff) working on the contract with as many as 2400 exposures per week during the peak of compressed air work. Compressed air working is still a versatile option to many construction problems, but increasingly the industry is finding ways of carrying out work without it. If it has to be used then by careful preplanning and a strict attention to detail contracts of the size and complexity of those in Cairo can be carried out without major problems. As shown on these contracts with the provision of proper training and supervision the local workforce can operate and maintain compressed air equipment correctly. Compressed air working however remains a system to be used sparingly and only when necessary. Because of the vast array of equipment required together with the potential for damage both to property and workers it is essential that the industry limits its application by the introduction of better methods. This indeed is exactly what CWC did on Contract 12/5A by applying the lessons learnt on Contracts 3 and 12 and improving the tunnelling methods and minimising the use of compressed air.

Acknowledgements

CWC would like to acknowledge the help and assistance of the client, the Organisation for the Execution of the Greater Cairo Wastewater Project of Greater Cairo (CWO) and the Engineer, American British Consultants (AMBRIC) comprising Binnie & Partners (UK), Black & Veatch International (USA), Camp Dresser & McKee (USA) and John Taylor & Sons (UK).

References

CIRIA (1982). Report 44. *Medical Code of Practice for Work in Compressed Air.* Third Edition.

Darling, P. (1991). Tunnelling alongside Egypt's heart-beat. *Tunnels and Tunnelling*, July 1991.

Elliot, I.H. and Spolton, R.L. (1985) Greater Cairo Wastewater Project: Design of East Bank Stage 1 Works Tunnels (1985). *Proceedings of the Institution of Civil Engineers*, Part 1, Design and Construction, Volume 78, August.

Flint, G.R. and Foreman W. (1991). Bentonite Tunnelling for the Greater Cairo Wastewater Project. *International Symposium on Tunnelling in Congested Cities.* Cairo, January 1991.

Haywood, D. (1991). Cairo Practice. *New Civil Engineer*, 28 November 1991.

McDonald, T. and Court, D.J. (1990). Challenges met on Contracts 3 and 12 in Cairo. *Tunnels and Tunnelling*, January 1990.

Roberts, D.G.M. (1992). Greater Cairo Wastewater Project. *World Tunnelling*, March 1992.

Wallis, S. (1987). Deciphering Cairo's Mysteries. *Tunnels and Tunnelling*, June.

8 Oakwood Beach Interceptor Sewer: A compressed air tunnel case history

R.J. Angarola, S.K. Jain and C.C. Chang
Sverdrup Corporation, New York, USA

Abstract
The Oakwood Beach Interceptor Sewer project consists of two pumping stations and approximately 13.7 km (8.5 miles) of sewer along Hylan Boulevard in Staten Island. The paper discusses the site geology and confined aquifer conditions along the alignment, various factors which necessitated the use of compressed air for tunnelling, and problems encountered during construction and their solution. Finally, the health-related safety guidelines exercised on the project are outlined.
Keywords: Geology, Tunnels, Compressed Air, Deep Well, Eductor, Dewatering, Venting, Health.

1 Site geology

The site (see Figure 1) is on the south shore of Staten Island in New York City, bordering Raritan Bay, within the Atlantic coastal plain. The geologic history of New York City Metropolitan area includes several events as follows:

- The deposition by the sea of the various sediments such as sand and clay which were slowly cemented in bedrock
- The uplift of these above sea level
- The prolonged weathering, erosion and removal of the surface rocks
- Glaciation.

Of these events, glaciation has had the most effect on the topography. Continental glaciers of the Pleistocene period moved over the land surface and, as they melted, deposited unconsolidated materials in the form of moraines. Glacial drift forms the near-surface material at the project site, and consists either of sand and gravel that was sorted and deposited as outwash by glacial melt water, or of till, a mixture of clay, silt, sand, gravel and boulders, which was deposited from the ice without the sorting action of water. Due to several ice advances in the area, the thickness and character of the glacial deposits varies substantially throughout the site.

Engineering and Health in Compressed Air Work Edited by F.M. Jardine and R.I. McCallum. Published in 1994 by E & FN Spon. ISBN 0 419 18460 0

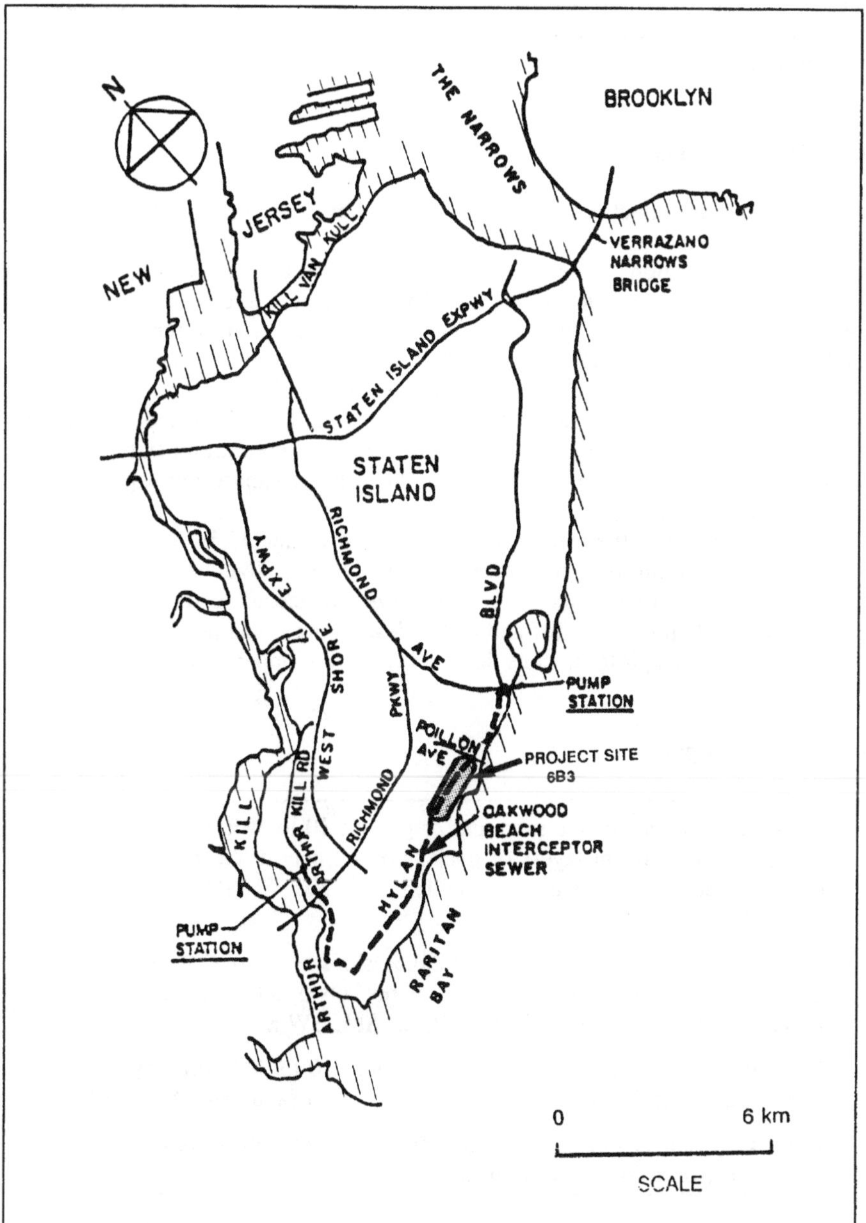

Fig. 1. Site location plan

The glacial soils at the site are underlain by Cretaceous deposits.

The interceptor sewer tunnel therefore had to be bored through strata containing diverse material ranging from Cretaceous hard clays to glacial sand and gravel mixed with boulders and till.

2 Subsurface conditions

The section/length of tunnel covered by this paper lies in a profile where glacial soils overlie the Cretaceous formation (see Figure 2). The tunnel invert is approximately at elevation -50 feet and the groundwater table varies from elevation +8 to -5 feet. Groundwater levels fluctuated several feet due to tidal variations in the nearby bay and site hydrologic conditions. All elevations are referred to Borough President of Richmond Datum, which is 3.192 ft above Standard Datum Mean Sea Level at Sandy Hook, New Jersey, U.S.A.

For clarity, the Glacial (G) and Cretaceous (C) strata are subdivided into groups with the same geological origin, but with significantly different grain size distribution and consistency. The four classes of soils relevant to the paper are described in Table 1 below.

The distinct feature of the area between Manholes 13 and 15 is the presence of glacial till (G1) underlain by a thick layer of coarse and relatively clean glacial outwash (G2) which is lying directly on single-size Cretaceous sands of stratum C2. The G2 and C2 soils presented difficulties for dewatering. The permeability of the G1 material is low enough to be considered as relatively impervious compared to the G2 soils.

Table 1 Soil description

G1	Red-brown coarse to fine sand and/or gravel; 10 to 50% silt/clay, occasional to frequent layers of pockets of silt/clay or clean sand; may contain nests of cobbles and/or boulders; SC, SM, SP-SM, GC, GM, GP-GM, SW-SM.
G2	Red-brown coarse to fine sand; 1 to 10% silt/clay, 1 to 35% coarse to fine gravel, occasional to frequent layers or pockets of silt/clay; may contain nests of cobbles and/or boulders; SP, SP-SM, SW-SM.
C1	Light gray, tan or pink medium to fine sand; 10 to 50% silt/clay, often micaceous; occasional layers or pockets of clean sand, SM, SC, SP-SM.
C2	Light gray, tan or pink medium to fine sand; 1 to 10% silt/clay, often micaceous; occasional layers or pockets of silt/clay; SP, SP-SM.

3 Selection of compressed air tunnelling method

The tunnelling at the 6B-3 site was initially planned to be carried out under free air in both the east and west headings from Manhole No. 18. The mining of the west heading was successfully accomplished in free air as planned. However, the tunnelling method of the east heading had to be changed to compressed air due to the following reasons:

1. It was difficult to drain the G2 soils effectively, particularly at the G2/C2 interface.
2. Large-scale field pumping tests and related data analysed by the contractor's geotechnical consultant demonstrated that a flow rate into the tunnel would remain between 0.3 to 0.6 l/s (5 to 10 gallons per minute) at the tunnel heading even with the most strenuous and elaborate dewatering efforts (e.g. with dewatering eductor wells at as close a spacing as 1.5 m (5 ft) on either side of the tunnel. Other measures to control groundwater such as grout injection, ground freezing or using a different type of tunnelling machine were considered, but would not have provided a viable cost-effective solution.

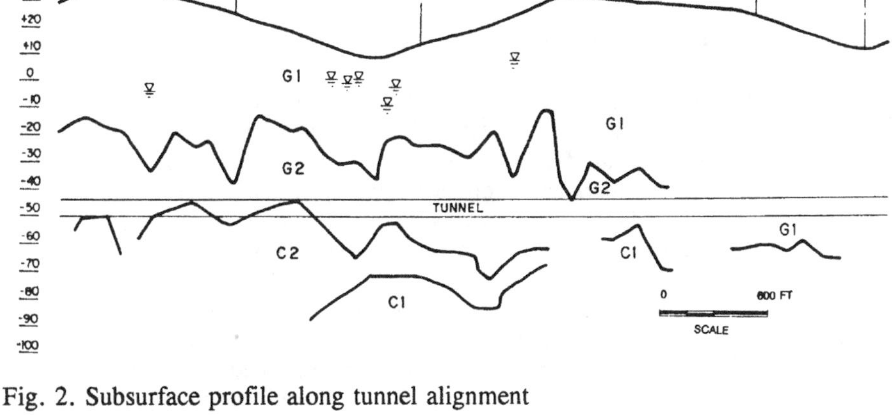

Fig. 2. Subsurface profile along tunnel alignment

Considering the potential of running sand conditions at tunnel face in the single-size sand of stratum C2, the contractor (Grow-Kiewit-Catapano) recommended to New York City Department of Environmental Protection (Owner) the use of a lower compressed air pressure in conjunction with partial dewatering by deep wells between Manholes 16 to 13 and closely spaced eductor wells between Manholes 18 to 16. However, Manhole 16 was deleted from construction due to design changes made by the Owner.

4 Dewatering well construction

Typically, the deep wells were constructed by drilling a 762 mm (30 in.) diameter hole extending to the bottom of the pervious water-bearing aquifer below the tunnel invert. The well screens of 406 mm (16 in.) diameter and of varying lengths were positioned in each well so as to intercept flow from pervious soil strata. The screen was Johnson galvanized steel wire-wrapped screen with 40-slot (0.04 in.) openings. The annular space of the well between the screen and the hole was filled with select sand and gravel filter. The well was then subsequently developed. Generally Layne deep-well 2-stage pumps were used. The horsepower of the pumps varies from 40 to 75.

Due to the low permeability of G1 soils, eductor wells were employed between Manholes 18 and 16 for dewatering the tunnel alignment. Typically, these wells were 152 mm (6 in.) in diameter. The well screen is Johnson PVC plastic screen with 40-slot (0.04 in.) openings. The eductors were installed by direct rotary method.

The deep dewatering system and vent wells were carefully and continuously attended during the entire tunnelling period. The pumped water from wells was collected in a header pipe leading to the storm sewer along Huguenot Avenue, near Manhold 17. The water was then disposed of into Raritan Bay in compliance with U.S. Environmental Protection Agency (EPA) regulations.

5 Tunnelling machine and compressed air equipment

The ground to be tunnelled through includes Cretaceous silts, clays and dense glacial soils mixed with boulders; thus the contractor chose an open-face excavator shield. This enables miners to work the face manually, if required, to break up and remove boulders. (Fortunately, boulders were not encountered.) The shield (see Figure 3) was designed and built by Elgood Mayo Corporation, Lancaster, Pennsylvania, U.S.A. The primary functions of the shield were hydraulically operated and controlled.

The shield is advanced by 12 rams. A hydraulic excavator, mounted at the centre of the shield under the hood, has the capability to move in and out, right, left, up and down and to rotate around the boom axis. The excavated soil is taken by conveyer to the muck cars. The tunnel was lined with a one-pass bolted concrete

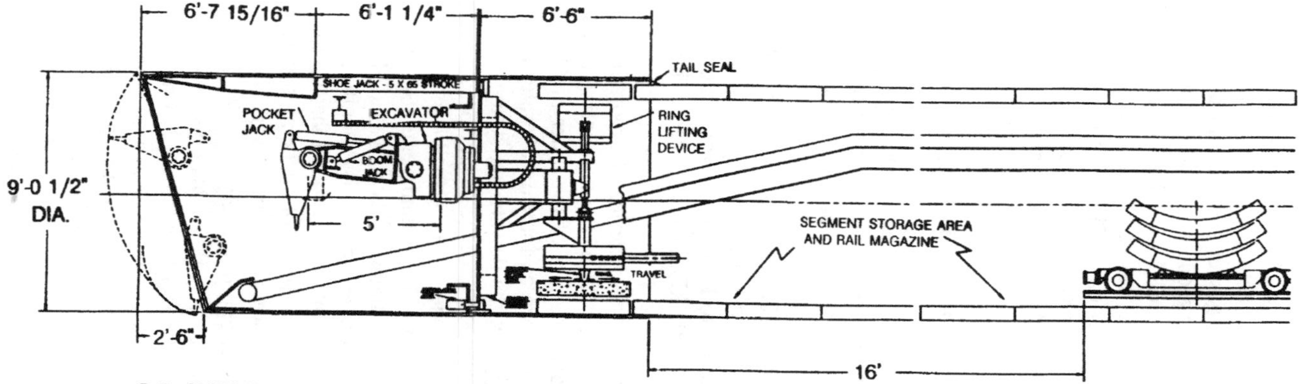

O.D. SHIELD	9'-1/2"
TAIL THK.	- 0'-2 1/2"
	8'-10"
I.D. TAIL-CLEARANCE	- 0'- 2"
O.D. SEGMENT	8'- 8"
SEGMENT THK. (6-1/2")	- 1'- 1"
I.D. SEGMENT	7'- 7"

Fig. 3. Tunnelling machine schematic

segmental lining 2311 mm (7'-7") I.D. Each of the six segments in the ring is trapezoidal in shape resulting in each segment acting as a key. The 1067 mm (42 in.) long segments have a full rubber sealing gasket and were installed by an erector in the tail of the shield. Either pea gravel or grout was used to fill the annular space, depending on ground conditions. Figures 4 to 7 show typical site photographs of the tunnelling operation and its various components. The average tunnelling progress rate was 7.6 m (25 ft) per day. There were six 4-hour shifts per day. The work was carried out five days per week. On weekends, the compressed air equipment and plant were continuously attended by the contractor's personnel.

6 Problems encountered during tunnelling at east heading and their solution

The dewatering wells were designed to lower the groundwater to approximately 3 m (10 ft) above the crown of the tunnel and the remainder of the water would be kept out of the heading by utilising compressed air with a pressure between 55 and 62 kPa (8 and 9 psig).

When the tunnel reached about 300 m (1000 ft) from work shaft No. 18, the tunnel broke out of a G1 till and into the permeable G2 soils. As tunnelling progressed the air pressure had to be raised to keep water from entering. Tunnelling stopped when pressures of approximately 103 kPa (15 psig) had to be used but water was still entering the tunnel.

Upon investigating the problem, Moretrench American Corporation, the dewatering sub-contractor, noticed that in several of the dewatering eductor wells in the area of the tunnel there existed a back pressure of the order of 41 kPa (6 psi).

The piezometers in the area had also shown an increase of water elevation of several feet. After several days of analysing the dewatering system it was recognised that escaping air was being trapped underneath the G1 soils, thus creating a back pressure in the aquifer and on the tunnel face. Usually under normal tunnelling conditions, the air that escapes into the ground vents itself to the surface. In the area where this problem was first encountered, there was a 24 m (80 ft) thick cap of impervious till preventing the air from escaping. By monitoring the water level in the piezometers along tunnel alignment, it was found that up to 274 m (900 ft) away from the tunnel heading there was evidence of air being trapped in the aquifer. (See Figure 8 for compressed air pressure in wells versus stations.) The figure suggests that when the shield was at approximately Sta. 12+25, the wells registered the effect of back pressure to Sta. 19+00. This effect was greater on wells closer to the shield as compared to the wells measured 213 m (700 ft) away from the shield. The back pressure in wells gradually decreased when the venting was applied. Subsequently the contractor was able to reduce the compressed air in the tunnel to 69 kPa (10 psi), as opposed to the 103 kPa (15 psi) previously required.

From the analysis it was suggested that when there was 41 kPa (6 psi) of back pressure in the aquifer and 69 kPa (10 psi) of air in the tunnel that there would be 71 kPa (10.3 psi) of pressure in the aquifer at the crown of the tunnel. This is

Fig. 4. Site view with dewatering wells and compressed air lines

Fig. 5. Site view with concrete liner segments

Fig. 6. Air lock

Fig. 7. Working shaft No. 18 and muck removal

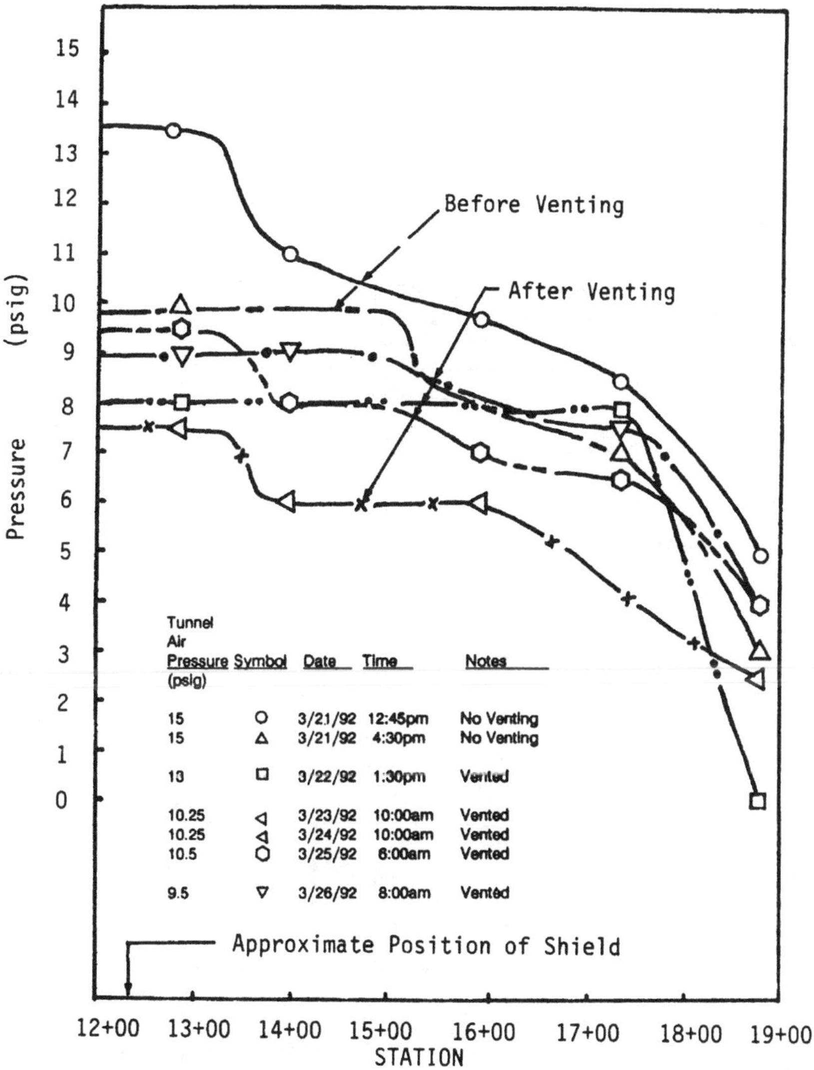

Fig. 8. Compressed air pressure in wells versus tunnel stationing

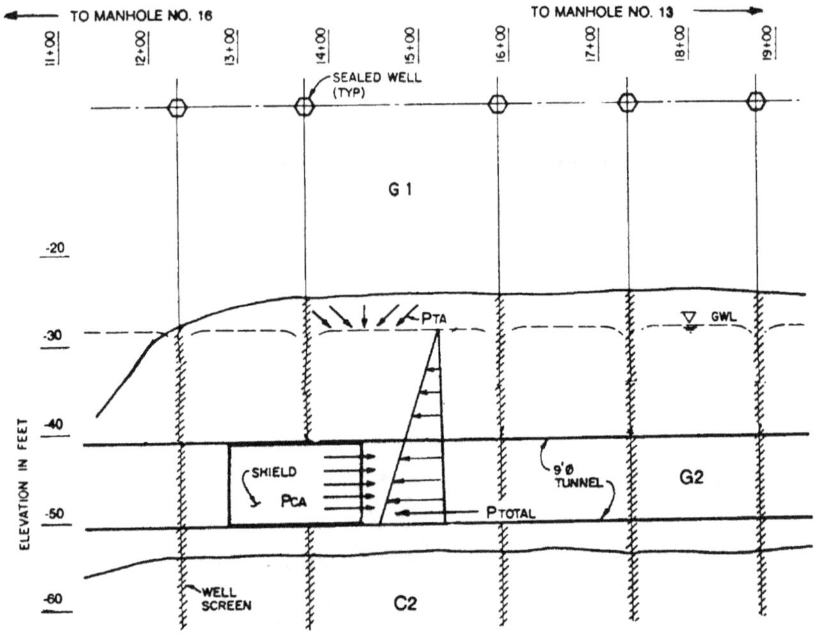

Fig. 9. Pressure on tunnel face without venting

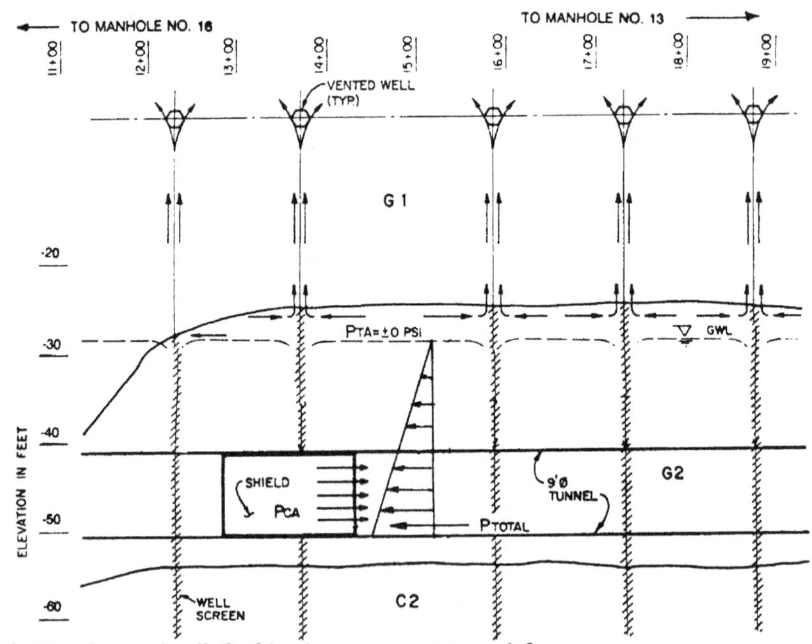

Fig. 10. Venting and relief of back pressure at tunnel face

Table 2 Pressure calculations with and without venting

Refer to Figures 9 and 10 for explanation of terms

P_{CA} = Pressure due to compressed air tunnelling
P_{TA} = Pressure due to trapped air in top of G2 level
P_{W} = Pressure due to limited dewatered level in tunnel
P_{TOTAL} = Total pressure exerted on face of shield

Before venting (assuming that water level is at 3 m (10 ft) above tunnel crown):

P_{CA} = 69 kPa (10 psi)
P_{TA} = 41 kPa (6 psi)
P_{W1} = 3 m (10 ft) of water = 30 kPa (4.3 psi) [at crown]
P_{W2} = 5.8 m (19 ft) of water = 57 kPa (8.2 psi) [at invert]
P_{TOTAL} [at crown] = 30 + 41 = 71 kPa (10.3 psi)
P_{TOTAL} [at invert] = 57 + 41 = 98 kPa (14.2 psi)

$P_{TOTAL} > P_{CA}$; water enters at invert level

After venting:

P_{TA} = 0
P_{TOTAL} = 30 to 57 kPa (4.3 to 8.2 psi)

$P_{TOTAL} < P_{CA}$; water no longer enters at invert level

illustrated in Figure 9. Whenever the tunnel air pressure was increased there was a corresponding increase in the back pressure in the aquifer; thus, when tunnel pressure increased the face dried for a short time but soon water would start to flow in again.

Once the problem was understood, the solution was to lower the back pressure of air in the aquifer. To achieve this, profiles of the alignment were analysed to find where the bottom of the G1 soils was at its highest elevation, i.e. the top of the domes of air. At these locations venting wells were drilled which included a screen at the top of the G2 aquifer. A valve at the top of the well riser allowed venting to be controlled. Valves were also installed in the dewatering wells, in order that there was no uncontrolled venting which might result in blowouts. On several wells extra valves were installed due to the greater volume of air that was encountered. Figure 10 shows the venting operation and the air flow path from the tunnel and up to the vents. Calculations showing the pressures before venting and after venting are presented in Table 2. Figure 11 shows the arrangement of vent wells.

Special precautions were also taken in the tunnel especially during periods when the face was not being worked or advanced. An impermeable membrane was put on

Fig. 11. Vent wells surrounded by hay to mitigate noise

the face of the tunnel so as to prevent air from escaping into the aquifer. This was especially important since the compressed air plant had a limited capacity of 4.7 m^3/s (10,000 cubic ft per minute) of air.

7 Health and safety guidelines

The work was carried out under strict supervision complying with United States Occupational, Safety and Health Administration (OSHA) guidelines. Some of the procedures followed at the site to protect the health and safety of the workers are summarised below:

- A licensed physician was retained who was familiar and experienced with the physical requirements and the medical aspects of compressed air work and the treatment of decompression illness.
- All employees who entered the compressed air environment were examined by the physician and reported by him to be physically qualified to engage in such work.

- A fully equipped first aid station was provided at the entry shaft.
- Each employee going under air pressure for the first time was instructed on how to avoid excessive discomfort.
- The site was also equipped with a medical lock facility to treat an employee for illness caused by compressed air operation.
- The decompression procedures used were in accordance with the Decompression Table 3 as shown below:

Table 3 Allowable continuous working hours in compressed air and decompression time

Air pressure (psig)	**Up to 12**	**12-16**
Maximum hours/shift	4	4
Maximum hours work under pressure per 24 hour day	4	4
Total decompression time (minutes)	3	7

Note: Work in compressed air was by six 4-h shifts per 24 hours, and no incident of decompression sickness was reported during the tunnelling period.

8 Conclusions

1. Complex and varying ground and groundwater conditions encountered along the tunnel alignment necessitated the change of tunnelling method of construction from free air to compressed air.

2. The confined aquifer conditions and its geometry entrapped the escaping compressed air from the tunnel face and created dewatering problems during tunnelling.

3. The phenomenon of entrapped air under a back pressure was carefully monitored and successfully overcome by venting the dewatering wells and installing additional vent wells along the alignment.

4. The field monitoring data indicated that entrapped air influenced wells as far as 274 m (900 ft) from the tunnel face.

5. It was found beneficial to cover the unworked face temporarily with an impermeable membrane to minimise escape of air.

9 Acknowledgements

Sverdrup Corporation and the authors would like to thank the New York City Department of Environmental Protection for the opportunity to work on this challenging project and allowing us to publish this paper.

The authors wish to thank the following individuals in particular for assisting in the review and preparation of this paper. Salvatore J. Vitale, Jr., Vice President and Deputy General Manager, Eastern Group; Dipak Shah, Resident Engineer; Charles Dillon, Assistant Office Engineer; all of Sverdrup Corporation; Mike McHugh, Project Manager, Moretrench American Corporation, Rockaway, New Jersey; Paul Dixit, Project Manager, Grow Tunnelling, New York, New York.

Discussion on Session 2: Construction projects – II

Mr P T Simpson presented the paper about compressed air tunnelling for the Hong Kong MTR on behalf of the authors, his colleagues, who could not be present. **Mr D J Court** presented his paper on Contracts 3, 12 and 12/5A of the Cairo Wastewater Project, Egypt. (The authors of the paper about the Oakwood Beach sewer tunnel could not be present). The session chairman, **Mr R S Hoad**, opened the questioning following the two presentations.

Mr R S Hoad. There are differences in the physiology and lifestyle of expatriate and indigenous workers in these overseas projects. How did this affect the ratios of decompression sickness in Cairo and in Hong Kong?

The presenters did not specifically answer this question but remarked on the different body sizes of European expatriates and the Chinese and Egyptians and that alcohol drinking could be a significant factor.

Dr J D King. What was the turnover of labour on the contracts?

Mr Court. On the Cairo contracts the turnover of Egyptian workers was relatively low; most stayed on for the later contracts.

Mr Simpson. There was an initial induction programme for Chinese workers.

Mr L Lofthouse. The turnover of Chinese labour on the Hong Kong Mass Transit 107 Contract (GKL Joint Venture) was not very high. Once the miners had overcome the fear of the 'magic air', they worked in a constant shift period.

Mr R J Reed. The indigenous workers in both Cairo and Hong Kong tend to be leaner and fitter.

Mr Simpson. There were also Korean workers in Hong Kong whose incidence rates of DCS were higher.

Mr F M Jardine. The difficulty of making comparisons of DCS incidence rates between different contracts or different races is separating all the variables. Overall DCS rates in terms of DCS incidents per exposure are of no value without at least defining pressures, working times and, I suggest, the number of exposures in distinct pressure/time combinations. Using this type of approach with some of the results of the analysis of the UK DCS Central Registry data carried out by Dr Evans, we put forward a way to represent DCS incidence in terms of pressure, p, and a parameter $B/tE^{0.5}$, where B is the number of DCS treatments resulting from E exposures of t hours at mean pressure p. For the Hong Kong contract 107 and for the Cairo contracts 3, 12 and 12/5A, for which papers give sufficient information, it is possible to represent their DCS rates in these terms. As can be seen in Figure 1, there is no obvious difference between the expatriate, the Chinese and the Egyptian DCS rates, and they are within the range found for the whole UK dataset. Perhaps, at an international conference, one might suggest that people are really the same the whole world over. (The above analysis is given in more detail in Part 8 of these Proceedings following the discussion on codes and regulations).

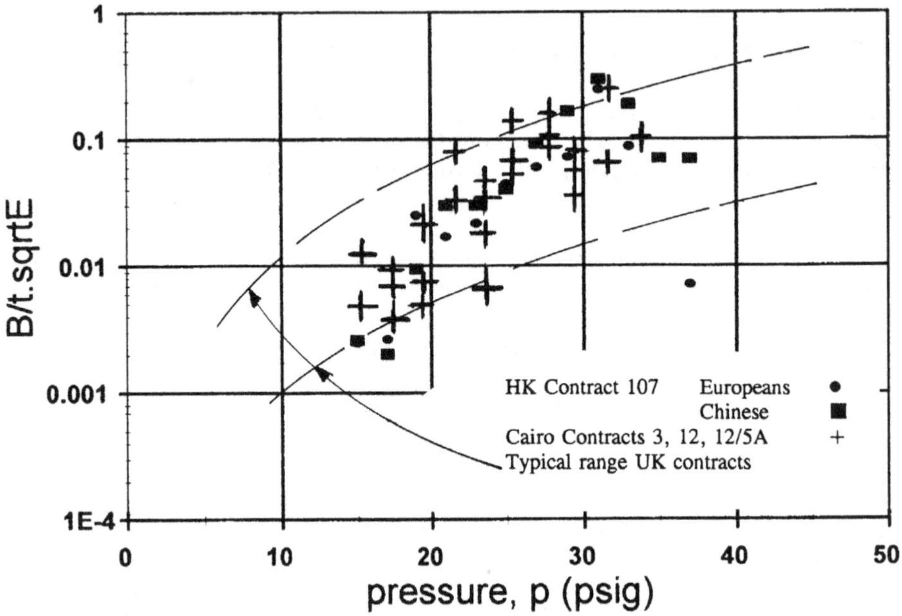

Fig. 1. Comparison of DCS incidence rates for different races on different contracts

Mr A J Howard. A question to Mr Simpson in relation to the Hong Kong work. Who decided what the tunnel air pressure should be; and on what basis? Was it always necessary to maintain full balance with hydrostatic; or was it possible to work satisfactorily at a lower pressure? How often was working pressure reviewed?
Mr Simpson. Mainly it was full hydrostatic balance. At crossovers, a different section was used and it was possible to reduce the range of pressures. This led to a damp sand face.
Mr Hoad. In Cairo, Mr Court suggested that the onset of decompression sickness frequently occurred later than the 1 hour minimum that workers are obliged to stay on site, perhaps recommending that this period needs re-examination. Can you please enlarge?

Mr Court. Looking at the reported times of the onset of DCS, only 34% were within 1 hour and 60% within a two-hour period. Does anyone know the origin of the one-hour rule? (Nobody in the audience was able to provide an answer).
Dr R W Hamilton. After stressful diving, most cases report within 1 hour, but long, shallow exposures tend to occur later.
Dr C E Lehner. With experiments on sheep, it was found that with short dives, the latency was relatively short, but increased for longer exposures.

Decompression sickness (DCS) or illness represents a number of clinically important forms of decompression injury. DCS has been described as either 'pain-only' Type I, or 'serious' Type II DCS.

Hyperbaric osteonecrosis is a clinically significant and disabling form of injury, and its earliest sign may be a persistent case of limb bends, so-called 'pain-only' bends. At this Conference, we report that persistent limb bends appears closely associated with the early development of hyperbaric osteonecrosis. This raises the question as to whether limb bends should continue to be referred to as 'pain-only' bends.

Dr L Greenbaum. Were workers penalised by reporting their DCS? There are problems about obtaining reliable figures as many cases go unreported.
Mr Court. There were, on the other hand, some unusually high incidences, and one was not sure if there was not over-reporting by workers fearing they might develop DCS. We had to rely on the doctor checking. Undoubtedly there were translation problems communicating to new workers.
Dr E P Kindwall. On a new contract, there are at first lots of DCS cases because of lack of acclimatisation or habituation.
Mr Court. In comparison with the tunnelling, shaft sinking is of relatively short duration, so pressures increase, say, from 0 to 2.3 bar within a week giving little time for acclimatisation.

(Reporters: J Buchanan, F M Jardine)

Part Three
Bone Necrosis

9 A preliminary survey of quantitative risk factors for bone necrosis in British compressed-air workers

A. Evans
Decompression Sickness Research Laboratory, University of Newcastle upon Tyne, UK

Abstract
The Newcastle Registry computer records have been searched to reveal personal and hyperbaric experience characteristics for two groups of compressed-air workers: cases (who developed radiological bone changes) and controls (who did not). *t* tests were used to identify any factors for which there was a significant difference between the groups. Age and experience of acute decompression sickness (DCS) were found to be the most important, though the separation achieved with a discriminant model based on these plus an important qualitative factor (the man's job) was disappointing. Several improvements to be incorporated at the next stage of the work are suggested.
Keywords: Compressed air workers, Bone lesions, Computer records, Risk factors, Statistical analysis.

1 Introduction

A serious long-term consequence for certain compressed-air workers and deep-sea divers of breathing gases at elevated pressure is the occurrence, years later, of aseptic necrosis in one of the long bones near a shoulder, hip or knee. Thus caisson disease of bone (as it was first described), which in the USA is known as dysbaric osteonecrosis, is an occupational hazard for the men who dig tunnels and sink caissons through waterlogged ground in a compressed air environment. The accompanying alterations to the mineral content soon make any such lesion detectable by radiography, so that early changes may be discovered by regular X-ray surveys of men at risk. The progress of this bone necrosis seems to be irreversible; thus when a lesion is present in the supporting structure beneath the surface of a major load-bearing joint there is a continual danger that the bone might crumble or shear, thereby disrupting the previously smooth geometry of the inner surface of the joint.

Consequently, a man with such a juxta-articular lesion may be concerned that it will develop and give him pain and limitation of movement of his shoulder or hip.

Engineering and Health in Compressed Air Work Edited by F.M. Jardine and R.I. McCallum. Published in 1994 by E & FN Spon. ISBN 0 419 18460 0

A severely damaged joint may require surgery to restore reasonable function, hardly satisfactory for an active working man. As there does not seem to be any reliable way of reversing the development of these bone lesions once they begin, it is clear that prevention must be the aim, though despite the introduction of the Blackpool decompression tables in 1966 giving longer total decompression times many men have developed bone damage since then. There is clearly still a problem; this paper describes a first attempt to identify those characteristics of the man and his hyperbaric experience which seem to precede bone damage.

2 The compressed air records

This study was only possible because of the existence of a large collection of records from the compressed-air work carried out at several important post-war British civil engineering tunnel and caisson contracts (1948 to 1977). Paper records which had been collected contemporaneously on behalf of the Medical Research Council Decompression Sickness Panel were concentrated at the former Decompression Sickness Central Registry at the University in Newcastle upon Tyne. A major funding initiative during the 1970s allowed the majority of these records to be transcribed by coders to one of several standard forms, from which cards were punched so that the data could be input for computer analysis and long-term storage.

These records consist of notes made at the initial medical examinations of identified men for fitness to engage in compressed-air work at specified construction sites and the results of any radiological surveys of their major joints carried out either at that time or subsequently. In addition there are day-by-day records of the workers' individual exposures to high-pressure compressed air from over a dozen projects including such major works as the Clyde road tunnel (1959 to 1963, where the statutory 1958 decompression tables were used) and the second Dartford road tunnel (1974 to 1977, where decompression was carried out according to the Blackpool tables). All these data are still accessible in the computer archive in Newcastle, making possible a rigorous investigation of many of the factors which could reasonably be expected to influence the development of the characteristic radiological abnormalities amongst active compressed air workers.

3 Cases and controls

The essence of the present survey is a comparison between two groups of compressed air workers. The first group, the 'Cases', is composed of selected men who developed bone necrosis, while some of their workmates who were employed at the same contracts but whose bones remained radiologically unchanged form the 'Control' group.

3.1 Radiographic results

The first stage was to examine the radiological results, so that each man who had experienced the usual full X-ray survey of shoulders, hips and knees on more than one occasion could be flagged. All men whose bones already exhibited some sign of aseptic necrosis of bone at first presentation were rejected, as was each man for which any of the individual radiological images (of which there are 8) from either the first or some later survey had been considered undiagnostic.

In this way it was possible to identify the complete set of men who were known to be both fit to work in compressed air and to have major joints giving normal radiological appearances on some day between about 1955 (when such surveys were introduced) and 1977, and who had participated in at least one further diagnostic X-ray survey the results of which had been recorded before Registry data entry ceased in about 1982. These men formed the basic pool from which both Controls and Cases would be drawn, by means of a carefully defined selection criterion.

3.2 Experience requirement for cases and controls

Before accepting a subject for inclusion into either category it was necessary to confirm that the Registry computer records appeared to cover that man's complete exposure history during a carefully defined 'vulnerable period'. This was intended to identify the part of his hyperbaric experience during which he might have sustained physiological damage which for a Case would result in the subsequent bone change.

As the natural history of this aseptic necrosis of bone is still poorly understood, it was only possible to make 'best guesses' for the bounds to this period of interest in advance, though having done so they were applied strictly to cases and controls alike. It was agreed for this trial that a bone lesion could be expected to take at least three months after any putative causative incident to become apparent on the radiograph, though it should not require more than a year to do so. Thus the subject experience window derived from and to be used in connection with any radiographic result pair was to begin 365 days prior to the earlier survey and end 90 days before the later or categorical examination.

Inevitably this method of selection resulted in a considerable range of window lengths, ranging from a few months to tens of years for both groups. Among the cases, the shortest sensitive period was 372 days, the mean was 1312 days and the longest was 6553 days. For the controls, the shortest experience window lasted 296 days, the mean was 1080 days, while the longest was 6499 days. Although the cases are seen to have marginally longer total time windows than the controls, this difference is found not to be significant.

3.3 Numbers of cases and controls

It was found that full hyperbaric work experience records were available in the Registry between the dates appropriate to 511 pairs of radiological results. The Registry coding scheme allowed for one of nine possible diagnostic results to be recorded for each joint survey, based on the diagnosis at the 'worst' single anatomical site. For each survey pair, the earlier result had by definition to be

Code 0 'Normal' or 2 'Irrelevant abnormality' (which means that nothing more serious than a radiological peculiarity not associated with aseptic necrosis of bone, such as a bone island, could be seen in bones covered by the films).

For the majority of the pairs the second result was also Code 0 or 2; 426 of these form the controls in the present study. Complete experience records were also available for 85 men who did develop some radiological appearance characteristic of of bone necrosis between surveys; these men are the Cases. All severities of lesion and specific radiological abnormality were accepted for a man to be included as a Case, ranging from the minimal 'suspected Head Neck or Shaft lesion' (result Code 3) through to the ultimate Code 9 'Surgical intervention following collapse of a joint surface'.

3.4 Personal attributes

For each case and each control, numerical information was compiled to define some of the subject's personal characteristics and to summarise his hyperbaric experience between the specified dates. Personal details extracted from the records of initial medical examinations included weight and height as recorded. These were also combined in the form of a Quetelet Index (Weight/Height squared), and the subject's stated date of birth was used to determine how old he had been at the later (diagnostic) survey.

3.5 Hyperbaric experience

Many cases and controls had worked according to the statutory British 1958 Regulations, which only required record-keeping when the working pressure reached 18 psig (1.24 bar). In order to achieve similarity in comparison between the pre-1966 experience of these men with that of men with experience under the Blackpool Tables (with which timed decompressions and so records are required for all pressures of 14 psig (0.97 bar) and above), it was decided that only exposures to pressures in excess of 17 psig would be considered when compiling the man's experience history. Thus in order to qualify, each exposure had to occur within the risk window and have been at a working pressure of 1.43 bar (18 psig) or more. For each such exposure details were extracted and compiled of:

- working pressure
- duration
- time of day (day, back or night shift)
- mode of decompression (i.e. direct or by decant)
- the nature of any ensuing acute decompression sickness (DCS).

A summary of this experience was prepared for each man from all relevant contracts to show the total number of qualifying exposures and their mean working pressure, together with their aggregate duration. A mean shift length could thus be calculated. The summary also includes the number of exposures which included decompression with decanting, and the total numbers of Type I (simple bends) and Type II (various symptoms other than bends) DCS events for which recompression therapy was given. The total length, in days, of the risk window was included, together with the total time in days which had elapsed between the first and the last of that man's qualifying exposures.

This full exposure summary might include exposures from 1.24 to 2.70 bar (18 to 39 psig) which might have lasted from as few as five minutes up to eleven hours. Although these features would to some extent be reflected in that man's mean working pressure or mean shift length, in the number count all decompressions are treated as equal. Some previous studies had indicated that longer working periods or higher pressures might be particularly injurious, so each exposure was also assigned according to length and working pressure to one of four quadrants which covered durations of more or less than four hours and pressures of more or less than about 3 bar absolute (29 psig). The numbers of exposures, decants, hours worked, and numbers of DCS events were then summarised separately for each quadrant. In all, 47 individual quantitative data items were compiled for each participating subject, some of whom were considered twice: first as their own control, and subsequently as a case.

4 Analysis and results

Values were computed from each numerical variable in turn to give the mean and standard error for all the cases and then again for all the controls, and a t test was conducted to assess the importance of any difference. The higher the value of t obtained with this test the lower the probability p that the observed distribution of values would occur by chance. For a difference to be considered at all significant the value of p must be no higher than 5% or 0.05, though much smaller values are required for high statistical significance. In some cases it was more appropriate to carry out the test on the logarithm of the natural value of the variable (plus unity to allow for possible zero values). It was found that the cases registered significantly larger values than the controls only for the fourteen variables shown in Table 1, where they are listed together with the pooled variance values for t and the associated 2-tailed probabilities p. For the remainder of the quantitative variables there was no significant difference between the groups.

4.1 Occupations

It was found that the highest t value was obtained with one of the few qualitative factors – a binary pseudo-variable based upon the subject's occupation. Men who had been employed as engineers, physicians or medical attendants were significantly less liable to develop bone damage than those in any of the other recorded job categories. No significant differences could be established however between the susceptibility of various other groups; foremen seemed to be just as liable to exhibit bone changes as miners or other tunnel workers or tradesmen.

Table 1. The only quantitative risk factors for which a significant difference was identified between the cases and the controls in this survey. 'VHP' means very high pressure (over 28 psig); 'long' means over 4 hours.

Risk factor	*t*	*p*
Log DCS from all VHP exposures	3.42	0.001
Age	3.36	0.001
Log DCS from all VHP long exposures	3.23	0.001
Log DCS from all exposures	2.90	0.004
Log total number VHP long exposures	2.81	0.005
Log total duration VHP long exposures	2.71	0.007
Quetelet index	2.68	0.008
Weight	2.63	0.009
Log total duration all VHP exposures	2.54	0.011
Log number all VHP exposures	2.28	0.023
Log number night shifts	2.25	0.025
Mean shift length	2.16	0.031
Log total duration all exposures	2.04	0.042
Log number decants from VHP long exposures	2.02	0.044

4.2 Personal factors

Clearly the most significant personal factor is the man's age. The controls were re-examined at a mean age of 32.5 years (with a standard error, s.e., of 0.4 years) while the mean for the cases was 35.7 (s.e. 0.9) years. For this difference t was found to be 3.36 (as shown in the Table), giving a probability of 0.001. This means that the age difference between these groups is highly statistically significant. The cases also yielded a higher mean value for the Quetelet Index than the controls, though the difference was less significant (t=2.68, p=0.008), and a similar result was obtained for bodyweight alone (t=2.63, p=0.009). This accords with the tradition that bone necrosis is more prevalent amongst older and heavier men.

4.3 Acute response to decompression

The association between bone damage and acute decompression sickness is strongly confirmed. Men who had been given recompression treatment for the symptoms of either kind of acute DCS were found to be much more liable to subsequent radiological abnormality than their workmates who had not reported problems. The strongest statistical correlation is seen to be with (the log of) the number of DCS events following any exposure at the highest pressure (i.e. over 2.0 bar or 29 psig), giving t=3.42 for which p=0.001, though the association with DCS events resulting just from the longer exposures (of 4 hours or more) at the highest pressures also gave a strong association (t=3.23; p=0.001). The cases are also seen to have reported more DCS events from any exposure irrespective of duration or pressure than the controls (t=2.90; p=0.004).

4.4 Other postulated risk factors
Direct measures such as exposure numbers or accumulated duration of high pressure experience did not give such a clear distinction between the groups. The (log of the) total number of decompressions recorded after long exposures to the highest pressures only gave t=2.81, p=0.005, and a similar result was obtained for the total time spent working long shifts at these pressures (t=2.71; p=0.007). Consideration of the aggregate duration of all highest pressure exposures irrespective of length gave t=2.54, p=0.011, though the total number of such experiences only gave a t value of 2.28 for which p=0.023. The aggregate duration of all qualifying exposures (the total time worked in air at 18 psig and over within the window) was slightly higher for the cases, to give t=2.04, p=0.042.

Three other factors derived from the exposure histories were found to be of little more than marginal importance. The (log of the) number of night shifts worked gave t=2.25 for p=0.025, even though the cases and the controls had not worked significantly different total numbers of exposures (of any length or pressure) for all three shifts combined. The average shift length (i.e. the mean duration of recorded work period) was found to be marginally greater for cases at 4.75 (s.e. 0.30) hours than for controls at 4.02 (s.e. 0.14) hours (t=2.16, p=0.031). Finally the (log of the) number of decompressions which had been conducted by decanting following long work periods at the highest pressures differed just enough for this factor to appear in the Table (t=2.02, p=0.044).

5 Discriminant analysis

Although some of the associations discussed above are individually highly significant, there is nothing in the t test analysis to indicate if all (or even most) of the important influences have been identified. To investigate this it is necessary to conduct a discriminant analysis in which an attempt is made to predict, on the basis of the individual values for the various risk factors for each subject, if he is likely to develop bone problems or not. All variables which appear to be significant are incorporated in some model. This specifies how they are to be combined, by means of some optimised discriminant function or rule, to provide a single numerical output for each combination of input parameters.

In the present investigation it was natural for the model to be based on the man's age. As the cases had been shown in general to be older, a large output from the formula will be associated with greater probability of bone change, and so other identified risk factors must be introduced in such a way that they increase the result. Experience of DCS following long higher-pressure work seemed to be equivalent to an extra 15 'years', though such experience without acute symptoms requiring treatment was only equivalent to an extra 3 'years'. Conversely following one of the least susceptible occupations was equivalent to a reduction of 40 years.

Application of these rules to all of the subjects results in a wide range of values, amongst both groups of men, from about -20 to over 60 equivalent years. A cut-off value is then found by trial which, for the postulated discriminant model, gives the clearest separation between cases and controls. However, setting the cut-off too high will ensure a low proportion of false positives though only by omitting a large proportion of the true positives, while too low a value will correctly identify many more true positives but also introduce many more false positives. An optimum value thus must be chosen; for the present model a level of 40 equivalent years was found to be best. Some 49 or 58% of the cases give values in excess of 40, while 301 or 71% of the controls yield lesser values.

6 Discussion

The analysis described in this paper represents a first attempt to apply these statistical techniques to the problem of bone necrosis. Although several highly significant associations have been identified, the degree of discrimination achieved with the first model is poor, and the model as it stands would appear to offer little guidance regarding the safe management of compressed air works. This may be because important determining factors do not yet feature correctly in the discriminant formula, despite the inclusion of all variables which showed highly significant differences between the groups. The minimal contribution of any direct measure of hyperbaric exposure is unsatisfactory considering that this bone necrosis is only seen in those who work in compressed gas environments. It is also odd to find that 11 of the 85 men selected to be cases seemed to have had no qualifying exposure (though it seems quite reasonable for 60 of the controls to be similarly unexposed).

Several reasons can be suggested for the poor predictive ability of this first discriminant model. No account has yet been taken of 'poor' decompressions, where either the decompression time was too short or the decant interval was too long, though this information is available for many thousands of man-decompressions. It is possible that personal factors in addition to weight and age are important, and that some others which were recorded in detail should be considered.

Some amendment to the definition of the experience window might also be appropriate. It is quite possible that a year is insufficient for radiological changes to become evident, and that the start of the window should extend back for two or more years before the first survey. Furthermore there may be no need for a preliminary negative X-ray result at all; it would be quite reasonable to assume that all young men develop to maturity with normal bones even if we do not happen to have radiological confirmation for every man. The exposure window might then be redefined to cover the subject's full hyperbaric experience from starting work to 3 months (or perhaps rather longer) before the diagnostic survey.

Such relaxation of the window definition would immediately allow many more men to be considered as potential cases, of which there have been less than a hundred using the original experience limits. Although any man claiming previous experience at his first recorded medical examination would not then be eligible to be a subject of either kind, a reasonable increase in the number of useful cases, as well as controls, may well result from these modifications without sacrificing the 'total exposure history' requirement which would remain a cardinal feature of the analysis;

the use of a subject as his own control would however be eliminated.

This first trial on the full Newcastle registry compressed-air dataset shows that useful correlations can be discovered, and has already provided some confirmation for the widely held beliefs that men may be more prone to develop bone necrosis as they grow older or fatter, and that an attack of the 'bends' today may indicate trouble in years to come. With the benefit of hindsight it has become clear how the investigation (which was supported by the Health and Safety Executive) could now be improved; the results from the next stage of this work should thus be of considerable interest to all who are responsible for the management and regulation of professional hyperbaric activities throughout the world.

10 The role of intramedullary pressure in hyperbaric osteonecrosis

E.H. Lanphier, C.E. Lehner, and S.E. Lemos
Department of Preventive Medicine and the Biotron, University of Wisconsin, Madison, Wisconsin, USA

This paper is dedicated to the memory of Albert R. Behnke, Jr., M.D., Captain, Medical Corps, United States Navy (Ret).

Abstract
Osteonecrosis appeared unexpectedly in sheep that we were using for decompression studies. Their dive history provided a recipe for an animal model of dysbaric osteonecrosis. Ficat and Arlet (1980) emphasized the role of high intramedullary pressure (IMP) in conditions with bone pain and osteonecrosis. This suggested that high IMP might cause both limb bends and dysbaric osteonecrosis. We found elevated IMP in limbs showing signs of bends. Artificially elevated IMP produced necrosis like that from long hyperbaric exposures with no-stop decompression. Fatty marrow can take up large quantities of nitrogen during exposure to compressed air. If bubbles form in the closed, rigid marrow space upon decompression, pressure in the space must rise. Elevated IMP may cause bone pain; blood flow may be impaired, producing a 'compartment syndrome' of bone. If high IMP and reduced blood flow are sufficiently prolonged, death of both marrow and calcified bone may occur. Prompt recompression of limb bends will probably prevent development of bone necrosis.
Keywords: Bone necrosis, Bends, Decompression, Intramedullary pressure, Recompression.

1 Introduction

Our own experience and related observations of others have given us a particular view of the mechanisms responsible for hyperbaric (dysbaric) osteonecrosis (Caisson disease of bone). The resulting hypothesis has a number of gaps, and parts of it may be in error. Nonetheless, we believe that it forms a useful framework within which greater understanding of this problem can be achieved.

Engineering and Health in Compressed Air Work Edited by F.M. Jardine and R.I. McCallum. Published in 1994 by E & FN Spon. ISBN 0 419 18460 0

2 Background observations

2.1 'Late limping'

In 1980, our laboratory was evaluating sheep and pygmy goats as substitutes for human subjects in proposed decompression studies. Both species seemed promising; but as the study went on, it was confounded by a phenomenon that we called 'late limping'. The characteristic limb-lifting sign of limb bends appeared in some of our sheep with no obvious relationship to pressure-chamber 'dives' that the animals had made.

2.2 Necrosis of bone

We considered dysbaric osteonecrosis as a possible cause of late limping, but the intervals were much too short according to the human timetable. Nevertheless, we obtained radiographs of our most pronounced limper. Bone necrosis was diagnosed, and bone scans with a gamma-emitting radionuclide (99mTC-MDP) were positive. The animal's dive history provided a recipe for an animal model of dysbaric osteonecrosis (Adams *et al.*, 1985). A series of 24-h exposures in the 2.5-2.9 atm abs. (22-28 psig) range, each followed by no-stop decompression, produced characteristic signs of bone necrosis. Four-hour exposures produced much less extensive changes.

2.3 Importance of intramedullary pressure (IMP)

In 1981, we discovered a slender volume, *Ischemia and Necroses of Bone* (Ficat and Arlet, 1980). The book's main thesis is that elevated intramedullary pressure (IMP) is a vital factor in conditions involving bone pain and osteonecrosis. This suggested that abnormally high IMP might be responsible both for the discomfort of bends and for ischemia leading to bone necrosis.

2.4 Measurement of IMP

We developed a transcutaneous method for measuring IMP in sheep bones (Lanphier *et al.*, 1990). A large hypodermic needle is inserted through skin and soft tissues to the surface of bone. A drill-tipped wire is introduced through the needle and is drilled through bone into the marrow cavity. The wire then guides a hollow needle that serves as both bone drill and pressure tap. In a pilot study, we compared pressures in long bones of limbs showing signs of bends with those in sheep that had not been exposed to increased pressure. We found no IMP above 40 mm Hg (5.3 kPa) in control animals, but there were instances of much higher pressures in bones of symptomatic limbs. The relationship of high IMP to discomfort seemed convincing enough to warrant shifting attention to elevated IMP and its role in osteonecrosis.

2.5 Artificial elevation of IMP

In anesthetized sheep, we utilized our IMP-measuring pressure tap to elevate IMP in a femoral shaft to the arterial level by infusion of Ringer's solution or normal saline into the marrow cavity. Very small amounts of fluid sufficed to maintain elevated

IMP, so we assumed that vascular stasis had been produced. The most likely mechanism seemed to be that of the *Starling resistor*: flow in a collapsible tube is reduced or stopped by increasing the pressure surrounding it. This principle is implicated in the ischemia of *compartment syndrome* elsewhere and in the effects of increased intracranial pressure.

Artificial elevation of IMP produced marrow necrosis indistinguishable from that produced by compressed-air exposure and decompression. Studies to date indicate that ischemia lasting 12 hours (or less?) can cause loss of viable osteocytes in calcified bone as well as marrow necrosis (Lemos *et al.*, 1990).

2.6 Loss of marrow blood vessels

Microangiography indicated that marrow necrosis includes massive destruction of marrow vasculature. However, remarkable ingrowth of new vessels is underway in two weeks, as is new-bone formation on the endosteum and dead trabeculae (Lemos *et al.*, 1990a).

2.7 Bone scans and magnetic resonance imaging (MRI)

We are currently using 99mTC-MDP bone scans to study the development of necrosis following decompression. In some instances, we have seen 'cold spots' that may reflect an early ischemic phase. It is now clear (Lehner *et al.*, 1994) that a single long exposure is sufficient to produce necrosis. Changes are readily confirmed by magnetic resonance imaging (MRI), but it is not yet clear whether MRI can reliably detect early ischemia (Marshall *et al.*, 1989). Detection early enough for effective remedial action seems particularly important in limb bends with omitted, delayed, or ineffective recompression – especially after long or deep exposures.

3 Hypothesis

The observations reported above, together with related findings in the literature, led us to think in terms of the hypothesis outlined here. It provides a framework for considering the elements that are presumably involved.

3.1 Importance of fatty marrow

We agree with the observation (Walder and Holloway, 1987; Walder 1990, 1991) that dysbaric osteonecrosis occurs only in bones in which fatty marrow is predominant. In commenting on animal studies in bone necrosis, Behnke (1968) said 'It is animals with white, fatty marrow in the long bones as the pig and the cow (certainly not the red-marrowed rodent) which are likely to yield the answer'.

3.2 Uptake of nitrogen

Nitrogen is about five times as soluble in fat as in 'aqueous' tissues. For this reason and because of low blood flow, fatty marrow is among the 'slowest' tissues in the body from the standpoint of saturation and desaturation with nitrogen. We believe that 4 hours in the neighborhood of 2 atm abs. is at the threshold for taking up enough dissolved gas to cause problems. Presumably, shorter times at greater pressures would be equivalent.

Swelling of adipocytes from elevated oxygen pressure, as reported by Pooley and Walder (1981), and the apparent decrease in bone perfusion reported by Walder and Holloway (1987) may increase the likelihood of osteonecrosis and the significance of 4 hours as the critical duration of exposure.

3.3 Importance of bubbles

Presence of expanding bubbles in the marrow cavity appears to be an essential part of the process leading to osteonecrosis. Bubbles or nuclei arriving in arterial blood would no doubt grow by uptake of additional nitrogen within the marrow space, but a very plausible source of bubbles is the great amount of nitrogen dissolved in marrow fat. We have no firm opinion as to where bubbles actually form. Whether they are within the sinusoids or elsewhere within the marrow cavity may not be important in terms of IMP.

3.4 Bends sensation

We do not know the location of relevant pain nerve endings, but we feel confident that elevation of IMP is sufficient to explain the discomfort of limb bends. Distention of sinusoids has been suggested as the mechanism (Walder, 1991), but we believe that the periosteum is capable of painful response to abnormal IMP, either high or low relative to ambient pressure. The possibility that bends is an example of ischemic pain (Cross, 1987) should not be neglected.

3.5 Vascular stasis

We favor the Starling resistor mechanism as the main factor in vascular stasis and tissue ischemia. Other possibilities include venous obstruction from bubbles formed (or grown) in sinusoids and unable to pass restrictive bony foramina. With any mechanism, stasis will probably be followed by thrombosis (Kawashima, 1991). Once stasis occurs, arterial pressure will presumably maintain IMP at a high level. The role of fat emboli in dysbaric osteonecrosis (Jones, 1968, 1985) is difficult to assess. Fat emboli from elsewhere could presumably obstruct marrow circulation, and it is plausible for high IMP to cause fat to be expelled from marrow into the venous return.

3.6 Marrow necrosis

Marrow is apparently much more susceptible to necrosis than calcified bone, so it may well be possible to have marrow necrosis without death of bone if the duration of ischemia is relatively short. With longer periods, loss of increasing numbers of osteocytes can be expected.

3.7 Loss of marrow vasculature

Destruction of vessels in the marrow (Lemos *et al.*, 1990a) must deprive the cortex of normal centrifugal blood supply even if IMP returns to normal. How much of this supply may be replaced by collateral flow from outside the marrow cavity is not certain. This factor is almost certainly variable and may account for some of the differences in response to decompression and to artificial elevation of IMP.

3.8 Bone death and new-bone formation

It is commonly assumed that death and disapearance of osteocytes is primarily due to ischemia. It is then supposed that the presence of dead bone leads to deposition of new bone and to ultimate replacement of necrotic structures. New bone produces the x-ray appearance that is considered diagnostic of osteonecrosis. Walder and Stothard (1978) have presented evidence that the presence of necrotic fat or its breakdown products is the initiating factor for new bone formation. In addition, Walder (1991a) has suggested that new bone may 'suffocate' old but still viable bone. Destruction of marrow circulation and the postulated effect of necrotic fat suggest that serious injury to marrow would almost guarantee eventual necrosis of calcified bone.

3.9 Recompression and limb bends

Universal experience that recompression relieves the pain of limb bends supports the importance of bubbles in that condition. This recalls an observation attributed to Behnke: that if we didn't know about bubbles, we would have to invent them to explain the phenomena associated with bends. If the pain of limb bends arises outside of bone, as with bubbles in tendons or articular structures, the efficacy of recompression would be easy to understand. If pain arises from bubbles in the marrow cavity and from resulting elevation of IMP, the course of events must be more complex.

As ambient pressure increases during recompression, displacement of blood into the gas-containing space might be expected; but no such movement can occur until ambient pressure (plus arterial pressure) exceeds IMP. When IMP and ambient pressures are equal, pain due primarily to pressure-difference between inside and outside of the bone should be relieved. If pain arises from sinusoids distended by bubbles, or distended by congestion from obstructing vascular bubbles, it may persist until displacement of blood into the cavity (with proportional elevation of IMP) causes significant compression of bubbles.

If the external pressure in recompression equals IMP and then rises more rapidly than movement of blood can equalize pressure within, a form of 'squeeze' will develop. Behnke (1945, 1968) attributed increased pain upon too-rapid recompression to this phenomenon. In the belief that this concept is important, we suggest calling it *Behnke's bone squeeze*.

Reduction of bubble size should permit reperfusion to occur unless significant thrombosis has developed during the period of stasis, or destruction of vessels has already occurred. Reperfusion injury, as reported following ischemia elsewhere, may further complicate the situation in bone (Jones, 1987).

3.10 Recompression and dysbaric osteonecrosis

An important area of uncertainty concerning hyperbaric bone necrosis is its relationship to limb bends. A high degree of correlation between sites of limb bends and sites of osteonecrosis has not been established, but we suspect that cases of necrosis having no such association are rare. In some instances, they may reflect differences in description and definition, for example as Cross (1987) in discussions of 'non-bend' pain. It now appears that this entity may be indistinguishable from the 'deep' form of limb pain (Cross, 1991) and similarly affected by recompression.

Walder (1987, 1991a) has presented a hypothesis of hyperbaric osteonecrosis involving adipocyte swelling due to elevated oxygen pressure. Ultimately, this also involves dissolved nitrogen and bubble growth. To that extent, it should also be amenable to recompression. The fact that osteonecrosis has been minimized in certain tunneling and diving operations may be attributed to conservative exposure limits, superior decompression tables, prompt treatment of bends (including 'niggles'), or to a combination of these (Behnke, 1968).

There is an urgent need to determine whether timely recompression of limb bends reliably averts development of dysbaric osteonecrosis following exposures otherwise likely to produce necrosis. We propose that recompression is important in preventing osteonecrosis and that it must be applied early: within 6 hours of the insult if possible; preferably much sooner.

3.11 Undesirable effects of treatment

Whether 'Behnke's bone squeeze', usually resulting from too-rapid recompression, can ever cause bone damage needs to be considered. How common this phenomenon may be and what factors favor its occurrence is also uncertain (Cross, 1991a). If overzealous recompression can cause or aggravate osteonecrosis, this certainly needs to be known.

Another possible problem concerns oxygen breathing in emergency management of decompression sickness (Bennett, 1991). Theoretically, breathing oxygen should hasten elimination of nitrogen and shrinkage of bubbles, meanwhile helping to relieve hypoxia. Actually, there is not much solid evidence of clinical improvement of limb bends with oxygen breathing at normal pressure. It is common knowledge that increased oxygen pressure causes vasoconstriction in some vascular beds; and reduced perfusion may occur for other reasons in bone and marrow fat (Walder and Holloway, 1987). In some situations, ill-effects might overshadow the expected benefit of oxygen breathing.

3.12 Prevention of dysbaric osteonecrosis

On the basis of his oxygen hypothesis, Walder (1991a) advocates a 4-h exposure limit as a means of preventing dysbaric necrosis. Whether time would have to be reduced further at higher pressures is not specified. Those who focus on nitrogen uptake as the crucial factor would also expect a 4-h limit to be preventive at moderate pressures, but they would offer longer decompression times as a means of extending useful workdays and safely reaching higher pressures.

Practical prevention probably depends upon conservative exposures, recognition of acclimatization or lack thereof, adequate decompression – perhaps with oxygen-breathing as suggested by Behnke (1968), and prompt and adequate recompression for any sign of bends.

Acknowledgment

This work was funded by the University of Wisconsin Sea Grant Institute under grants from the National Sea Grant College Program, National Oceanic and Atmospheric Administration, U.S. Department of Commerce, and from the State of Wisconsin. Federal grant NA9OAA-D-SG469, project R/DP-6.

References

Adams, W.M.C., Lehner, C.E., Dubielzig, R.R., Wilson, J.W., Palta, M., and Lanphier, E.H. (1985) Dysbaric osteonecrosis in sheep: a model of aseptic bone necrosis. *Physiologist*, **28**, 308 (abst 39.19).

Behnke, A.R. (1945) Decompression sickness incident to deep sea diving and high altitude ascent. *Medicine*, **24**, 381-402.

Behnke A.R. (1968) *Medical Aspects of Work in Pressurized Tunnel Operations*. Transit Insurance Administrators, San Francisco, Calif.

Bennett, P.B. (1991) (Discussion on oxygen in treatment of bends), in: *What is Bends?* (eds. I. Nashimoto and E.H. Lanphier), Undersea and Hyperbaric Medical Society, Bethesda, Maryland, pp. 136-137.

Cross, M.R. (1987) Is bone necrosis in diving painful? in: *Proceedings of the 9th International Symposium on Underwater and Hyperbaric Physiology* (eds. A.A. Bove, A.J. Bachrach, and L.J. Greenbaum, Jr.) Undersea and Hyperbaric Medical Society, Bethesda, Maryland, pp. 1143-1153.

Cross, M.R. (1991) (Discussion on increased pain upon recompression), in: *What is bends?* (eds. I. Nashimoto and E.H. Lanphier), Undersea and Hyperbaric Medical Society, Bethesda, Maryland, pp. 145-146.

Cross, M.R. (1991a) (Discussion on 'non-bend' pain), in: *What is bends?* (eds. I. Nashimoto and E.H. Lanphier), Undersea and Hyperbaric Medical Society, Bethesda, Maryland, p. 149.

Ficat, R.P. and Arlet, J. (1980). *Ischemia and Necroses of Bone*. (ed. D.B. Hungerford), Williams & Wilkins, Baltimore, Maryland.

Jones, G.R.N. (1987) Dysbaric osteonecrosis (Caisson disease of bone): are active oxygen species and the endocrine system responsible, and can control of the production of free radicals and their reaction products confer protection? *Free Rad. Res. Comms.*, **4**, 397-406.

Jones, J.P., Jr. (1968) Avascular necrosis of bone, in: *Medical Aspects of Work in Pressurized Tunnel Operations*. (ed. A.R. Behnke), Transit Insurance Administrators, San Francisco, Calif, pp. 53-63.

Jones, J.P., Jr. (1985) Chapter 86, Osteonecrosis, in: *Arthritis and Allied Conditions*, Tenth Edition (ed. D.J. McCarty), Lea & Febiger, Philadelphia, pp. 1356-1373.

Kawashima, M. (1991) (Discussions concerning platelet thrombosis), in: *What is Bends?* (eds. I. Nashimoto and E.H. Lanphier), Undersea and Hyperbaric Medical Society, Bethesda, Maryland, pp. 65 and 74.

Lanphier, E.H., Lehner, C.E., Kanikula-Ives, T., Anderson, L.C. and Lemos, S.E. (1990) Measurement and modification of bone marrow cavity pressure. *Undersea Biomed. Res.* **17** (suppl.), 58 (abstract 67).

Lehner, C.E., Lin, T.F., Lanphier, E.H., Wilson, M.A., Dueland, R., and Dubielzig, R.R. (1994) Early pathogenesis of hyperbaric osteonecrosis. in: *Engineering and Health in Compressed Air Work* (eds. F.M. Jardine and R.I. McCallum), E & F N Spon.

Lemos, S.E., Lehner, C.E., Brunson, D.B. and Lanphier, E.H. (1990) Elevated intramedullary pressure causes bone marrow necrosis. *Undersea Biomed. Res.*, **17**, (suppl.), 59 (abstract 68).

Lemos, S.E., Lehner, C.E., Nordheim, E.V., Brunson, D.B. and Lanphier, E.H. (1990a) Ischemic change in cortical osteocytes following exposure to elevated intramedullary pressure, *Undersea Biomed. Res.*, **17**, (suppl.), 59-60 (abstract 69).

Marshall, D.G., Steinberg, M.E., Dalinka, M.K., Rao, V.M., Fallon, M. and Kressel, H.Y. (1989) Magnetic resonance imaging of the ischemic hip: alterations within the osteonecrotic, viable, and reactive zones. *Clin. Orthopaed.*, **244**, 60-77.

Pooley, J. and Walder, D.N. (1981) Changes in cell volume following hyperbaric exposure: a manifestation of oxygen toxicity, in: *Underwater Physiology VII: Proceedings of the Seventh Symposium on Underwater Physiology* (eds. A.J. Bachrach and M.M. Matzen), Undersea Medical Society, Bethesda, Maryland, pp. 45-53.

Walder, D.N. and Stothard J. (1978) Bone necrosis: Reimplantation of anoxic autologous marrow. *Undersea Biomed. Res.*, **5**, 39-40.

Walder, D.N. and Holloway I. (1987). Role of oxygen in dysbaric osteonecrosis, in: *Proceedings of the 9th International Symposium on Underwater and Hyperbaric Physiology* (eds. A.A. Bove, A.J. Bachrach and L.J. Greenbaum, Jr.) Undersea and Hyperbaric Medical Society, Bethesda, Maryland, pp. 613-626.

Walder, D.N. (1990) Chapter 17, Aseptic necrosis of bone, in: *Diving Medicine*, Second Edition, (eds. A.A. Bove and J.C. Davis), W.B. Saunders Company, Philadelphia, pp. 192-199.

Walder, D.N. (1991) The pain and mechanism of bends, in: *What is bends?* (eds. I. Nashimoto and E.H. Lanphier), Undersea and Hyperbaric Medical Society, Bethesda, Maryland, pp. 58-62.

Walder, D.N. (1991a) The links between the bends and dysbaric osteonecrosis, in: *What is bends?* (eds. I. Nashimoto and E.H. Lanphier), Undersea and Hyperbaric Medical Society, Bethesda, Maryland, pp. 68-71.

11 Optimum schedules for caisson decompression

E.P. Kindwall
Medical College of Wisconsin, Milwaukee, Wisconsin, USA

Abstract
Evidence accumulated over many years indicates that all the current caisson decompression schedules in use are inadequate. The reported incidence of decompression illness (DCI) is considerably lower than the actual. Most previous decompression schemes for tunnel workers have been extrapolated from naval schedules, albeit navies have had very little experience with long-duration dives at shallow depths. Oxygen decompression tables developed in Milwaukee are based on commercial data and eliminate nitrogen several times more effectively than existing schedules. Oxygen decompression appears to be the only viable method of decompressing tunnel workers on a daily basis. These tables appear to be a great deal safer than existing tables, but regulating bodies in the United States, Great Britain and Japan have failed to keep pace with recent developments in decompression physiology. In the past, dysbaric osteonecrosis has been the most feared complication of tunnel decompression. New evidence from Milwaukee, Wisconsin, indicates that there may be the possibility of central nervous system damage secondary to improper decompression. Tables from Japan, the United States, Great Britain, Germany and France are compared with the new oxygen decompression tables. Data regarding central nervous system changes in the brain of compressed air tunnel workers as seen with magnetic resonance imaging are reported.
Keywords: Decompression illness, Oxygen decompression, Caisson decompression, Compressed air, Magnetic resonance imaging, CNS, Dysbaric osteonecrosis.

1 Introduction

Caisson and compressed air work differ from diving in several important ways. Caisson exposures are usually carried out at quite modest pressures never exceeding the equivalent of 50 psig (345 kPa). Bottom times or shift lengths, however, almost always range between 4 and 8h on a daily basis with 6 to 8h exposures being quite common. Furthermore, each exposure tests the decompression schedule being used

Engineering and Health in Compressed Air Work Edited by F.M. Jardine and R.I. McCallum. Published in 1994 by E & FN Spon. ISBN 0 419 18460 0

almost to the limit, as a new schedule is provided for each additional 2 psig (14 kPa). This is done for economic reasons as every effort is made to avoid extending decompression times unnecessarily. For this reason, about four out of five caisson exposures test the table in use to its maximum limit whereas only three in ten water dives tax the table to a similar extent. Diving tables are usually set up in larger 3 m (30 kPa) increments.

Caisson workers' bodies are not supported by water while exposed. In addition to supporting their own weight throughout the exposure, they must carry their tools and the burdens they bear. The air in caissons and tunnels usually contains more contaminants than are present in diver's air. While working at pressure, tunnellers are rarely chilled but are often exposed to heat during concrete pours. On the other hand, divers are typically chilled during their working exposure. In tunnels, chilling takes place during decompression when the workers are seated in a dry chamber. Balldin *et al.* (1969) show that there is 30% less nitrogen eliminated in those decompressing in the dry as opposed to those decompressing while immersed. Finally, despite the fact that caisson exposures are much more numerous than diving exposures, relatively few good data are available regarding the efficacy of the decompression tables used compared to diving decompression schedules.

1.1 The reported incidence of decompression illness

When decompression illness (DCI) rates are tabulated in caisson work, it is usually only treated cases which are reported. These are the official rates which generally give little indication of the actual occurrence of symptoms. Caisson tables have been so extremely bad for so many years that compressed air workers tend to accept symptoms of DCI as an inescapable element of their jobs. For this reason, they rarely report symptoms unless they are unbearable or incapacitating. If they were to report for treatment too frequently, they would be removed from more remunerative compressed air work. Using an anonymous system of reporting on more than one tunnel job in Milwaukee, we found that up to 26% of a shift might be suffering symptoms of DCI on any given day, although none came in for treatment. On one project which had pressures ranging between 19 and 31 psig (131-214 kPa) anonymous reporting revealed that DCI was present on 42.5% of the working days. During that period of time the 'official' decompression illness incidence, representing cases treated, was 1.44%. The present United States Occupational Safety and Health Administration (OSHA) enforced decompression schedules were the tables used when these observations were made.

1.2 Dysbaric osteonecrosis

I consider dysbaric osteonecrosis to be another form of decompression illness as bone is simply another target organ along with spinal cord and brain. Although there have been other theories as to its causation, bone disease appears to be caused by improper decompression. Repetitive human experience has demonstrated that decompression time is the only independent variable in the production of bone necrosis. Short decompressions produce necrosis; long decompressions for the same exposures produce less or none.

1.3 The split shift

The split shift was formalised in the United States when the 1922 New York code was adopted. It was hoped that the shorter exposures would decrease the incidence of DCI. The split shift embodied two rather short shifts separated by a one-hour rest interval on the surface during the middle of the day. Decompressions for both shifts were identical which did not take into account the residual nitrogen from the first shift. The one-hour surface interval was woefully inadequate to allow release of meaningful amounts of nitrogen present in the body. The split shift also exposed the worker to the trauma of two decompressions per day instead of just one. Unfortunately, the split shift is still in use in Japan. It produces a very high rate of decompression illness but contractors have been reluctant to report this to the authorities as they might be subject to disciplinary action should the government become officially aware they had not reported these high rates in the past (Nashimoto, 1986). Figure 1 shows a comparison between the Japanese split shift and the US Navy repetitive dive schedule. The Japanese schedule is totally unphysiologic.

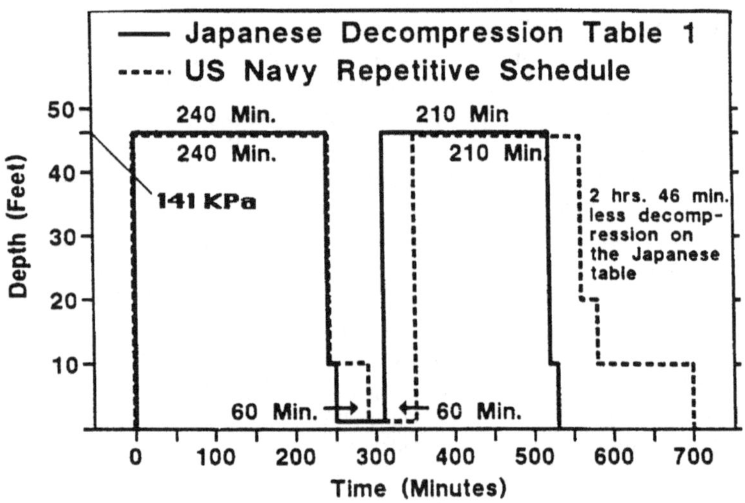

Fig. 1. Comparison between Japanese split shift and USN repetitive dive schedules

2 Shift length

Working shift lengths are typically divisible into 24 hours as tunnelling usually proceeds around the clock. Eight, six and four hour shifts are used, the last at the higher pressures. Because of table inadequacy, the French have drastically shortened permissible shift length, as pressures rise, to increase safety. In Britain, current practice permits work shift lengths to be longer, as decompression time may be added to a full 8 h work shift. The 1982 edition of CIRIA Report 44 stipulates only that there must be a 12 h surface interval between exposures and that the total of the combined work and decompression exposures be limited to 10 hours per day. My own view is that the time between shifts should be at least 16 h and that 12 h is too short a period for adequate denitrogenation. Even a 16 h hiatus is probably too short, if present theories regarding the half time of the slowest tissues are correct. This implies that employees are not completely free of work-absorbed nitrogen when re-entering compressed air after the first shift of the week. In the US, all work, including decompression, must be completed within an 8 h shift, guaranteeing a 16 h surface interval.

Historically, continuous decompression rather than stage decompression was the norm in compressed air exposures. This emanated from decompression simply being a matter of opening a valve and waiting until the lock pressure equalised with atmospheric pressure. When it become understood that DCI was caused by rapid decompression, the valves were not opened fully, and slower continuous ascent was provided. The present US OSHA decompression tables still utilise continuous decompression at increasingly slower rates for this reason. This complicates matters, as it is very difficult to control such decompressions by hand and a computer or cam-controlled exhaust mechanism is required. Using stage decompression there is no such requirement. Additionally, continuous decompression is also inefficient and wasteful. For example, if the last stage from 4 psig (28 kPa) to the surface were to take one hour, at least one half of the time is spent at a pressure less than 2 psig (14 kPa) which provides no meaningful bubble suppression. In UK, France, Germany and Brazil, the more efficient stage decompression has been adopted.

Behnke (Kindwall *et al.*, 1983) has suggested that saturation habitats be constructed in tunnels with storage pressures for the workman up to 22 psig (152 kPa). This would be the pressure limit because of the cumulative effect of oxygen toxicity from compressed air. At that pressure, the effective oxygen level would be equivalent to 53% at the surface. Much deeper 8 h excursions to working pressure could be made without a decompression obligation. A major problem with saturation exposure is the space required for the living quarters and securing the cooperation of unions for these exposures. Saturation would be feasible only on the largest projects. To this date, it has not been utilised.

3 Modern decompression tables

In 1963, Duffner devised the Washington State tables which have since become the federally enforced OSHA tables in the USA (U.S. Bureau of Labor Standards, 1971). These tables replaced the split shift but retained continuous decompression as opposed to stage decompression.

Because the British 1958 tables were found to produce a 19% incidence of aseptic necrosis on the Clyde tunnels (McCallum, 1967), Hempleman (1973) devised the Blackpool tables in 1966. The Blackpool tables were much longer, adopted stage decompression and required decompression for all exposures greater than 14 psig (97 kPa), as do the American tables.

Duffner was unable to adopt stage decompression in his tables because of the objections of contractors and the unions who refused to stray from tradition (Duffner, 1983). The American OSHA tables also have another major flaw in that for each 2 psig (14 kPa) pressure range a decompression time is provided for shift length 'over 8 h'. These decompression schedules would be disastrous if ever used, as conceivably a foreman could work two 8 h shifts back-to-back in an emergency, totalling 16 h. Were he working at 20 psig (138 kPa), the decompression requirement using the OSHA code would be only 113 minutes. In modern saturation diving with air, such an exposure calls for over 14 h decompression.

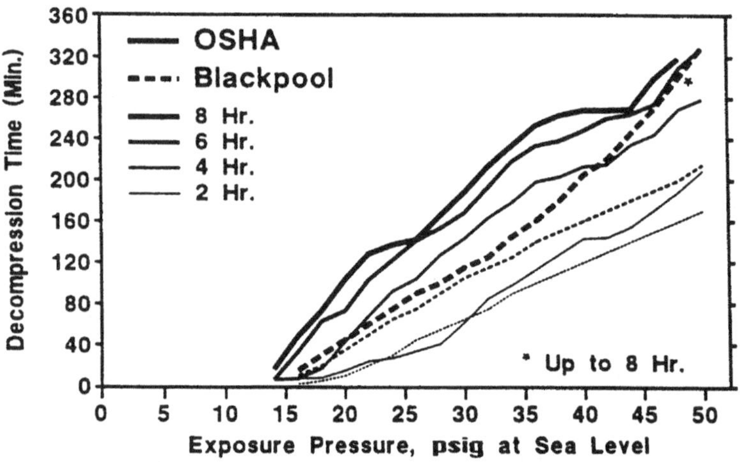

Fig. 2. Comparison of OSHA and Blackpool Tables decompression times

As seen in Figure 2, the OSHA tables call for longer decompressions over most of their range than do the Blackpool tables and therefore should be inherently safer. The steeper the rise of the curve, the greater the safety. Note also the internal inconsistencies in the OSHA schedules which show that at 26 and 44 psig (179 and 303 kPa), the decompression time is the same for both 6 and 8 h exposures. This occurs at several other places in the tables. When these discrepancies were shown to Duffner, he was unable to explain them as they were not so calculated. They are probably due either to typographical errors which crept into the transcription of the tables or resulted from the slight displacement of the ruler which was placed across a nomogram to find the decompression times.

Although the OSHA tables are safer than the Blackpool tables on inspection, they still are not as safe, for the practical exposure times which would be used, as the US Navy Exceptional Exposure Air Tables for which a comparison is shown in Figure 3.

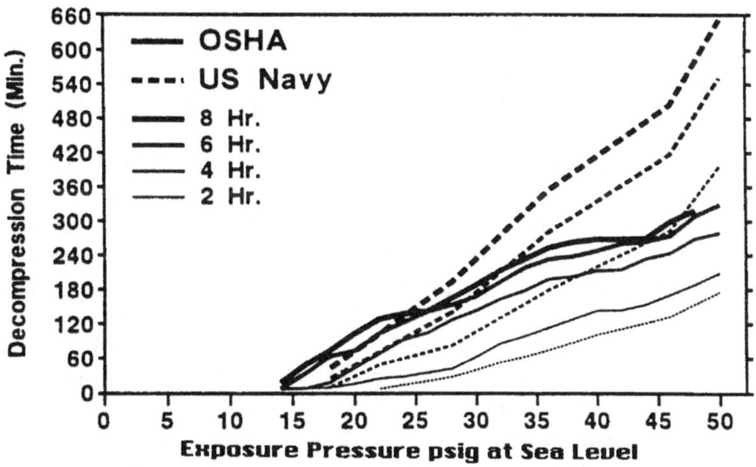

Fig. 3. Comparison of OSHA and USN Exceptional Exposure Air tables decompression times

Thus, it would appear that the US Navy Exceptional Exposure Air Tables would be safest for tunnel work as they provide the longest decompression times over most of their range. However, the US Navy Exceptional Exposure Tables were only tested at the 140 fsw (965 kPa) level before being promulgated to the fleet. On each test there were never more than six men in the chamber and on every occasion at least one man suffered DCI and sometimes two. Realising that these tables showed a trend for producing a DCI incidence of between 17 and 33%, they are now reserved *for emergency use only*. Thus it would appear that in the United States and Great Britain, contractors routinely expose their workman to daily decompressions that are more dangerous than decompressions the United States Navy approves only for dire emergencies.

Considering the theoretical parameters outlined above, we would be concerned about the actual results of using these tables in the field. Experience has shown these concerns to be valid. Previously, the statistics were given for DCI when the OSHA tables were in use. We do not have similar statistics for the Blackpool tables as no anonymous reporting of symptoms has been employed when the tables were in use. A possible exception is Yau's personal communication (1987) that 83% of the men working on the Hong Kong subway project reported decompression illness in association with the Blackpool schedules.

Bone necrosis for all practical purposes appears to be unknown at pressures less than 17 psig (117 kPa) (Walder, 1971). When the OSHA tables were first employed in Washington State, they were used up to a maximum of 36 psig (248 kPa). Ten years later, Sealey (1975) found four individuals with dysbaric osteonecrotic lesions in the shafts of the proximal tibia. No other lesions were found and the tables were felt to be safe from the standpoint of dysbaric bone disease. Subsequent use on the Bay Area rapid transit project was for the most part at pressures never reaching 17 psig (117 kPa) except for one two-week period where the pressures peaked at 36.5 psig (251 kPa) under the Embarcadero. No aseptic necrosis was found on yearly x-ray of the workers but no follow-up examination of them has been carried out. The present OSHA tables were adopted by emergency order in 1970 in Wisconsin and initially no new cases of aseptic necrosis were seen on yearly x-ray of the workmen. However, two years after a tunnel job had been completed at a maximum pressure of 43 psig (296 kPa) 33% of the workmen were found to have dysbaric osteonecrosis when surveyed (Kindwall *et al.*, 1982). Thus, at pressures over 36 psig, the OSHA tables were as ineffective at reducing dysbaric osteonecrosis as were the old split-shift tables used in Milwaukee which produced a 35% incidence of bone disease (Nellen and Kindwall, 1972). Trowbridge (1977) reported an 8% incidence of bone disease when the Blackpool tables were used on the Dungeness B Power Station cooling water tunnels. Bone necrosis has also been reported on the Blackpool tables from the Hong Kong subway (Yau, 1987). It would seem only reasonable that the Blackpool tables should produce as much or more aseptic necrosis as the OSHA tables, as they provide shorter decompression times and longer shifts. They have physiologic advantages over the OSHA tables, however, in that they take deeper stops and use stage decompression.

Looking at experience on the continent, we find that the French air tables are less conservative than the OSHA tables, but that exposure at the high pressures is severely limited (Figure 4). It appears that the French have recognised the inadequacy of the air decompression tables at the higher pressures and have simply curtailed the work shift length and limited the maximum pressure.

The German air schedules are also less conservative than the OSHA schedules and are more limited in shift length and pressure than either the Blackpool or OSHA schedules (Figure 5).

Comparison between the exposure limits of the various tables is made in Figure 6. Notice that the Blackpool tables have an area of diagonal lines between four and eight hours and over one bar (14.7 psig, 101 kPa). This indicates that decompression is the same for any exposure over four hours and depends only on the pressure at that point. The Blackpool tables cut off at eight hours and the US Navy tables cut off at a 12 hour exposure. Note however that OSHA tables are indefinite as they do give schedules for 'any shift over eight hours'. In theory they have no maximum exposure limit. The Navy Exceptional Exposure Air Tables are not shown above 50 psig (345 kPa) where they are not comparable to tunnelling.

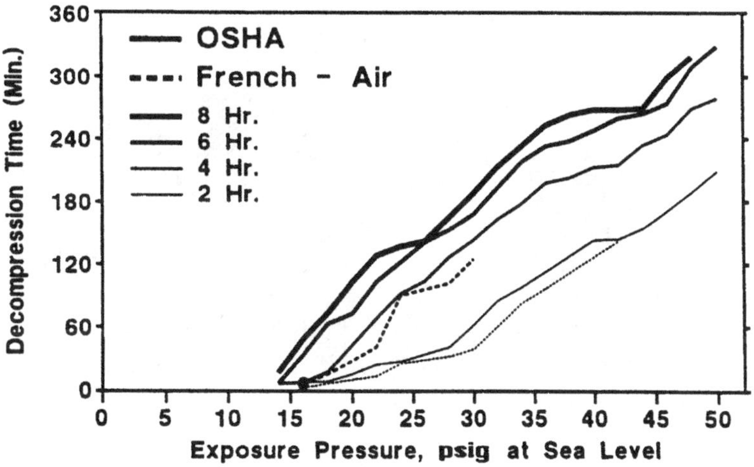

Fig. 4. Comparison of OSHA and French decompression times

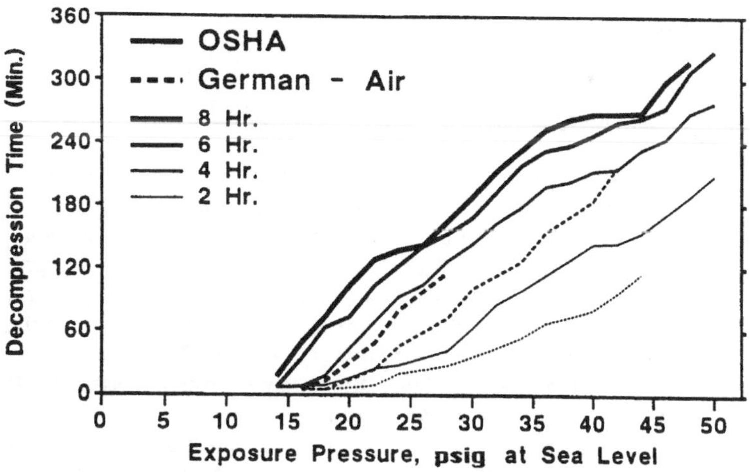

Fig. 5. Comparison of OSHA and German decompression times

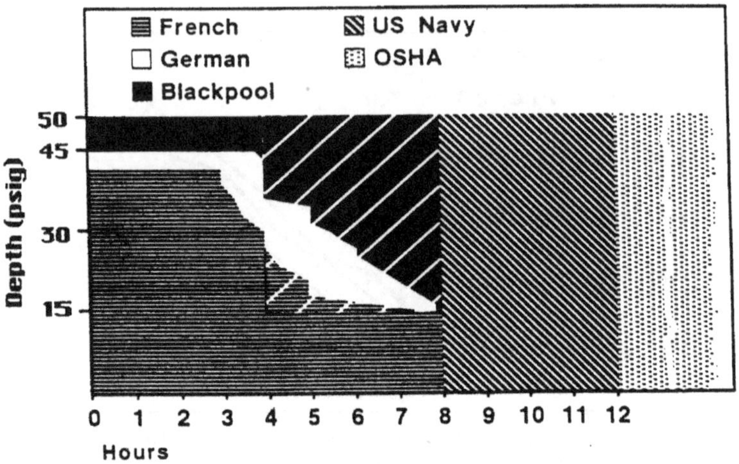

Fig. 6. Comparison of exposure limits

I do not have access to data with regard to the incidence of DCI in workers using the French and German air tables but unless an anonymous reporting system were used to collect those data, they would not be indicative of the true incidence. Since these tables are less conservative than the OSHA tables, we can assume that the incidence of DCI is high.

4 A new approach to development of decompression schedules

Decompression tables for compressed air workers currently in use have relied chiefly on naval experience and some sort of mathematical model, usually Haldanean. It appears that all our efforts to produce tables using this methodology have been flawed. As Duffner himself pointed out to me over 25 years ago, all decompression schedules can be designed to be more or less adequate over their intended working range but fail when extrapolated to deeper depths and longer times. This implies that our system of mathematical modelling is flawed. I do not know of any decompression tables, whether they be Navy Diving tables or compressed air tables that become safer as the shifts grow longer and the pressures increase.

Because the OSHA tables were producing aseptic necrosis at an alarming rate at the higher pressures, we sought funds from the National Institute of Occupational Safety and Health to devise safer decompression tables for compressed air workers. This work was commenced in 1979 and finished in 1983 (Kindwall *et al.*, 1983). As previous attempts at mathematical modelling had failed and naval experience was inadequate on which to base long-term exposures, I sought the help of Peter Edel in New Orleans, Louisiana. Edel had a computer with data pertaining to 15 years of successful and unsuccessful commercial dives in the Gulf of Mexico and elsewhere, as well as some experimental data. Some of these data were his own and included additional experimental material from commercial as well as naval sources.

It was evident from the start that any decompression tables would have to be economically viable as well as physiologically sound. Unlimited decompression in the name of safety would render the tables unusable in the tunnelling industry. In an effort to meet these requirements, some extremely long duration data were initially excluded from consideration, and the computer was asked essentially to draw a line between the successful and unsuccessful decompressions for increasingly longer times and pressures of up to 50 psig (345 kPa). Gaps in the available data were filled in using a proprietary neo-Haldanean method of calculating tissue saturation and de-saturation. It was assumed that a worst-case situation would be workers decompressing from the second shift of the week when they still had residual gas from the first shift, despite a 16 hour hiatus between shifts.

The first set of decompression tables so produced was labelled Autodec 2 and they were subjected to test by volunteers in our laboratory. The volunteers consisted of six males and one female who ranged in age from 26 to 59 years of age. Each decompression schedule was tested after exposure to the longest shift time which could be accommodated in an eight-hour work day when the workshift and decompression times were combined. In practice, these would be the only schedules which would be used in the field. They would also be the most stressful. The subjects walked on a treadmill up a 3% grade at a speed of three miles per hour for 10 minutes. Every 10th step the subject would raise a five pound weight held in each hand from a position at his side to above his head. Each 10 minutes of exercise was followed by a 10 minute rest. This was to approximate a moderate work effort throughout exposure to compressed air in an effort to simulate the work load of a tunnel labourer. The Autodec 2 schedules were initially spot tested at various pressures up to 34 psig (234 kPa). However, Autodec 2 decompression produced chokes and decompression illness in multiple joints in more than one subject and finally produced a positive bone scan which indicated pathologic activity in the proximal tibia. Further testing of Autodec 2 was abandoned. It had been hoped that Autodec 2, which used decompression times similar to the present OSHA schedules would be adequate, as stage decompression was employed. It was hoped that this alone would provide the necessary safety. In practice, it proved to be a vain hope.

The Autodec 3 schedules were then generated using all of the data available to the computer. This included material from extremely long exposures. The results proved disappointing from the standpoint of economics as the decompression schedules generated by the computer were prohibitively long, although obviously much safer. An example can be given: the decompression requirement for a 4 h work period at 44 psig (303 kPa) is 4 h using the present OSHA schedule. The Autodec 3 schedule required 10 h and 46 min. If this and comparable schedules were the ultimate price for tunnelling at pressures greater than 18 psig (124 kPa), commercial compressed air work as we know it was doomed. Since physiology is unalterable, another way had to be found to rid the body of excess nitrogen in a more timely fashion.

5 Oxygen decompression

Oxygen decompression is nothing new and has been used successfully by the navies of the world since the 1930s. It is a matter of routine in commercial diving. Oxygen decompression carries with it certain risks, not the least of which is fire. For many years the US Navy did not permit oxygen to be used in the treatment of DCI because it was not 'sailor proof'. This was despite the fact that Behnke had shown it to be efficacious in DCI as far back as 1939. However, it was only in 1964 when research disclosed that air treatment of decompression sickness produced a 41.7% failure rate on the initial recompression that research was again directed toward the use of oxygen. By 1967, oxygen treatment of decompression sickness had been approved for use in the US Navy and the failure rate on initial recompressions had dropped to 3.6% (U.S. Navy, 1967). Needless to say they were the same kind of dive supervisors, treating the same kind of divers, using the same chambers, and breathing the same oxygen that had been available in 1939. Suddenly in 1967, oxygen had become sufficiently 'sailor proof'. The only differences were in the recognised requirement and proper training of the personnel.

Oxygen decompression was the first used in compressed air tunnelling in 1938-39 in connection with the construction of the Queen's Midtown Tunnel in the city of New York (Jones *et al.*, 1940). Some 14,000 experimental oxygen decompressions were carried out without mishap on that job despite an oxygen delivery system of poor design with no overboard dump system. Additionally, the oxygen breathing apparatus had very high breathing resistance. Despite these inadequacies, there were no fires and there were also no serious cases of DCI in those workers who breathed oxygen.

The next use of oxygen in compressed air work was in 1959 in Japan. Unfortunately, the workers had not been adequately trained and one of them apparently lit a cigarette that resulted in a fire which killed six men undergoing decompression (Nashimoto, 1967). Despite that unfortunate accident, oxygen decompression was put into use on a regular basis in Germany in 1972 when the first oxygen tables for caisson workers were promulgated. The Germans were pleased with the results and reported that 'the rates of DCI decreased remarkably in spite of longer shifts' (Altner, 1986). Recent German experience with new oxygen tables developed by Faesecke (1991) has been excellent. Le Péchon (1987) reports that the French have had oxygen decompression tables for tunnel workers since 1974, but these remain to be tested on a large project. Ribeiro (1985) has also pointed to very successful results of oxygen decompression used in 1976 in the construction of the São Paulo subway. To date, the German, French and Brazilian experience has been ignored by British and American regulatory bodies and there has been extreme reluctance to adopt oxygen decompression, much as the US Navy was reluctant to accept oxygen treatment of DCI.

6 The Milwaukee oxygen tables

From the physiologic standpoint, it became apparent to us that the only available means of removing nitrogen from the tissues of caisson workers subject to daily decompressions was the use of oxygen. For this reason, Edel's computer was instructed to calculate an oxygen variant of the long Autodec 3 table. From the economic point of view, these new oxygen tables looked very promising indeed. Now, 4 h work periods at 44 psig (303 kPa) required only 3 h 21 min when oxygen was breathed intermittently during decompression. This schedule and the schedules at the lower pressures save considerable time over the present OSHA table (Figure 7).

The new oxygen table was tested in the same way as Autodec 2 at maximum shift length at 2 psig (14 kPa) intervals up to 46 psig (317 kPa). No DCI was seen in any of the test subjects and all of the bone scans were negative six months later. This represents the only set of decompression schedules for compressed air workers ever pre-tested in the laboratory. Obviously, it is clear that two or three tests of the table at each pressure level, even with different subjects, is totally inadequate to project an incidence of DCI. However, DCI did not appear, even at the higher pressures and at maximum shift length. At least, catastrophic error can be ruled out. Decompression times for a 4 h exposure using various decompression tables (both air and oxygen) are compared in Figure 8.

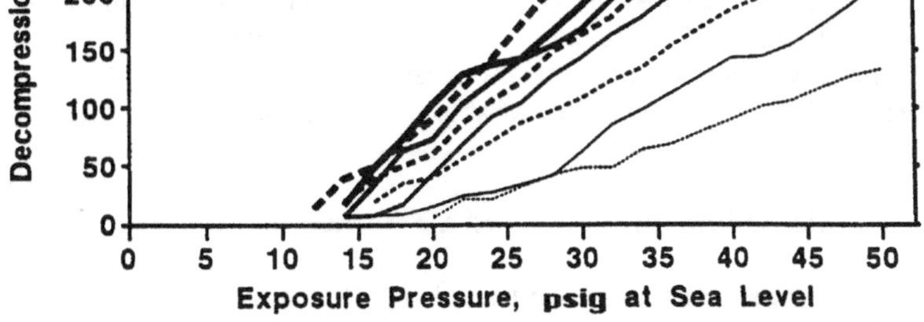

Fig. 7. Comparison of OSHA and Milwaukee oxygen tables decompression times

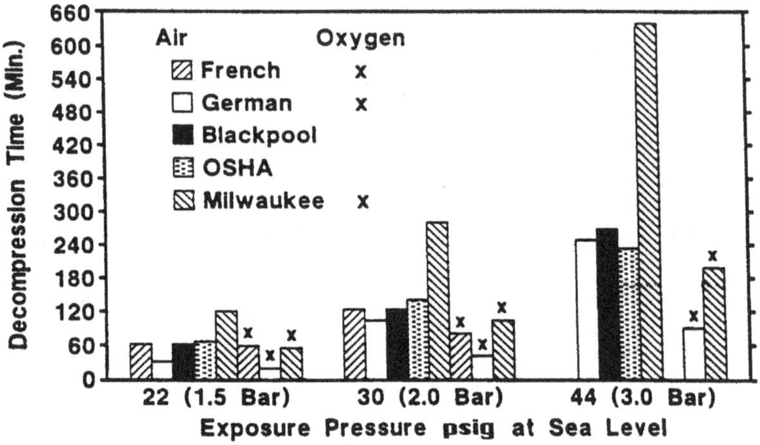

Fig. 8. Comparison of decompression times for 4 h of compressed air work

Oxygen toxicity is not a matter to be taken lightly and its occurrence cannot be ruled out when these schedules are used on a daily basis. However, its appearance is unlikely due to the intermittent nature of the oxygen breathing periods and should oxygen toxicity occur, it is easily reversible. This is not the case with bone and neural damage.

Oxygen decompression offers other advantages beyond simply expanding the 'oxygen window' for nitrogen elimination. Raising the pO_2 in venous blood lessens blood sludging (End, 1939), reduces the stickiness of white cells, and reduces the no-reflow phenomenon (Zamboni, 1990), renders red cells more flexible in passing through capillaries (Mathieu, 1984), and tends to counteract the 81% decrease in the deformability and filterability of white cells after exposure to air at 50 fsw (153 kPa) for 4 h, as demonstrated by Pimlott *et al.* (1987).

We can no longer be cavalier about 'minor bends symptoms' or 'simple limb pain'. Gorman *et al.* (1987) have shown that patients suffering from so called 'Type I' DCI with negative neurologic examinations later demonstrated pathologic psychological testing, a 40% incidence of abnormal EEG findings and brain atrophy.

7 Pathologic brain findings in compressed air workers

Using magnetic resonance imaging, we have demonstrated a very high incidence of unidentified bright objects (UBOs), probably representing lacunar infarcts and gliosis in tunnel workers in Milwaukee, Wisconsin (Fueredi *et al.*, 1991). We compared 19 subjects who had been exposed to various degrees of compressed air with 11 age-matched controls who belonged to the same labour union but who had never been exposed to hyperbaric air. Foci of increased T-2 intensity (UBOs) deep within white matter tracts were evaluated as to their number and location. Psychometric testing was carried out in both groups to exclude pre-existing brain disease. The 19 subjects in the compressed air group had a statistically higher number of white matter lesions (more than 152) than the control group (22 lesions) ($p=0.05$). Thirty-seven percent of the compressed air group had more than 20 white matter lesions each, while only 18% of the controls had 10 to 11 such lesions each. The experimental group had a five times higher risk than the control group of having high grade lesions and a high statistical correlation was found between the number and severity of lesions in the compressed air group as compared with the control group when linear trend analysis was performed ($p=0.02$).

We are aware that these findings are at variance with those of Brubakk (1991) and Todnem *et al.* (1991). They found more UBOs in the controls than in the divers, but it must be remembered they were studying divers. We were examining compressed air workers whose decompression histories were considerably more traumatic. Although UBOs are thought to represent lacunar infarcts and possibly gliosis they are normal findings in an ageing population. In our series, however, the compressed air group and the controls were closely matched as to age. Our compressed air workers scored 'lower' on psychological testing than the controls but the difference was extremely slight and was not statistically significant.

8 Conclusions

It is time that we took a long hard look at our present decompression schedules. It must be noted, however, that the Health and Safety Executive in the UK has been against change in the present tunnelling decompression procedures until there is assurance that 'no new risk is added'. In the United States, the Office of Variance Determination of the Occupational Safety and Health Administration has now admitted that oxygen decompression may be feasible. In 1988, Dr Ralph Yodaiken, Director of the Office of Occupational Medicine of OSHA stated in a memorandum to his staff that the present OSHA tables are 'by modern standards flawed' and that 'the OSHA tables have failed any reasonable test of adequate performance over the past 16 years'. He further noted that 'oxygen decompression can be very safe as proven by the German, French and Brazilian experience'. He concluded 'in summary, oxygen decompression is long overdue in caisson work'. Thus, it is obvious that oxygen decompression will be coming into general use.

As the 21st century approaches, it is clear that the construction industry and its regulators must abandon trying to solve high-tech problems using 19th century tradition.

References

Altner, A. (1986), Chief Safety Engineer. Fachausschuss, Tiefbau, Munich. Personal Communication

Balldin, U.I. and Lundgren, C.E. (1969) The Effect of Immersion on Inert Gas Elimination. *Acta Physiological Scand.* (Suppl.), **330**: 83

Brubakk, A.O., Rinck, P., deFrancisco, P., *et al.* (1990) Central Nervous Changes in Professional Divers as Evaluated by MRI. (Abstract only) *Undersea Biomedical Research* (Suppl.), **17**: 88

Duffner, G.J. (1983) Personal Communication

End, E. (1939) The Physiologic Effects of Increased Pressure. *Proceedings of the Sixth Pacific Science Congress*, University of California, 6: 91-97

Faesecke, K. (1991) Landesgewerbeartz State Occupational Health Authority, Kiel, Personal Communication

Fueredi, G.A., Czarnecki, D.J. and Kindwall, E.P. (1991) MR Findings in the Brains of Compressed Air Tunnel Workers: Relationship to Psychometric Results. *American Journal of Neuroradiology*, **12**: 67-70, January/February

Gorman, D.F., Edmonds, C.W., Parsons, D.W., Beran, R.G., Anderson, T.A., Green, R.D., Loxton, M.J. and Dillon, T.A. (1987) Neurologic Sequelae of Decompression Sickness: A Clinical Report. In: *Underwater and Hyperbaric Physiology IX*, pp. 993-998, (Eds. Bove, Bachrach and Greenbaum). Undersea Hyperbaric Medical Society, Bethesda

Hempleman, H.V. (1973) In: *Medical Code of Practice for Work in Compressed Air.* Construction Industry Research and Information Association Report 44, 1st Edition

Jones, R.R., Crosson, J.W., Griffith, F.E., Sayers, R.R., Schrenk, H.H. and Levy, E.P. (1940) Administration of Pure Oxygen to Compressed Air Workers During Decompression: Prevention of the Occurrence of Severe Compressed Air Illness. *Journal of Industrial Hygiene and Toxicology*, **22**: 427-444

Kindwall, E.P., Nellen, J.R. and Spiegelhoff, D.R. (1982) Aseptic Necrosis in Compressed Air Tunnel Workers Using Current OSHA Decompression Schedules. *Journal of Occupational Medicine*, **24**(10): 741-745

Kindwall, E.P., Edel, P.O. and Melton, H.E. (1983) *Safe Decompression Schedules for Caisson Workers*, National Institute of Occupational Safety and Health, Research Grand #5R 01 OH 00947-03, Final Report

Le Péchon, J.C. (1987) Personal Communication

Mathieu, D., Coget, L., Vinkier, F., *et al.* (1984) [Erythrocyte Filterability and Hyperbaric Oxygen Therapy], *Med. Sub. Hyp.*, **3**(3): 100-104 (French)

McCallum, R.I. (1967) Aseptic Necrosis of the Bone in Compressed Air Workers, pp 138-144. In: *Decompression of Compressed Air Workers in Civil Engineering* (Ed. R.I. McCallum). The Oriel Press Limited, Newcastle-upon-Tyne

Nashimoto, I. (1967) The Use of Oxygen During Decompression of Caisson Workers. pp 217-227. In: *Decompression of Compressed Air Workers in Civil Engineering*, (Ed. R.I. McCallum). The Oriel Press Limited, Newcastle-upon-Tyne

Nashimoto, I. (1986) Personal communication

Nellen, J.R. and Kindwall, E.P. (1972) Aseptic Necrosis of Bone Secondary to Occupational Exposures to Compressed Air: Roentgenologic Findings in 59 cases. *Amer. J. of Roentgenology, Radium Therapy and Nuclear Medicine*, **65**(3): 512-523

Pimlott, J., Ormsby, P.L. and Cross, M.R. (1987) The Effect of White Cells on Blood Filterability at Pressure. *Proceedings of the European Undersea Biomedical Society Annual Scientific Meeting*, Palermo

Ribeiro, I.J. (1985) Letter to Dr Richard Vann, Duke University

Sealey, J.L. (1975) Aseptic Bone Necrosis Survey in Compressed Air Workers. *Journal of Occupational Medicine*, **17**(10): 666-667

Todnem, K., Skeidsvoll, H., Svihus, R., *et al.* (1991) Electroencephalography, Evoked Potentials and MRI Brain Scans in Saturation Divers. An Epidemiological Study. *Electroenceph Clin. Neurophysiol*, **79**: 322-329

Trowbridge, W.P. (1977) Bone Necrosis in British Compressed Air Workers. In: *Aseptic Bone Necrosis. Proceedings of the Symposium of the European Undersea Biomedical Society*, Newcastle-upon-Tyne

Walder, D.N. (1971) Personal Communication

U.S. Bureau of Labor Standards (1971) Department of Labor: Safety and Health Regulations for Construction. *Federal Register*, **36**: 7395-7404 (April 17th)

U.S. Navy (1967) Oxygen Breathing for Decompression Sickness and Air Embolism, *Bumed Instruction 6420.2, Bumed 74*, August 22, 1967. Washington D.C. Bureau of Medicine and Surgery

Yau, K.P. (1987) Retained Physician, Hong Kong Subway Project. Personal Communication

Zamboni, W.A. (1990) The Effect of Hyperbaric Oxygen Treatment on the Microcirculation of Ischemic Skeletal Muscle. *Undersea Biomedical Research*, (Suppl.), **17**: 26 (Abstract only)

12 Early pathogenesis of hyperbaric osteonecrosis

C.E. Lehner, T.F. Lin, E.H. Lanphier, M.A. Wilson, R. Dueland and R.R. Dubielzig
Departments of Preventive Medicine and Radiology, Medical School, Departments of Surgical Sciences and Pathobiological Sciences, School of Veterinary Medicine, Department of Veterinary Science, and the Biotron, University of Wisconsin, Madison, Wisconsin, USA

Abstract
Hyperbaric osteonecrosis is a form of non-traumatic bone necrosis associated with prolonged or repeated hyperbaric exposures such as experienced by tunnel and caisson workers and by diving fishermen. Hyperbaric osteonecrosis, also called dysbaric osteonecrosis, can be induced experimentally in sheep long bones by hyperbaric exposures similar to those which cause osteonecrosis in humans. Osteonecrosis was induced in five sheep by 12-13 twenty-four hour hyperbaric exposures to compressed air pressures of 2.5-2.9 atm abs with no-stop decompression. At 26-36 weeks after the first hyperbaric exposure, lesions included extensive fatty marrow necrosis, endosteal thickening in bone shafts and juxta-articular osteonecrosis. Even single 24-h hyperbaric exposures of sheep to compressed air pressures of 2.2-2.4 atm abs with no-stop decompression also induced osteonecrosis. Bone scans with ^{99m}Tc-methylene diphosphonate revealed 'hot spots' two weeks after hyperbaric exposure in those limbs which were affected by persistent limb bends. Roentgenographic films of these limbs at eight weeks revealed endosteal thickening and medullary opacities at 'hot spot' sites, and MRI at eight weeks indicated fatty marrow infarction. Necropsy at nine weeks confirmed extensive fatty marrow and bone necrosis. In our view, the early pathogenesis of hyperbaric osteonecrosis involves a bone compartment syndrome initiated by N_2 bubble formation in the fatty marrow of long bones after decompression from prolonged hyperbaric exposure.
Keywords: Bone Necrosis, Dysbaric Osteonecrosis, Decompression Sickness, Pathogenesis, Compartment Syndrome

1 Introduction

Hyperbaric osteonecrosis, also known as dysbaric osteonecrosis, is a disabling form of bone necrosis which occurs in humans exposed to prolonged hyperbaric pressures (Davidson, 1976, 1989; Gregg and Walder, 1986; McCallum and Harrison, 1982). Among caisson and tunnel workers and diving fishermen, the prevalence of

Engineering and Health in Compressed Air Work Edited by F.M. Jardine and R.I. McCallum. Published in 1994 by E & FN Spon. ISBN 0 419 18460 0

hyperbaric osteonecrosis is highest for those who breathe compressed air for extended periods of time and whose decompressions have been rapid. The causal association between hyperbaric osteonecrosis and prolonged hyperbaric exposure is essential to our understanding the etiological and developmental basis of this form of bone disease.

Prolonged hyperbaric exposure and a rapid decompression can provoke a high incidence of decompression sickness (DCS) or illness, especially in the form of limb bends (Elliott and Kindwall, 1982). The strong association between prolonged hyperbaric exposure and a chronic form of limb bends, which often persists for several days, supports a model for an important pathogenic mechanism in the development of hyperbaric osteonecrosis. The proposed mechanism of a bone compartment syndrome, with elevated tissue pressure and ischemia, appears consistent with ischemic bone and marrow necrosis which develops in hyperbaric osteonecrosis (Zizic and Hungerford, 1985; and Lanphier *et al.*, 1992).

Experimental findings of bone necrosis in sheep long bones after hyperbaric exposure provide a basis for a proposed etiological and pathogenic model of hyperbaric osteonecrosis. Early stages of hyperbaric osteonecrosis induced in sheep correspond closely with the same signs and symptoms in humans. Consequently, the sheep model offers a means for the study of this and other forms of non-traumatic osteonecrosis (Catto, 1976; Jones, 1985; Mankin, 1992). Moreover, the apparently beneficial role of early recompression treatment suggests therapeutic approaches which can significantly reduce the prevalence of human hyperbaric osteonecrosis.

2 Etiological basis of hyperbaric osteonecrosis

2.1 Prolonged hyperbaric exposure: a risk factor

Marked differences in the prevalence of hyperbaric osteonecrosis among various human populations which have been exposed to compressed air provide clues about the etiological basis of hyperbaric osteonecrosis. A high prevalence of hyperbaric osteonecrosis, greater than 10%, can occur in tunnel workers (Gregg and Walder, 1986; Kindwall *et al.*, 1982), commercial divers (Decompression Sickness Central Registry and Radiological Panel, 1981), and diving fishermen (Amako *et al.*, 1974; Kawashima *et al.*, 1978; Wade *et al.*, 1978). Hyperbaric exposures of diving fishermen, often exceeding four hours cumulative duration per day, are typically repetitive, with several closely-spaced dives per day (Lehner, 1992). Recreational divers appear to have avoided bone necrosis presumably due to their relatively brief hyperbaric exposures.

2.2 Recreational scuba divers

Very few cases of hyperbaric osteonecrosis have been reported among recreational scuba divers, although many recreational dives exceed the maximum pressure in a tunnel worker's hyperbaric exposure. With the advent of the dive computer, the recreational diver can freely conduct several multi-level dives per day over a period of several days (Lang and Vann, 1992). Such dive profiles sometimes resemble the hyperbaric exposures associated with a high risk of bone necrosis in tunnel workers and diving fishermen. Indeed, the cumulative effect of closely-spaced repetitive dives can involve a significant risk of bone necrosis as is prevalent in diving fishermen (Lehner *et al.*, 1992).

2.3 Diving fishermen and tunnel workers

Diving fishermen conducting severe repetitive dives with little regard for conventional decompression tables have sustained some of the highest reported prevalence rates of hyperbaric osteonecrosis, exceeding 50% in the surveyed Japanese (Kawashima *et al.*, 1978) and Hawaiian (Wade *et al.*, 1978) divers. Hard-hat shellfish diving in Asia involves the use of compressed air provided by an air line from the surface. Prolonged hyperbaric exposures conducted by these diving fishermen often exceed six hours of cumulative hyperbaric exposure in their multiple dives to depths of 18-20 m (Lehner *et al.*, 1992).

Reports of bone necrosis in British tunnel workers and divers indicate a positive correlation between the duration of hyperbaric exposure and the prevalence of hyperbaric osteonecrosis (Golding *et al.*, 1960; Decompression Sickness Central Registry and Radiological Panel, 1981). This pattern in bone necrosis prevalence is evident in a comparison of shift tunnel workers who work longer hours and have a higher frequency of bone necrosis than non-shift tunnel workers who typically remain under hyperbaric pressure for less than four hours (Golding *et al.*, 1960). However, the strength of this association between an increased duration of hyperbaric exposure and greater risk of bone necrosis, based on tunnel worker data (McCallum and Walder, 1966), remains unclear.

2.4 Epidemiological patterns in hyperbaric osteonecrosis

A high prevalence of hyperbaric osteonecrosis in populations of diving fishermen and tunnel workers exposed to compressed air points to prolonged hyperbaric exposure as an important risk factor in hyperbaric osteonecrosis. In contrast, vastly greater numbers of recreational divers, approximately 2 million scuba divers in the United States alone, exposed to compressed air during comparatively brief dives have sustained an extremely low incidence of hyperbaric osteonecrosis (Eric Kindwall, personal communication). Such benign outcomes in recreational diving after brief 'bounce' dives would change dramatically if the frequency of closely-spaced, severe repetitive diving, which would inexorably favor hyperbaric osteonecrosis, were to increase.

2.5 Bone blood flow, bubbles and bone necrosis

Low blood flow rates in bone offer a physiological explanation for the relationship between the risk of hyperbaric osteonecrosis and the duration of hyperbaric exposure. Nitrogen uptake studies in fatty marrow of goat long bones by Campbell and Hill (1933) and similar experiments by Thomas *et al.* (1982) point to the low rates of blood flow in the long bones. Not coincidentally, the long bones are the principal sites of hyperbaric osteonecrosis (Walder 1990).

A bone compartment syndrome is proposed in the early pathogenesis of hyperbaric osteonecrosis (Figure 1). The developmental basis for this hypothesis of bone necrosis induction is discussed in a companion paper by Lanphier *et al.* (1994), also presented in this conference.

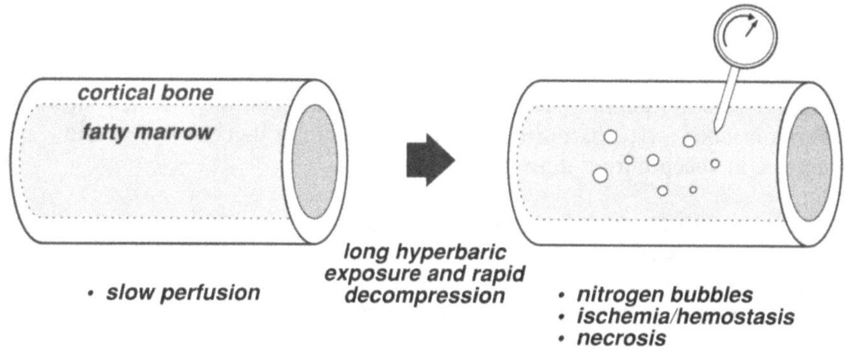

Fig. 1. Proposed mechanism of hyperbaric osteonecrosis induction. Decompression from prolonged hyperbaric exposure triggers bubble formation which initiates a bone compartment syndrome.

2.6 Proposed model of hyperbaric osteonecrosis induction

Bone blood flows are notoriously slow. Such low perfusion rates presumably contribute to the long bones' vulnerability to decompression injury, particularly hyperbaric osteonecrosis (Lehner, 1992; Lehner *et al.*, 1992). Because of the low perfusion in bone and marrow tissues, prolonged hyperbaric exposure, typically an exposure greater than four hours, is usually necessary for significant quantities of dissolved N_2 or other physiologically-inert gases to accumulate in the fatty marrow. N_2 is about five times more soluble in fat than in water. With air as the breathing gas mixture, fatty marrow and other lipid-rich tissues can become important sources of dissolved N_2 and potential sites of bubble formation. The cortical bone surrounding fatty marrow establishes a semi-closed compartment with a complex vasculature potentially vulnerable to vascular obstruction due to bubble formation (Behnke, 1945).

In addition to bubble formation, other mechanisms have been implicated in the early pathogenesis of hyperbaric osteonecrosis. Some of these mechanisms include circulatory embarrassment associated with fat cell swelling in the presence of elevated O_2 levels (Pooley and Walder, 1984) and fat emboli (Jones, 1985; Kitano and Hayashi, 1981), together with blood factors involved in intravascular coagulation (Jones, 1992).

3 Experimental hyperbaric osteonecrosis in sheep

3.1 Induction of non-traumatic osteonecrosis in animals

Previous animal studies have induced hyperbaric osteonecrosis in the mini-pig (Stegall and Smith, 1976; Gregg *et al.*, 1980) and in a strain of obese mice (Chryssanthou, 1976) after numerous hyperbaric exposures. Based on these reports and other studies, the experimental induction of non-traumatic osteonecrosis in animals, particularly by hyperbaric exposure (Reeves *et al.*, 1972), was considered difficult. Large animals offer advantages over small animals in anatomical and physiological scaling (Schmidt-Nielsen, 1984), especially considering their slower blood flows in bone. At Wisconsin, we have successfully induced hyperbaric osteonecrosis in sheep during three recent studies.

3.2 Bone necrosis after 24-h hyperbaric exposures

Bone necrosis was experimentally induced in sheep with a protocol based on information drawn from various sources. Previous reports of human bone necrosis, especially bone necrosis in tunnel workers (McCallum and Walder, 1966), pointed to prolonged hyperbaric exposures as those being most likely to induce hyperbaric osteonecrosis. Saturation hyperbaric exposure serves to maximize the level of N_2 supersaturation in decompressed tissues. After near-saturation hyperbaric exposures of 24 hours and no-stop decompression, we had previously observed that sheep which developed persistent limb bends later developed bone necrosis. In fact, this observation in sheep was remarkably similar to the delayed onset of limb bends observed in Japanese diving fishermen (Hayashi *et al.*, 1978) who are also known for their high prevalence of bone necrosis (Amako *et al.*, 1974; Kawashima *et al.*, 1978). We interpreted persistent limb bends as an early clinical manifestation of an underlying disease process which leads to bone necrosis. This interpretation suggested that prolonged hyperbaric exposures and rapid decompression would induce bone necrosis in sheep.

Five mature sheep underwent 12 or 13 hyperbaric exposures of 24-hour duration in a two-month period (Adams *et al.*, 1985). Hyperbaric exposures included compression at 0.7 atm/min with maximum pressures of 2.5-2.9 atm abs, equivalent to 51-62 feet of sea water pressure. Sheep underwent rapid no-stop decompression at 1.1 atm/min to 0.97 atm abs, mean atmospheric pressure at Madison. A high rate of ventilation in the pressure chamber maintained CO_2 levels below 0.2 %.

All five sheep developed hyperbaric osteonecrosis in their long bones (Lanphier and Lehner, 1990). Bone lesions were visualized in serial, plain x-ray films taken at two, five and nine months after the first hyperbaric exposure. Progressive development of endosteal thickening and medullary opacities in the shafts and necks characterized the induced hyperbaric osteonecrosis seen with x-ray films. These and other lesions were confirmed at necropsy conducted nine months after the first hyperbaric exposure.

At necropsy, extensive fatty marrow necrosis was observed in gross sections of the long bones. Areas of necrotic fatty marrow were typically located in the center of the marrow cavity, and these lesions often extended to much of the marrow.

Microangiography revealed widespread neovascularization of the fatty marrow surrounding the fatty marrow infarcts.

Bone lesions were confirmed by standard histological techniques with hematoxylin and eosin staining. Bone necrosis in the long bones occurred with a loss of osteocytes in cortical bone and trabeculae of the spongy, cancellous bone. Endosteal thickening observed with x-ray films represented viable appositional new bone which formed on the dead inner surface of the bone shafts. Calcification of necrotic fatty marrow presumably represented the medullary opacities previously visualized in plain x-ray films.

Juxta-articular, subchondral necrosis occurred in some long bones. Hyperbaric osteonecrosis in the ends of long bones is particularly important, because subchondral necrosis represents a potentially disabling form of bone injury. If a juxta-articular fracture and collapse occurs in humans, a site of previously 'silent' osteonecrosis then becomes painfully symptomatic and disabling (McCallum and Walder, 1966). In contrast, shaft lesions remain almost exclusively asymptomatic (Golding, 1966).

3.3 Bone necrosis after brief hyperbaric exposures

In another study, sheep underwent hyperbaric exposures of one-half hour to four hours in duration and with no-stop decompression. Comparatively mild lesions, again mostly focused in the long bone shafts, were induced by this series of brief hyperbaric exposures (Lehner *et al.*, 1992). Fatty marrow lesions were characterized by calcification and granulation tissue at the necrotic foci and endosteal thickening of the long bone shafts (Kitano *et al.*, 1992).

3.4 Bone necrosis after a single 24-h hyperbaric exposure

In our most recent experiments, we induced hyperbaric osteonecrosis in all six mature sheep which underwent a single hyperbaric exposure. Now that successful bone necrosis induction from a single exposure has been achieved, the early pathogenesis of hyperbaric osteonecrosis can be followed.

Six mature sheep each underwent a single 24-h hyperbaric exposure at moderate pressures of 2.2 to 2.4 atm abs and with no-stop decompression. Non-invasive diagnostic techniques of roentgenography, bone scanning, and magnetic resonance imaging (MRI) were used to detect the early developmental stages of hyperbaric osteonecrosis. Plain x-ray films of the long bones were taken one week before hyperbaric exposure and at two, four and eight weeks after hyperbaric exposure. Bone scans were frequent, with static bone scans taken one week before and two days, and two, four and eight weeks after hyperbaric exposure. MRI was used in one sheep at eight weeks. All sheep underwent euthanasia and necropsy at nine weeks.

3.5 Early pathogenesis of hyperbaric osteonecrosis

After a single 24-h hyperbaric exposure, sheep developed persistent clinical signs of limb bends within a few hours of decompression. These classic signs of limb bends, mostly mild limb-lifting and non-weight-bearing, typically persisted for more than 24 hours. We used transcutaneous Doppler ultrasound to detect circulating bubbles in the venous return. Circulating venous bubbles preceded clinical signs of chokes or respiratory DCS in the sheep (Atkins *et al.*, 1988), and these bubble signals indicated massive bubble formation in the decompressed sheep. Sheep observed with high levels of bubble signals later developed labored breathing, a characteristic sign of respiratory DCS. All cases of respiratory DCS, usually mild, resolved uneventfully.

Excellent bone scans of the sheep humerus and femur were difficult to obtain. In sheep, the femur and humerus, lie within large muscles and near large bones, and so only the distal long bones, the tibiae and the fused radius-ulna were routinely

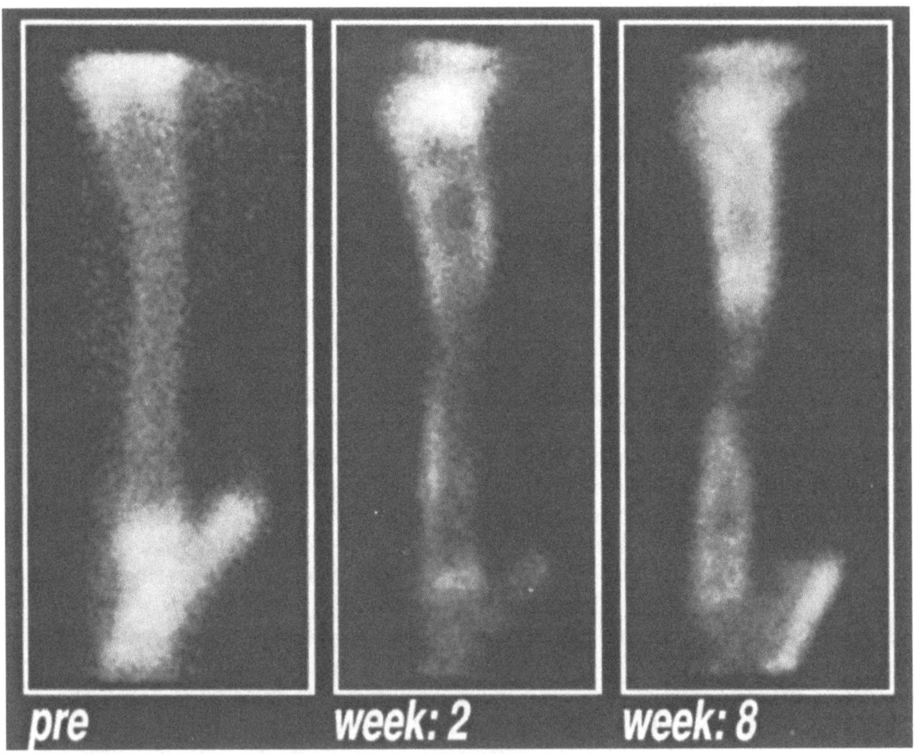

Fig. 2. Progressive development of hyperbaric osteonecrosis visualized in serial bone scans (^{99m}Tc-methylene diphosphonate) of the left tibia in the representative sheep, #1810. A bone scan taken several days before (pre) hyperbaric exposure represents the normal condition. Two weeks after 24-h hyperbaric exposure to 19 psig (2.3 atm abs), heterogeneity and distinct 'hot spots', white areas, were visualized in the cortical bone shaft. At eight weeks, tracer uptake suggested reparative calcification in the fatty marrow.

scanned. Sheep were injected intravenously with 25-35 mCi of ^{99m}Tc-methylene diphosphonate two to four hours before imaging with a gamma camera. Long bone lesions first appeared as mild but distinct 'hot spots' two weeks after the sheep underwent hyperbaric exposure (Figure 2).

'Hot spots' were especially evident in the shafts of the radii and tibiae and became progressively more intense at four and eight weeks. The left tibia represented in Figure 2 had numerous 'hot spots' in the shaft and neck regions. Long bones of this sheep, whose tibiae were scanned in Figure 2, serve to characterize staging of the repair process in the early pathogenesis of hyperbaric osteonecrosis.

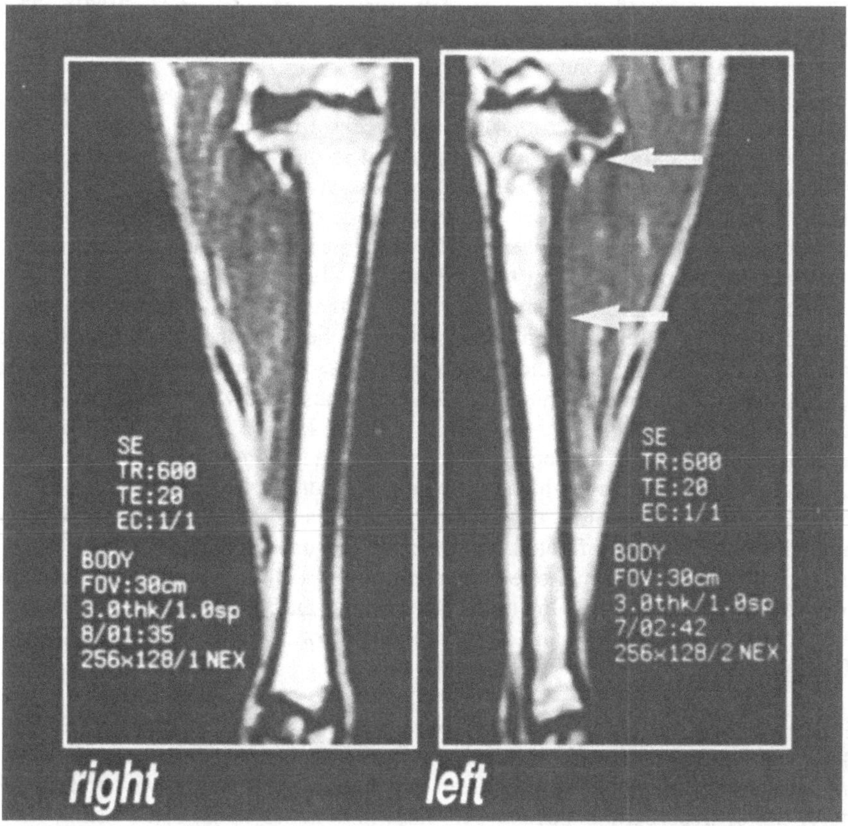

Fig. 3. Magnetic resonance imaging (MRI) of Sheep #1810 tibiae eight weeks after a 24-hour hyperbaric exposure to 2.3 atm abs, 43 feet of sea water, and no-stop decompression. The right tibia appeared normal: uniformly light areas represented fatty marrow, and black areas represented bone. The left tibia revealed marrow lesions: arrows point to neck and shaft lesions in the fatty marrow. Infarcted fatty marrow sometimes appeared surrounded by grey rings or bands (upper arrow) which represented inflammatory repair.

MRI of the previous sheep tibiae confirmed the presence of extensive fatty marrow lesions in the left tibia and the normal appearance and absence of fatty marrow lesions in the right tibia (Figure 3). The tibiae of Sheep #1810 were imaged in a GE Signa MRI scanner (General Electric, Waukesha, WI) with a 1.5 Tesla magnet. The sheep rested on a wooden pallet which was placed in the body coil. In these T1-weighted images (TR 600 msec/TE 20 msec), normal fatty marrow appeared white, and bone appeared black. Marrow lesions produced intermediate signal intensities visualized as various shades of grey. A fatty marrow lesion in the proximal neck of the left tibia was particularly interesting. This lesion was partially surrounded by a dark ring signifying the hyperemic, inflammatory repair characteristically associated with infarcted fatty marrow (Mitchell *et al.*, 1989). The left tibia also contained shaft lesions which were located where 'hot spots' previously appeared in the eight-week bone scan (Figure 2).

Serial radiography of the sheep long bones before and after hyperbaric exposure revealed only subtle changes (Figure 4). Eight weeks after hyperbaric exposure, endosteal thickening and opacities became apparent in the shaft and proximal neck of the left tibia, although medullary opacities were less distinct.

Necropsy was conducted on the sheep eight weeks after hyperbaric exposure, and gross sectioning of the tibiae of Sheep #1810 confirmed the presence of extensive bone and marrow lesions in the left tibia (Figure 5). Extensive fatty marrow necrosis occurred in the proximal neck and shaft of the previously imaged left tibia. In Figure 5, hyperemic regions in a gross section of the left tibia, which appeared as dark reddish areas, spatially coincided with 'hot spots' previously visualized in bone scans, endosteal thickening and medullary opacities observed with x-ray films, and grey areas of fatty marrow lesions visualized with MRI.

The left tibial neck from the representative sheep had a longitudinal, light-colored area of infarcted fatty marrow which was nearly surrounded by a reddish area of hyperemic neovascularization. Similarly in a distal area of the tibia, hyperemic tissue surrounded a small area of infarcted fatty marrow (Figure 5). The modified appearance of fatty marrow in the left tibia contrasted with the normal appearance of fatty marrow in the right tibia. In control sheep, fatty marrow of the long bones appeared sparsely vascularized and colored light yellow as in the right tibia of Sheep #1810 (Figure 5).

3.6 Stages of early hyperbaric osteonecrosis in sheep

The stages of clinical signs and diagnostic findings in hyperbaric osteonecrosis were remarkably consistent among the five sheep which underwent a single hyperbaric exposure. After hyperbaric exposure, persistent limb bends signs, often lasting several days, presaged the development of bone necrosis in the long bones of those same limbs.

Lesions in the sheep bone shafts were easily visualized with bone scans. Bone scans of the distal long bones at two days revealed few abnormalities, although there were occasional 'cold spots'. Persistent limb bends and the occasional 'cold spot' suggest either that a bone compartment syndrome, with elevated tissue pressure and ischemia, continued for at least two days after hyperbaric exposure in some long bones, or that post-ischemic tissues were inactive or dead. Such an ischemic insult in the long bones would be expected to induce bone necrosis.

Bone scans at two weeks frequently detected 'hot spot' lesions, a prominent manifestation of early hyperbaric osteonecrosis. 'Hot spots' indicate high rates of bone and marrow blood flow associated with hyperemia, angiogenesis, bone formation, and tissue calcification which occur in the repair process.

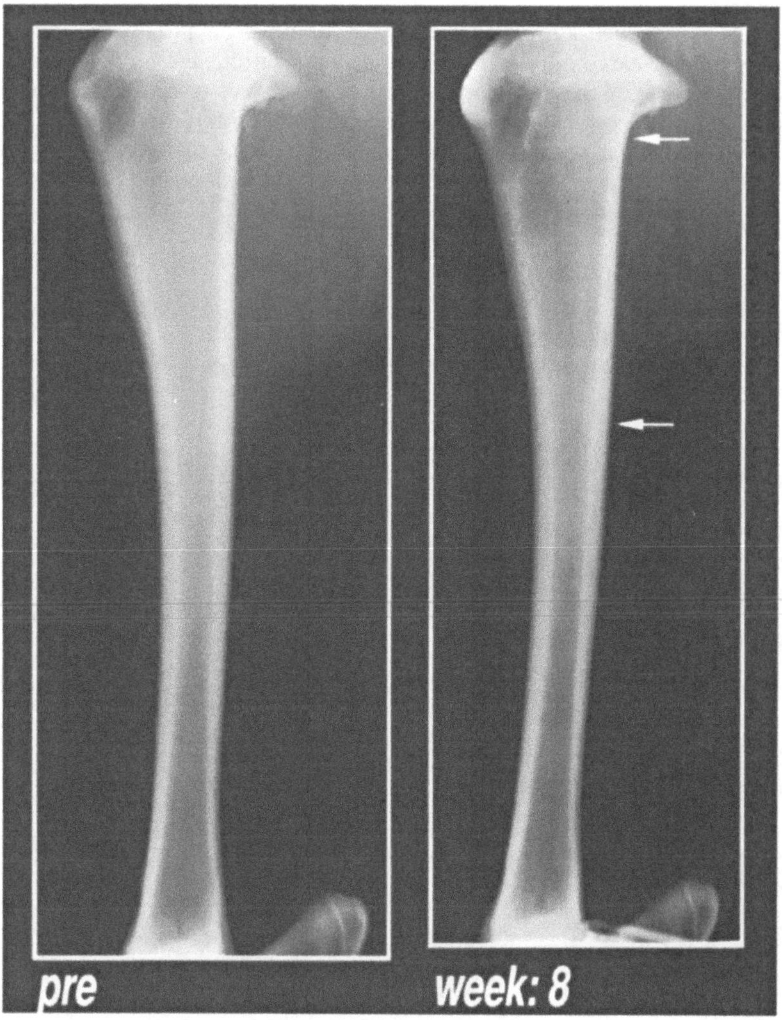

Fig. 4. Plain x-ray films of Sheep #1810 left tibia in lateral view. In the eight-week film, arrows point to subtle thickening of the endosteum (mid-shaft & neck) and to opacities (neck) which represent reparative calcification.

Radiographic findings in sheep hyperbaric osteonecrosis were mostly subtle, even at eight weeks. Only at eight weeks did the lesions appear distinct with serial x-ray films, although subtle endosteal thickening could be detected at four weeks. The earliest reported radiographic findings of hyperbaric osteonecrosis in humans were three to four months after injury (Gregg, 1989).

MRI at eight weeks demonstrated prominent fatty marrow necrosis in the single sheep (#1810) previously described. Earlier MRI of sheep long bones was not performed.

Necropsy at nine weeks revealed extensive areas of repair in fatty marrow of many long bones. Large fatty marrow infarcts were surrounded by regions of hyperemic repair. Lesion sites observed post-mortem were consistent with the lesion sites visualized with non-invasive diagnostic techniques. Moreover, the extent of hyperbaric osteonecrosis lesions was best predicted by bone scans and MRI, because the lesions were considerably more extensive than seen with plain x-ray films.

4 Hyperbaric osteonecrosis in humans and sheep

4.1 Etiological findings in sheep

Sheep experiments induced hyperbaric osteonecrosis after controlled hyperbaric

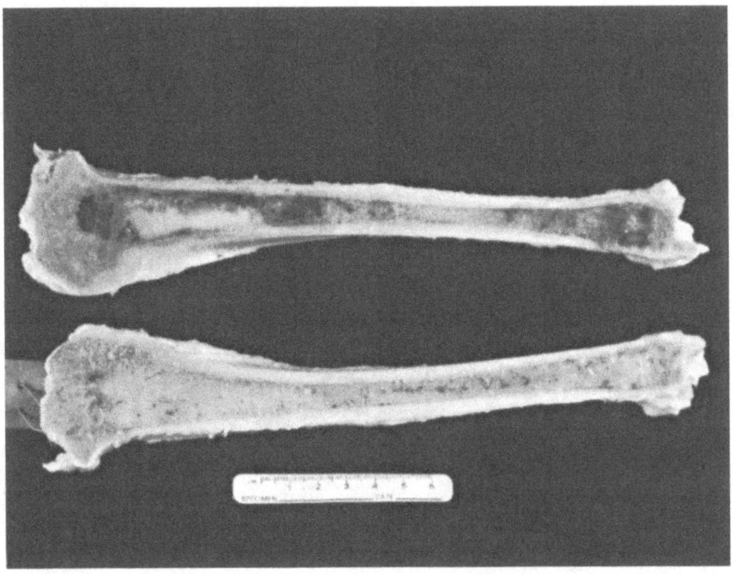

Fig. 5. Gross sections of the tibiae from Sheep #1810. The right tibia (near ruler) appears normal, but the left tibia contained large regions of hyperemic tissue (dark areas) which indicate active, ongoing repair of fatty marrow lesions.

exposures, and these findings are particularly useful for understanding the etiological basis of human hyperbaric osteonecrosis. In sheep experiments, the severity and frequency of bone lesions pointed to the relative risk of hyperbaric osteonecrosis across a broad spectrum of hyperbaric exposures with no-stop decompression (Lehner *et al.*, 1992), but the most severe lesions occurred after 24-h hyperbaric exposures.

In humans, most hyperbaric osteonecrosis lesions occur in bone shafts (Davidson, 1976), and this was also the case in sheep. Multiple hyperbaric exposures, each not greater than four hours in duration, induced mild bone lesions in some sheep, while multiple hyperbaric exposures of 24-h duration induced moderate to severe bone lesions in all five sheep (Lehner *et al.*, 1992). Single hyperbaric exposures of 24-h duration also induced moderate to severe bone lesions in all six sheep. In humans, single hyperbaric exposures can also induce severe bone necrosis, with disabling juxta-articular collapse (James, 1945).

Sheep findings, decompression theory, and epidemiological trends in the prevalence of hyperbaric osteonecrosis in tunnel workers and divers indicate that increased duration, pressure, and rate of decompression in hyperbaric exposures are all correlated with increased risk and severity of hyperbaric osteonecrosis.

4.2 Etiological basis in human populations

We view the mechanisms responsible for the experimental induction of hyperbaric osteonecrosis in sheep to be the same as those which cause hyperbaric osteonecrosis in humans. Epidemiological trends of bone necrosis in humans strongly support the importance of duration and pressure of hyperbaric exposure as etiological factors in hyperbaric osteonecrosis.

The prevalence of hyperbaric osteonecrosis in caisson and tunnel workers demonstrates an overall pattern in the risk of hyperbaric osteonecrosis. Risk increases with the increasing duration and pressure of hyperbaric exposure.

4.3 Limb bends and hyperbaric osteonecrosis

There appears to be a strong correlation between persistent limb bends and hyperbaric osteonecrosis. Although limb bends is sometimes referred to as 'pain-only' bends, results from the sheep experiments and other observations suggest otherwise. Hyperbaric osteonecrosis is a potential outcome faced by the diver and tunnel worker with persistent limb bends.

4.4 Therapeutic recompression and bone necrosis

Prompt recompression therapy can likely prevent or block the development of hyperbaric osteonecrosis in persistent limb bends cases. Increased ambient pressure in recompression will reduce the volume of any bubbles within the marrow cavity and decrease tissue pressure in the long bones. A more normal tissue pressure would promote reperfusion of the bone tissues. In fact, the relief of limb bends pain may directly reflect a more normal tissue pressure in bone, as occurs in osteoarthritis with the relief of pain when intraosseous pressure is reduced.

In a preliminary study, recompression treatment was tested with a modified U.S. Navy Table 1A to determine whether bone necrosis induction could be blocked in decompressed sheep. After recompression, there was a prevailing lack of limb bends signs in the sheep, and bone scans were mostly negative, except for a minimal reaction in the shaft of a tibia at four weeks. MRI revealed no fatty marrow lesions. Gross sections of the long bones at necropsy confirmed the absence of fatty marrow lesions and a minimal cortical bone reaction in the tibia which previously had shown a mild, transient 'hot spot'.

These preliminary results in two sheep suggest the possible role of recompression therapy in blocking the induction of bone necrosis. A formal study is now in progress. These findings also suggest that recompression, within a few hours of decompression, offers an important therapeutic approach for preventing the development of hyperbaric osteonecrosis in tunnel workers and divers with early limb bends symptoms.

5 Conclusions

Prolonged hyperbaric exposures carry a risk of hyperbaric osteonecrosis. Prolonged hyperbaric exposures can involve both single and closely-spaced exposures which permit a substantial accumulation of dissolved N_2 in the fatty marrow of the long bones. Decompression may form bubbles in the semi-closed compartments of the long bones. If significant numbers of bubbles form, then tissue pressures can rise and ischemia may result. Persistent limb bends may reflect elevated bone marrow pressure and ischemia in the long bones. If the ischemia is sustained, then hyperbaric osteonecrosis may result.

Persistent ischemia brought about by the proposed mechanism of a bone compartment syndrome, triggered by rapid decompression and bubble formation, appears to be the most likely pathogenic mechanism responsible for hyperbaric osteonecrosis. Preliminary recompression experiments with sheep indicate the importance of early recompression therapy in potentially stemming the development of hyperbaric osteonecrosis in the limb bends patient.

Acknowledgment The authors wish to thank Dr. Arthur DeSmet for performing MRI of the sheep, and Dr. Chess Adams for radiographic evaluations. We also thank Mr. James Schwarz, Ms. Joan Capelle, Mr. Glen Moen, and Ms. Dana Rhode for their assistance in these studies.

This work was funded by the University of Wisconsin Sea Grant Institute under grants from the National Sea Grant College Program, National Oceanic and Atmospheric Administration, U.S. Department of Commerce, and from the State of Wisconsin. NA9OAA-D-SG469, project R/DP-6.

6 References

Adams, W.M.C. Lehner, C.E. Dubielzig, R.R. Wilson, J.W. Palta, M. and Lanphier, E.H. (1985) Dysbaric osteonecrosis in sheep: a model of aseptic bone necrosis. *Physiologist*, 28, 308.

Amako, T. Kawashima, M. Torisu, T. and Hayashi, K. (1974) Bone and joint lesions in decompression sickness. *Seminars Arthr. Rheum.*, **4**, 151-190.

Atkins, C.E. Lehner, C.E. Beck, K.A. Dubielzig, R.R. Nordheim, E.V. and Lanphier, E.H. (1988) Experimental respiratory decompression sickness in sheep. *J. Appl. Physiol.*, **65**, 1163-1171.

Behnke, A.R. (1945) Decompression sickness incident to deep sea diving and high altitude ascent. *Medicine*, **24**, 381-402.

Campbell, J.A. and Hill, L. (1933) Studies in saturation of the tissues with gaseous nitrogen. I. Rate of saturation of goat's bone marrow in vivo with nitrogen during exposure to increased atmospheric pressure. *Quart. J. Exp. Physiol.*, **23**, 197-210.

Catto, M. (1976) Pathology of aseptic bone necrosis, in: *Aseptic Necrosis of Bone* (ed. J.K. Davidson), Elsevier, New York, pp. 147-212.

Chryssanthou, C.P. (1976) Dysbaric osteonecrosis in mice. *Undersea Biomed. Res.*, **3**, 67-83.

Davidson, J.K. (1976) Dysbaric osteonecrosis, in: *Aseptic Necrosis of Bone* (ed. J.K. Davidson), Elsevier, New York, pp. 147-212.

Davidson, J.K. (1989) Dysbaric disorders: aseptic bone necrosis in tunnel workers and divers. *Baillière's Clin. Radiol.*, **3**, 1-23.

Decompression Sickness Central Registry and Radiological Panel. (1981) Aseptic bone necrosis in commercial divers. *Lancet*, ii, 384-388.

Elliott, D.H. and Kindwall, E.P. (1982) Manifestations of decompression disorders, in: *The Physiology and Medicine of Diving* (eds. P.B. Bennett and D.H. Elliott), Baillière Tindall, London, pp. 461-472.

Golding, F.C. (1966) Radiology and orthopaedic surgery. *J. Bone Joint Surg.*, **48B**, 320-332.

Golding, F.C. Griffiths, P. Hempleman, H.V. Paton, W.D.M. and Walder, D.N. (1960) Decompression sickness during construction of the Dartford Tunnel. *Br. J. Indust. Med.*, **17**, 167-180.

Gregg, P.J. (1989) Osteonecrosis, in: *Imaging Techniques in Orthopaedics* (eds. C.S.B. Galasko and I. Isherwood), Springer-Verlag, London, pp. 363-370.

Gregg, P.J. and Walder, D.N. (1986) Caisson disease of bone. *Clin. Orthop.*, **210**, 43-54.

Gregg, P.J. Walder, D.N. and Rannie, I. (1980) Caisson disease of bone: A study of the Göttingen mini-pig as an animal model. *Br. J. Exp. Path.*, **61**, 39-54.

Hayashi, K. Kitano, M. Kawashima, M. Torisu, T. and Matsuoka, S. (1978) Studies of decompression sickness in Japanese diving fishermen, in: *Underwater Physiology VI* (eds. C.W. Shilling and M.W. Beckett), FASEB, Bethesda, Maryland, pp. 547-554.

James, C.C.M. (1945) Late bone lesions in caisson disease. Three cases in submarine personnel. *Lancet*, ii, 6-8.

Jones, J.P. Jr. (1985) Fat embolism and osteonecrosis. *Orthop. Clin. North Am.*, **16**, 595-633.

Jones, J.P. Jr. (1992) Intravascular coagulation and osteonecrosis. *Clin. Orthop.*, **277**, 41-53.

Kawashima, M. Hayashi, K. Torisu, T. and Kitano, M. (1978) Pathological review of osteonecrosis in divers. *Clin. Orthop.*, **130**, 107-117.

Kindwall, E.C. Nellen, J.R. and Spiegelhoff, D.R. (1982) Aseptic necrosis in compressed air tunnel workers using current OSHA decompression schedules. *J. Occup. Med.*, **24**, 741-744.

Kitano, M. and Hayashi, K. (1981) Acute decompression sickness. *Acta Pathol. Jpn.*, **31**, 269-276.

Kitano, M. Kawashima, M. Taya, Y. Lehner, C.E. and Lanphier, E.H. (1992) Histopathological analysis of dysbaric osteonecrosis experimentally induced in sheep, in: *Proceedings of the Eleventh Meeting of the United States-Japan Cooperative Program in Natural Resources, Panel on Diving Physiology and Technology*, September 26-27, 1991, Hakone, Japan, Japan Marine Science and Technology Center (JAMSTEC), Yokosuka, pp. 36-47.

Lang, M.A. and Vann, R.D., eds. (1992) *Repetitive Diving Workshop*, Proceedings of the American Academy of Underwater Sciences, (Duke University, March 18-19, 1991). American Academy of Underwater Sciences, Costa Mesa, California.

Lanphier, E.H. and Lehner, C.E. (1990) Animal models in decompression, in: *Man in the Sea* (eds. Y.C. Lin and K.K. Shida), Best Publishing Co., San Pedro, California, pp. 273-295.

Lanphier, E.H. Lehner, C.E. and Lemos, S.E. (1994) The role of intramedullary pressure in hyperbaric osteonecrosis, in: *Engineering and Health in Compressed Air Work* (eds. F.M. Jardine and R.I. McCallum), E & FN Spon.

Lehner, C.E. (1992) Dive profiles and adaptation: pressure profiles target specific tissues for decompression injury, in: *Repetitive Diving Workshop* (eds. M.A. Lang and R.D. Vann), American Academy of Underwater Sciences, Costa Mesa, California, pp. 203-217.

Lehner, C.E. Lanphier, E.H. Wilson, M.A. Kawashima, M. Kitano, M. Mano, Y. and Taya, Y. (1992) Dive profiles control the risk of dysbaric osteonecrosis: Validation in Japanese diving fishermen and sheep, in: *Proceedings of the Eleventh Meeting of the United States-Japan Cooperative Program in Natural Resources, Panel on Diving Physiology and Technology*, September 26-27, 1991, Hakone, Japan, Japan Marine Science and Technology Center (JAMSTEC), Yokosuka, pp. 85-101.

Mankin, H.J. (1992) Nontraumatic necrosis of bone (osteonecrosis). *N. Eng. J. Med.*, **326**, 1473-1479.

McCallum, R.I. and Harrison, J.A.B. (1982) Dysbaric osteonecrosis: Aseptic necrosis of bone, in: *The Physiology and Medicine of Diving* (eds. P.B. Bennett and D.H. Elliott), Baillière Tindall, London, pp. 488-506.

McCallum, R.I. and Walder, D.N. (1966) Bone lesions in compressed air workers. With special reference to men who worked on the Clyde Tunnels 1958 to 1963. *J. Bone Joint Surg.*, **48B**, 207-235.

Mitchell, D.G. Steinberg, M.E. Dalinka, M.K. Rao, V.M. Fallon, M. and Kressel, H.Y. (1989) Magnetic resonance imaging of the ischemic hip. *Clin. Orthop.*, **244**, 60-77.

Pooley, J. and Walder, D.N. (1984) The effect of compressed air on bone marrow blood flow and its relationship to caisson disease of bone, in: *Bone Circulation* (eds. J. Arlet, R.P. Ficat and D.S. Hungerford), Williams & Wilkins, Baltimore, pp. 63-67.

Reeves, E. McKee, A.E., Stunkind, J.A. and Schilling, P.W. (1972) Radiographic and pathologic studies for aseptic bone necrosis in dogs incurring decompression sickness. *Aerospace Med.*, **43**, 61-66.

Schmidt-Nielsen, K. (1984) *Scaling, Why is Animal Size so Important?* Cambridge University Press, Cambridge.

Stegall, P.J. and Smith, K.H. (1976) The etiology and pathogenesis of decompression sickness: radiographic, hematologic and histologic studies in swine, in: *Underwater Physiology V. Proceedings of the Fifth Symposium on Underwater Physiology* (ed. C.J. Lambertsen), FASEB, Bethesda, MD, pp. 141 152.

Thomas, I.H. Evans, A. and Walder, D.N. (1982) Saturation of mini-pig bone marrow with nitrogen during exposure to compressed air. *Lancet*, ii, 127-129.

Wade, C.E. Hayashi, K. Cashman, T.M. Jr. and Beckman, E.L. (1978) Incidence of dysbaric osteonecrosis in Hawaii's diving fishermen. *Undersea Biomed. Res.*, **5**, 137-147.

Walder, D.N. (1990) Aseptic necrosis of bone, in: *Diving Medicine* (eds. A.A. Bove and J.C. Davis), W.B. Saunders, Philadelphia, pp. 192-199.

Zizic, T.M. and Hungerford, D.S. (1985) Avascular necrosis of bone, in *Textbook of Rheumatology* (eds. W.N. Kelly, E.D. Harris Jr., S. Ruddy, and C.B. Sledge), W.B. Saunders, Philadelphia, pp. 1689-1710.

13 Report of seventeen cases with delayed occurrence of dysbaric osteonecrosis

L.-D. Zhang, J.-F. Kang and H. Zhang*
Yangpu District Central Hospital, Shanghai, China
**Diving Medical Centre, Shanghai Salvage, Shanghai, China*

Abstract
Dysbaric osteonecrosis (DON) has increasingly become a public health problem recognised by scientists of many countries, but the delayed occurrence of the disease in patients who left hyperbaric work has not received sufficient attention. From 1962 to 1988, 1140 hyperbaric workers were examined in our hospital. Of these, 137 workers who left hyperbaric work were followed up. We found 17 patients developed DON at 4 to 19 years, an incidence of 12.4%. These patients had one or more of the following characteristics: (1) decompression not performed according to the decompression table after hyperbaric work: (2) acute decompression sickness not treated completely: (3) most of the patients were found to have had bone lesions before, and new lesions tended to occur more easily at other spots in these patients: (4) most of these patients drank no alcohol. Therefore, the authors recommend that decompression after hyperbaric work be adhered to more stringently: acute decompression sickness should be treated completely: dysbaric osteonecrosis be treated actively: X-ray evaluation of bones and joints after leaving hyperbaric work should be made once every 5 years, especially for those who have a long hyperbaric work experience, high pressure exposure, no-stop decompression or irregular decompression after diving, and acute decompression sickness. Examinations should be every year for suspicious lesions.
Keywords: Dysbaric osteonecrosis, Radiography.

1 Introduction

Dysbaric osteonecrosis (DON) is an ischaemic disease of the long bones caused by inadequate decompression after hyperbaric exposure: gas bubbles and blood coagulation obstruct and compress the vessels in the long bones. With the development in diving and its associated high frequency of DON of 10 to 50%, the subject has been studied by scientists in many countries (MRC, 1981; Xue *et al.*, 1966; Kang *et al.*, 1980; and Ohta and Matsunage, 1974), but the long-term occurrence of the disease in patients who leave hyperbaric work has not received

Engineering and Health in Compressed Air Work Edited by F.M. Jardine and R.I. McCallum. Published in 1994 by E & FN Spon. ISBN 0 419 18460 0

Table 1. Seventeen cases with delayed DON

Case No.[1]	Hyperbaric experience (years)	Decompression	DCS	Alcohol	Stage & Site of DON	X-ray examinations interval (years)						Incubation period (years)
						1	2	3	4	5	6	
1	2	regular	severe	no	I-right femur head	6	6	7				12-19
2	23	no-stop	no	no record	I-left humerus head	4	7					4-11
3	29	no-stop	moderate	no record	I-right humerus head	4	3					4-7
4	26	no-stop	moderate	no	II-both lower femurs	1	2	8				3-11
5	6.5	no-stop	moderate	no	I-left lower femur	6	10					6-16
6	11	no-stop	moderate	no	I-right lower femur	10		6				10-16
7	13	no-stop	moderate	yes	I-left humerus head	1	3	1	2	6		7-13
8	2	no-stop	mild	no	I-right lower femur	12		6				12-18
					I-left upper tibia	12		6				12-18
9	31	irregular	moderate	no	I-right lower femur	1	4					1-5
10	14	irregular	moderate	no	I-left humerus head	2	3	1				5-6
11	16	irregular	no	yes	I-both humerus heads	1	1	3	1	6	6	12-18
12	20	irregular	moderate	no	I-right humerus head	3	6					3-9
13	24	irregular	moderate	no	I-left upper tibia	3	2					3-5
14	17	irregular	moderate	no	III-left femur head	3	1	7				4-11
15	31	irregular	moderate	no	I-left upper tibia	4	3					4-7
16	2.5	irregular	severe	no	II-both lower femurs	7	2	2				9-11
17	0.67	irregular	moderate	no	I-right humerus head	7	3	3	1			13-14

Notes:

1 All cases were divers except No. 16 (a caisson worker) and No. 17 (a tunnel worker)

2 Case No. 1 experienced severe decompression sickness following a blow-up accident, but decompressions were otherwise regular

sufficient attention.

From 1962 to 1988, 1140 hyperbaric workers in diving, caisson work and tunnelling were examined in our hospital: of these, 152 individuals were diagnosed with DON, an incidence of 13.3%. The 137 workers who left hyperbaric work were followed up. These workers had one of more of the following characteristics:

1. Long hyperbaric work experience
2. Decompression not in accordance with the decompression table
3. A history of decompression sickness
4. They had been examined regularly for DON.

Among the 137 persons, we found 17 cases of DON in follow-up examinations from 4 to 19 years after their leaving hyperbaric work, an incidence of 12.4%. Description of these changes and some of the relevant clinical data form the basis of this report.

2 General description

All 17 patients are male, from 38 to 62 years old, with occupational exposure to hyperbaric conditions from 8 months to 31 years (av. 15 years and 9 months). Most of the 17 patients started hyperbaric work before 1960 in the Yangtze River and Huangpu River in China. At that time, diving equipment was very simple. The typical working depth was shallow, about 20 metres (204 kPa), and the deepest was 68 metres (693.6 kPa). The occurrence of DON was observed 4 to 19 years after leaving hyperbaric work (av. 11 years and 8 months). Individual case records appear in Table 1.

Ours is a retrospective examination, there was not a previous design plan. The number of times and intervals of X-ray examination differ from case to case. The interval period of X-ray examination in some cases is one year, but in others ten years.

In Table 1, the cases have been classified in the following terms:

Decompression schedule (Gong, 1985)

- Regular decompression: decompression performed after diving completely in accordance with the decompression table.
- No-stop decompression: ascending to surface in 2 to 5 minutes after diving, no stop-station.
- Irregular decompression: decompression not fully in accordance with the decompression table for various reasons.

Decompression sickness (PRC National Standard GB8782-88)

- Mild: itch and marble rash of skin, mild muscle and joint pains.
- Moderate: severe muscle and joint pains and other symptoms, such as vomiting, dizziness etc.

- Severe: in addition to the above symptoms, CNS disorder and shock etc.

Alcohol

- Yes: consuming 100 to 150 ml alcohol which contains 40 to 60 per cent ethanol every day.
- No: consuming alcohol a little or occasionally.

DON Stages in terms of Chinese Standard (Xue, 1988) method

Stage I: the total size of localised dense areas (segmental, linear, or irregular opacities; translucent areas or cysts as projected on radiographs) is less than one third of the humerus or femur head.

Stage II: the total size of the above mentioned lesions projected on the radiograph is larger than one third of the humerus or femur head.

Stage III: lesions that extend to articular destruction.

3 Discussion

1. The occurrence of DON in these patients correlates with the length of hyperbaric work experience, exposure pressure, the decompression schedule and the history of DON (Kang *et al.*, 1980). All 17 cases underwent no-stop decompression or irregular decompression (see note 2 for case 1). Acute decompression sickness was experienced by 15 of the 17 cases, an incidence of 88.2%. To protect workers against the delayed onset of DON, existing decompression procedures must be adhered to more stringently and acute decompression sickness should be treated completely.

2. Among the 17 clinical cases evaluated in this paper, the description of their bone lesions is that 14 are in Stage I, 2 in Stage II, and 1 in Stage III. Fourteen of the 17 cases initially had bone lesions in various degrees in other parts of bone. Four were in Stage I, 7 in Stage II, and 3 in Stage III. There are 8 cases with Stage II or Stage III whose lesions had worsened and spread; while among Stage I cases, only one case had worsened and the other three cases improved. These data show that the more severe the disease is, the more likely new lesions will be observed in other bones. Therefore to decrease or delay the occurrence of late-onset DON, patients should undergo treatment for the initial bone lesions, for example, by treatment with hyperbaric oxygen and Traditional Chinese Medicine (Xue, 1984).

3. It has been reported that alcohol consumption after diving protects against acute decompression sickness (Zhang *et al.*, 1989 and 1991) and also DON (Zhang *et al.*, 1987). Of the 17 DON cases, 13 cases did not consume alcohol, 2 reportedly consumed alcohol: it is not recorded for the other two. Alcohol consumption after diving might be protective against late-onset DON.

4. McCallum has recommended that the X-ray evaluations ought to be conducted regularly for two years after the patients leave hyperbaric work (McCallum, 1983). The 17 patients of this paper reported clinical symptoms of DON after leaving the hyperbaric work, their symptoms appearing after 5 years in 29.4% of the cases;

after 10 years − 41.2%; and after 15 years − 29.4%. This high incidence and long incubation period of DON should receive more attention in hyperbaric medicine. The Authors recommend that hyperbaric workers should be examined and undergo X-ray evaluation of bones and joints regularly after leaving hyperbaric work, so as to find early lesions. It is appropriate that examinations should be once every 5 years, especially for those individuals of long hyperbaric work experience, high pressure exposure, and who have experienced no-stop decompression or irregular decompression after diving, and acute decompression sickness.

4 References

Gong, J-H. *Diving Medicine*. 1st Edition. Beijing: People's Army Physician Press, 1985: 138-156

Kang, J-F. *et al.* Investigation of dysbaric osteonecrosis in 650 divers. *Chinese Journal of Industrial Hygiene and Environmental Medicine* 1980: **3**: 11-14

McCallum, R.I. *ILO Encyclopedia of occupational health and safety*, 3rd Edition. Geneva: International Labour Office, 1983: 596

MRC Decompression Sickness Central Registry and Radiological Unit. Aseptic bone necrosis in commercial divers. *Lancet* 1981: **2**: 384-388

Ohta, Y. and Matsunage, H. Bone necrosis in divers. *J. Bone Joint Surg.* 1974: **56B**: 3-16

Peoples' Republic of China: National Standard *Diagnostic Criteria and Principles of Management of Occupational Decompression Sickness*. GB 8782-88: 1-3

Xue, H-L. *et al.* Bone and joint diseases in divers. *Chinese J. Surg.* 1966: **14**: 219-222

Xue, H-L. Stage III dysbaric osteonecrosis and its management. A report of 10 cases. *Chinese Journal of Orthopaedics* 1984: **4**(2): 106-109

Xue, H-L. Dysbaric osteonecrosis and its radiographic classification in China. *Undersea Biomedical Research* 1988: **15**(5): 389-395

Zhang, L-d, *et al.* Study in the ethanol treatment of decompression sickness. *Ocean Engg.* (in Chinese) 1987: **5**(3): 85-91

Zhang, L-d, *et al.* Ethanol treatment for acute decompression sickness in rabbits. *Undersea Biomedical Research* 1989, **16**(4): 271-274

Zhang, L-d, *et al.* Ethanol treatment for acute decompression sickness in 6 cases. *Undersea Biomedical Research*, 1991, **18**(1): 64-65

Discussion on Session 3: Bone necrosis

The session chairman, **Mr J Pooley**, who had introduced Professor D N Walder, thanked him for giving so clear and challenging a keynote lecture.

In the session, **Dr A Evans** presented his paper giving some of the results of the survey of risk factors associated with bone necrosis in British compressed-air workers. **Dr C E Lehner**, speaking on behalf of the Rev. Dr E H Lanphier, whose illness prevented him from attending, introduced the paper about the role of intramedullary pressure in bone necrosis. **Dr E P Kindwall** spoke to his paper on optimum decompression schedules. **Dr Lehner** then presented the paper of which he is lead author on the early pathogenesis of bone necrosis observed in sheep. The authors of the other paper of the session, Drs Zhang and Kang, were unable to attend the conference. Mr Pooley invited questions from the audience both on the papers and on Professor Walder's lecture.

Mr V J Belman. I note Professor Walder's theory that bone necrosis commonly occurs in that quadrant defined by pressure above 2 bar for a time exceeding 4 hours, whereas Dr Evans's research does not support this – in fact some 15% of those investigated had not been in pressures above 17 psig.

Is the explanation that it is not certain that bone necrosis, in all cases, is caused by circumstances attributable to compressed air? or, as Professor Walder states, is there something wrong with Dr Evans's analysis?

Dr G C Voigt. Is my understanding correct that the radiographic findings consistent with dysbaric osteonecrosis are not specific for that entity? That is to say that such x-ray abnormalities are seen in bone necrosis due to many other causes.

Professor Walder replied that there are difficulties with the recorded data available to the analysis carried out by Dr Evans. Although there are uncertainties of radiographic interpretation in some cases, the abnormalities seen in the bones of young men who have been working in compressed air are not likely to be from other causes.

Mr Pooley raised the question of oxygen at raised partial pressures in the development of bone necrosis and commented that bone blood flow has to be equivalent to that of skeletal muscle or much greater.

Dr Lehner commented that bones with fatty marrow have a lower flow rate.

Dr R D Vann. Have measurements of intramedullary oxygen tension been made at elevated inspired oxygen partial pressure? These measurements might provide insight into the potential for oxygen toxicity and the swelling of fat cells.

Mr Pooley. Such measurements have not been made yet but should be made.

Dr E P Kindwall. With the measurements of intramedullary pressure, when did they rise? and when were they back to normal?

Dr Lehner. Typically they rose after some 4-6 hours, and reverted to normal about 24 hours after decompression.

Mr F M Jardine. Dr Lehner referred to no higher intramedullary pressures than 40 mm Hg in control animals, but much higher pressures in bones of symptomatic limbs. What pressures did you measure in these?

Dr Lehner. We did not do many experiments, but recorded pressures of more than 100 mm Hg.

Mr Pooley commented that capillary pressures are much less than this. He referred to his work with Professor Walder on fat cells swelling under increased partial pressure.

Professor D H Elliott. The pathophysiology is focused upon shaft lesions which have no clinical significance: what is the evidence that the pathology of shaft and subchondral necrosis is the same?

Mr Pooley. Both lesions show the same mechanism. There is a difference in the distribution between compressed air workers and divers, i.e. in divers the lesions are in the shoulders. Their work exercises are different so that could account for the different locations, the link perhaps being the decreased perfusion in the bone on exercise.

Dr Lehner. The location of the lesions could reflect blood flow rates.

Professor Walder. Oxygen in decompression may be acceptable; it is a somewhat different application from the role of oxygen at raised partial pressures on fat cells; it is not subjecting a worker to high partial pressures of oxygen for 4 to 5 hours. Questions for Dr Lehner: (1) if they were increasing the intramedullary pressure to arterial for 12 hours, what effect was there on the bones? (2) The exposure of 24 hours at 2.2 atm with a very rapid decompression over 1m/sec was very severe, why was this procedure used? (3) If bends pain comes from stretching of the periosteum, was there evidence of this from radiographs?

Dr Lehner referred to work of Dr M Cross in relation to periosteal reaction and three autopsy cases. He suggested that the push of marrow in the bone acted on pain nerve fibres. They chose the exposure regime for the following reasons: bone necrosis is less likely at lower pressures, 24 hours is needed to approach maximum saturation and cost was one of the factors controlling the decompression time.

Professor R B Philp. With respect to the question of prolonged exposures to elevated oxygen tension, there is one population which has not yet been alluded to. I refer to the habitat saturation divers (e.g. those who used Hydrolab Habitat in the Bahamas) who lived at 25 psig for 5-7 days with daily 4-6 hour diving (and swimming) exposures to 40-60 psig. In about 2,500 person-days of saturation diving followed by prolonged decompression, the bends incidence was near zero and no cases of bone necrosis have been reported. This does not argue for an oxygen role in the pathogenesis of bone necrosis.

Mr J C Le Péchon. Concerning oxygen breathing during decompression, to be efficient in reducing hypoxia associated with bone marrow ischemia, it must be administrated before bubble formation occurs there at a 'deep' stop compatible with oxygen toxicity. We shall need that information for further discussion on proper management of oxygen in decompression.

Dr J D King. An important implication for engineers that comes from this discussion of the pathogenesis of bone necrosis is the need for more surveys of tunnel workers.

Dr R W Hamilton. It is important to keep in mind that Professor Walder's references to oxygen being implicated in the mechanism of dysbaric osteonecrosis and the use of oxygen for decompression are not the same. Professor Walder's hypothesis that oxygen may be involved in the mechanism of the bone necrosis is interesting and meaningful, but it does not mean that oxygen is not beneficial in decompression. Professor Walder made the same comment himself.

Dr H V Hempleman. Bone blood flow is not well understood, it varies considerably in different regions.

(Reporters: J Buchanan, F M Jardine)

Part Four
Engineering Design

14 Compressed air tunnelling between Somerset Station and Newton Station, Singapore

A.R. Gaba and D.B. Parkes
*Soil Mechanics Associates, Wokingham, UK**
Ove Arup & Partners, London, UK

Abstract
Twin bored tunnels, each 6.4 m in diameter with an aggregate length of some 3.5 km, were constructed using a full-face tunnel boring machine under compressed air between Somerset and Newton Stations for the Singapore Mass Rapid Transit System. The tunnels were bored beneath Orchard Road and Scotts Road close to some of Singapore's most prestigious and substantial structures. This paper examines certain problems encountered during the tunnelling works, in particular the settlements generated by the work and their effect on neighbouring structures and utilities. The settlements were principally due to consolidation of deep deposits of soft compressible clay encountered by the tunnels. The paper shows how the magnitude of the compressed air pressure had a major effect on piezometric water levels in the clay and hence consolidation settlements. The methods adopted to minimise these settlements are described. The paper also discusses the factors involved in determining an adequate provision of compressed air supply during tunnelling for the Singapore Mass Rapid Transit System.
Keywords: Compressed Air, Tunnelling, Settlement, Consolidation, Air Pressure, Air Supply Provision.

1 Introduction

The Singapore Mass Rapid Transit (MRT) System comprises 42 stations over a 67 km route. Much has been written about the system: notably by Hulme *et al.* (1989a) on its planning and implementation and by Copsey *et al.* (1989) and Hulme *et al.* (1989b) on the design and construction of the System respectively.

Contract 105 of the Singapore MRT involved the construction of twin bored tunnels between Somerset Station and Newton Station beneath the busy tourist districts of Orchard Road and Scotts Road, see Figure 1.

---

* Formerly Ove Arup & Partners

Engineering and Health in Compressed Air Work Edited by F.M. Jardine and R.I. McCallum. Published in 1994 by E & FN Spon. ISBN 0 419 18460 0

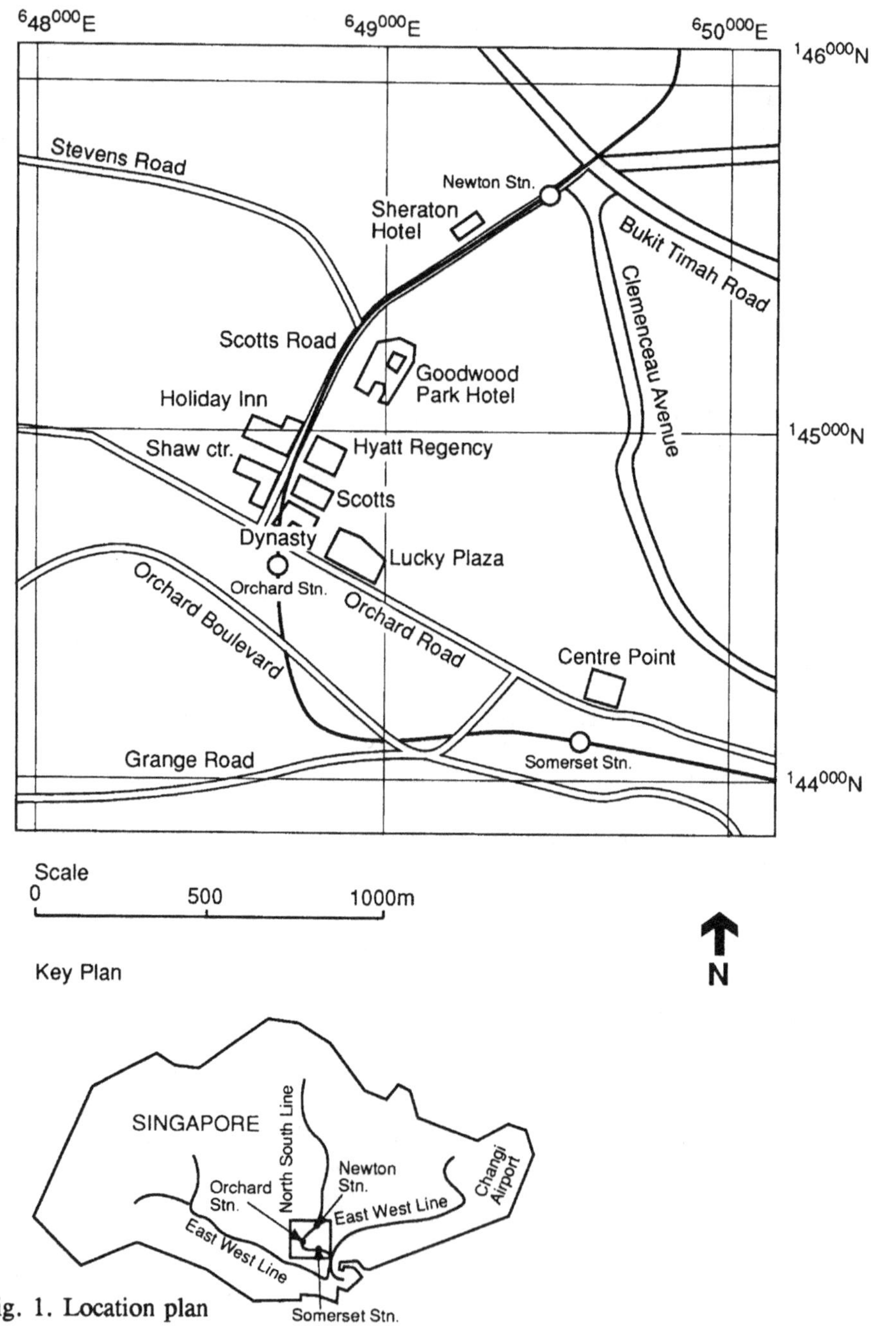

Fig. 1. Location plan

A major concern in the design and construction of the tunnelling works in this relatively sensitive urban environment was the effect of the works on adjacent properties and utilities. Extensive settlement monitoring was therefore carried out during the underground work. This was done to identify the causes of the settlements and to assist in deciding how to minimise them. Analysis of the results of this monitoring also provided a useful guide to the prediction of the effects of future work.

This paper presents some of the results of this monitoring and discusses the methods adopted to minimise settlements. The tunnels were constructed under compressed air. Some engineering aspects of compressed air working are also discussed.

2 Ground conditions

2.1 Geological setting

The geology of Singapore can be divided into the predominantly soft clays and loose sands of the Kallang formation and the older generally stronger soils and rocks. The Kallang formation is recent, of Holocene and late Pleistocene age, and was laid down in valleys eroded into the underlying rocks during periods of low sea levels. The most common material within the Kallang formation is a normally to lightly overconsolidated clay of marine origin. Lenses of fluvial sand and clay occur within the marine clay, particularly at the interface of an older, late Pleistocene member and a younger, Holocene member. The marine clay has a low coefficient of permeability, about 10^{-9} m/s, and is compressible. Inland, the soft clay tends to have a higher organic content, probably because it was laid down under estuarine or marshy conditions. This estuarine clay is more compressible than the marine clay.

The main rock types in Singapore are the igneous rocks of the Bukit Timah Granite and the Gombak Norite formations and the sedimentary rocks of the Jurong formation. Bukit Timah Granite is used to describe a suite of granitic rocks and is more typically a granodiorite than a true granite. Completely weathered rocks of these formations tend to be of relatively low permeability.

The geological environment through which the Contract 105 tunnels traverse is shown in Figure 2. The geology along the route of the tunnels can be conveniently divided into three areas or zones as discussed below.

2.2 Zone A ground conditions

Zone A ground conditions were found to occur for a distance of about 300 m west of Somerset Station and for a similar distance north of Orchard Station, see Figure 2. The typical stratigraphy in this area is shown in Table 1.

A typical geological section representing this area is shown in Figure 3.

The Contract 105 tunnels through Zone A ground conditions were constructed within the completely decomposed granite (G4).

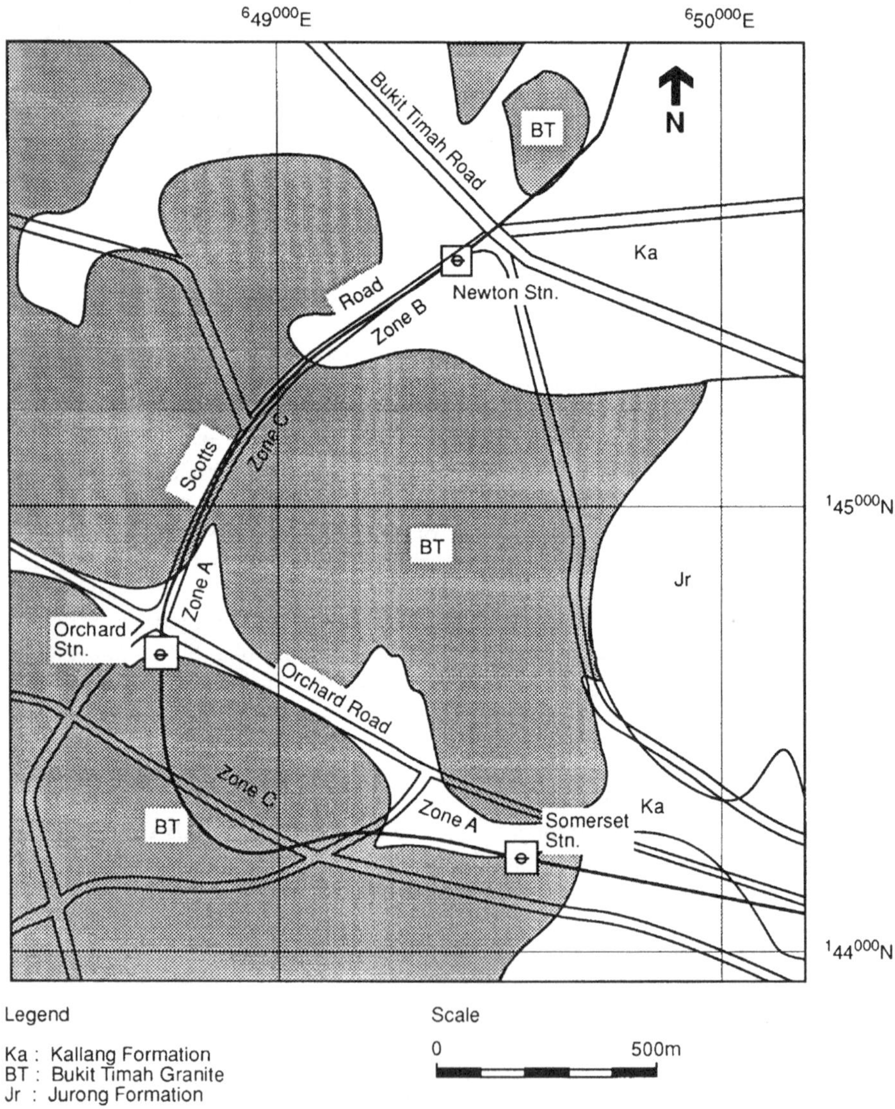

Fig. 2. Geology

Table 1 Typical stratigraphy: Zone A ground conditions

Stratum	Elevation of top	Thickness
MADE GROUND	103 m to 104 m	1 m - 4 m
KALLANG FORMATION	100 m to 102 m	
- Estuarine Clay (E) very soft to soft peaty clay		1 m - 6 m
- Fluvial Sand (F1) loose to medium dense sand		1 m - 2 m
BUKIT TIMAH GRANITE complete decomposed granite (G4)	95 m to 96 m	
- stiff silty sandy clay		1 m - 5 m
- dense clayey silt		greater than 18 m

2.3 Zone B ground conditions

Zone B ground conditions were encountered over a distance of some 300 m south of Newton Station, see Figure 2. The typical stratigraphy in this area is shown in Table 2.

Table 2 Typical stratigraphy: Zone B ground conditions

Stratum	Elevation of top	Thickness
MADE GROUND	103 m to 104 m	0 - 2 m
KALLANG FORMATION	101 m to 104 m	
- Upper Marine Clay (M1) very soft to soft silty clay		7 m - 10 m
- Fluvial Sand (F1) loose to medium dense sand		0.5 m - 1.5 m
- Lower Marine Clay (M2) soft silty clay		3 m - 10 m
BUKIT TIMAH GRANITE	93 m to 81 m	
completely decomposed granite - dense clayey silt		greater than 12 m

A typical geological section representing this area is shown in Figure 4.

The Contract 105 tunnels through Zone B ground conditions were constructed entirely within the Kallang Formation and intercepted the layer of Fluvial Sand (F1) between the two units of Marine Clay.

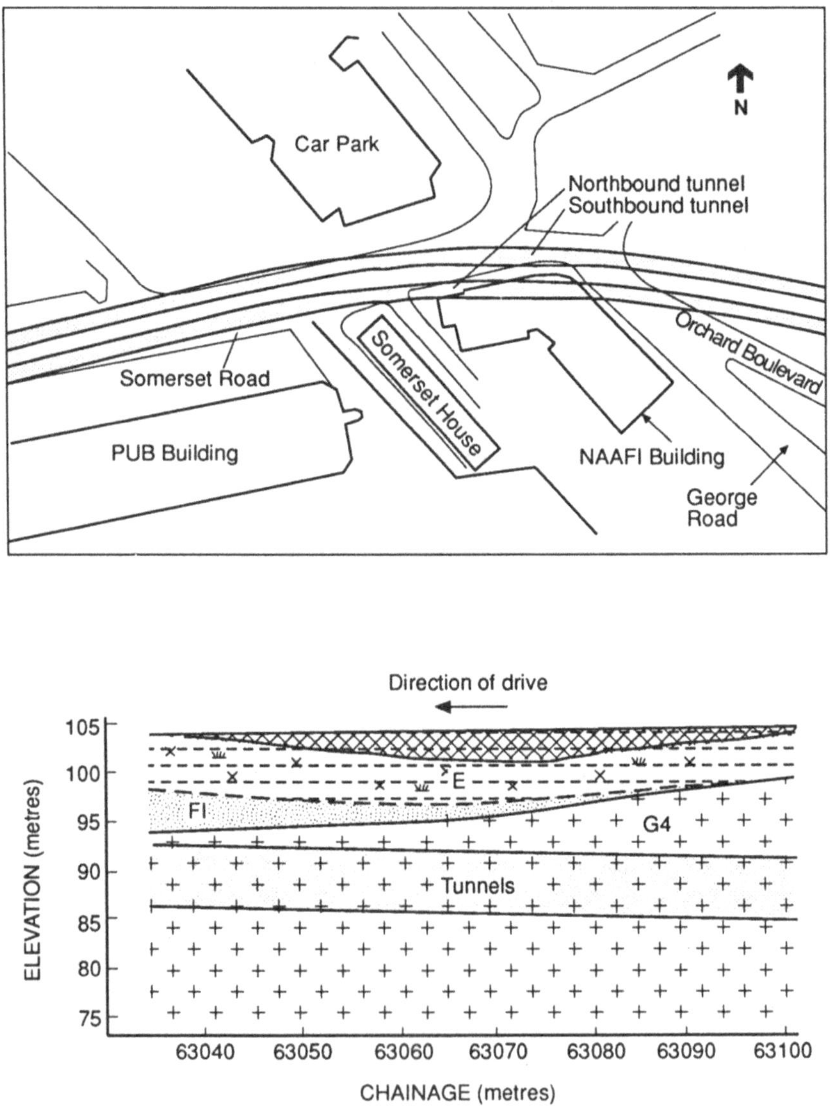

Fig. 3. Zone A: Typical ground conditions

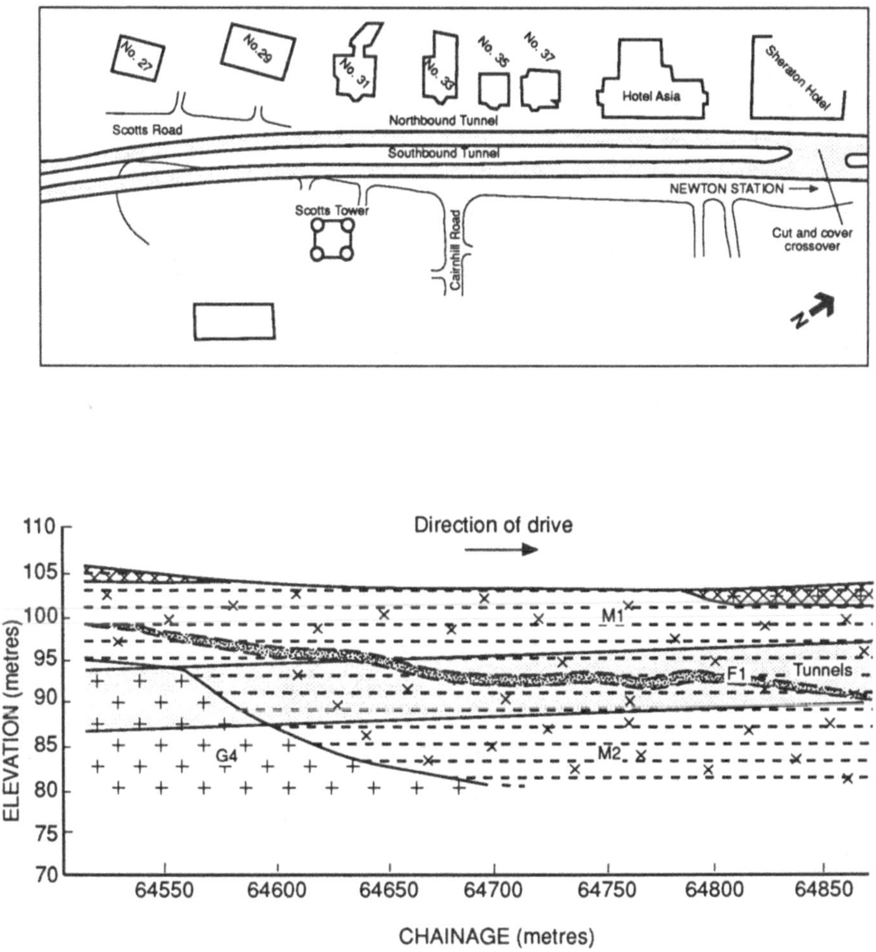

Fig. 4. Zone B: Typical ground conditions

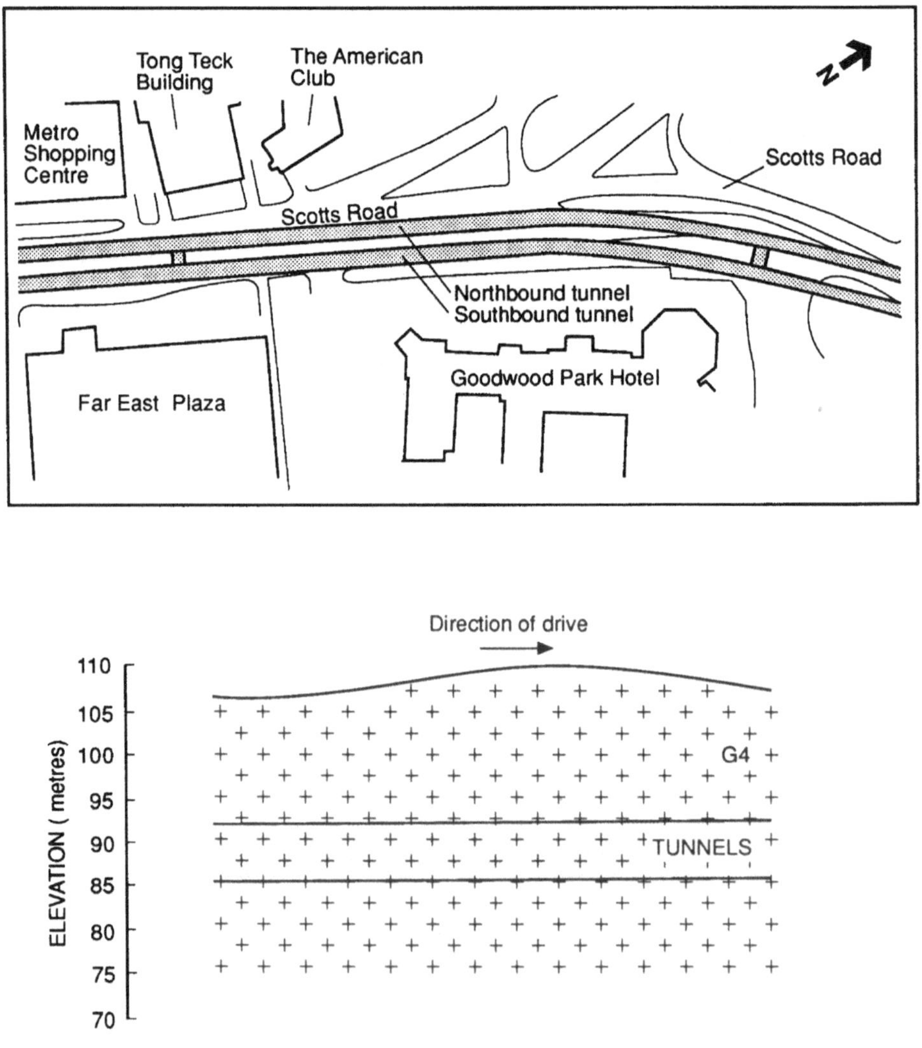

Fig. 5. Zone C: Typical ground conditions

2.4 Zone C ground conditions
Zone C ground conditions were encountered over a distance of some 600 m south and east of Orchard Station and beneath Scotts Road between the Zone A and Zone B areas, see Figure 2. The typical stratigraphy in this area comprised Bukit Timah Granite from ground level to a depth in excess of 35 m. A typical geological section representing this area is shown in Figure 5.

2.5 Groundwater
Site investigation data prior to the commencement of tunnelling operations indicated a high groundwater table to be present along the route of Contract 105. Groundwater was encountered at depths varying from ground level (at Zone B) to about 2 m below ground level (at Zone C) with a hydrostatic profile with depth below ground level.

3 Tunnel construction

The Contract 105 tunnels were all driven using the intermediate Orchard Station as a working site with the shafts incorporated into the station box.

The majority of the tunnelling was expected to be in completely decomposed granite (Zones A and C ground conditions) with some bouldery areas and a length of about 300 m of the drives being in soft clay (Zone B ground conditions). It was considered that all four tunnel drives could be completed with one full-face fully mechanised shield (TBM) using a steel rib and timber lagging method of temporary support. A permanent cast-in-situ reinforced concrete lining was installed after completion of excavation. Compressed air was adopted for substantial lengths of the drives. In the event, only three of the tunnels were completed using the full-face mechanised shield. A second shield, a Greathead type with a full circle backhoe, was used for the fourth drive.

Access to the face of the mechanised shield was very restricted which added to the problems of controlling the excavation. Thus the shield proved disappointing in bouldery areas where blasting ahead of the shield became necessary to break up boulders. In the soft clay areas, there were initial difficulties in controlling excavation where, because of the sensitivity of the clay, the shield was being operated as a blind shield. This led to some initial heave and settlements, which are discussed further in Section 4.5 below.

Compressed air was used and was found to be successful in maintaining stable conditions at the tunnel face. However, there were several locations where the ground conditions were such that air pressures could not be held at desirable levels because relatively permeable sands connected the tunnel face to the surface. These were mostly in the Zone C ground conditions. On occasions air vents occurred up to 30 m from the face.

The basic features of the geology and progress in the construction of the south-bound tunnels are shown in Figures 6 and 7.

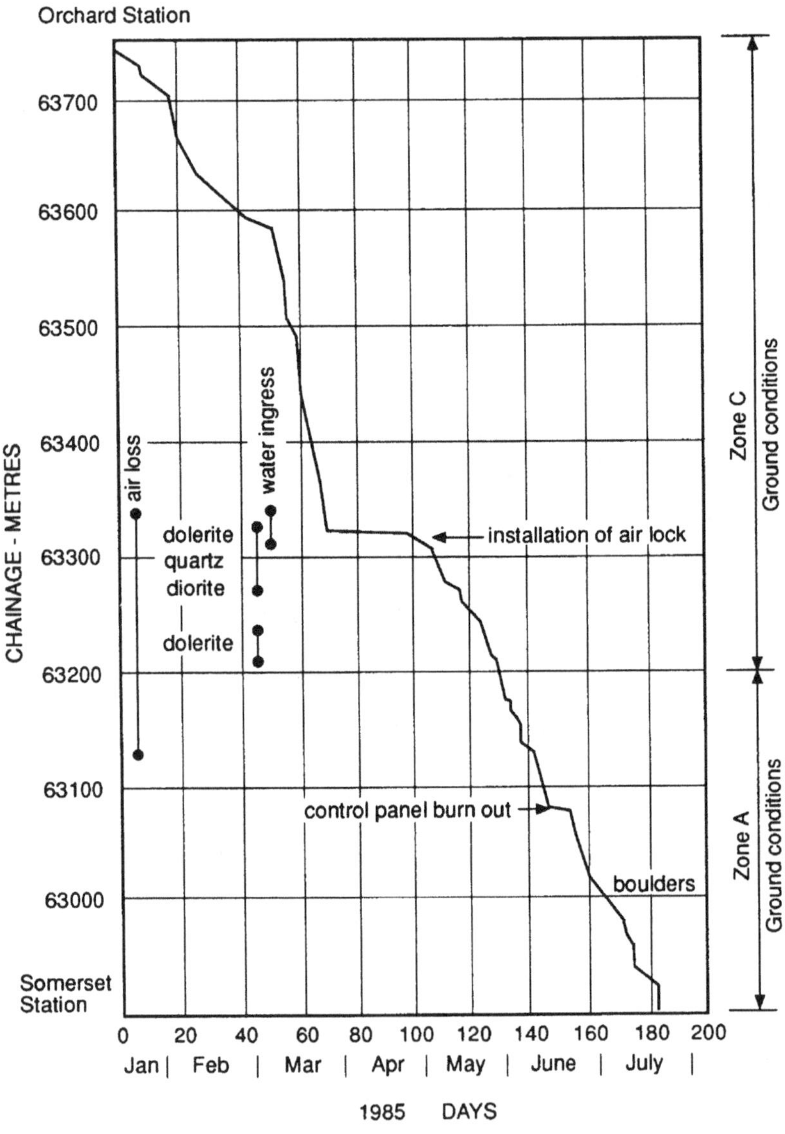

Fig. 6. Southbound tunnel: Orchard Station to Somerset Station

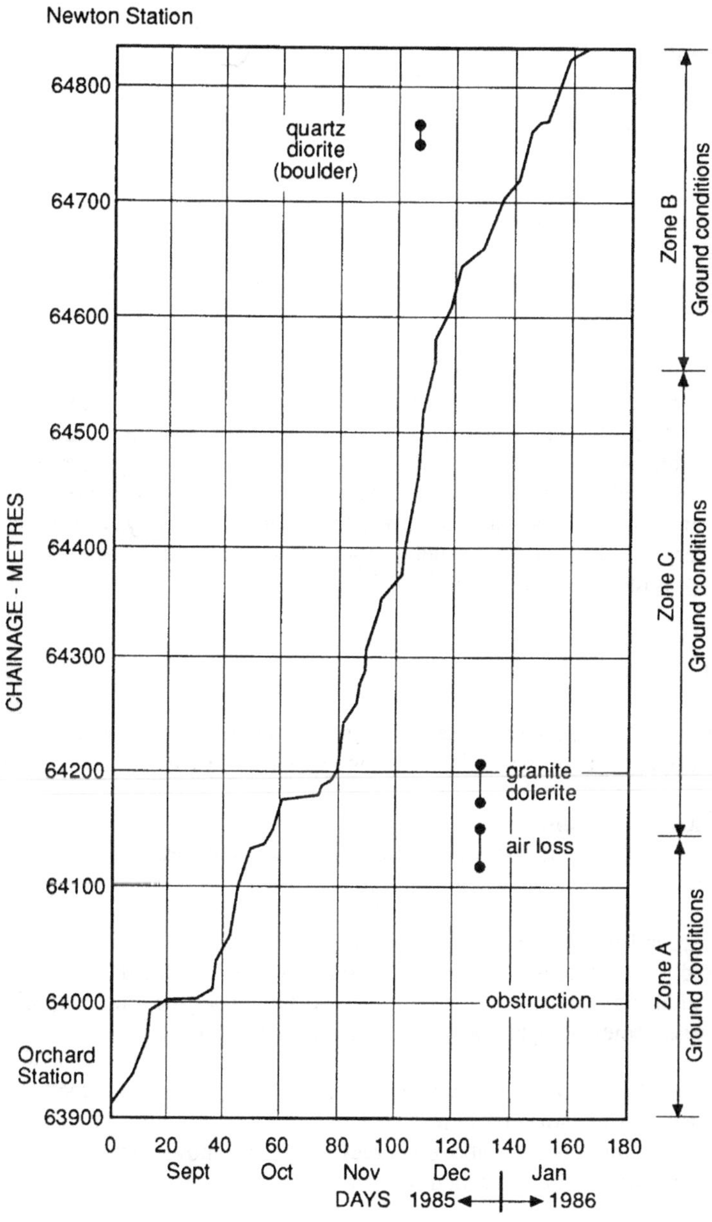

Fig. 7. Southbound tunnel: Orchard Station to Newton Station

Table 3 Typical TBM progress rates

Ground conditions	Typical progress (m/week)
Zones A & C	
- completely decomposed granite G4 (no boulders)	65
- completely decomposed granite G4 (boulder areas)	35
Zone B	
- marine clay	50

The presence of obstructions in the tunnel face caused some delay. Boulders were either split hydraulically or by explosives. The limited access through the head made these operations slow and difficult. Progress in the completely decomposed granite (G4) was good, and in the marine clay area, the machine was used in blind mode and progress was also good. However, the balance necessary to prevent surface heave or settlement when extruding material was difficult to achieve. Typical TBM progress rates are given in Table 3.

4 Settlement

There are two major components of settlements from bored tunnelling: immediate or ground-loss settlements and consolidation settlements (Shirlaw and Doran, 1988).

4.1 Ground loss settlements

Settlement because of ground loss during tunnelling typically results from the following sources:

- yield of ground into the face of the heading
- inward radial movement of the ground around the tunnel shield
- inward radial movement of the ground behind the tail of the shield, prior to the erection and grouting of the permanent lining.

The magnitude of these movements depends on the tunnelling method, the equipment used and the standard of workmanship.

Ground loss resulting from the above sources of ground movements can be expressed in terms of a relative ground loss, $\Delta V/V$,

$$\frac{\Delta V}{V} = \frac{\text{Volume of ground lost into the tunnel}}{\text{Volume of tunnel excavation}}$$

The magnitude of relative ground loss depends on the stability of the tunnel face and other factors such as overcut for steering and the timing of back-grouting. The stability of tunnels in soft clay may be expressed in terms of a stability number, N, defined as :

$$N = \frac{(\gamma z - \sigma_i)}{c_u}$$

where γ = bulk unit weight of soil
z = depth to tunnel axis
σ_i = tunnel compressed air pressure
c_u = average undrained shear strength above axis level.

Compressed air was adopted to provide support to reduce the stability number to less than 4 to provide a reasonable margin of safety and to limit ground losses during tunnel excavation.

For tunnelling in clay in undrained conditions, predictions of relative ground loss can be made assuming that ground deformations take place with no volume change. Clough and Schmidt (1981) considered the relationship between ground loss and stability number. Their relationship depends on the ratio (E_u/c_u). Empirical correlations between relative ground loss and stability number have also been established based on a study of case history records. Correlations were proposed by Glossop (1977) and by Attewell *et al.* (1986) on this basis. Mair (1983) reported the results of a series of centrifuge model tests on tunnels in clay and developed a relationship between ground loss and stability number. All of these correlations and relationships are shown on Figure 8.

Case history data from tunnels driven using compressed air in soft clay in Singapore (reported by Shirlaw and Copsey, 1987) have also been added to Figure 8. These data show relative ground losses generally between 1% and 3% for a stability number between 3 and 4.

For the Contract 105 tunnels, relative ground loss was typically found to be about 3%, at the upper bound of the data presented by Shirlaw and Copsey (1987) for the Singapore MRT.

Figure 8 shows that ground losses can vary over a wide range depending on construction techniques, workmanship and ground conditions. There appears to be a tendency for the theoretical curves to overpredict the ground losses observed in practice, particularly at the higher stability numbers.

From a study of many case histories it has been found that the surface settlement trough usually takes the form of a Gaussian normal distribution. The trough width is defined in terms of the trough width parameter, i.

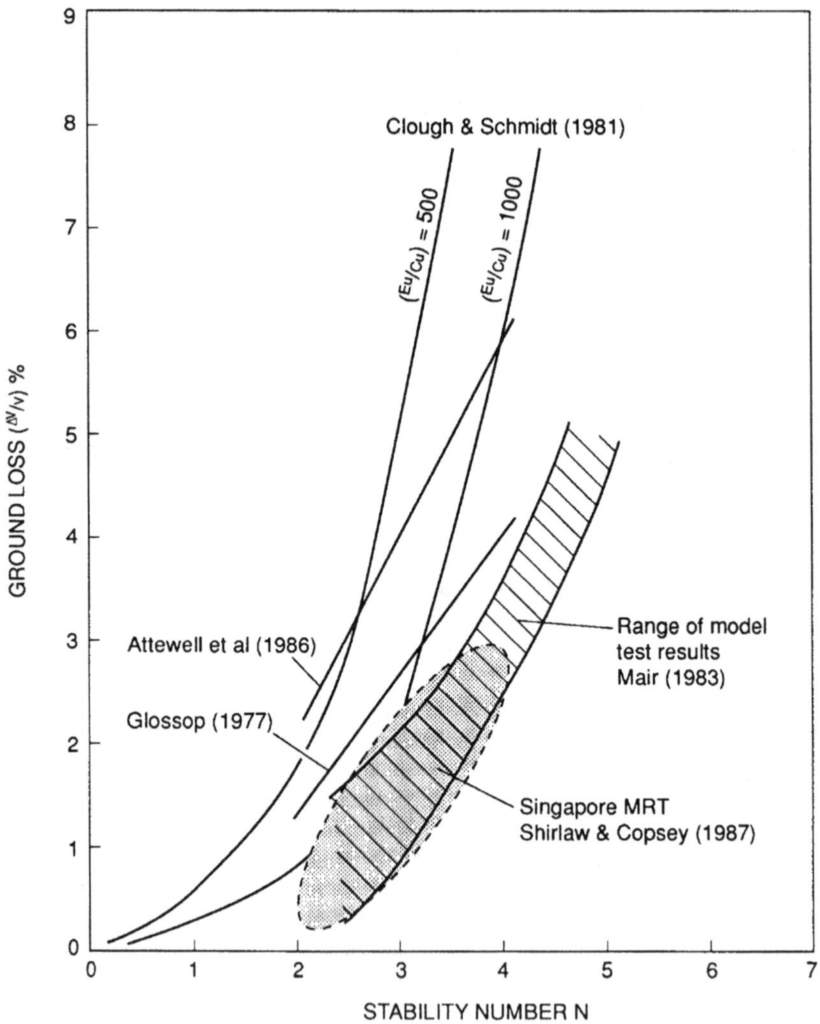

Fig. 8. Relationship between stability number and ground loss

Clough & Schmidt (1981) proposed the correlation:

$$(i/a) = (z/2a)^{0.8}$$

where a = tunnel radius
z = depth to tunnel axis.

The value of i tends to be larger for soft clays than for stronger soils, and from a review of tunnelling in UK, O'Reilly and New (1982) suggested that:

$$i = Kz$$

Settlement monitoring data show that typically $K = 0.5$ for the ground conditions at Contract 105.

4.2 Consolidation settlements

The excavation of a tunnel under compressed air results in changes to both the total and effective stresses in the ground around the tunnel. The pore pressures generated depend on the total stress changes and pore pressure parameters.

In soft clays a small rise in pore pressure as the tunnel machine passes followed by a reduction in pore pressure can be observed. The long-term pore pressures are dependent on the rate of water seepage into the tunnels. The long-term consolidation settlement therefore depends on the watertightness of the tunnel lining and the compressibility of the surrounding ground. The importance of the watertightness of the tunnel lining was recognised from the outset, the objective being to limit water inflow into the tunnels to a minimum both during construction and in service.

The contractor chose to adopt a steel rib and timber lagging method of temporary support to the tunnels.

The temporary lining to the Contract 105 tunnels consisted of 100 mm x 100 mm hardwood laggings inserted between 125 mm x 125 mm rolled steel ribs and expanded behind the shield tail. A packing piece or 'dutchman' was installed on both shoulders after expansion. Rib spacing was generally 1.2 m, but a 0.9 m spacing was adopted in marine clay areas (Zone B ground conditions). Following initial unsatisfactory experience with water inflow, a polythene sheet was placed behind the timber laggings to reduce water ingress. It was also necessary to grout behind the laggings.

Without significant grouting, the timber lagging primary lining was observed to be relatively ineffective in restricting water seepage into the tunnels and the method was considered inappropriate for a sensitive urban environment with difficult ground conditions. A fourth tunnel drive was completed using precast concrete lining with better results.

In addition to measures to stem water inflow to the tunnels, groundwater recharge wells were installed along the route of the tunnels to maintain groundwater levels in the relatively more permeable Fluvial deposits located within and below the soft clay areas to minimise drawdown and consolidation settlements in the soft clays.

4.3 Settlement monitoring

The primary objective of any settlement monitoring system is to identify the moment when the works begin to affect adjacent structures. Having ascertained that movement has occurred, the aim is then to determine the magnitude and direction of this movement and to identify any trends. With this in mind, an early warning system was established, consisting of precise level marks on all roads and structures along the route of the tunnels. Structures located within 70 m of tunnel centre were monitored. The location and number of marks depended on the age, condition and type of structure. A minimum of two points on the front and back of buildings were established to detect differential settlement. Care had to be taken with cladded buildings and, where necessary, holes were drilled through the cladding and a sleeve inserted into the structure. Measurements could then be taken to a bolt inserted into the sleeve to determine movement.

In addition, use was made of inclinometers, piezometers and extensometers to separate the various components of settlement.

Ground settlement monitoring marks were established in arrays along the route of the tunnels, typically 25 m apart. Each array extended to a distance of some 70 m perpendicularly from the alignment of the tunnels. Marks were typically located at 7 m spacing within each array.

Initial levels to all monitoring marks were taken using precise levelling techniques. Automatic levels fitted with parallel plate micrometers were normally used together with invar staves. Initial values were agreed between the Mass Rapid Transit Corporation (MRTC) and the contractor prior to the commencement of work and monitored at regular intervals. During tunnelling activities, structures were normally monitored at least weekly with sensitive structures being monitored daily. Techniques capable of detecting 2 mm of movement between consecutive readings were used.

Ground settlement data were plotted on a plan and interpolated to produce settlement contour lines. If these indicated a shallow depression, then this was indicative of consolidation settlements. A more localised steeper settlement profile indicated settlements due to ground loss.

4.4 Typical settlements in Zone A ground conditions

Typical ground surface and building settlements and piezometric groundwater pressures monitored during the construction of the southbound tunnel in the vicinity of Somerset House and the NAAFI building (Zone A ground conditions) are shown in Figure 9. This figure illustrates the relative effects on the development of ground surface and building settlements of:

- the passing of the tunnel face
- the removal of compressed air pressure
- the activation of the groundwater recharge wells
- the installation of the permanent tunnel lining.

The figure also shows the variation in the piezometric groundwater pressures in the soft Estuarine clay (E), the underlying Fluvial sand (F1) and the completely decomposed granite (G4). The tunnel was constructed using a compressed air pressure of 0.9 bar (12.5 psig).

Figure 9 shows that the passing of the tunnel face and the removal of the compressed air pressure had a significant effect on the development of the ground surface settlements and reduction in piezometric pore pressure. The activation of the groundwater recharge wells helped to stabilise groundwater conditions in the F1 sand and the Estuarine clay and this also helped to curtail further consolidation settlements. The installation of the permanent tunnel lining led to a gradual recovery in the groundwater pressures.

Total ground settlements were measured to be of the order of 80 mm. The settlement of Somerset House, which is founded on relatively shallow foundations supported by bakau (short timber) piles, closely followed the settlement of the ground surface around the building. The settlement of the NAAFI building, which was supported on groups of 350 mm diameter driven precast piles, was much less than the ground surface settlement around the building. This led to some problems being experienced with utility connections to the building.

A similar pattern of settlement development and groundwater variation was observed in the Zone A ground conditions north of Orchard Station, except that all of the adjacent buildings were supported on deep piled foundations and did not undergo any noticeable settlement. Some difficulties, however, were experienced becauses of the separation of unpiled elements, principally utility connections which settled with the ground.

4.5 Typical settlements in Zone B ground conditions

Typical ground surface settlements and piezometric groundwater pressures monitored during the construction of the southbound tunnel in the area south of Newton Station (Zone B ground conditions) are shown in Figure 10. This figure exhibits similar characteristics to those of Figure 9. Groundwater recharge wells were installed to maintain piezometric pressures in the Fluvial sand (F1) within the marine clay deposit and were relatively successful in achieving this. Piezometric levels in the upper marine clay were generally maintained as a result of artificial and natural recharge of the F1 sand layer and grouting behind the temporary lining of the tunnel.

Piezometric pressures in the G4 underlying the marine clay dropped, particularly following the passage of the tunnel face and the removal of compressed air pressure.

It is important to note that heave was experienced ahead of the tunnel face. This was a result of difficulties in controlling face pressure in the soft relatively sensitive marine clay. Positive pore pressures were generated in the marine clay at the tunnel face because of pressures being greater than the 'at rest' earth pressures. The subsequent dissipation of these positive pore pressures led to further consolidation settlements following the passage of the shield and the release of compressed air pressure.

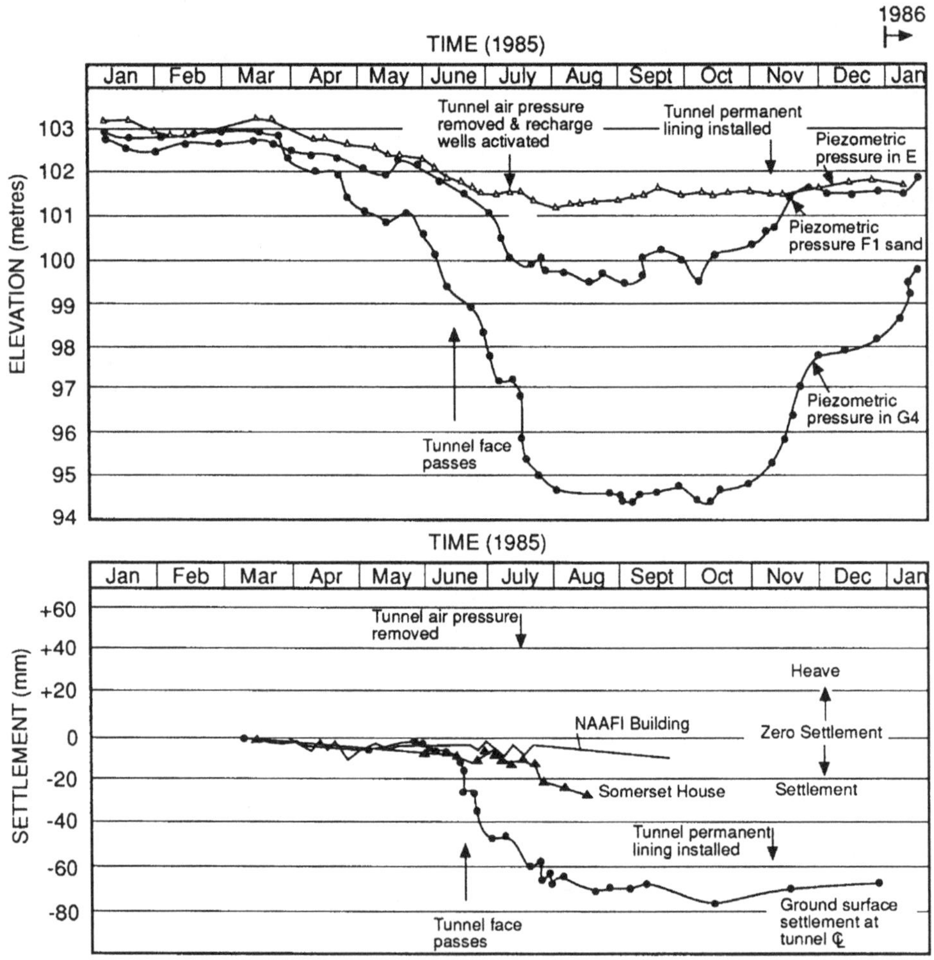

Fig. 9. Zone A: piezometric pressures and settlements

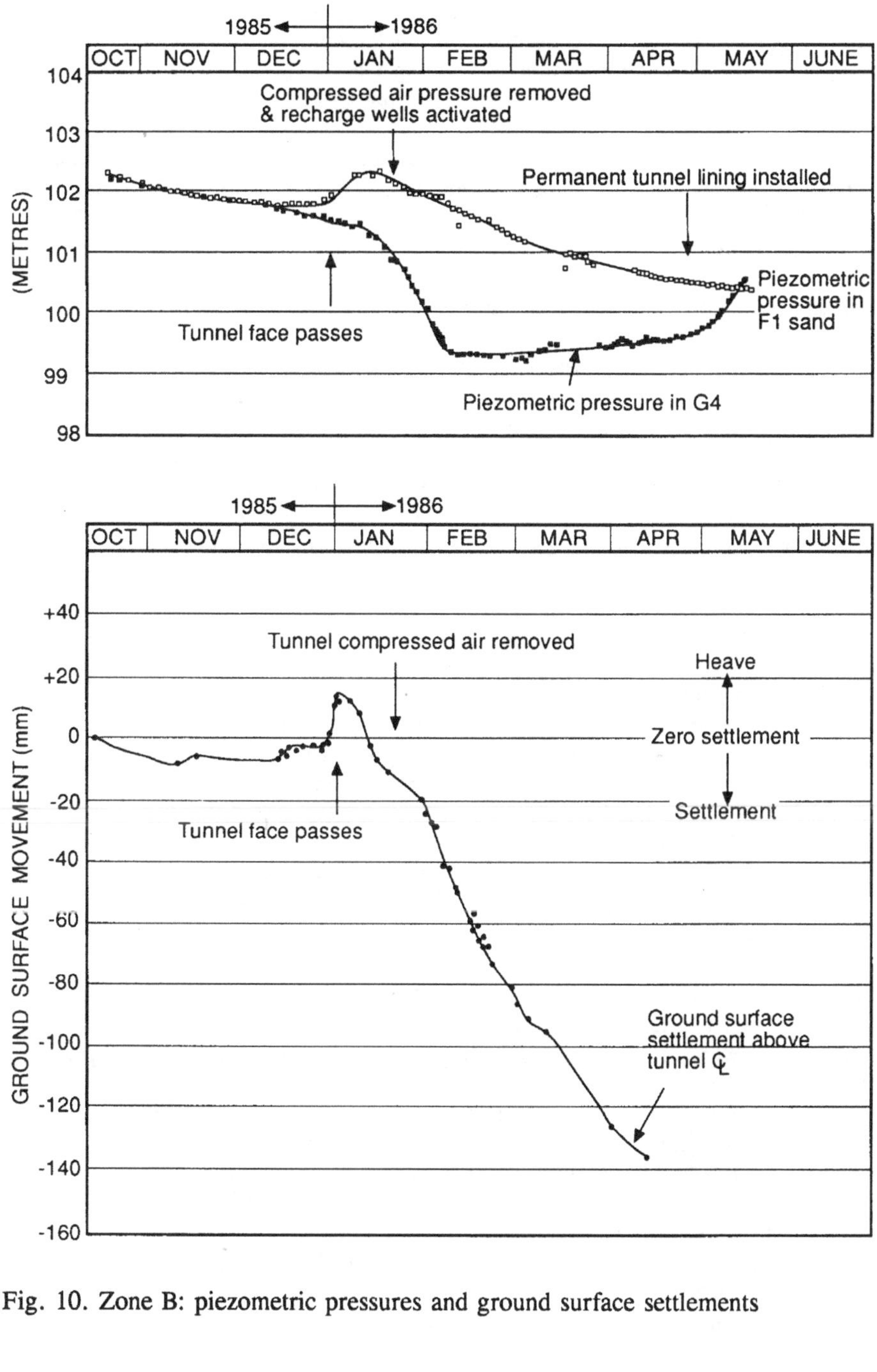

Fig. 10. Zone B: piezometric pressures and ground surface settlements

The magnitude and rate of consolidation settlements monitored on the ground surface was found to be related to the thickness of the compressible soft clay and the degree of piezometric drawdown in the underlying G4.

The construction of the permanent lining to the tunnel resulted in a recovery of piezometric levels in the G4.

Nearby adjacent low-rise structures were all founded on shallow foundations and were observed to settle in accordance with the ground surface settlement around the structures. This resulted in some slight damage being experienced in some of the structures from differential settlements between the front and back of the structures.

4.6 Typical settlements in Zone C ground conditions

Typical ground surface settlements resulting from ground loss due to tunnelling amounted to about 20-30 mm. However, the total settlement was in places much greater than this where there were problems related to loss of compressed air and to slow tunnelling progress on encountering granite boulders and dykes that required blasting techniques at the tunnel face. Ground surface settlements of up to 200 mm were observed south of Orchard Station, the result of local tunnel face instability and the consolidation of a deposit of Estuarine clay overlying the granite. Fortunately, few structures were located within the zone of influence of the ground movements.

5 Working in compressed air

5.1 The effect of ground conditions

The majority of the drives for Contract 105 were carried out with compressed air at less than one bar overpressure. In common with present practice in the UK, contractors working on the Singapore MRT sought to limit working pressures to less than one bar to avoid the need for decompression procedures and the attendant increased risk of decompression sickness. However, some driving was carried out at higher pressures in order to stabilise water-bearing F1 materials at greater depth. Decompression procedures were carried out under local regulations which were based on the Hong Kong 1975 regulations and utilised the Blackpool tables from CIRIA Report 44.

In efforts to limit both the consumption and the pressure of air used, various measures were taken to improve the ground, including jet grouting and chemical injection. These were only partially successful, and on occasion both consumption and pressure attained the maxima predictable in the absence of treatment. In some areas existing services and traffic congestion prevented full protective measures. In others, imperfections in the protective grouting, possibly connected with unidentified and thus untreated permeable lenses, may have given rise to problems. These occurrences serve to underline the necessity to provide sufficient plant to achieve those maxima in addition to treating the ground.

In Singapore the F1 material appears in lenses and pockets in the Kallang formation soft clays and at the interface with the underlying formations. The material varies from a coarse to a silty fine sand, with permeabilities from 10^{-4} to

10^{-6} m/s. The strata are impersistent, but can be up to 10 m thick. These uncertainties are not uncommon on compressed air tunnelling projects and underline the need to make adequate provision where the safety of the works is concerned.

5.2 The provision of compressed air equipment

Three distinct criteria must be satisfied in the establishment of compressed air equipment:

- maximum pressure
- maximum volume rate of supply
- standby capacity.

Where the overriding consideration is the safety of the works, the provision should be based on maximum possible values and not on probable values.

The power supply must also be secured by standby alternatives.

5.3 Determination of maximum pressure

The maximum pressure required for permeable ground is determined from $\sigma_i = \gamma_w h_i$, where h_i is the maximum possible hydrostatic head at the invert through the drive. In soft clays the required reduction in the stability number of the material in the tunnel face may be the criterion, that is, $\sigma_i > \gamma z - 3c_u$. An upper limit will be imposed by the ability of the ground to resist the uplift. If this limit is less than the maximum requirement then special measures may be necessary, either to reduce the maximum pressure requirement or to increase the sustainable uplift.

The desire to increase air pressures in an attempt to improve conditions in the invert of the tunnel face in permeable materials must be tempered with the need to limit air losses through the top of the face so that the ground is not disrupted with resulting ever higher air losses or catastrophic escape. It is, however, important that the volume capacity of the provision should be adequate at the maximum pressures that are envisaged. This is particularly important where compressor equipment has alternate serial or parallel operation modes depending on the required delivery pressure.

5.4 Determination of maximum consumption

Schenck and Wagner in 1963 set out a method of calculation for the flow of air through porous ground, taking compressibility into account. They concluded that the approximations involved made this treatment no more precise than an approximate equation for the prediction of air losses subsequently quoted by Szechy (1973):

$$Q = 70\,kif$$

where Q	=	air flow through the face at mean pressure
70	=	air/water permeability ratio
k	=	coefficient of permeability to water
i	=	hydraulic gradient
f	=	gross area at the face.

This was based on work in the temperate climate of northern Europe. In tropical climates, however, where soil temperatures are higher, the permeability ratio is somewhat lower than 70. This is due to the fact that the viscosity of water reduces more rapidly than the viscosity of air with increasing temperature. For example, the ratio is 44 at a ground temperature of 29°C.

It follows that 'the rule of thumb' assessments may be reduced by about one third in situations where comparisons are being made on the basis of the recorded permeability of the ground between temperate and tropical soils. The hydraulic gradient is taken as the pressure drop between the face and the nearest point of escape, divided by the distance. The face area is taken as the excavated area plus a nominal annulus to the rear of the order of one metre long to allow for losses around the shield, the tail, etc.

If we take a 6.03 metre diameter tunnel face at 20 m depth below the groundwater table in Singapore, the parameters are as follows:

Assuming a full face of sand,	$k = 10^{-4}$ m/s
Maximum air pressure	2 bar (3 bar absolute)
Flowpath	20 m
Area of ground exposed	47.5 m^2

then Q, the volume flow, is 12.5 m^3/min through the face area at a mean absolute pressure of (3+1) - 2 = 2 bar.

and Q_a, the volume flow at atmospheric pressure, is 25 m^3/min.

5.5 Determination of the level of provision

Total air consumption, in addition to face losses, will include mechanical loss (pipeline losses and cooling), and functional use (locking, snorer, extra ventilation). It is assumed that losses through the lining are minimal.

These may be calculated from simple volumetric factors or as percentages, and a figure arrived at for a 6 m diameter tunnel is of the order of 30 m^3/min.

The overall total is thus 25+30, i.e. 55 m^3/min or just less than 2 $m^3/min/m^2$ of the net face area.

The volumes are quoted as 'free air' values, that is, at atmospheric pressures.

In addition to the volumetric capacity to maintain this total consumption, standby capacity is also necessary. This is required not only for routine maintenance and breakdown cover, but against any emergency which might eventuate and give rise to additional requirements for air. There is no standard set for the level of standby provision, although 50% standby is often provided. Breakdown and maintenance can be covered by a provision of one or more additional machines. Beyond that the risk of requiring more capacity may be assessed, possibly with the assistance of risk analysis. Where two or more tunnels are working concurrently, the risk of both requiring maximum supply at the same time would have a relatively low probability. If 50% standby is assumed then the requirement for a 6 m diameter tunnel would be 3 $m^3/min/m^2$ of net face area.

Empirical rules of thumb for determining the level of provision are quoted by various authors, and modified by such comments as 'judging from previous experience'. The figures quoted vary from 3.6 to 7.2 $m^3/min/m^2$ of tunnel face area depending on the openness of the ground. These refer to use in the temperate climates of Europe and North America where initial experience was gained.

Table 4 shows the actual provision made on the Singapore contracts:

Table 4 Air provision on Singapore contracts

Contract No.	**Capacity provided** ($m^3/min/m^2$ of net excavation area)
104	5
105	9
107	5
108	7.5
109	5

The 5$m^3/min/m^2$ of net excavation area provisions made on contracts 104, 107 and 109 are consistent with a provision for moderately open ground in a temperate climate. On Contract 108, there was some risk of encountering more permeable ground in areas where there was no protective grouting. This required increased provision. On Contract 105, the higher provision took account of the intended use of steel rib and timber lagging temporary lining. As far as the authors are aware, these provisions, which include standby allowance, were sufficient in practice.

6 Conclusions

The data presented in this paper, consisting of field measurements made during the construction of bored tunnels for Contract 105 of the Singapore MRT shows the effect of the magnitude of the compressed air pressure on piezometric water pressures and ground surface settlements and illustrates the importance of maintaining relatively high compressed air pressures for tunnel stability and in limiting ground settlements.

Generally the settlements due to the excavation of tunnels caused less building and utility damage than the settlements due to consolidation. The settlements from consolidation tended to be widespread and have a relatively flat profile and did not produce angular distortions which can cause major structural damage. However, ground settlement did result in the separation of utilities from buildings founded on piled foundations.

Acknowledgements
The authors are indebted to Ms J Arnold of Ove Arup & Partners for her assistance in producing the figures for this paper. Thanks are also expressed to Mrs A Dixon of Soil Mechanics Associates for typing the manuscript.

References

Attewell, P.B., Yeates, J., and Selby, A.R. (1986) *Soil movements induced by tunnelling and their effects on pipelines and structures.* Blackie.

CIRIA (1982). *Medical code of practice for working in compressed air.* Report 44, third edition.

Clough, G.W. and Schmidt, B. (1981) Design and performance of excavations and tunnels in soft clay. In: *Soft clay engineering*, Elsevier, Amsterdam.

Copsey, J.P., Hulme, T.W., Kraft, B. and Sripathy, P. (1989) Singapore Mass Rapid Transit System: design. *Proc. Instn. Civ. Engrs.*, Part 1, 1989, 86, Aug., 667-707.

Glossop, N.H. (1977) *Soil deformation caused by soft ground tunnelling* PhD thesis. University of Durham.

Government of Hong Kong (1989). *Factories and industrial undertaking (work in compressed air) regulations 1975.* Revised 1978, 1981, 1982, 1989.

Government of Singapore (1985). *The Factories Act 1973 (Building Operations and works of Engineering Construction). Regulations 1985 - Part XII Excavation and tunnelling works.*

Hulme, T.W., Chapman, P.H.J. and Foo, P.S. (1989a) Singapore Mass Rapid Transit System: planning and implementation, *Proc. Instn. Civ. Engrs.*, Part 1, 1989, 86, Aug, 627-665.

Hulme, T.W., Nicholls, D., Burchell, A.J. and Stewart, D.F. (1987) Tunnel construction for the Singapore MRT. *Proc. Singapore MRT Conf.*, Institution of Engineers, Singapore, 1987, 261-278.

Hulme, T.W., Potter, L.A.C. and Shirlaw, J.N. (1989b). Singapore Mass Rapid Transit System: construction, *Proc. Instn. Civ. Engrs.*, Part 1, 1989, 86, Aug., 709-770.

Mair, R.J. (1983). Geotechnical aspects of soft ground tunnelling. *Proc. Int. Seminar on Construction Problems in Soft Soils*, NTI, Singapore.

O'Reilly, M.P. and New, B.M. (1982). Settlements above tunnels in the United Kingdom - their magnitude and prediction. *Proc. Conf. Tunnelling '82*, Instn. Min. Metall. London.

Schenck, W.R. and Wagner, H. (1963). Air consumption and cover when driving tunnels with compressed air. *Die Bautechnik*. February 1963. (In German)
Shirlaw, J.N. and Copsey, J.P. (1987). Settlements over tunnels in Singapore Marine Clays. *Proc. 5th Int. Seminar*, NTI, Singapore.
Shirlaw, J.N. and Doran, S.R. (1988). Ground movements and settlements caused by tunnelling for the Singapore Mass Rapid Transit system. *Proc. Conf. Tunnelling '88*, Institution of Mining and Metallurgy, London, 295-314.
Szechy, K (1973). *The Art of Tunnelling*. Akademiaikiado Budapest.

15 Estimation of compressed air losses

R.S. Hoad and G.P. Gittoes
Sir William Halcrow & Partners Ltd, London, UK

Abstract
This paper considers the pre-estimation of compressed air losses or consumption for tunnels driven in soft ground. Causes of air consumption are discussed and empirical methods of forecasting are compared with theoretical methods. The importance of permeability and the influence of other factors which can introduce errors in estimation are discussed. Some recent case histories are described and tentative conclusions and recommendations are given.
Keywords: Compressed Air Losses, Permeability,

1 Introduction

Whenever compressed air working is employed for tunnelling an estimate of compressed air loss must be made to identify plant and standby plant requirements. Regulations provide for a minimum exchange of air in the working environment for health reasons, but usually loss of compressed air into the ground determines the final plant size.

Contractual claims have sometimes been based on the difference between the actual air losses and those which might reasonably have been estimated by an experienced contractor at the time of tender. The issue has even been used in an attempt to justify reimbursement for additional compressor plant or to attempt to verify changed physical conditions leading to additional reimbursement for time delays, extra grouting, etc. These contractual issues may be contentious and are outside the scope of this paper; it is sufficient to say that pre-contract and post-contract estimation of compressed air losses may be important for either party to a contract.

The nature of the ground, essentially its permeability, is a major factor influencing compressed air losses. Experience of the previous construction of other tunnels in the same location or in the same geological strata is invaluable when estimating air losses. However, the form of construction also plays a part.

Engineering and Health in Compressed Air Work Edited by F.M. Jardine and R.I. McCallum. Published in 1994 by E & FN Spon. ISBN 0 419 18460 0

In soft ground conditions where compressed air working is applied, the ground is rarely uniform and in any case, soil particles, inter-granular pores, interstices, fissures, etc. all vary. A wide range of permeabilities is commonly found from *in-situ*, theoretical and laboratory methods. Air losses are also time dependent as the surrounding ground is dried and the particles loosened by air movement. In these circumstances it is not surprising that all methods for estimating compressed air losses must be labelled as imprecise.

The capacity of compressors is always stated in terms of the volume of air per minute measured at atmospheric pressure (free air). In the United Kingdom imperial units are still commonly used for this plant and a compressor may be rated at, for example, 1000 cubic feet per minute (cfm) of free air delivered at 2 bar pressure. In this paper air consumption related to the area of the tunnel face is given both in SI units as $m^3/min/m^2$ of tunnel face and for convenience also as cfm/m^2 of tunnel face. Throughout this paper air consumptions are given in terms of the volume of free air.

The method of measurement of compressed air losses can vary and depends on the type of compressed air plant used.

It is normal practice to apply a compressed air pressure to balance the hydrostatic pressure at the invert or thereabouts, in order to minimise loss of ground at the invert. In some circumstances, such as to minimise the risk of a blow-out or where the invert stratum is stable, a reduced pressure may be applied. In the formulae in this paper, pressure balance is assumed at the invert; i.e. the excess air pressure at the tunnel crown is equivalent to the tunnel diameter D.

2 Causes of air consumption

The volume of air required to maintain the pressure in the tunnel depends on the amount of leakage, arising principally from:

- the area and the permeability of the tunnel face, especially the crown
- the area around a tunnel shield
- joints in the segmental tunnel lining for the length of tunnel in-bye the lock
- use of men and muck locks.

Loss of air through the face is an important cause of losses in most circumstances. Empirical and semi-theoretical formulae are commonly used for estimating face losses.

Around the tunnel shield losses are more difficult to quantify and can be appreciable if the shield cutting edge is not buried and/or if there is no tail seal between the segmental lining and the shield tail.

Loss of air through the completed tunnel lining will depend on the design and working methods employed. Factors with a major influence include delayed grouting or caulking, and use of neoprene gaskets or strips of hydrophilic compound to seal joints between segments.

Loss of air through locks can readily be calculated from the size of the lock, pressure and frequency of use but is not generally significant when compared with face losses occurring in open ground.

3 Empirical methods

The usual means of estimating the capacity of plant required is by means of empirical formulae based on previous experience.

3.1 Hewett and Johannesson

Hewett and Johannesson (1922) studied a number of early tunnels lined with cast iron segments and then developed the following formulae, related to the area of the tunnel face. These quick formulae introduced more than 60 years ago are still in general use.

In the original Imperial units:

$Q = 12\,D^2$ for fairly average open ground
$Q = 24\,D^2$ for open sand and gravel

where Q is the capacity in cfm (cubic feet per minute)
D is the external diameter of the tunnel in feet.

Alternatively, in SI units:

$q = 3.66d^2$ for fairly average open ground
$q = 7.32d^2$ for open sand and gravel

where q is the capacity in m^3/min
d is the external diameter of the tunnel in metres.

Their experience showed that in adverse conditions rather more compressed air capacity was needed. It led to their general recommendation, still widely employed, that 50% additional reserve compressed air plant capacity should be provided above the maximum expected to be needed for a tunnel drive.

3.2 Other empirical formulae

More recently other references have suggested a variety of formulae and descriptions of ground conditions including Kirkland (1984), in Imperial units:

$Q = 5\,D^2$ for pre-treated ground
$Q = 15D^2$ for sands and silts
$Q = 30D^2$ for coarse ballast.

A sample of the ground descriptions and the appropriate losses is shown in Table 1, derived from the above and other quoted references.

Table 1. Empirical formulae

		Expected loss per unit of face area	
Reference	Ground description	$m^3/min/m^2$	cfm/m^2
Hewett & Johannesson (1922)	fairly average open ground	4.7	164
	open sand and gravel	9.3	328
Richardson & Mayo (1941)	gravel (worst conditions)	19	670
Megaw & Bartlett (1981)	moderately open ground	3.6 to 6.0	130 to 210
Kirkland (1984)	pre-treated ground	1.9	68
	sands and silts	5.8	205
	coarse ballast	11.6	410

4 Semi-theoretical formulae

4.1 Schenck and Wagner

Schenck and Wagner (1963) provided semi-theoretical formulae for estimating air losses based on Darcy's filter law and modified by the Hagen-Poiseulle law.

Although capillaries are irregular making a full mathematical assessment impractical, Schenck and Wagner developed equations to take account of the air flow process through ground as opposed to liquid flow. Once capillary stress is overcome and air begins to move, water is driven out with increasing speed, and eventually particles are loosened and the permeability and consequent air loss increases. The equations apply only for laminar flow which can be expected in material where $k_w < 1 \times 10^{-3}$ m/s, k_w being the water coefficient of permeability of the strata.

The full equation takes into consideration the following factors in calculating air consumption:

- permeability of strata to air movement
- excess air pressure in terms of balancing hydrostatic pressure at the invert
- reduction in the air entry cross sectional area in relation to the capillary entry resistance and the form of the air flow field.

The full Schenck and Wagner formula is:

$$Q = w.\frac{k_L}{\gamma_w}.\frac{P_1-P_2}{L}.\frac{P_1+P_2}{2}.\frac{F}{P_1}.\frac{D-h_{KAP}}{D}. \tag{1}$$

where Q = compressed air loss at tunnel pressure in m^3/min

w	=	correction factor based on experimental results and set at 2
k_L	=	coefficient of permeability of the ground to air. Based on Darcy's Law k_L approximates to $70k_w$.
P_1	=	absolute pressure at the tunnel face which is equivalent to the absolute pressure generated by the water table at the tunnel axis
P_2	=	absolute pressure at the surface
L	=	length of air flow field, which is taken from the crown of the tunnel to the point where air can flow freely.
F	=	area of excavated face
D	=	excavated diameter of the tunnel face
h_{KAP}	=	the capillary rise in metres in the pores of the soil. Values used in calculating air consumption need to be based on typical values relative to soil particle size
γ_w	=	unit weight of the water

Additionally a conversion factor is needed to change the volume of air lost at absolute tunnel pressure to absolute atmospheric pressure (free air).

A simplified form of their equation is quoted by Schenck and Wagner as:

$$Q = 2k_L.i.F \tag{2}$$

where i = hydraulic gradient equivalent to $(P_1\text{-}P_2)/\gamma_w L$.

Based on Boyle's Law in order to convert the volume rate of air loss at absolute tunnel pressure to a free air volume (i.e. at absolute atmospheric pressure), Q should be multiplied by P_1/P_2.

4.2 Modifications to Schenck and Wagner

Modifications to the Schenck and Wagner expression have been reported by Buchanan (1984) based on Krabbe (1968), providing additional empirical factors relating to the ground and the form of construction. These modifications were admitted as being of an arbitrary nature although they provide a first estimate which takes both the form of construction and the length of tunnel into account. The equation is most simply shown as:

$$Q = 2.k_w 70.\frac{(P_1-P_2)}{\gamma_w.L}.F.\frac{P_1}{P_2}(n_a+n_b+n_c) \qquad (5)$$

where n_a relates to the tunnel face, with a value between 1 for soft ground, also described as homogeneous, and 0 for hard ground

n_b relates to the tail seal of the shield, with a value of 0 in hard ground or if there is a plenum chamber at the face; in soft ground, $n_b = 0$ with tail seal or between 1 and 2 without a tail seal.

n_c relates to a typical 500 m length of tunnel behind the face in-bye the lock. $n_c = 0$ to 1 for grouted and caulked lining and $n_c = 1$ to 2 for grouted lining without caulking.

Taken in an extreme instance this modification could increase the estimate of air loss very significantly, perhaps by a factor of 5 for a short length of tunnel.

5 Permeability measurement

In soft ground, hydraulic permeabilities can be obtained by several methods including:

- *in-situ* falling head tests
- *in-situ* pumping tests
- laboratory constant head tests
- theoretical prediction based on particle size, such as Hazen's formula used on sands
- Laboratory consolidation tests.

The most representative permeability test is generally recognised to be *in-situ* pumping but this often requires large-diameter closely spaced boreholes. The cost of such special pumping tests may be significant. Frequently, however, there are several reasons for attempting to determine permeability especially in urban areas where inevitable settlement of structures needs to be assessed.

Terzaghi and Peck (1967) included tabular information on the range of coefficients of permeability for various soils and the most appropriate methods for performing the permeability tests on these soils.

It is not unusual for permeabilities based on laboratory tests or on *in-situ* falling head tests or even on theoretical predictions, to vary widely, and frequently ground conditions and hence permeabilities vary within a short distance along a tunnel drive.

Pumping tests may well be the most reliable method of investigating 'mass' transmissivity of an aquifer, but attention also needs to be paid to the less permeable strata and the degree to which they are continuous and can act as a barrier to compressed air losses. Expert geotechnical opinion should be sought in order to determine a range of permeabilities for a given tunnel drive. These values of permeability can then be used for estimation of compressed air losses.

Where mixed face conditions occur a further judgement has to be made, but it is the nature of the ground at the crown which will have the greatest influence on air losses.

6 Factors introducing errors in estimation

If a permeable stratum lies close to the tunnel this will form a sink attracting compressed air losses. Although the presence of such a layer may be suspected it is not usual to attempt to confirm this by probing because additional sealing difficulties will result.

As an example, Figure 1 shows a situation where very high compressed air loss would be expected.

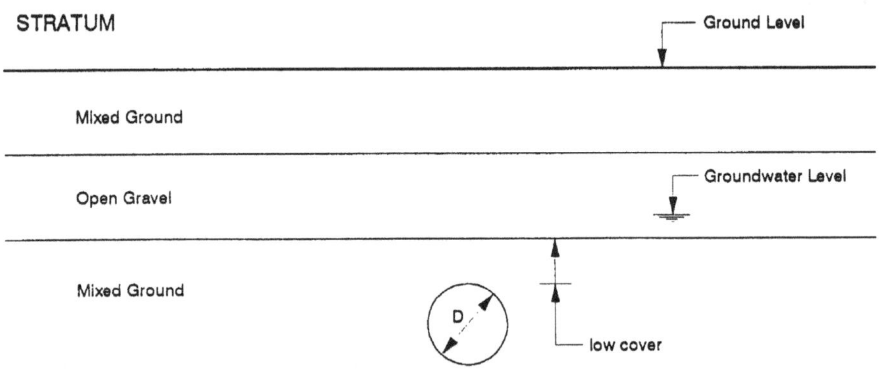

Fig. 1. Situation where high air loss would be expected

Inadequately sealed boreholes or buried utilities, sewers or even basements can form a short route for air escape.

Modern tunnel boring machines are much longer than hand shields and the area of ground exposed to tunnel pressure is therefore much larger than with hand shields.

Delaying back-grouting to the end of a shift exposes more of the ground for longer, particularly with rapidly advancing TBMs. In some instances, the air escape loosens capillaries and results in further instability, increasing air losses and leading to more difficulty with grouting and sealing.

Air can escape through the joints in a segmental tunnel lining. The problem is similar to water inflow to a completed tunnel. Leakage may occur in local patches mainly on open circumferential joints where there is an unfavourable combination of the build of the lining and imperfect grouting.

Arbitrary rules of thumb without any allowance for the diameter, joint width or the number of joints have been suggested as:

For uncaulked tunnel: 0.14 to 0.2 m^3/min (5 to 7 cfm) per ring
For caulked tunnel: 0.06 to 0.08m^3/min (2 to 3 cfm) per ring

We found no published material to substantiate these figures. Clearly, use of neoprene gaskets in a shield drive might be expected to produce a very well sealed tunnel at least as good as caulking. Hydrophilic material would probably not be so successful in the dry condition while the air was turned on. Longitudinal joints between tunnel segments would tend to be closed and may not represent a significant source of loss. However, losses through the circumferential joint between tunnel rings are dependent on such factors as the joint width, the nature of the circumferential joint packing, the alignment, and the large shove forces required to advance a heavy TBM.

CIRIA Report 81 *Tunnel Waterproofing* (1979) provided a basis for a different approach to attempting to assess losses from a completed tunnel. By assuming a simple ratio of 70 between permeability to air and permeability to water, the wettest tunnel classification 'E' with water leakage 100 litre/day/m^2 would be equivalent to an air loss of 7000 litre/day/m^2. In, for instance, a 2.9m ID tunnel with 3.2 OD this approximates to 140 cfm per 100 m length of driven tunnel or 0.8 cfm per ring, assuming 0.6 m long rings. This cannot be judged as a precise comparison but it seems to discount the arbitrary rules of thumb above.

For all aspects of estimation a careful study of case histories is needed in order to attempt to improve reliability.

7 Recent case histories

7.1 London Docklands - Royal Docks Drainage Phase 4

Compressed air working has been used for sections of several recent drainage contracts in London Docklands, although slurry TBMs were usually chosen for the main drives in water-bearing ground.

On Royal Docks Drainage Phase 4, compressed air working at around 22 psig was used for the construction of approximately 1400 m of 1.8 m ID precast concrete segmental tunnel. A description of this contract was provided by Ferguson *et al.* (1989) and ground conditions in the area are described by Howland (1991). The Contractor was Miller Construction Ltd. and the Engineer was the London Borough of Newham with advice from Sir William Halcrow & Partners Ltd.

Ground conditions ranged through three distinct phases:

Upper Chalk
Thanet Sands
Woolwich and Reading Beds.

The Contractor employed a back-hoe excavator mounted in an 8 m long articulated shield with a conveyor.

On this contract there are several interesting features which have a bearing on compressed air losses. However, in this case history detailed examination of compressed air losses is restricted to the length of drive in Thanet Sands where relatively uniform conditions apply and good records are held.

The Thanet Beds in the South East of England consist mainly of sands but with some clay and silt content especially towards the base of the formation. The sand is quartzose with some glauconite and where the beds are argillaceous the clay is evenly distributed. Although subdivisions within the Thanet Beds have been described elsewhere, at the Royal Docks Phase 4 site the majority of the boreholes show Thanet Beds to be a very dense grey silty fine sand, with some zones of fine sands without silt. The Thanet Beds themselves exhibited areas of low silt content, but the gradings of samples taken from the face remained within the range given by the borehole samples.

Falling head permeability tests as part of the site investigation generally gave values ranging from 4.4×10^{-7} to 8.9×10^{-7} m/s for slightly silty fine sand with a value of 1.6×10^{-6} m/s in another borehole where the description was very silty fine sand.

Empirical formulae for estimation of compressed air losses give the following results:

1. Hewett and Johannesson, using D as 7 feet and the multiplier of 12 for fairly average open ground gives 590 cfm, or 16 m^3/min.
2. Schenck and Wagner's simplified formula with permeability of 1.6×10^{-6} m/s, P_1 approximately 2.5 atm. abs., and L as 14 m, gives 5 cfm. In terms of loss per face area this is equivalent to 1.3 cfm/m^2 or 0.04 m^3/min/m^2.

During the early stages of the tunnel drive, compressed air losses were held below 600 cfm (170 cfm/m^2). As the drive progressed so the losses increased peaking at 3,300 cfm (921 cfm/m^2) despite additional grouting at weekends and use of bentonite as a further grouting seal. There are several factors to explain these increases as follows.

The tunnel was driven in Thanet Sands but with approximately only 2m cover over the crown to the Floodplain Gravels, which have a permeability two or three orders of magnitude greater than the Thanet Sands.

Air would move slowly through the relatively impermeable Thanet Beds until it reached the permeable Flood Plain gravels where it would be released. Piping systems could then form and the compressed air losses would rise, the air simply bubbling through the groundwater within the gravel. This is taken into account in the Schenck and Wagner formula by the factor L. For instance with a permeable gravel layer only 2 m above the crown instead of 14 m to the surface, air loss estimates then increase by a factor of 7.

The use of an 8 m long tunnelling shield compared with a conventional hand shield meant that a much greater area of the ground than usual was exposed. This is illustrated in Table 2.

Table 2. Comparison between exposed ground area for hand shield and TBM of 2m outside diameter and 0.6m long rings.

	Hand shield	**TBM**
Length (m)	2	8
Progress per shift (rings)	5	10
Face area (m^2)	3.1	3.1
Shield cylindrical area (m^2)	12.6	50.4
Circumferential area per shift (m^2)	18.9	37.8
Overall area exposed (m^2)	34.6	91.3

Hence, when grouting is done only at the end of a shift the area of ground exposed during a shift using the rapid advance rate of the back-hoe TBM system is almost three times that of a conventional hand shield.

In the confined space of 1.8 m diameter tunnel with an overhead conveyor feeding rail skips, there is virtually no room to grout over the tunnel crown until rings emerge beyond the conveyor. Even if perfect grouting is achieved beyond the conveyor, there is still a significant length of ground exposed where piping can arise over, say, a weekend stoppage. Grouting at weekends makes it difficult to analyse many of the site records, but where weekend grouting was not performed, losses over a weekend increased by around 5 cfm per hour from the rate on the Friday night (Table 3).

These values represent an average rate of increase in losses of about 4.6 cfm per hour when adjusted for short weekends.

Hydrophilic compound in the form of strips glued to the segments before erection was introduced into the circumferential joint between rings in order to attempt to restrict air losses. However, in an air-drying situation this may not have been very effective as a water/air seal until air pressure was lowered. Pointing was carried out to restrict air losses, but analysis shows a rate of approximately 2.6 cfm per ring of driven tunnel; this is for an uncaulked and probably less than perfectly grouted condition. This is equivalent to approximately 0.46 cfm per m of circumferential joint (Table 4).

Table 3. Increase in rates of air loss over weekends

	Losses recorded (cfm)		
Week no.	**Start of week***	**End of week***	**Weekend loss**
56	-	1033 (Sat)	627
57	1660	2730 (Sun)	175
58	2905(Tues)	1532	128
59	1660	2760 (Sat)	40
60	2800	2000 (Sat)	223
61	2223	2086	404
62	2490	2573	145
63	2718		

*These records were generally taken during the weekends.

Table 4. Compressed air losses per ring of driven tunnel

Week no.	Start of week* cfm	End of week* cfm	Rings driven	Loss per ring cfm
55	585	830	69	3.6
56	850	1033	105	1.7
57	2124	2490	85	4.3
59	1660	2282	91	6.8
60	2124	2158	87	0.4
61	2112	2124	115	0.1
62	2490	2656	77	2.2

Average over 629 rings is 2.6 cfm per ring

*These records were generally taken during the working shifts

This analysis is of course a simplification because it assumes that the loss from the face remains constant.

Table 5. Compressed air consumptions for UK road tunnels

Project (Dates)	**Drive**	**Ext dia. m**	**Provision m^3/min ($m^3/m^2/min$)**	**Use m^3/min ($m^3/m^2/min$)**	**Pressure bar**	**Note**
Blackwall (1892-97)	Main	8.23	283 (5.32)	283 (5.32)	2.55	A
Rotherhithe (1904-08)	Main	9.17	472 (7.15)	- (-)	2.41	B
Dartford (1956-63)	Pilot	4.11	227 (17.04)	227 (17.04)	2.48	C
	Main	9.30	595 (8.76)	127 (1.88)	1.86	
Clyde (1957-64)	Main 1	9.86	810 (5.31)	198 (1.30)	2.07	D
	Main 2	9.86	810 (5.31)	85 (0.56)	1.86	
Blackwall 2 (1961-67)	Pilot	2.59	- (-)	265 (25.14)		E
	Main	9.30	560 (8.34)	150 (2.29)		
Tyne (1961-67)	Main	10.21	421 (5.14)	57 (0.69)	2.90	F
Dartford 2 (1972-73)	Pilot	4.00	- (-)	35 (2.78)	2.34	G
	Main	10.11	- (-)	85 (1.06)	1.06	

Notes to Table 5.

General. The capacities quoted are measured at atmospheric pressure.

A. The Blackwall tunnel was driven through open gravel. Whilst passing through the gravel the whole capacity of the plant was used.

B. The Rotherhithe tunnel was driven mostly through close ground. No important leakage occurred and the plant was usually only run at half its capacity.

C. For the Dartford pilot tunnel, the maximum losses were encountered in an area of broken chalk under the Kent river bank. The main tunnel was driven through ground treated with cement grout. The compressed air plant was never

(Notes to Table 5 continued)

run at full capacity. Ground conditions included silty clay with interlayered peat, Thames gravel and chalk.

D. The ground conditions for the Clyde tunnel comprised silt, boulder clay, sandstone and shale. The full capacity of the plant was required during a blow. Ground treatment was carried out from the surface and the pilot tunnel.

E. The ground conditions prior to ground treatment for the Blackwall tunnel are as given below:

	Permeability (m/s)
sand and gravel	1×10^{-3}
silty sands & Woolwich Beds	5×10^{-5}
clay silt layers	1×10^{-7}
fine sands	1×10^{-4}

Permeabilities were required to be reduced to an average value of 3×10^{-6} m/s by ground treatment as described by Kell and Ridley (1966).

F. The ground conditions for the Tyne tunnel comprised alluvial deposits, boulder clay and coal measures. The highest loss of compressed air occurred on the pilot tunnel in fissured ground of clay and decomposed shales. Ground treatment was carried out from the pilot tunnel for the main drives.

G. The ground conditions for the Dartford 2 tunnel comprised chalk, alluvial deposits and gravels. The pilot tunnels were driven through ground which had in part been treated by injection grouting to reduce air consumption, water inflow and to stabilise alluvial deposits.

7.2 Other UK records

Often compressed air losses are small and so not recorded, as was the case on several other recent contracts in London Docklands and on the Stansted Airport Rail Link tunnel. However, Table 5 has been compiled to give an indication of the difference between capacity provided and capacity used on major road tunnel contracts.

7.3 Cairo Wastewater Project: Contracts 3 and 12

The Cairo Wastewater Project is described by Elliott and Spolton (1985). The Contractor for Contracts 3 and 12, Cairo Wastewater Consortium (CWC), employed Sir William Halcrow & Partners for technical advice, including preparation of Schenck and Wagner estimates. The following analysis of compressed air losses is based on site records which were recently made available. The major compressed air tunnel drives are shown in Table 6.

The ground was characterised in terms of soil units as follows:

A made ground

B1 silty clay and clayey silts
B2 interbedded clay, silt and sand
B3 very silty sand (15 to 33% fines)
C1 sand (15% fines)
C2 sand/gravel (15% gravel)

Boundaries between the soil units were not always distinct, but for most of the tunnel drives the tunnel crown lies in unit C1 which dictated compressed air losses. This soil unit C1 has a range of D_{10} size from 0.03 to about 0.3mm and therefore varies from soil which would cause little loss of air to soil which can cause significant air loss.

Part of the drive from shaft 20 was in rock. Around shaft 16, soil type C1 was absent and the tunnelling was in types B2 and B1.

Table 6. Compressed air drives: Cairo Contracts 3 and 12

Contract 3 Shaft Numbers	**Rings no.**	**Length (m)**	**Finished ID (m)**	**Tunnel ID (m)**	**OD (m)**
5A2-5A3-5A4-5A5	1773	1076	1.8	2.20	2.64
5A2-5A2VI	183	120	1.2	1.62	2.06
4A1-4	160	111	2.8	3.17	3.62
3VI-3	172	114	1.2	1.62	2.06
5A1-5	194	150	1.8	2.87	3.32
5A1-5A2	489	350	1.8	2.87	3.32
Contract 12 Shaft Numbers					
20-19*	154	125	4.0	4.57	5.15
19-18*	1007	810	4.0	4.57	5.15
16-15*	525	420	4.0	4.57	5.15
16-17-18*	762	620	4.0	4.57	5.15
19-19A1*	671	420	1.2	1.62	2.06
17VI-17*	528	325	1.2	1.62	2.06

* Tunnel drives where gaskets were used.

Median permeability estimates (m/s) for soil type C1 were:

	Contract 12	Contract 3
In-situ falling head tests	1.5×10^{-6}	4×10^{-6}
Laboratory constant head tests	3×10^{-5}	3×10^{-4}
Hazen's formula	1.3×10^{-4}	3×10^{-4}

The *in-situ* falling head test results have been discounted in this analysis because they could be expected to be less reliable than other methods (see Terzaghi and Peck). Using geotechnical advice, 2 x 10^{-4} m/s has been used in the analysis for Contract 12, and 3 x 10^{-4} m/s for Contract 3.

Table 7. Summary of Schenck and Wagner estimates for Cairo tunnels

Shaft	Tunnel OD(m)	Soil Unit	Permeability (m/s)	Pressure (bar)	Losses in: m^3/min	$m^3/min/m^2$	cfm	cfm/m^2
15*	5.15	B3-C1	$2x10^{-4}$	1.4	62.4	2.99	2200	105
16	5.15	B2	$2x10^{-6}$	1.3	0.5	0.03	17	0.8
17	5.15	C1	$2x 10^{-4}$	1.3	82.2	3.95	2900	140
18	5.15	C1	$2x10^{-4}$	1.3	83.1	3.99	2930	140
19*	5.15	C1/B2	$2x10^{-4}$	1.4	30.1	1.45	1060	50
20	5.15	Rock	$2x10^{-6}$	1.4	0.8	0.04	28	1.3
17	2.06	C1	$2x10^{-4}$	1.0	8.7	2.60	310	93
17VI	2.06	B3	$3x10^{-5}$	0.9	0.2	0.06	7	2.1
19	2.06	C1	$2x10^{-4}$	1.0	8.7	2.61	310	93
19VI	2.06	B3	$3x10^{-5}$	0.5	0.2	0.06	7	2.1
19V2	2.06	A	$3x10^{-5}$	0.4	0.2	0.07	7	2.1
19A1	2.06	A	$3x10^{-5}$	0.5	0.2	0.06	7	2.1
19A2	2.06	A	$3x10^{-5}$	0.4	0.2	0.07	7	2.1
5	3.32	C1	$3x10^{-4}$	1.3	35.7	4.13	1260	145
5A1	3.32	C1	$3x10^{-4}$	1.3	42.1	4.86	1490	170
5A2	3.32	C1	$3x10^{-4}$	1.2	48.2	5.56	1700	200
5A2	2.64	C1	$3x10^{-4}$	1.2	29.1	5.31	1030	190
5A3	2.64	C2	$4x10^{-4}$	1.0	35.9	6.56	1270	230
5A5	2.64	C1	$3x10^{-4}$	0.4	7.0	1.28	250	46

* Reduction in face area has been assumed in order to take account of mixed tunnel face conditions.

Example calculation for Shaft 18. Using the following values in equation (1):

$w = 2$
$k_L = (70).(2.10^{-4}).(60)$ (m/min)
$\gamma_w = 9.81$ (kPa)
$P_1 = 234$ (kPa $\equiv$ 1 + 1.3 bar)
$P_2 = 101.3$ (kPa $\equiv$ 1 bar)
$L = 8.9$ (m)
$F = \pi\, 5.15^2/4$ (m²)
$D = 5.15$ (m)
h_{KAP} is taken as 0.28 (m)

gives at tunnel pressure $Q = 35.97$ m³/min
correcting for free air pressure $Q = 35.97 \times 234/101.3 = 83.1$ m³/min

Table 8. Summary of actual losses

Near Shaft	Tunnel OD	Max.* cfm	Min.* cfm	General increase in losses				Rate cfm/ ring
				Ring	cfm	Ring	cfm	
15	5.15	10,500	6,000	198	5,000	497	7,800	9.3
16	5.15	5,000	1,200	15	2,000	50	3,200	20
17	5.15	5,000	3,000	475	3,600	522	4,500	19
18	5.15	10,500	8,000	685	8,000	762	9,500	19
18	5.15	15,500	9,000	932	12,600	997	14,500	29
19	5.15	4,000	3,000	70	3,200	121	3,600	7.8
19	5.15	6,000	5,000	582	4,500	627	5,500	22
20	5.15	2,500	700	169	400	500	2,000	4.8
17	2.06	3,500	400	161	700	492	3,000	6.9
17VI	2.06	1,500	200	20	700	81	600	Nil
19	2.06	2,000	400	536	800	615	950	1.9
19A1	2.06	2,000	200	60	500	262	850	1.8
5	3.32	9,000	3,000					Nil
5A1	3.32	9,200	4,000	25	4,000	146	6,500	21
5A1	3.32	8,500	1,000	51	6,300	88	7,200	24
5A2	3.32	9,500	4,300	390	4,200	473	4,700	6
5A2	2.64	4,000	300	20	600	84	2,500	29
5A3	2.64	7,000	4,000	561	3,000	653	4,000	11
5A4	2.64	2,500	1,000	732	2,500	891	6,000	22
5A5	2.64	7,000	4,300	1594	4,500	1727	5,500	7.5

* These are general records; localised maxima and minima are not shown

Soil unit B2 occurs in the crown of drive near Shaft 16 and a permeability of $2x10^{-6}$ m/s has been used in the analysis for tunnel drives 16-17 and 16-15.

The range of estimates resulting from use of the Schenck and Wagner equation is shown in Table 7, whereas Table 8 shows actual losses and a general rate of increase in loss with length of tunnel driven.

It should be noted that a significant degree of subjective judgement has been applied to arrive at the values in Table 8 in order to try to avoid the more exceptional localised results. For instance pocket holing was used near shaft 5A3 and ground treatment near most shaft locations may also have affected losses.

In Contract 3, where gaskets were not used, the rate of increase in losses averaged about 15 cfm/ring whereas in Contract 12 (with gaskets) the average rate of increase in losses was 12 cfm/ring. However, the average permeabilities where these results were taken, was in the ratio of approximately 2:1 for Contract 3 compared with Contract 12. With this ratio in mind when the circumference of the tunnels is taken into account, the benefit of gaskets is not apparent, i.e.

Contract 3:	without gaskets, average 1.65 cfm per metre of circumferential joint
Contract 12:	with gaskets, average 0.81 cfm per metre of circumferential joint.

In any given tunnel drive the maximum value for compressed air losses is a sum of both losses at the face and losses from the completed length of tunnel. Variations between the maximum and minimum amounts also depend on the timing and efficiency of grouting. The minimum, i.e. that after backgrouting, can be expected to be closest to a theoretical estimate like the Schenck and Wagner equation.

Krabbe's factors can be applied to Schenck and Wagner predictions and are compared with actual losses in Table 9.

A further relationship has been assumed in the last column of Table 9. From the above average losses per metre of circumferential joint for Contract 3 and Contract 12 an overall figure of 1 cfm per metre of circumferential joint has been assumed for simplicity. Depending on the distance from the shaft that the maximum loss occurred, the extra losses through the circumferential joint were then calculated and added to the Schenck and Wagner estimate.

Clearly there is a very wide range of both the actual and the estimated losses (which depend on assumed permeability estimates). It is not reasonable to draw conclusions over such a narrow band of assumed permeabilities. Nevertheless Figure 2 has been completed to show the results in terms of best fit graphs. This illustrates a similarity between the actual minimum losses and the Schenck and Wagner estimates. When the losses from the completed length of tunnel are taken into account, the estimates approach the actual maximum.

Table 9. Comparison between actual and estimated losses

			Loss in cfm/m² of face area				
Near shaft	**Drive No**	**Permeability m/s**	**Actual min.**	**Actual max.**	**Schenck & Wagner**	**Krabbe(1)**	**S&W modified**
15		$2x10^{-4}$	290	500	105	520	470
16		$2x10^{-6}$	57	240	0.8	3	94
17	1	$2x10^{-4}$	140	240	140	700	350
18		$2x10^{-4}$	380	740	140	700	910
19	1	$2x10^{-4}$	140	190	50	250	93
20		$2x10^{-6}$	34	120	1.3	4	340
17	2	$2x10^{-4}$	120	1000	93	460	1060
17VI		$3x10^{-5}$	60	450	2.1	8	180
19	2	$2x10^{-4}$	120	600	93	370	1190
19A1		$3x10^{-5}$	60	600	2.1	11	190
5		$3x10^{-4}$	350	1000	145	730	220
5A1		$3x10^{-4}$	120	980	170	680	370
5A2	1	$3x10^{-4}$	500	1100	200	1000	690
5A3		$4x10^{-4}$	180	460	230	1150	670
5A5		$3x10^{-4}$	780	1300	46	230	1200
5A2	2	$3x10^{-4}$	55	730	190	760	380

(1) From equation (3) using: $n_a = 1$ except for shaft 20, where $n_a = 0$;
$n_b = 2$;
$n_c = 1$ at start of drive, or $n_c = 2$ at completion

(2) To the value derived from the Schenck and Wagner equation, an addition of 1 cfm per metre of circumference has been added.

8 Conclusions

Simple empirical methods of estimating compressed air losses have been in use for some time, and they are still useful for a quick first estimate.

More theoretical methods such as that of Schenck and Wagner depend on an assumed permeability of the ground. Soft ground tunnels are commonly driven through variable ground, and permeability to water, let alone air, cannot be determined with certainty. Hence these methods can only give an order of losses; and this is often rather less than actual losses.

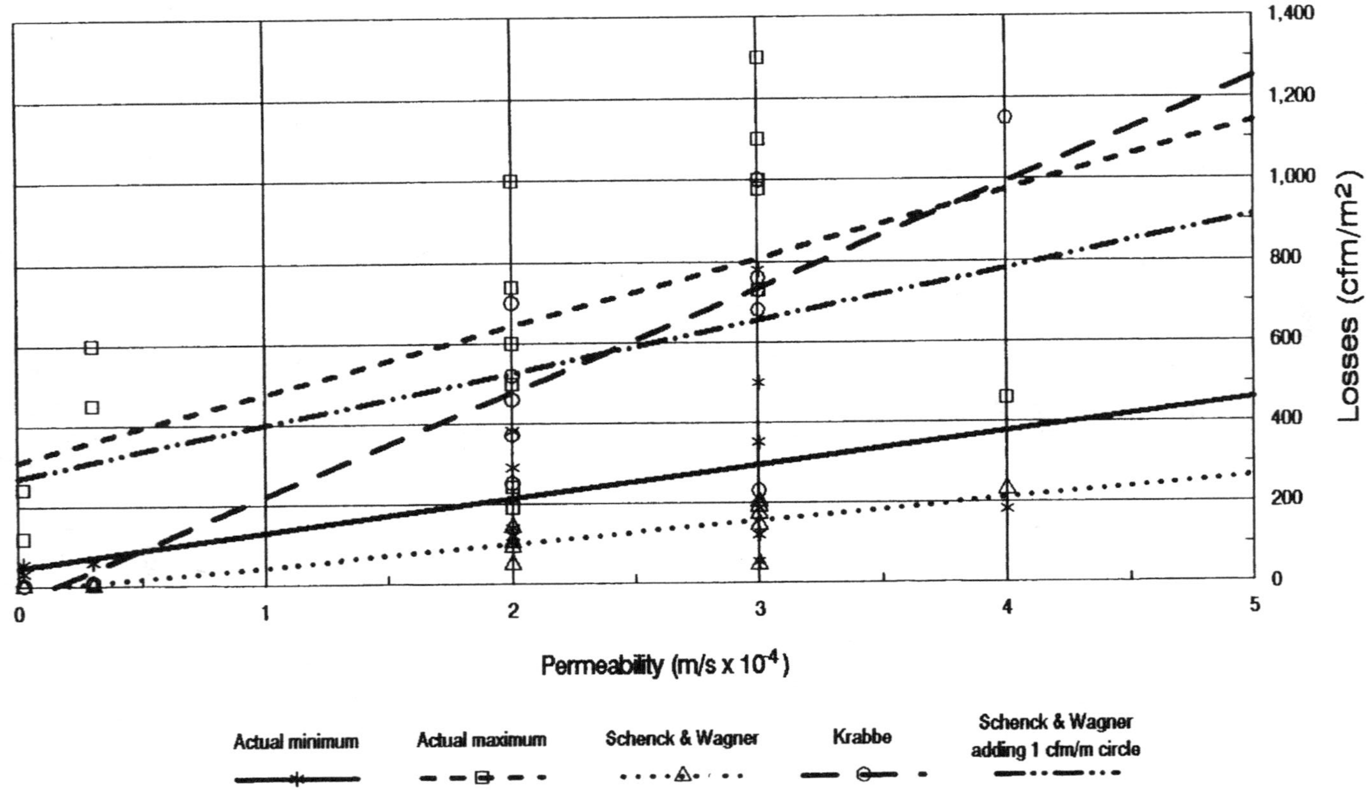

Fig.2. Comparison between actual and estimated losses

It would be very difficult to find uniform ground conditions where theory can be tested in the field. The case histories studied in this paper are not ideal. In Royal Docks Phase 4, the tunnel was driven in relatively uniform Thanet Beds but the compressed air losses were increased both by the proximity of more permeable Flood Plain Gravels and by the working methods. In Cairo Wastewater Contracts 3 and 12 the C1 type ground (sand) had a variation in D_{10} value suggesting that permeability and compressed air losses could vary widely. For Contracts 3 and 12, there is no verification that the ground actually encountered was the same as for the mean permeability value from the site investigation. In practice the actual compressed air losses changed rapidly during the course of tunnel drives and it was difficult to determine reasonable maxima and minima for analysis.

The method of construction can have an influence on compressed air losses. For instance, relatively long TBMs achieving rapid progress rates but with limited access for back-grouting, result in a large area of ground being exposed for a long duration. This can provide an opportunity for piping systems to be set up increasing the rate of air losses and making it more difficult to seal the tunnel by conventional back-grouting.

As well as losses occurring at the tunnel face, losses are apparent in the completed but uncaulked tunnel and in the case histories studied above ranged from 0.5 to 1.6 cfm per metre of circumferential joint. Use of gaskets did not appear significantly to reduce air losses on Cairo Contract 12 compared with Contract 3.

For the Cairo records, the Schenck and Wagner formula gives estimates which approximate to the actual minimum losses in tunnel drives. The addition of an allowance for losses in completed lengths of tunnels at a rate of 1 cfm per m of circumferential joint produced a resultant similar to the actual maximum recorded. However, there is a very wide scatter of individual results even for a narrow range of permeability. The permeability assumed in this analysis was not verified during construction.

9 References

Buchanan, J. (1984) *The use of compressed air.* British Tunnelling Society course on soft ground tunnelling, September 1984

Elliott, I.H. and Spolton, R.L. Greater Cairo Wastewater project: design of east bank stage 1 works tunnels (August 1985), *Proceedings of the Institution of Civil Engineers*, Part 1, Design and Construction, Volume 78

Ferguson, P.A.S., Crabtree, G., Talbott, T. and Hughes, C. (1989) Difficult Docklands contracts prompt creative solutions. *Tunnels and Tunnelling.* Volume 21, No. 11, November, Morgan Grampian (Construction Press) Ltd, London

Hewett, B.H.M. and Johannesson, M, (1922) *Shield and Compressed Air Tunnelling.*
McGraw-Hill Book Company Inc, New York

Howland, A.F. (1991) London's Docklands: engineering geology. *Proceedings of Institution of Civil Engineers*, Volume 90, Part 1.

Kirkland, C.J. (1984) Design and construction of urban tunnels - compressed air. *Advances in Tunnelling Technology and Subsurface Use*, Volume 4, No 4, Pergamon Press Ltd
Kell, J, and Ridley, G. (1966) Blackwall Tunnel Duplication. *Proceedings of the Institution of Civil Engineers*, Volume 35, October, Session 1965-66
Krabbe, W. (1968) Die Wasserhaltung beim Schildvortrieb durch Grundwasserabsentung und Druckluft, *Proceedings of Baugrundtagung 1968*, Hamburg. Editor Deutsche Gesellschaft für Erd - und Grundbau ev Essen
Megaw, T.M. and Bartlett, J.V. (1981) *Tunnels: Planning, Design, Construction.* Ellis Horwood Ltd, Chichester
Richardson, H.W. and Mayo, R.S. (1941) *Practical Tunnel Driving.* McGraw-Hill, New York
Schenck, W-R. and Wagner, H. (1963) Lufterbrauch und Uberdeckung beim Tunnelvortrieb mit Druckluft. *Die Bautechnik*, February
Terzaghi, K. and Peck, R.B. (1967) *Soil Mechanics in Engineering Practice.* John Wiley & Sons, Inc, New York
CIRIA (1979) *Tunnel Waterproofing*, Report 81, Construction Industry Research and Information Association, London

16 Tunnelling in compressed air – Necessity or option?

G.B. Shutter
Mott MacDonald, Croydon, UK

Abstract
This paper considers the current use of compressed air in tunnel construction suggesting circumstances where it is necessary and discussing alternative construction methods. Specific examples are described. The assessment of risk is considered.
Keywords: Compressed Air, Tunnels, Tunnel Boring Machines, Ground Treatment, Ground Freezing, Risk.

1 Introduction

In the first half of the nineteenth century Marc Brunel's epic struggle to build the first Thames Tunnel took twenty years. There was a clear need for compressed air, but no such option existed. Compressed air, however, has now been used in bored tunnel construction for well over a hundred years and during this time it has made a major contribution to satisfying demands for tunnels, particularly in adverse ground conditions in urban areas and subaqueous locations.

Its limitations have in turn been a constraint on tunnel construction, affecting size and alignment, time and cost, and health and safety.

In recent years there have been substantial advances in reducing the dependence on use of compressed air, particulary the development of the closed-face tunnel boring machine (TBM) in the forms of slurry shield and earth pressure balance machine (EPBM). In these the excavation area at the face of the machine is pressurised whilst the operation and service area at the rear together with the completed tunnel remain in free air.

The use of ground treatment and ground freezing have also enabled reductions in the use of compressed air. These various techniques may be used alone or in conjunction with compressed air, but all have their own restrictions and risks.

The assessment of necessity or option for using compressed air remains a matter for experienced judgement preferably involving client, consultant and contractor.

Engineering and Health in Compressed Air Work Edited by F.M. Jardine and R.I. McCallum. Published in 1994 by E & FN Spon. ISBN 0 419 18460 0

2 Major options

Cut and cover tunnels, immersed tube tunnels, and bridges may often be alternatives to compressed air bored tunnels.

A particular example of the evolution of such options on a major project is the highway crossing of the River Thames at Dartford, now part of the London orbital motorway. Three crossings have been constructed during the last half century. Following a compressed air pilot tunnel (1936 - 38), the first crossing was a Greathead shield driven tunnel (1959 - 61) using balancing air pressure up to 2.5 bar (35 psig) combined with large-scale ground treatment.

When planning the second crossing in the late 1960s the advance in immersed tube tunnel techniques made such a tunnel a serious option. Despite comparable estimated costs a bored tunnel was considered to involve less risk and a shield-driven bored tunnel together with cut-and-cover approaches was constructed from 1972-77. Extensive ground treatment was used with the intention of restricting air pressures to a maximum of 1.5 bar (20 psig). At this pressure whilst there was no risk to the ground stability for tunnelling, the water inflow was considered by the Contractor to have an adverse effect on the performance of men and plant in the tunnel. Pressure was therefore increased to 2.6 bar (36 psi), but the subject of increased pressures became a matter for protracted contractual dispute.

The third crossing was planned and constructed on a design-and-build basis in the 1980s. Technological advances dictated that a large diameter EPBM was considered instead of a traditional shield for the bored tunnel option. This option would have restricted the use of compressed air to maintenance and emergency requirements. An immersed tube tunnel and bridge were also options that were offered at tender stage. In the event a bridge crossing was chosen and whilst apparently at greater cost than a bored tunnel it must be recognised that the construction time of 2½ years is unlikely to have been matched had a bored tunnel solution been selected.

Large-diameter Greathead shields with compressed air are unlikely to be used again for similar works. Their use for the second Dartford Tunnel was probably the last of a long line of notable subaqueous crossings which have included such tunnels as Blackwall, Clyde, Hudson and Tyne.

3 Bored tunnel options

When bored tunnels in adverse ground conditions are required the closed-face TBM is now the likely choice when there is sufficient tunnel length to justify the high cost of such machines. However additional measures for ground support must usually be adopted at the start and finish of these drives, and also for the shafts, adits, cross passages and sumps that are often associated with such tunnels. Additional measures are likely to be ground treatment, freezing, compressed air or a combination of these.

Ground treatment by grouting is widely used, but is limited by difficulties encountered in penetrating soils of low permeability such as silts and fine sands.

Compressed air is usually very effective in such soils and eliminates the risk of incomplete coverage, an aspect to which ground treatment is susceptible.

Ground freezing is an important option but is usually costly and time consuming. Whilst generally reliable, difficulties arise where there is groundwater movement; and the nature of frozen soil presents its own excavation problems. Freezing is particulary beneficial for high hydrostatic heads which would otherwise require high compressed air pressures. Installation for freezing is usually carried out from ground level as it can be both difficult and hazardous when installed from an existing tunnel.

A wide range of ground conditions, varying from water-bearing gravel to hard rock, perhaps in boulder form, may be encountered in a single tunnel face. Despite the advances in TBM technology the universal machine to cope with such variations remains an elusive objective. The facility for pressurised manned intervention in the face is often needed to overcome problems of ground conditions in addition to maintenance requirements. This can be provided by a man-lock in the bulkhead of the TBM or with a separate bulkhead behind the TBM.

The use of closed-face TBMs has encouraged tunnel construction at depths where the hydrostatic head is greater than the traditional limit of 3.4 bar (50 psig) for compressed air construction. Whilst every opportunity is taken to minimise such use, it has moved tunnel construction into the discipline of diving regulations.

4 Effectiveness of compressed air

There are no simple criteria for ensuring the effectiveness of compressed air.

The application of a single air pressure at any one time in order to counteract the natural variation of hydrostatic head is essentially a contradiction. Compromise is required when determining an appropriate pressure and the use of pressures that are too high can be as dangerous as the use of those that are too low.

The ranges into which air pressures are categorised are dictated by health and safety rather than technical considerations. Air pressures below 1 bar, accepted as minimising health risk, are adequate for probably the majority of compressed air tunnels.

The ability of the ground to minimise air volume loss is dependent on the soil properties, particularly the grading and permeability. In simple terms, medium sands and finer soils have a sufficiently low permeability. Coarser soils are likely to need additional measures to reduce air loss. Many soils, however, are not so easily categorised and may be heterogeneous combinations of gravels, sands, silts and clays. Examples of such soils are found in glacial deposits and complex measures may be required to achieve adequate stabilisation.

The application of compressed air to a soil not only changes the ground water regime, but may also induce soil particle movement, thereby varying the soil properties and, in turn, affecting strength and stability.

Subject to the above limitations compressed air is a powerful and effective means of stabilising soil.

5 Risk

Assessment of options is often an assessment of risk; and risk analysis is becoming an increasingly accepted tool for tunnel construction. Such analyses are only as reliable as the input data which are often limited by lack of availability over the whole range of risks inherent in tunnel construction.

The assessment of risk, to a great extent and necessarily, will continue to be based upon experienced judgement and should involve client, consultant and contractor.

An example of a tunnel where there was thought to be no risk of encountering bad ground is in Toronto, Canada. When the Yonge Subway Northern Extension was planned a very extensive site investigation was carried out. 23 boreholes were sunk along a 1200m length of twin subway tunnel directly below Yonge Street. A perched water table was identified from the boreholes and both tunnels, at the same level, were aligned vertically to provide apparently safe cover below the base of the water table. The tunnels were planned for construction in free air and one tunnel was so constructed without incident. The second tunnel however, only 7m distant from the first, encountered water in the crown. The base of the stratum dipped some 4m across the width of the street and over a length of some 60m. When the water was encountered the options of ground treatment by grouting, freezing and compressed air were considered. The permeability of the soil was too low for effective ground treatment. Freezing from the street was considered too disruptive, and from the tunnel impractical. It was also the most expensive option. Compressed air was selected, installed and successfully used, the overall delay to the work being some three months. Whilst this incident took place over 20 years ago the problem and its resolution would probably take a similar course today.

There are situations where it is prudent to plan for every reasonable eventuality. Such a situation is the construction of the step-plate junction for the London Underground's Northern Line safety measures work at London Bridge Station, being undertaken with the Jubilee Line Extension works. A new station tunnel is required and the geometry of the associated south-bound running tunnel diversion is such that the north junction with the existing running tunnel is located below the River Thames. The junction consists of a bored tunnel encirclement of the existing tunnel through which train services must be maintained with maximum safety and minimum risk of flooding. Boreholes and records indicate construction in London Clay with adequate cover at the tunnel horizon, implying safe tunnelling in free air. However, immediate protection is required in the event of unforeseen water conditions. This cannot be provided by ground treatment or ground freezing. A standby compressed air installation is therefore specified to be available for immediate use if required. Fortuitously the existing tunnel can also be pressurised in an emergency as floodgates exist for tunnel protection on both sides of the River Thames.

The risk to health from high-pressure compressed air tunnelling is a major consideration and an emotive issue. Comprehensive assessment of alternatives for pressures over 1 bar is essential. Below 1 bar, however, the option is not perhaps exercised as frequently as is desirable. The cost of providing standby installation is often modest when compared with the cost of delay resulting from emergency measures needed during the course of the works.

6 Ground movement

Compressed air in tunnelling can reduce the extent of ground movement and the associated settlement of adjacent structures. However, relatively high pressures are required to make significant reductions, and use of compressed air solely for this purpose is usually considered unjustified.

There are new alternatives for controlling ground-surface and building settlement such as compensation grouting, the pressure injection of grout into fractures created in the soil. The experience of such methods is at present limited, but the use is likely to increase as the problem of settlement becomes an increasingly important issue.

7 Conclusions

The extensive development and use of closed-face TBMs in recent years has dramatically reduced the need for compressed air tunnel construction. However there remain particular circumstances where other alternative methods such as ground treatment and ground freezing are not necessarily preferable and where compressed air is needed to minimise risk and achieve safe construction.

The facility to plan, specify and use compressed air remains an important and necessary aspect of tunnel construction. The efforts of those concerned should be directed towards ensuring its acceptability.

8 References

Kell, J (1963) The Dartford Tunnel. *Proceedings Institution of Civil Engineers*, 26, 359-371.

Shutter, G.B. and Bell, G.A. (1979) Design and construction of the second Dartford Tunnel, in: *Tunnelling '79*, The Institution of Mining and Metallurgy, London.

17 Modelling compressed air losses from a tunnel in sand

C.P.M. Snee
University of Bradford, UK

Abstract
Adoption of a tunnel pressure, and the required volume of air to maintain that pressure, is currently based on judgement or oversimplified formulae. The risks associated with leakages of air from tunnels are not addressed by these methods. Such risks include surface settlement, uplift of structures and development of air paths to other structures or excavations. Furthermore, it is considered that standard ground investigations do not provide the parameters necessary for carrying out analysis if a method did exist. A series of experiments has been carried out on a small-scale physical model of a tunnel in sand to demonstrate the mechanisms of air flow through soils, including the behaviour of the ground during a blow-out, the settlement of the surface and the changes in soil condition after a blow-out. The observations are presented and a more rigorous method for the prediction of tunnel pressure, air consumption and the zone influenced by the air flow is discussed.
Keywords: Compressed air, Air loss, Ground investigation, Permeability to air, Modelling air loss.

1 Introduction

Tunnels driven through unconsolidated water-bearing strata can be pressurised with air to prevent collapse and inundation. The appropriate air pressure in the tunnel must be applied to balance the groundwater pressure and, because of leakage of air from the tunnel, a constant recharge of air must be supplied to maintain that pressure. If air leakages increase, a greater volume of air must be supplied or measures taken to reduce the air losses.

Compressed air acts to enhance the stability of a tunnel in three ways (Attewell *et al.*, 1986):

1. By balancing the external head of water and resisting its entry into the tunnel.
2. By providing a direct reaction to the field forces attempting to displace the soil particles into the tunnel.

Engineering and Health in Compressed Air Work Edited by F.M. Jardine and R.I. McCallum. Published in 1994 by E & FN Spon. ISBN 0 419 18460 0

3. By drying out soil at the tunnel face and so increasing its effective strength.

Karlsson (1973) suggested that the air pushes the water away from the tunnel until an equilibrium surface between water and air is formed in the soil.

This paper presents an overview of existing methods for calculating the appropriate air pressure and the likely air consumption. The main problems with using site investigation data for design purposes are examined and the observations of a series of small-scale laboratory model tests are presented and discussed.

Although there has been extensive use of this tunnelling method there is still a lack of understanding of how the air, water and soil interact. For example, seepages of water into the tunnel can result in groundwater lowering and settlement, over-pressuring can cause ground heave, high air pressures or large quantities of air can fracture or disturb the soil and reduce its strength. Also, the air can travel laterally to an adjacent excavation. Records of the Dartford Tunnel (Glossop, 1976) refer to high air losses issuing 800 m from the face and a blowout 100 m from the face. These situations are of immediate concern to those constructing the tunnel, but the threat to deep excavations supported by temporary works, dewatering projects, basements, cuttings, other tunnels and surface structures must be considered.

2 Current design methods

Adoption of a tunnel pressure is based on judgement or empirical formulae; usually the aim is to balance the water pressure at or slightly above the tunnel invert. In addition, engineers must determine the plant capacity to supply a sufficient volume of air to maintain that pressure against leakages from the tunnel. Typical sources of air loss are leaks in pipework, airlocks, segment joints, around the shield and through the face of the tunnel.

In recognition of the practical considerations, several empirical methods for calculating tunnel pressure P_1 (see Table 1) and the quantity of air required, Q, (see Table 2) have been proposed.

Table 1. Formulae for determining compressed air pressure.

Reference	Formula
Anon.	$P_1 = \gamma_W(H_{inv} - a)$
Anon.	$P_1 = \gamma_W(H + 2/D3)$
Anon. (USA)	$P_1 = 0.76\ H_{inv}$
Atkinson & Mair (1981) cohesive soils	$P_1 = \sigma_s - c_u\ T_c/F_s + 0.5\gamma_W\ (1 + 2C/D)$
Atkinson & Mair (1981) cohesionless soils	$P_1 = \gamma_W\ D\ T_\gamma$

Where

P_1	is the air pressure in the tunnel	T_c, T_γ	are stability parameters
γ_W	is the unit weight of water	D	is tunnel diameter
H_{inv}	is the head of water at tunnel invert	C	is depth of cover
a	is an on-site adjustment for ground conditions	c_u	is cohesive strength
H	is head of water at tunnel crown	F_s	is the factor of safety
σ_s	is the surcharge.		

Table 2. Formulae for determining quantities of air required

Reference	Formula
Schenk & Wagner (1963)	$Q = 70\ kiA$
Schenk & Wagner (1963)	$Q = \dfrac{\omega k_a}{\gamma_w} \cdot \dfrac{P_1 - P_2}{L} \cdot \dfrac{P_1 + P_2}{2} \cdot \dfrac{A}{P_1} \cdot \dfrac{D - h_k}{D}$
Hewett & Johannesson (1922)	$Q = 3.7D^2$ to $7.2D^2$ (m^3)

Where

k	is the coefficient of permeability for water
i	is the hydraulic gradient
A	is the face area of the tunnel
k_a	is the coefficient of permeability for air
ω	is a correction factor
h_k	is the capillary rise.

Calculations of P_1 are generally based on the head of water above the tunnel crown with an additional pressure to account for a proportion of the tunnel face. The formulae for Q comprise three basic components; permeability for water, hydraulic gradient, and tunnel dimensions. These design methods do not account for the soil permeability to air or the effect air might have on the state of the soil. The flow of air through soil can present a risk to structures and nearby excavations by displacing water, washing out fines and encouraging settlement. This is a particular problem in an urban environment where there are often complex changes in cover, groundwater level and imposed stresses such that overpressuring the tunnel is difficult to avoid.

The consequences of simplifying the mechanics of compressed air tunnelling may be illustrated by recent tunnelling projects that encountered difficulties. The claims, which were for changes in method to reduce the permeability of the soils to air and extension of time, could be apportioned to the following:

1. The site investigations did not provide any specific parameters for the design of a tunnel driven under compressed air.

2. The possibility of a reduction in soil strength because of air losses and the risk to structures was not considered.
3. Monitoring of air consumption (and so air losses) and ground conditions was insufficient to identify the proximity of more permeable soils and so enable appropriate action to be taken.

3 Limitations of ground investigation

Ground investigations are sometimes inappropriately designed for tunnelling projects because there is a tendency to use general suites of tests. Adopting a general philosophy for works as specialised as compressed air tunnelling can introduce significant risks. For example, it is common practice to use coefficients of permeability derived from water injection or abstraction tests to assess the suitability of the ground to compressed air tunnelling. The permeability values are the basis for judgements on whether the soil will contain the air or whether there is a low-permeability horizon above the tunnel that will restrict the flow of air to surface. In addition to the sources of inaccuracy in these test procedures and analysis, which are well documented, there are problems that relate more specifically, though not exclusively, to the use of the data for the design of compressed air tunnels, i.e.

Region tested	---	Lateral permeability measured vertical permeability also required
Measurement limit	---	Low permeability materials may control real flows
Borehole analogy	---	Different permeant Different drawdown Scale
Testing conditions	---	Loss of fines Settlement Consolidation
Untestable concepts	---	Change of soil properties during tunnelling.

Two fundamental inadequacies with the existing procedures are that the permeant, water, is some 100 times more viscous than air, and it is not compressible. The scope of the ground investigation should be increased to determine the air permeability as well as the water permeability.

4 Parameters affecting air flow

The quantity of a fluid that can flow through a soil is normally described by or derived from Darcy's law, $Q = kAi$. However, the coefficient, k, referred to as the effective permeability, is not an absolute value but one which is dependent on fluid viscosity, which, in the case of Darcy, is water. Therefore, the parameter is not suitable for other fluids, particularly those that are compressible. The relationship between permeability for compressible and incompressible fluids is discussed in detail by Peck (1990). The absolute coefficient of permeability, K, is independent of fluid characteristics and is related to the effective permeability for water by the formula;

$$k = \gamma_w K/\mu \tag{1}$$

Where γ_w is the unit weight of water and μ is the viscosity of water. However, converting values for water to air is not simple because the permeability for compressible fluids such as air changes inversely with pressure.

The degree of saturation is another significant control over air permeability (Olson and Daniels, 1981) in that the water in the pore spaces impedes the flow of air. There have been limited studies of the relationship between air permeability and degree of saturation, S_r. Yoshimi and Osterberg (1963) carried out studies of compression of partially saturated soils and the role of air permeability but the findings were limited to low permeability cohesive soils which are not usually problematic in compressed air tunnelling.

The equations presented above provide a means of estimating the required air pressure and the quantities of air likely to be consumed during tunnelling under idealised conditions. However, they do not address the real condition because of three basic inadequacies:

1. Site investigation data generally relate to horizontal permeabilities whereas vertical permeability is the dominant control of flows of air.
2. The degree of saturation is very difficult to determine using conventional site investigation techniques.
3. As air travels from the tunnel to the surface it expands. If this expansion has sufficient energy it can disturb the soil. This is referred to as a hydrodynamic condition.

The last point is believed to be a significant omission in current design practice. Observations in the field (Snee, 1989) have shown that the hydrodynamic condition is not only associated with blow-outs, but also occurs under what might appear in the tunnel to be stable conditions. Yet there is no analytical method currently available in soil mechanics to describe the hydrodynamic effects of compressed air tunnelling. There are studies of gas migration from organic sources in offshore sediments (Wheeler, 1988) but this involves small quantities of gas and relatively low velocities. Several alternatives for the preliminary research were considered including computer modelling, field measurements and physical modelling. As the mechanics of the problem are poorly understood and vary with time, computer methods are of limited use. There are practical and contractual constraints on field work in tunnels. Therefore, the physical model was considered to be the most useful method.

5 Small-scale model experiment

5.1 Objectives of modelling

The experiments were designed to demonstrate the following phenomena:

- the air pressure required to exclude water from the tunnel
- the pressure at which catastrophic air losses (blow-out) occurs
- the effects of consecutive blow-outs
- the necessary reduction in pressure to stop blow-out conditions
- the development of air paths through the soil mass
- the effects on surface structures.

The model comprised a rectangular glass channel containing the water-bearing soil and a perspex cylinder with one perforated end to simulate the open face of the tunnel. Air was supplied to the cylinder by a plastic pipe *via* a manometer; the pressure range was 0 to 50 kN/m^2. The soil samples were prepared to a specific particle size distribution by sieving and placed loosely into the glass channel. Three particle size distributions were used; a 0% silt, 100% fine to coarse sand; a 10% silt, 90% fine to coarse sand; and a 20% silt, 80% fine to coarse sand. The observations were made for all three soil samples.

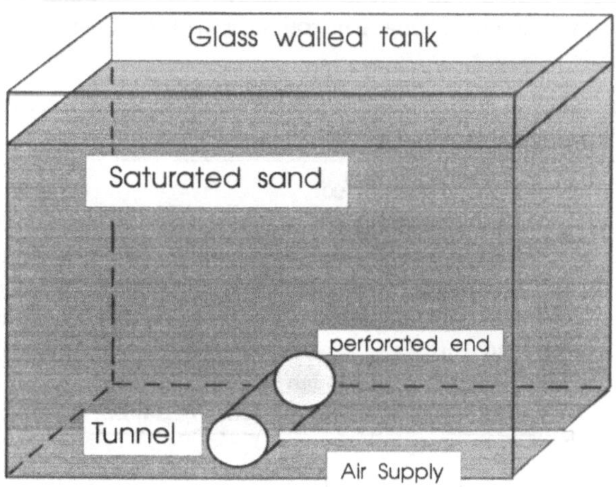

Figure 1. Physical model of compressed air tunnel.

In each experiment the controlled variant was the air pressure in the tunnel, P_1, and the constants were the head of water or water pressure, P_w, and the cover. The observations are related to the ratio of two quantities, P_a/P_w, referred to as the air pressure coefficient, P_c which is 1 when the applied air pressure is equal to the water pressure above the invert of the tunnel.

The following accounts of the experiments use descriptive terms which are defined as follows:

Term	Definition
Dewatering	The water in the tunnel is displaced by the compressed air. The water flows through the perforated end of the tube into the soil.
Partial dewatering	The water level in the tunnel falls as a result of the applied compressed air.
Soil disturbance	The grains of sand move because of the migrating air and displaced water but are not transported through the soil.
Air pipes	Vertical pipes in the sand containing only air extending from the tunnel to the surface.
Voids	Elliptical voids in the sand containing only air.
Blow-out	Extensive disturbance of the soil mass, violent bubbling at surface, particles transported vertically through pipes and voids, slight inflow of water to tunnel.
Stable condition	No soil disturbance and limited movements of fine particles, whilst the tunnel remains dry.

5.2 Experiment 1: Time-dependent effects at balance pressure

Procedure: the air pressure was increased in stages until the water was expelled from the tunnel. Changes in soil state with time were observed at balance pressure.

P_c	Time (min)	Observations
0.29		No apparent effect
0.63		No apparent effect
1.00		Partial dewatering
1.10	0	Complete dewatering. No disturbance of the soil mass.
	10	Air bubbles appear at surface. P_c fluctuates between 0.94 and 1.10
	20	No air bubbles at surface. P_c stable
1.04		Tunnel remains dewatered. P_c stable

5.3 Experiment 2: Effect of a blow-out on soil state and determination of the required pressure to re-establish a stable condition.

Procedure: The air pressure was increased in stages until a blow-out occurred. The pressure was reduced until the condition stabilised.

P_c	Time (min)	Observations
1.02		Tunel starts to dewater
1.32		Tunnel completely dewatered. Soil disturbed.
1.48	0	Blow-out occurs. Air bubbles appear at surface concentrated directly above the tunnel. Air pipes develop temporarily
	20	Elliptical voids up to 30 mm across develop
	24	Voids and pipes collapse
	42	Voids and air pipes reappear
This cycle of opening and collapse of air paths was repeated for a 3-h period		
1.65		Permanent air paths develop accompanied by violent bubbling at surface. Air bubbles appear at surface up to 8 diameters from the tunnel line
1.00	30	Blow-out features persist for approximately 30 min P_c reading fluctuates between 0.98 and 1.11
0.90		Stable condition re-established

5.4 Experiment 3: Behaviour of surface structures above a blow-out.
Procedure: Two weights (5.2 kg each) with a base area of 0.014 m^2 were placed on the surface to represent footings. One was directly above the tunnel and the other at the observed outer limit of the blow-out. The air pressure was increased in stages to a blow-out. The air pressure was decreased to establish a stable condition. Movements were monitored by dial gauges.

P_c	Time (min)	Observations
1.00		Tunnel only partially dewatered
1.44		Tunnel dewatered. Stable condition
1.88	10	Blow-out. Air paths and voids develop everywhere except below the weights
	20	Weight directly above tunnel settles by 5 mm
	30	Weight directly above tunnel settles differentially up to 20 mm. Second weight settles differentially by 5 mm
1.52	40	Blow-out continues. Settlement stops
1.06	45	Blow-out continues for 15 minutes. P_c fluctuates between 1.02 and 1.18
	60	Stable condition
	24 hours	Weight above tunnel settles by 24 mm. No change to second weight

5.5 Summary of observations

It must be emphasised that at this stage in the research the model only produces qualitative information with the exception of P_c. On the basis of these tests it is possible to summarise the observations as follows.

1. An air pressure greater than the head of water above the invert is required to keep the tunnel completely dry. This P_c is a function of the fines content of the soil and ranges between 1 for 20% silt and 1.07 for 0% silt.
2. The air pressure required to cause a blow-out is dependent on the silt content and varies from a P_c of approximately 1.2 for 0% silt to 1.35 for 10% silt and 1.45 for 20% silt.
3. During a blow-out, air paths such as pipes and voids up to 30 mm across appear temporarily. Eventually, the voids and pipes interconnect and collapse then reappear in approximately the same positions. This cycle is repeated although air continues to issue at the surface throughout the cycle.
4. At blow-out fluidisation occurs which results in a settlement trough causing differential settlement of simulated footings. The settlement increases with time.
5. During fluidisation, the upward velocity of the air is sufficient to move and sort particles in the soil mass by size. The greater the volume of air supplied, and so the greater the energy of the air, the larger the particles that are transported. The lighter or smaller particles, with the lower settling velocities, are transported to the surface and are deposited in pockets of fairly uniform grain size.
6. After a blow-out, when the air pressure is slightly above balance pressure, the phenomena associated with a blow-out persist for a limited period. During this period the air pressure in the tunnel fluctuates around the set air pressure. After this period, stable conditions are re-established at an air pressure slightly above that originally set and air bubbles continue to appear at the surface.
7. Repeated blow-outs result in a progressively lower pressure needed to establish a stable condition.
8. When the air pressure is reduced to zero the tunnel floods almost immediately. The soil structure returns to a stable state but there is clear evidence of sorting of particles by settling velocity with the finer material appearing at the surface and the coarser material concentrated around the tunnel.

6 Discussion

The available formulae for tunnel pressure and compressed air capacity simply provide a tool for determining approximate values to corroborate judgements. They do not provide a means of assessing the risks, a procedure for dealing with air loss problems, or a procedure for controlling a blow-out. There is no way of assessing the likely impact of losses from compressed air tunnelling on other structures or excavations even though it is well known that air travels substantial distances in the ground, fluidisation occurs and the soil state changes.

A more rational approach would be to assess and use realistic parameters and actual conditions, not oversimplified ones. An important step would be the use of a coefficient of permeability for air and not water. As air is compressible and the degree of saturation changes in the presence of air the air permeability will change with space and time. Furthermore, the assumption of static conditions would seem to be invalid. The physical model shows that the flow of a compressible fluid through a granular soil on a steep hydraulic gradient generates velocities in excess of the settling velocities of the particles, causing sorting. The change in particle size distribution introduces another time-dependent change in permeability. Furthermore, quite substantial voids and pipes can develop in a friction-only material.

It is considered that a more accurate design method should be developed. The existing analytical methods based on Darcy's law consider the important parameters to be the permeability, the hydraulic gradient and the diameter of the tunnel face. An improved method must consider the hydrodynamic aspects of compressible fluids which entail turbulent flow, temperature effects and phase relationships. Such a method is currently being developed at the University of Bradford. Mechanisms of flow of air are taken from the gas laws and thermodynamics. The distribution of air is measured in the model by tracing the zone around the tunnel that is affected by the air and a curve-fitting method is used to predict the zone of influence for different model parameters. It has been found that a fairly simple power relationship exists between the quantity of air leaving the tunnel and the migration of the air through the soil. The main parameters are the viscosity of air, the hydraulic gradient and the voids ratio of the soil. The method will be sufficiently developed for publication after further experiments using a redesigned model.

The redesigned model will be much larger than the prototype and will be used to examine the effects of particle size distributions, settling velocities, flow mechanisms of compressible fluids and soil/structure interaction for different groundwater levels, cover and air pressures. The model will contain instrumentation for measuring parameters such as pore pressure and strain at several depths below the surface.

References

Atkinson, J.H. and Mair, R.J (1981) Soil mechanics aspects of soft ground tunnelling, *Ground Engineering*, July.

Attewell, P.B. Yeates, J. and Selby, A.R. (1986) *Soil movements induced by tunnelling and their effects on pipelines and structures*, Blackie, Glasgow.

Glossop, R. (1976) The invention and early use of compressed air to exclude water from shafts and tunnels during construction, *Géotechnique*, **26**, No. 2, pp 253-280.

Hewett, B.H.M and Johannesson, S. (1922) *Shield and compressed air tunnelling*, McGraw Hill, New York.

Karlsson, N.G. (1973) A complicated tunnel driving method, *Tunnels and Tunnelling*, Vol 5, No 8 September, pp 452-457.

Olsen, R.E and Daniel, D.E. (1981) Measurement of the hydrostatic conductivity of fine grained soils, in: *Permeability and Groundwater Containment Transport*, ASTM STP746, pp 18-64.

Peck, J.G. (1990, *Finite element modelling of blow-outs in compressed air tunnels*. Unpublished, MSc. thesis, University of Durham.

Schenck, W. and Wagner, H (1963) Luftuerbrauch v. Uberdeckung beim tunnelvortrieb mit druckluft, *Bautechnik* 2.

Snee, C.P.M. (1989) The use of tunnel based grouting to control compressed air losses beneath London's Docklands, Paper to the British Tunnelling Society and awarded the Harding Prize, reported in *Tunnels and Tunnelling*, Vol. 21, No 7.

Wheeler, S.J. (1988) A conceptual model for soils containing large gas bubbles, *Géotechnique*, **38**, pp 389-397.

Yoshimi, V and Osterberg, J.O. (1963) Compression of partially saturated cohesive soil, *J. Soil Mech. Found. Div.*, ASCE, **89**, SM4, pp 1-24.

18 Developments in low-pressure compressed air equipment for tunnelling

R.T. Slocombe
Specialist Plant Associates, Wellingborough, UK

Abstract only

Low-pressure compressed air is an important option for tunnelling. The presentation described the development of transportable, quickly assembled compressed air systems which provide additional flexibility and economy in use. The systems are suitable both when compressed air is the basis of the tunnelling method and when it is a standby or emergency facility. The components of the systems and typical installations were illustrated by slides. The planning needed to achieve the greater advantages from these systems was discussed. The points particularly stressed were that:

1. A high proportion of tunnelling is for infrastructure works at shallow depth for which low-pressure (< 1 bar) compressed air is sufficient.
2. At these pressure there are not the health hazards of higher pressures.
3. Many of these projects are relatively small in scale and the headworks are set in places awkward of access and restricted in area.

The compressed-air systems described can be tailored to suit specific situations.

Engineering and Health in Compressed Air Work Edited by F.M. Jardine and R.I. McCallum. Published in 1994 by E & FN Spon. ISBN 0 419 18460 0

Discussion on Session 4: Engineering design

The session chairman, **Mr R F Jennion**, introduced the presenters of the papers. **Mr A R Gaba** described aspects of compressed air tunnelling in Singapore. **Mr R S Hoad** presented a review of methods for estimating air losses and results from case histories. **Mr G B Shutter** discussed factors affecting the adoption of compressed air methods in tunnelling. As **Dr C P M Snee** had to teach on a geology field course, **Mr Jardine** presented some of the results of model tests to investigate air losses given in Dr Snee's paper. Mr Jardine also drew attention to a geotechnical point in the paper 5 by Angarola *et al.*, i.e. that of the geological structure of the ground trapping escaping air above the groundwater table effectively increasing the water pressure at tunnel level. The steps taken to overcome this were to install venting wells to release the air back-pressure. **Mr R T Slocombe** presented examples of the rapidly assembled, readily transportable, sectional units which can be used to provide a convenient compressed air system, particularly suitable for smaller contracts. The discussion that followed these presentations is given below.

Mr V J Belman. As a comment on tunnel driving into trapped water, there is an alternative to venting which is to carry on with reduced air pressure, accepting greater seepage into the tunnel.

A comment on Dr Snee's paper; the modelling in sand gives a useful understanding of the phenomenon of air exit through the face. However since the face was static it did not show that with an advancing face in real tunnels the 'blow' will occur to the rear of the face having been weakened by an air passage previously. This experiment serves to emphasise that when in doubt pressure should be reduced.

With regard to the paper by Hoad and Gittoes on the estimation of compressed air losses, these seldom occur at the face – this can be confirmed by generating smoke to assess the velocity of the air entering the tunnel, and where the velocity thus shown reduces is where air is escaping. There is on the market a pocket air speed indicator for those who do not smoke! Hewett and Johannesson's method should be discarded – a tunnel 2m in dia. and 1,000 rings long will have 50,000 grommets, 10,000 metres of caulking and 4,000 square metres of concrete; all of which can let out compressed air. The only criterion to determine air losses is the area of gravelly surround (after caulking and grouting is complete).

Mr R P Lewis. I would like to raise the subject of air locks fitted to tunnel boring machines (TBMs). It is a matter which impacts on the subject matter of all this morning's papers. Very often these airlocks on machines are specified by people purchasing TBMs, but I would ask whether they are safe to use. The size of the air reservoir into which the worker must enter is very small and the danger of sudden pressure loss is high. It may not be through the ground, it could be through the tail seal or tunnel lining joints. Secondly, the size of the air lock will be very small and it will only be on very large machines that locks which allow stretcher handling and other mandatory requirements will be capable of being fitted. The question is then raised; how do you maintain the machine? There is no magic solution, but I would suggest that the safe answer is to pressurise the tunnel as we have just heard from Roy Slocombe how simple and rapid this can be.

Mr D R Lamont. HSE is concerned about the problems of rescuing a casualty from the relatively small chambers and working compartments now being found on earth-pressure balance and slurry machines.

Mr R S Hoad. In many cases it is still convenient to make an initial estimate with the Hewett and Johannesson method but to add 50%. A difficulty with Schenck and Wagner's method is estimating the ground permeability to air. Certainly, air losses depend on the length of tunnel, and our paper suggests a simple rule of thumb approach.

Professor D N Walder. I endorse what Mr Slocombe said that low pressure work, i.e. at pressures less than 14 psig is not a problem area for bone necrosis.

Professor R B Philp. Mr Slocombe indicated that 90% of compressed air work in UK is at pressures of less than 17 psig, yet yesterday we had figures suggesting that up to one third of British tunnel miners have radiological evidence of osteonecrosis. What matters epidemiologically is not what proportion of work is done at low pressures, but what percentage of miners are exposed to high pressures, since it takes one bad exposure to set in motion the sequence of events which may lead to osteonecrosis. Would you clarify for me, please, the frequency of high pressure exposures in UK tunnel miners.

Dr G C Voigt. On a recent US tunnelling contract on which compressed air was used, the air drove volatile contaminants from the ground into nearby homes. The contaminants included dry-cleaning fluids.

Mr F M Jardine. I understand that at Cairo air escapes manifested themselves at considerable distances from the tunnel. Could Mr Court comment on this?

Mr Hoad. There had also been a problem in Los Angeles tunnelling with volatile contaminants in the ground. The use of gaskets between segments may not be any more effective in reducing air losses than grouting.

Mr J C Le Péchon. In the French regulations, there is a pressure limit of 1.2 bar. Everyone going into the pressure has to be trained and they are not allowed to undergo higher pressures without additional training. In regard to the size of locks and their position, e.g. at the machine (TBM) or at ground level, a minimum size is specified, but the choice depends on the number of interventions: each case has to be considered separately.

Mr D J Parks. The use of man locks on a TBM is successful on the Storebaelt Project in Denmark for maintenance and repair to the cutterhead. Despite the sophistication of these earth-pressure-balance TBMs, it has proved the most effective method for this intermittent work, and the contractor has gone to considerable trouble to ensure the work is carried out safely and effectively. However, the question which needs to be asked of the industry as a whole is whether we make sufficient provision on the complex TBMs needed for more difficult jobs to treat the ground ahead to minimise interventions of this type in relatively small chambers, given that decisions of this sort at an early stage may be extremely difficult.

Mr H R Davies. In regard to the modelling of air losses described in Dr Snee's

paper, further improvements could be to place a pressure on the 'sand' to simulate the natural consolidation of the geological strata. Additional loading could also be applied to increase the K_0 (coefficient of earth pressure at rest) to simulate overconsolidated strata. In order to increase the accuracy of scale, the sand should be graded down to silt, as it appears that the present experiment used sand which, to scale, would react more like looser graded gravels – a most unlikely scenario.
Mr Slocombe noted the importance of having provision of a sufficient back-up air volume.

(Reporter: J Buchanan)

Part Five
Occupational Health and Safety

19 Medical aspects of compressed air tunnelling – The Singapore experience

J. How, A. Vijayan and T.M. Wong
Republic of Singapore Navy, Singapore

Abstract
In the Singapore Mass Rapid Transit Project (MRT), 11 km of underground tunnels were built between 21 September 1984 and 17 April 1987 using compressed air. 1737 compressed air workers (CAWs) were employed in the project. They underwent 188,538 man decompressions at the various compressed air work-sites. 160 CAWs developed Type I decompression sickness (DCS) and 4 developed Type II DCS. Knee, elbow and shoulder pain were the predominant presenting symptoms of Type I DCS. Two Type II cases showed peripheral and central nervous system dysfunction. The third case presented with vestibular disturbance and the fourth with 'chokes'. The overall incidence of DCS was 0.087%. The adoption of strict medical selection, strict adherence to decompression procedures and the provision for acclimatisation of new starters contributed greatly to this low incidence. Prompt treatment of DCS accounted for the low relapse rate. The clinical presentation of DCS is discussed here. Prevention of DCS by health education and monitoring of the worksite environment and work practice are advocated.
Keywords: Decompression sickness, Compressed air worker, Mass-transit project, Singapore, Injuries.

1 Introduction

The use of compressed air in tunnelling work on a large scale was seen for the first time in Singapore during the Mass Rapid Transit (MRT) Project. Prior to this, compressed air had been used on a small scale in some local sewage works. Singapore had until then no experience in managing medical problems associated with this type of engineering work.

Compressed air is used to prevent seepage and flooding of tunnels in areas where there is a high water table. This was a technical problem faced by the engineers in the Central Business District of Singapore which has a very high water table. Underground railway lines make up 20 of the 65.8 km of the MRT route. Compressed air was used in the construction of 11 of these 20 km underground.

Engineering and Health in Compressed Air Work Edited by F.M. Jardine and R.I. McCallum. Published in 1994 by E & FN Spon. ISBN 0 419 18460 0

2 The Singapore experience

In 1982, the Singapore Government announced its decision to build the Mass Rapid Transit System after the State and City Planning Study had examined the country's land use and transportation policy. The Protem Committee of the MRT Project subsequently approached the Republic of Singapore Navy (RSN) in August 1982 to assist in providing medical support for the Project. The Diving and Hyperbaric Medical Centre (DHMC) of the RSN was the only agency in Singapore with the facilities and experience in the selection of divers and in the management of diving-related illnesses such as DCS, barotrauma and dysbaric osteonecrosis. In view of the similarity of medical conditions suffered by divers and compressed air workers (CAWs), DHMC was in the best position to help in the MRT Project.

The use of compressed air in tunnelling is not new. Its first practical use in tunnelling was described by Triger (1841). Since then it has been used in many large-scale construction projects. With extensive use of compressed air new clinical entities such as decompression sickness (DCS) and dysbaric osteonecrosis (DO) were described and their pathophysiology studied. The etiological basis of both these conditions is similar. They result from the mechanical and biochemical consequences of nitrogen bubbles in the vascular and extravascular tissues. These nitrogen bubbles are produced within the tissues when the drop in ambient pressure exceeds the supersaturation levels of dissolved nitrogen.

In order to prevent supersaturation of nitrogen within the tissues, CAWs are decompressed slowly in accordance with rates set out in decompression tables or schedules. Several tables have been used in the past. Hempleman's Blackpool Tables in the CIRIA Report 44 (3rd Edition) have been used with a low incidence of DCS in Hong Kong. These tables were employed in the Singapore MRT Project. However, with information garnered from diving research, refinements to the Blackpool Tables were added to increase their safety and to reduce the incidence of DCS in the repetitive nature of the compressed air shift work.

In this paper, the authors present their experience in the management of DCS in the MRT Project from 1984 to 1987.

3 Organisation of medical support

Under the MRT tender specifications, DHMC was given the sole responsibility of providing medical support for the compressed air tunnelling. This situation was ideal. It allowed for strict control of the medical selection standards, proper record keeping, systematic medical surveillance and rapid response to the treatment of decompression sickness and other illness related to dysbarism.

Medical support was organised under an operational HQ which coordinated the activities of several site medical centres (SMCs) located at the worksites.

4 Operational headquarters

DHMC of the RSN served as the medical operations headquarters; its roles were to:

(a) Serve as a central Medical Selection Centre for all new potential CAWs.
(b) Conduct medical screening for firemen from the Singapore Fire Service and inspectors from the Mass Rapid Transit Corporation and Industrial Health Division, Ministry of Labour.
(c) Monitor and coordinate appointments for periodic medical examination, x-rays, blood and other medical investigations.
(d) Provide structured training to CAWs, medical-lock attendants, man-lock attendants, and medical personnel (i.e. doctors, nurses and medical orderlies) in various health and procedural aspects of compressed air work.
(e) Monitor the physical environment of the tunnel such as working pressure, humidity, temperature, noise level and levels of oxygen, carbon dioxide, carbon monoxide and other contaminants through the SMCs.
(f) Centralise all documentation and data collection to expedite data processing and information retrieval.
(g) Run a MRT Medical Operations Room 24 hours a day with a medical team in readiness for any emergency. Medical lock personnel and safety supervisors at the various worksites could telephone this Operations Room and request medical assistance. The Duty Medical Officer would then take charge of the management of the cases.

To back up the SMCs, DHMC installed a larger recompression chamber in order to provide better second-line treatment. A high-dependency room was also set up to manage serious cases of DCS.

5 Worksite medical support

SMCs were established after the various companies signed contracts with the Ministry of Defence. In all, six contracts were drawn up with the contractors to formalise the agreement to provide medical support. These were with:

Tobishima-Takenaka Joint Venture (TTJV) – Contract 104
Bocotra Construction Pte Ltd (Bocotra) – Contract 105
Kajima-Keppel Joint Venture (KKJV) – Contract 107
Taisei-Shimizu-Marubeni Consortium (TSM) – Contract 108
Ohbayashi-Gumi/Okumura Joint Venture (OOJV) – Contract 109
Nishimatsu-Lum Chang Joint Venture (NLJV) – Contract 301.

Each SMC was run by a Naval Medical Officer (MO) trained in the medical management of CAWs and related industrial injuries. A medical-lock attendant, was hired by the contractor, to assist the MO in the consultation room and in patient administration. The SMCs were each equipped with a medical therapeutic recompression chamber or medical lock which the medical-lock attendant was trained to operate. Diagnostic equipment and a small dispensary were also available in this set-up.

The objectives of the worksite medical support were to provide:

(a) Periodic medical screening and recertification of fitness to work in the compressed air environment.
(b) Medical surveillance of the safety practices of the worksite and the compressed air tunnels in order to prevent fire and other accidents.
(c) Surveillance of the physical environment of the worksite as well as the tunnel pressures to ensure compliance of the contractors with regard to safe work habits and decompression schedules prescribed in the Ministry of Labour legislation.
(d) Healthcare facilities to CAWs who required specialised treatment and management of even simple ailments like rhinitis which could disqualify CAWs from entering the compressed air tunnels because of the high risks of barotrauma, thus reducing labour availability.
(e) On-site emergency recompression treatment facilities to treat all cases of decompression sickness or pulmonary embolism arising from the tunnelling work.

6 Categories of compressed air workers

For planning purposes, the CAWs were categorised into three groups in order to provide specific supervision and training. They were the:

(a) Manual shift workers: these included skilled and semi-skilled workers who were involved in heavy work of lifting, use of vibrating and pneumatic tools. They usually perform shifts of specified duration, depending on the working chamber pressure.
(b) Supervisory category: this group included those supervisors who worked in compressed air almost daily, but did not have to perform heavy manual tasks.
(c) Special category: these include workers with special skills, e.g. electricians, blasting experts, who were usually non-shift employees. The medical personnel, men of the Fire Service and Industrial Health Division were also included in this category.

7 Training

Much of the planning would be useless without trained personnel to do the tasks effectively. As such training was high on our priority list. It was necessary to train the various categories of CAWs in the medical aspects of compressed air work. Manual shift workers were given lectures to increase their awareness of dysbaric-related illnesses and on preventive measures. Among other things, they were taught to equalise ear pressures properly and identify the early symptoms of DCS. In addition to these lectures, supervisors were lectured on environmental hazards in compressed air tunnels and the monitoring and management of such hazards. The supervisors were certified upon completion of this training package by the Ministry of Labour. A number of courses were also run for other groups of people. These included compressed air medical courses for medical officers, nurses and medical orderlies. The Industrial Health Division of the Ministry of Labour certified these courses. Special courses certified by the National Productivity Board were also run for the medical-lock and man-lock attendants. Medical training was also given to the inspectors from the Industrial Health Division so that they could work safely in the compressed air environment.

8 Medical selection objectives and rationale

DHMC, with the approval of the Ministry of Defence, assisted the Ministry of Labour to draft new regulations for the safe conduct of compressed air work. This was based on CIRIA Report 44, but adapted for our conditions and work-force. These new regulations were required because prior to 1982, compressed air work had not been used in Singapore. The new regulations were published by the Ministry of Labour as an amendment to the Republic of Singapore Factories Act 1973, Building Operations and Works of Engineering Construction (1985).

Medical selection of CAWs started in April 1984 prior to the commencement of compressed air work at Shan Road worksite. The objectives of medical selection were to exclude those workers with medical conditions which would be detrimental to health if exposed to compressed air work. Medical standards for medical selection were set and the regulatory of such tests were specified within the Factories Act.

The manual shift workers were usually in the younger age-group. They were selected based on stringent criteria which included age, degree of body fat and freedom from pulmonary and ear, nose and throat (ENT) pathology. However, the selection standards had to be reviewed on a case-to-case basis, especially the experienced supervisors of the older age group as they manifest the usual age-related physical impediments like obesity, poorer lung function and physical fitness. In addition, if these supervisors had previous experience with compressed air, they would also have a greater probability of suffering from dysbaric osteonecrosis.

Paton and Walder (1954) showed that men over 40 had a greater risk of DCS. Age restriction was thus implemented. CAWs should be fit individuals of age 18 years to 40 years. Those over 40 years were medically cleared on a case-to-case basis with cardio-vascular stress testing to ensure their fitness as well as the limitations of their exposure to compressed air.

Nitrogen is 5 times more soluble in fat than in water (Bennet and Elliot, 1969). A man whose body weight comprises 30% fat will have 2,000 ml more nitrogen than a lean man who has only 10% body fat, for every 1 bar change in pressure. This predisposes the obese to a greater risk of decompression sickness on account of the larger nitrogen content. CAWs are therefore preferably lean individuals of less than 20% body fat, as measured by the skin-fold method. Individuals who had between 20 to 24% body fat were given limitations on the duration and pressure at which they could work.

The medical selection process was controlled and performed at DHMC by trained doctors. A comprehensive medical history was taken to identify any condition, especially asthma, pulmonary tuberculosis or ENT conditions, which could place the worker at greater risk of getting barotrauma. Experienced CAWs were also screened for past histories of decompression sickness as well as problems of dysbaric osteonecrosis. Their height, weight and body fat were measured. They were assessed on their cardio-vascular health and an audiometric examination was performed.

Investigations included a full blood count, urine testing, eyesight testing, serum lipids and cholesterol levels, chest and long-bone x-rays. Cardio-vascular stress testing on a bicycle ergometer was required for a worker over the age of 40.

Entrapment of air under pressure within the lungs because of secondary lung pathology can result in serious pulmonary embolism. Congenital cysts, scar tissues, and emphysematous bullae are possible sites of air entrapment. Many physicians feel that the stethoscope is not efficient in the detection of small lesions. The routine examination, therefore included spirometry and the measurement of peak expiratory velocity or equivalent quantitative tests of pulmonary function. Large chest x-rays were taken prior to commencement of work in compressed air and this was repeated yearly. Potential CAWs were also put through a trial of compression and decompression in a recompression chamber under constant monitoring to ensure that they were able to equalise the pressure in their ears, sinuses and lungs.

A total of 2,392 potential compressed air workers were seen for pre-employment and medical examination. Out of these, 1,737 (72.7%) passed and 654 (27.3%) failed the medical examinations. Table 1 shows the causes of failure and the incidence of the causes.

Table 1. Reasons for failing medical examination

Reasons	No.	%
Ear	285	43.6
Cardiovascular	102	15.6
Respiratory	102	15.6
Recompression chamber test	36	5.5
Nose	32	4.9
Paranasal sinuses	19	2.9
Endocrine	14	2.1
Musculoskeletal	10	1.5
Dysbaric osteonecrosis	3	0.5
Others	51	7.8
Total	654	100.0

Among the failures, ear conditions stood out as the commonest cause, accounting for 43.6%.

These ear conditions included perforation of the tympanic membranes and chronic infection of the ears. Cardio-vascular and pulmonary problems together contributed 31% to the failures. Common pulmonary conditions were asthma and pulmonary tuberculosis. Hypertension and vascular heart conditions were the common cardio-vascular problems.

It is interesting to note that three cases were rejected because of dysbaric osteonecrosis. All three had previously worked in compressed air in other parts of the world.

9 Monitoring of worksite and tunnel safety

In addition to running a daily clinic for CAWs at the worksite, the worksite medical officer was also responsible for the monitoring of worksite and tunnel safety. In this, he was assisted by the site safety supervisor and the man-lock attendant. His functions were to:

(a) Monitor fitness to work in compressed air environment.
(b) Monitor proper working of man-locks and medical-locks.
(c) Monitor tunnel pressures.
(d) Ensure proper recording of man-lock registers.
(e) Monitor that decompression schedules were properly followed.
(f) Monitor gaseous constituents in tunnel atmosphere.
(g) Monitor general work safety and occupational hygiene.

Tunnel pressures were recorded by barograph.

Inspection of the man-locks revealed that some of their chambers leaked especially in the last 0.3 bar of decompression. Immediate rectification of these inadequacies was carried out.

Man-lock attendants were fined if wrong decompression schedules were carried out. CAWs who were found in infringement of safe work practice, like tampering with barographs to reduce their decompression time or smoking within the compressed air tunnels, were strictly dealt with. Disciplinary actions included levying of fines, suspension or even dismissal from the workforce.

10 Our experience

The medical support team was tested in its provision of treatment to the CAWs on several occasions. The response times were good, meeting our set goals of immediate response during working hours and within 1 hour of recall after working hours. This was achieved with doctors available at the worksites during office hours and two doctors on standby after office hours to take care of any decompression emergency. One hundred and sixty cases of Type I DCS and four cases of Type II DCS were seen. Our set-up with the site recompression chambers and medical centres ensured that first-line treatment was quickly and efficiently carried out. The back-up recompression chambers located at DHMC were on standby 24 hours a day as well.

The population at risk was a multinational cohort of 1,737 compressed air workers. There were Thai, Korean, Japanese, Chinese, Indian, Malay and Caucasian workers of various age groups. Both shift and non-shift workers were used in the contracts. The non-shift workers included engineers, supervisors, electricians and fitters who entered the tunnel for specific tasks. They spent a variable time in the tunnel and had to be decompressed accordingly.

For ease of reporting and presentation, the cases of DCS were classified into two types after Golding *et al.* (1960), i.e.

Type I ('pain only')	-	symptoms and signs are mild and present as musculoskeletal pain, swelling due to lymphatic obstruction and skin involvement.
Type II ('serious')	-	symptoms and signs are severe and attributable to disorder of the nervous, pulmonary and cardio-vascular system.

Various factors were analysed including the host characteristics, duration of onset, age, length of exposure, pressure and number of episodes of DCS. Overall incidence, physical environment of the worksite, treatment methods and outcome were also analysed.

Data on the incidence of dysbaric osteonecrosis were obtained by review of the long-bone x-ray reports performed yearly as required by legislation. Factors with regard to pressures of exposure and other host characteristics like obesity, race and number of episodes of DCS were analysed.

11 Decompression sickness

There were 164 cases of DCS. 160 were of the Type I category while 4 were of the more serious Type II category. Nine of these cases were the result of explosive decompressions in contracts 104 and 105 following accidents.

In contract 104 (TTJV) a runaway rail car smashed open the muck-lock doors of the tunnel. Working pressure was 1.45 bar gauge. The 15 CAWs who had been working within for 7½ h were exposed to explosive decompression. Four developed symptoms of Type I DCS. One developed symptoms of Type II DCS involving the pulmonary vasculature (chokes). One worker working just outside the muck-lock was thrown about 10 m by the out-rush of the compressed air and suffered multiple injuries.

The second incident occurred at contract 105 (Bocotra) under almost similar circumstances. A runaway rail car smashed open the muck-lock doors of the pressurised tunnel. 15 men working in a pressure of 1.88 bar were exposed to explosive decompression. All of them developed aural or sinus barotrauma. Four developed Type I DCS.

11.1 Type I DCS

The commonest presentation of Type I DCS in this series was joint pain (154 cases or 96.3%). 84 cases (55%) presented with monoarticular pain and 70 cases (45%) with polyarticular pain. In 76% of the cases, the joints of the lower limb were involved. The knee joint was the commonest joint involved (108 cases) followed by the shoulder joint and the elbow joint (Table 2).

Table 2. Sites of Type I symptoms

Site	Pain	Skin rashes	Lymphatic manifestations
Head	-	1 (Macular rash)	-
Shoulder	19	-	-
Elbow	16	-	-
Wrist	2	-	-
Abdomen	-	2 (Papular rash)	-
		1 (Marbling rash)	-
Inguinal region	-	-	1
Hip	4	-	-
Knee	108	-	-
Lower leg	-	1 (Itchy, macular)	-
Angle	5	-	-
Total	154	5	1

There were five (3.1%) cases of cutaneous DCS. Two cases presented with papular rashes over the abdomen. Two cases had macular rashes over the forehead and both shins respectively. One case presented with marbling rash over the abdomen. One case of lymphatic DCS was seen involving the inguinal lymphatics and presenting as swelling of the penile skin (Table 3).

The characteristic nature of the joint pain noted was that it was deep in nature, aggravated by movement (Table 3). There was also a limitation of the range of joint movement and altered sensation, including numbness, around the joints.

11.2 Type II DCS

Table 4 illustrates the clinical presentation of the four Type II cases seen on the MRT project. Four of them were exposed to compressed air for periods exceeding 8 h. All of them presented within 2 h of decompression. Case number 4 developed pulmonary DCS following an explosive decompression of the tunnel when a runaway locomotive smashed open the muck-lock door.

Table 3. Clinical presentation of Type I DCS

Symptomatology		**No. of cases**
Pain:	Deep pain	133
	Superficial	7
	Constant	24
	Throbbing	5
	Radiation of pain	3
	Limitation of movement	16
Rashes:	Erythematous, papular, itchy	2
	Erythematous, macular, itchy	2
	Marbling rash over abdomen	1
Lymphatic swelling (Penile skin)		1
Itch		5
Warmth around joints		8
Numbness around joints		16

11.3 Onset of symptoms

94.5% of the Type I and II cases developed symptoms within 12 hours of decompression (Table 5).

11.4 Episodes of DCS

Of 1,737 people who worked in compressed air from September 1984 to April 1987, 136 (7.83%) suffered from DCS. Of these, 114 (83.8%) of them suffered from a single episode of DCS while 22 (16.2%) had two or more. Table 6 shows the distribution of DCS episodes amongst the 136 affected. The maximum number of episodes of DCS in any one man was 4.

11.5 Incidence and exposure pressure and duration

A total of 188,538 man-decompressions were performed in the MRT project. There were 160 cases of mild (Type I) DCS and 4 cases of severe (Type II) DCS, giving the overall incidence of DCS at 0.087%. There were 64,059 man decompressions over 1 bar gauge, with 154 cases of DCS, giving an incident rate of 0.240%.

A study of the compressed air exposure time of the 164 cases showed that the majority, 125 cases (76.2%), occurred at exposure times exceeding 8 h as seen in Table 9.

The incidence of DCS from exposures at less than 1 bar was 0.008%. There were 10 cases in 124,479 man decompressions. Table 7 shows in detail the cases of DCS at less than 1 bar pressure, and Table 8 summarises the symptoms and the treatment.

Table 4. Clinical presentation of Type II DCS

Case No.	**Working pressure (bar)**	**Exposure time**	**Onset of symptoms following decompression**	**Clinical presentation**	**Treatment**
1	1.75	8 h 5 min	1/2 < 1 h	Pain - left knee and hip joints. Loss of sensation L3,4 bilateral bp 110/70. 2 previous eposides of Type 1 DCS	CIRIA 2
2	1.53	8 h 23 min	1/2 < 1 h	Pain - right knee. Loss of sensation to pinprick over right half of body	CIRIA 2
3	1.45	8 h 32 min	1 < 2 h	Felt weak and giddy 70 min after leaving manlock. Noted to be staggering and vomiting but no nystagmus, visual or auditory symptoms. Bp 110/70. Relapsed with giddiness and low bp of 90/50 and vomiting on sitting	CIRIA 2 and Table 6A
4	1.4	6 h 25 min	< 1/2 h	Sudden decompression accident. Pain - both knees. Aural barotrauma bilateral. Chest pain and dyspnoea (chokes). Bp 140/80 pulse 84. Chest x-ray normal	Table 6

Table 5. Onset of symptoms of DCS

Time (h)	**<1**	**1<4**	**4<6**	**6<12**	**12<24**	**>24**	**Total**
DCS Type 1 (No.)	32	68	32	19	1	8	160
DCS Type 2 (No.)	3	1	0	0	0	0	4

Table 6. Episodes of DCS

Episodes of DCS (No.)	1	2	3	4	Total
CAWs affected (No.)	114	17	4	1	136
% affected	83.8	12.5	2.9	0.8	100%

Table 7. Cases of decompression sickness (DCS Type I) at less than 1 bar pressure

Case	Race	Contract	Type of worker	% Body fat	Age	Working pressures (bar gauge)	Date of incident	Exposure time (h:m)	Time of onset after decompression
1	Thai	109	CAW	11	32	0.80	1.2.85	12	A few hours
2	Thai	109	Asst. Surv.	6	28	0.80	1.2.85	10	A few hours
3	Thai	109	Asst. F/man	20	39	0.80	8.2.85	12	A few hours
4	Thai	109	CAW	9.4	18	0.95	23.12.85	12	1½ hours
5	Thai	109	CAW	15	24	0.95	25.12.85	11:45	3 hours
6	Thai	109	CAW	11.3	31	0.95	29.12.85	12	1½ hours
7	Thai	109	CAW	12	25	0.95	29.12.85	12	3 hours
8	Indian	109	CAW	15	29	0.75	16.2.86	10:55	4½ hours
9	Korean	301	CAW	19	31	0.90	21.1.87	12:22	6 hours
10	Korean	301	CAW	14.2	35	0.94	16.3.87	12:22	28 hours

*including decompression time of not more than 10 min

Table 8. Summary of symptomatology and treatment used for DCS occurring at below 1 bar

Case	Shoulder		Elbow		Hips		Knees		Treatment table used
	UNI	BI	UNI	BI	UNI	BI	UNI	BI	
1								✓	61*
2						✓		✓	61
3	✓					✓		✓	61
4							✓		CIRIA 1
5		✓							61
6							✓		61
7							✓		CIRIA 1
8				✓		✓		✓	CIRIA 1
9			✓				✓		CIRIA 1 and 62*
10				✓					CIRIA 1

Notes:
1. UNI - Unilateral, BI - Bilateral
2. In all cases the outcome of the treatment was a complete cure
3. Tables 61 and 62 are the Royal Navy oxygen recompression therapy tables. Treatment with Table 61 takes 2 h 15 min while that with Table 62 takes 4 h 45 min.

The incidence of DCS at pressures from 1 to 2 bar was 0.221% with 138 cases in 62,350 exposures. In this pressure range, for durations less than 4 h, the results were significantly lower ($p < 10^{-6}$) than the mean values. At exposure times of greater than 4 h, however, the number of cases of DCS observed were significantly higher than values extrapolated from the incidence of DCS for exposures shorter than 4 h (both p values $< 10^{-6}$) (see Table 9).

The incidence of DCS occurring at pressures greater than 2 bar was 0.936%, with 16 cases in 1,709 exposures. Above 2 bar pressure, with duration of exposure less than 4 h, there were no cases of DCS. However, the observed increase in the number of cases of DCS was statistically significant for exposures of between 4-8 h ($p < 0.03$) and exposures greater than 8 h ($p < 10^{-6}$) (see Table 9).

11.6 Incidence of DCS, heat and humidity

The incidence of DCS did not appear to correlate significantly with the observed high temperatures and humidity of some tunnels. The highest temperatures were recorded at Contract 105 where temperatures reached a high of 42°C. Most tunnels had relatively high humidity, i.e. above 80%. At Contracts 107 and 108, there were occasions where humidity reached 100%.

Table 9. Incidence of DCS with pressure and duration of exposure

Maximum working pressure (bar gauge)	DCS type	Decompression sickness incidence in % (No.)			DCS incidence by type (%)	Total incidence (%)
		<4 h	4-8 h	>8 h		
<1	I	0.00 (0)	0.00 (0)	0.027 (10)	0.008 (10)	0.008%
	II	0.00 (0)	0.00 (0)	0.00 (0)	0	
No. of man-decompressions		60,976	27,080	36,423	124,479	
1-2	I	0.016 (4)	0.307 (32)	0.359 (98)	0.215 (134)	0.221%
	II	0.00 (0)	0.010 (1)	0.011 (3)	0.006 (4)	
No. of man-decompressions		24,596	10,423	27,331	63,250	
2-3	I	0.00 (0)	1.786 (2)	1.724 (14)	0.936 (16)	0.936%
	II	0.00 (0)	0.00 (0)	0.00 (0)	0.00 (0)	
No. of man-decompressions		785	112	812*	1,709	
Total	I	0.005 (4)	0.090 (34)	0.193 (122)	0.085	0.087%
	II	0	0.003 (1)	0.005 (3)	0.002	

* Exposure time exceeded Blackpool Tables for working pressure. Extensions to the tables were made to allow safe exit from the tunnels.

Table 10. DCS and percentage body fat

Percentage of body fat	No. of DCS events	No. of man decompressions	Incidence
< 20%	155	180,744	0.086%
20-30%	6	6,655	0.090%
>30%	3	1,139	0.263%

Table 11. Incidence of DCS by occupation

Occupation	No. of DCS events	No. of man-decompressions	Incidence (%)
CAW	110	110,904	0.099
Engineer	15	51,050	0.029
Supervisor	23	12,883	0.179
Others	16	13,701	0.117
Total	164	188,538	Average: 0.087%

Table 12. Incidence of DCS by contract

Contract	Maximum pressure (bar)	Number of DCS events			Number of man decompressions		Incidence of DCS					
							Overall %			Over 1 bar %		
		Type I	Type II	Total	Total	Above 1 bar	Type I	Type II	Total	Type I	Type II	Total
104	1.50	36	1	37	79,363	39,064	0.045	0.001	0.046	0.092	0.002	0.094
105	1.43	3	0	3	27,976	937	0.011	0	0.011	0.320	0	0.320
107	1.60	31	0	31	8,757	2,679	0.354	0	0.354	1.157	0	1.157
108	1.95	26	1	27	19,520	6,666	0.133	0.005	0.138	0.390	0.015	0.405
109	1.50	28	2	30	38,110	5,550	0.073	0.005	0.078	0.505	0.036	0.541
301	2.35	36	0	36	14,812	9,163	0.019	0	0.019	1.746	0.0	1.790
Total	-	160	4	164	188,538	64,059	0.085	0.002	0.087	0.250	0.006	0.256

Table 13. Incidence of DCS by race

Race	No. DCS events	No. of man-decompressions	Incidence (%)	*p* value*
Chinese	39	45,008	0.0867	0.91
Malay	14	15,849	0.0883	0.89
Indian	34	39,064	0.0870	0.90
Japanese	20	23,035	0.0868	0.90
Korean	18	20,539	0.0876	0.89
Thai	35	40,239	0.0870	0.90
Caucasian	4	4,804	0.0830	0.90
Total	164	188,538	0.0897	0.896

* This denotes the average *p* value obtained for each ethnic group when compared against the other groups individually, using the Chi-square test on a computerised statistical programme.

11.7 DCS and obesity

The percentage body fat of CAWs was measured using the Durnin and Womersley (1974) method. The number of CAWs who were obese (percentage body fat of more than 24%) were few. However, there were five cases of DCS occurring amongst CAWs who had body fat between 24 and 30% and three cases of DCS amongst the CAWs who had more than 30% body fat during initial medical clearance (Table 10).

11.8 DCS and occupation

The distribution of DCS among the various occupations is shown in Table 11.

11.9 Incidence of DCS by contracts

The maximum working pressure in the various contracts varied, the lowest being 1.43 bar in Contract 105, and the highest, 2.35 bar in Contract 301. The incidence of DCS by contracts is given in Table 12.

11.10 DCS and race

There appeared to be no significant difference between the different races in incidence of DCS ($p > 0.89$) (see Table 14). No correlation between DCS incidence and race has been noted in our literature search.

11.11 Treatment of DCS

Treatment primarily consisted of recompression with supportive drug therapy. Recompressions in most instances were carried out at the medical-locks of the various worksites. Following the recompression therapy, 100% surface oxygen were administered ensuring no relapse of symptoms.

The CIRIA air therapeutic tables alone using the pressure + technique were used in most (75%) of the cases. Not all the medical locks at the worksites were equipped with built-in oxygen breathing systems. In resistant and relapsed cases, Workman and Goodman oxygen therapeutic tables were used by bringing into the chamber oxygen breathing apparatus or by transporting the patient to DHMC. Table 14 lists the different recompression tables used in the MRT project.

Of the 164 cases treated, four (2.4%) cases relapsed after the first treatment and had to be treated with oxygen. None of the 36 (22%) cases treated with oxygen in the first instance, had a relapse.

12 Dysbaric osteonecrosis

All potential CAWs underwent a pre-employment long-bone x-ray survey (LBXR). These were reported by a radiologist. Radiological findings of DO were classified in terms of the classification developed by the Medical Research Council's Decompression Sickness Panel (Panel Report 1966). Of the 1,737 CAWs who were certified fit to work in compressed air, 32 had previous Type B lesions of DO while 11 had other benign findings like bone islands (Table 15). The CAWs with Type B

Table 14. Recompression treatment used

Recompression table	Contracts						Total
	104	105	107	108	109	301	
CIRIA 1	35	-	26	21	19	14	115
CIRIA 2	1	-	1(1)	2(1)	2	-	6(2)
Table 61	-	3	4	3	9	1	20
Table 62	(1)	-	-	-	-	15	15(1)
CIRIA 1 followed by Table 61	-	-	-	-	-	1	1
CIRIA 1 followed by Table 62	-	-	-	-	-	3(1)	3(1)
Total	37	3	31	27	30	36	160(4)

Note: Numbers in brackets denote cases of Type II DCS

Table 15. Abnormal long-bone x-rays of CAWs in the SMRT project

Contract	Abnormal entry LBXR		Abnormal exit LBXR	
	Type B	Other ortho.⁺ conditions	Type B	Other ortho.⁺ conditions
MTF*	1	2	1	2
109	0	0	0	0
108	0	0	0	0
104	0	0	0	0
107	2	2	0	0
MRTC	8	0	3	0
105	2	1	0	0
301	19	6	5	6
Total	32	11	9#	8#

\+ Orthopaedic

* MTF: MRT Task Force, comprising Firemen from the Singapore Fire Service

\# Only 643 CAWs had an exit LBXR. No new cases of DO were seen. The abnormal exit LBXRs listed here belonged to those CAWs who had existing lesions noted during the pre-employment examination.

lesions were allowed to work in compressed air, with limitations of exposure to not more than 2 bar. Follow-up yearly LBXRs showed no new cases of DO and of those CAWs who had pre-existing DO, further changes were not seen. Six-hundred and forty-three CAWs had exit LBXRs at the end of their contract. The remaining 1,094 CAWs either had their annual LBXR just before terminating their job or they did not turn up for their LBXR appointment before terminating their job. None from the group that had exit LBXR or near-exit annual LBXR had new findings of DO (Table 15).

13 Non-decompression sickness cases

From 21 September 1984 to 17 April 1987 (31 months), the 6 SMCs saw a total of 2,033 cases of non-decompression sicknesses. This gives an average monthly reporting sick rate of 65.6 patients. The breakdown for each contract of sick consultations is shown in Table 16.

The majority of cases (69.6%) were non-pressure-related illnesses followed by injuries (15.7%). Barotrauma contributed to 14.7% of non-DCS cases (Table 16).

There was one attendance at the medical centre for every 787.9 man hours worked. For every 5,005.8 man hours worked, there was one reported injury case seen.

13.1 Barotrauma

The number of barotrauma cases reported for the project was 299. The incidence of barotrauma based on the total number of man decompression was 1.59 per 2,000 man-decompressions.

Aural barotrauma accounted for 57.8% of all cases of barotraumas. The remaining 42% arose as a result of sinus barotrauma. No cases or pulmonary barotrauma was reported (Table 17). Dental barotrauma from poor filling of tooth cavities was also not reported.

13.2 Diseases

Table 18 shows the distribution of the different non-dysbaric diseases. Upper respiratory tract infection was the commonest (59.5%) medical condition seen. Afflictions of the skin contributed 8.3% and vasomotor rhinitis and rhinitis another 5.4% of the disease. 4.2% of those who reported sick suffered from otitis externa and media. This is not an uncommon conditions seen amongst divers and compressed air workers who work in high humidity environments. This environment could also have contributed to the relatively high incidence of skin problems.

13.3 Non dysbaric injuries

These accounted for 320 or 15.7% of the occasions when people reported sick at the various SMCs over the period of 31 months. Sprains and strains of muscles and ligaments accounted for 71.3% of the overall injuries. This is not surprising as most

Table 16. Non-decompression sickness conditions amongst CAWs of MRT Project by Contract

Contract	Barotrauma	Non-pressure related		No. of cases	% contract	Duration of CAW (months)	Number of man decompressions	Average cases per month
		Diseases	Injuries					
104	57	310	70	437	21.5	18	79,363	24.3
105	76	310	51	437	21.5	17	27,976	25.7
107	21	105	29	155	7.6	6	8,757	25.8
108	27	116	21	164	8.1	7	19,520	23.4
109	53	217	80	350	17.2	17	38,110	20.6
301	20	163	17	200	9.8	10	14,812	20.0
Others (MRTC, SFS, minor contractors)	45	193	52	290	14.3	31	-	1.7
Total	299	1414	320	2033	100.0	31*	188,538	65.5+
Percentage of non-DCS conditions	14.7	69.6	15.7	100%				

* This indicates the duration of CAWs in the MRT project as a whole

+ Average value over 31 months.

Table 17. Barotrauma amongst all CAWs of MRT Project

Barotrauma	**No.**	**%**
Aural	173	57.9
Sinus	126	42.1
Pulmonary	0	0
Total	299	100

of the workers handle heavy machinery or were employed on heavy labour. Contusions form 12.5% while cuts form 8.1% (see Table 19).

14 Discussion

14.1 Incidence of DCS in Singapore

The incidence of DCS on the Singapore MRT project was low when compared with compressed air work done elsewhere in the world (Table 20).

The main reason for the low incidence was that the pressures were low. We also believe that the intense and comprehensive medical training programme contributed significantly to achieving a low incidence. The highest pressures were at Contract 301, which had a maximum tunnel pressure of 2.35 bar.

From the start, no compromise was made with regard to the selection of men working in compressed air. Workers who were susceptible to DCS were either certified unfit or given a conditional clearance to work in compressed air. This included the obese (> 24% body fat by skinfold measurement), those with a high incidence of DCS in the past and CAWs with extensive DO. The obese and those with Type B DO were given conditional clearance. They were allowed to work at a limited pressure and for a limited duration only.

The control of proper decompression of the CAWs was also another key element in the low DCS incidence. The man-lock and medical-lock attendants were required to attend certification courses conducted by DHMC in conjunction with the National Productivity Board and the Ministry of Labour. In additional all the CAWs were thoroughly briefed on the safety aspects of compressed air work. They were taught the importance of proper decompression and the signs and symptoms of early decompression. Some CAWs, when found tampering with the emergency exhaust valves in the manlocks to release themselves early, were disciplined. The strict control of the decompression procedure paid off and minimised DCS occurring from negligence.

Table 18. Non-dysbaric diseases amongst CAWs

Condition	No.	Percentage
Upper respiratory tract infection (including pharyngitis, bronchitis, sinusitis)	842	59.5
Fungal and bacterial skin infection and other skin rashes	118	8.3
Otitis media/externa	60	4.2
Gastro-enteritis	39	2.8
Conjunctivitis and other conditions of the eye	32	2.3
Viral fever	35	2.5
Vasomotor rhinitis/rhinitis	77	5.4
Gastritis	18	1.3
Bites and stings	9	0.6
Pneumonia	3	0.2
Malaria	2	0.1
Heat exhaustion	2	0.1
Other injuries	177	12.5
Total	1,414	100.0

Table 19. Injuries in CAWs

Injuries	No.	%
Cuts/abrasions	26	8.1
Burns	4	1.3
Contusions	40	12.5
Chest injuries	2	0.6
Strains/sprains	228	71.3
Head injuries	4	1.3
Backache	3	0.9
Fractures	13	4.0
Total	320	100

Table 20. Comparison of various compressed air contracts

Contract	Period of compressed air (mths)	Total No. of CAWs	Maximum pressure (bar)	No. of man decompressions	No. of DCS cases (overall)	DCS incidence overall (%)	DCS incidence > 1 bar (%)
East River Tunnel NY 1914-21	84	-	3.26	1,360,000	680	0.05	-
Nowrah Bridge India 1938	6	509	2.72	12,400	353	2.8	-
Lincoln Tunnel NY 1955-56	18	704	2.31	138,000	44	0.03	0.07
Dartford Tunnel 1957-59	24	1,200	1.90	122,000	685	0.56	0.97
Blackwall Tunnel 1960-64	44	1,536	2.65	81,000	863	1.1	1.09
Tyne Road Tunnel 1960-65	38	650	2.86	44,800	711	1.6	1.74
Hong Kong Island line 1982-85	36	3,966	2.85	443,430	2,003	0.46	0.52
Singapore MRT 1984-87	31	1,737	2.35	188,538	164	0.087	0.26

The legislative framework determined the requirement for the control of the time of exposure. In addition, we believed that a period of acclimatisation for new starters did much to reduce the incidence of DCS. This was incorporated in the legislation for all new CAWs as well as for those who had been away for more than 12 consecutive days. Paton and Walder (1954) noted a high incidence of DCS in newly introduced workers, but with acclimatisation the incidence fell. As a result of the regulations, there were very few cases of new starters who experienced DCS.

The manlock attendants had to abide with the decompression schedules which were laid down in the Factories Act. The control of proper decompression was further ensured by frequent checks by the safety supervisors and the worksite medical officer.

The Blackpool Tables, designed by Hempleman, have been well tested in the UK and Hong Kong, with an overall incidence rate of less than 2%. For exposures less than 1 bar gauge, the Tables call for the CAWs to be decompressed to the surface at a rate not exceeding 0.4 bar/min, without any stops. However, in the Singapore MRT Project, a stop at 0.2 bar for 5 min was included in the regulations to reduce the rate of ascent even more, as we believed that it would have been too risky to ascent immediately to surface, especially after more than 8 h of exposure.

CAWs exposed to pressures of more than 1 bar gauge had to follow the decompression schedules according to the Blackpool Tables. These Tables required the CAWs to remain at the surface for at least 12 h in every 24 h. This was because at the end of decompression, residual nitrogen still remains in the body. This will become accumulative without an adequate rest period at atmospheric pressure.

Multiple entries into the compressed air tunnels were allowed for the supervisors. By law they were allowed to enter the chambers only 5 times in a 24-h period at pressures not greater than 2 bar, for not more than half an hour on any one occasion and with a minimum surface interval at normal pressure of more than 1.5 h.

There were instances where CAWs were exposed to compressed air for longer than the time limits imposed by the Blackpool Tables, either by misinformation or because of exigencies of the work. This posed a problem of decompression for two reasons. First, the Blackpool Tables are designed for a maximum exposure of 8 h and, secondly, nitrogen saturation of tissues in general may take as long as 3 to 4 days to complete. In order for the CAWs to work these long hours, extension of the Tables was required and additional stops had to be included as a safety measure. The incidence of DCS was lowered when compared with the incidence following the decompression with the unmodified Blackpool Table (see Table 9).

Among the 164 cases of DCS, 76.2% of them occurred with exposures exceeding 8 h. A higher incidence of DCS was also noted with increasing pressures (Table 9). This was observed with the incidence of DCS rising from a low of 0.008% at less than 1 bar exposures, to 0.936% for exposures at 2-3 bars. With the increase in duration of exposure, there was an overall increase in the DCS incidence.

The higher incidence of DCS noted at the extremes of exposure indicates that far greater risks are associated with long exposures with the use of the Blackpool Tables in spite of the adequacy of control. It is likely that with longer working hours, the workers had absorbed greater amounts of nitrogen and had also been subjected to greater physical stress such as the prolonged use of vibrating tools and lifting heavy loads over long distances in the tunnels. These cumulative factors could have resulted in the development of DCS in some cases. They appeared to be more susceptible to developing DCS than their peers who had been similarly exposed to the same pressures and performed the same type of work.

The overall incidence of the four cases of Type II DCS was 0.002% compared with 1 case (0.001%) reported by Lam and Yau (1988) in the Hong Kong MTR series. Three of the Type II cases developed DCS after more than 8 h exposure at 1.45 bar and above. The fourth case developed DCS as a result of sudden decompression, when the muck-lock doors were smashed open by the runaway rail cars. Three out of the four developed symptoms within 1 h of decompression, while the fourth case developed symptoms within 2 h.

The incidence of DCS among the obese CAWs was higher than the other CAWs (Table 10). Nitrogen has been shown to be five times more soluble in fat than in lean tissues like muscles. With long exposures, it is expected that there is near saturation of the fatty tissue by nitrogen. This results in a nitrogen load five times greater during decompression. Therefore it was not surprising that despite conditionally passing the obese CAWs and allowing them to work for limited time and pressure, this group still showed a higher incidence of DCS.

The supervisor category of personnel had a greater overall incidence of DCS when compared with the other categories of workers (Table 11). This is related to the nature of their work. They were required to enter and exit from tunnels up to five times a day, increasing their risk of DCS. The CAWs, in comparison, although they were performing heavier work over a longer period of time, exited only once per shift.

The incidence of DCS among the various races was comparable, with no significant difference being noted.

14.2 Clinical presentation

The commonest presentation of Type I DCS in our series was that of joint pain (Tables 2 and 3). This occurred in 96.3% of cases, with 76% of cases having pain in the lower limbs. Pain was mostly around the joints of the long bones. 84 cases (55%) presented with monoarticular pain while 70 cases (45%) presented with polyarticular pain. 76% had pain the their lower limbs compared with 87.82% in the Hong Kong MRT Project (Lam and Yau, 1988). The knee was commonly affected in our series (108 cases - 70.1%), followed by the shoulder joint (19 cases - 12.3%) and the elbow (16 cases - 10.4%).

A detailed analysis of the onset of symptoms revealed that 82.9% of cases had symptoms within 6 h of decompression, with 95.1% of cases presenting before 24 h. This is comparable with Lam's Hong Kong MRT series, with 93.6% presenting within 6 h.

There were five cases of cutaneous Type I DCS presenting as rashes. These cases proved difficult to diagnose. Two cases presented as swelling of the lymphatics and one who had lymphatic swelling over the inguinal region. These symptoms subsided with recompression therapy. In the Hong Kong series, four cases had skin mottling,

39 had non-specific symptoms of headache, nausea and vomiting. The other 94.6% (749) cases presented with joint pains.

14.3 Treatment

Early recognition of symptoms and prompt treatment were ensured by making it compulsory for CAWs to remain at the worksite for 2 h following decompression from the tunnel. This probably accounted for the fairly low relapse rate of 2.44% (4 cases). By comparison, 8.2% (64 cases) required a second treatment and 1.4% (11 cases) required third treatment in the Hong Kong series. Oxygen recompression therapy was used on 39 cases. This limited use was because not all site recompression chambers had built-in breathing systems and over-board dump systems for oxygen.

Adjuvant therapy included fluid replacement. This included Dextran, in addition to normal saline and Hartman's Solution. Aspirin was not given to the CAWs because of the possibility of masking the symptoms.

The objective in fluid therapy was to replace depleted blood volume, to restore haemocrit, prevent blood sludging and to improve tissue perfusion. It is common to find patients with acute DCS to have reduced blood volume, and many animal studies have confirmed this finding. Dextran has advantages in restoring intravascular fluid loss by increased vascular permeability during DCS (Cockett and Nakamura, 1964), although there are those who are more concerned with the potential of Dextran for creating an acute volume overload and further lung congestion. Our experience has shown us that judicious use of Dextran, especially in the young, robust CAW or fisherman divers with Type I and delayed Type II DCS, reaps benefits of improvement in the symptoms of our patients. We advocate the combined use of fluid therapy consisting of colloids and crystalloids with aspirin, dexamethasone, and recompression therapy in the management of serious cases of DCS.

14.4 DCS at less than 1 bar

In our experience, seven cases presented with DCS after exposure times of 12 or more hours, including the time required to decompress. The other three cases occurred after exposures of between 10 h 45 min and 11 h 45 min. By regulations, at pressures less than 1 bar, CAWs were only required to be decompressed to the surface at not greater than 0.4 bar per minute.

None of the cases was obese. The oldest worker was a 39 year old assistant foreman. Other than a surveyor, the rest were manual workers. The lowest working pressure from which we saw cases of DCS was 0.8 bar. All presented with deep joint pain, with tenderness in 1 worker, limitation of movement (3) and numbness (1). 60% of them presented within 6 h, but 50% of these were not sure of the exact onset time. One case presented with characteristic symptoms 28 h after the last decompression. The diagnosis was confirmed by recompression therapy.

14.5 Prevention

Monitoring and safety checks by medical officers and safety officers at the worksite were conducted regularly to ensure that there were no unsafe work practices. Daily checks of the barographs and the man-lock registers were made to ensure compliance with the prescribed Blackpool Decompression Schedules. They also ensured that workers would not exceed the prescribed working hours for the exposure pressure.

Inspection of man-locks revealed that some of them leaked especially in the last 0.3 bar of decompression. This was because of poor seals and door alignment. This problem was further compounded as some man-locks were not level. This meant that the workers were rapidly decompressed when the doors suddenly opened. As there was a real danger that this would precipitate DCS amongst the CAWs, measures were taken by the contractor to modify the chamber doors by including various locking devices. These held the doors shut until decompression was completed. Leaks found in some of the pipe-work were rectified promptly.

With the high humidity and temperatures in some of the tunnels, the contractors were told to install water coolers within the tunnels to provide the CAWs with ample cool potable water to prevent dehydration.

14.6 Accidents

Two major accidents, one each in Contract 104 and Contract 105, have been described above. Both cases involved explosive decompression resulting in DCS and barotrauma.

In both instances, the MRT medical operations room coordinated the emergency medical support. DHMC's recompression facility was thrown into full alert. Triage at the site was carried out and the casualties appropriately dealt with.

The operations room maintained open lines of communications between the SMCs, DHMC's recompression facilities and other relevant external agencies.

The accidents highlight the necessity for a back-up recompression facility with multiple chambers capable of treating all the workers of a single shift should an accident occur. The presentation of the casualties can vary and because of the differing severity, a single recompression chamber alone cannot provide the standard of treatment required for such a mass casualty situation. The number of cases of DCS would have been less (145 instead of 164) if not for these two accidents.

15 Conclusion

Decompression sickness is the bane of compressed air work. Though not totally preventable, much can be done to reduce its incidence. The prevention of this illness is to a large extent dependent on the recognition of the hazards involved and in the application of recognised medical and environmental control measures. The DCS incidence of 0.087% in the Singapore experience was low and the medical team involved in the project can look back at the months of compressed air works with much satisfaction. The attention to detail during planning and the adoption of strict medical standards in the selection of men and in safety control ensured that very few cases of DCS occurred out of ignorance or poor compliance with safety regulations. The preventive measures undertaken enabled a fairly good safety record to be achieved.

16 Acknowledgements

The authors wish to thank all the medical officers, nursing staff, medical orderlies and administrative staff of DHMC who have contributed to the completion of the Singapore Mass Rapid Transit Project.

17 References

Bennet, P.B. and Elliott, D.H. (1969) *The physiology and medicine of diving*; Ballière Tindall, London, 162-164

Cockett, A.T.K. and Nakamura, R.M. (1964) New concept in the patho-physiology of experiement dysbaric - decompression sickness. *Am. Surg.*, 30447-451

Durnin and Womersley (1974) Body fat assessed from total body density and its estimation from skin fold thickness on 481 men and women. *Br. J. Nutrition*, **32**, 77-92

Golding, F.C., Griffiths, P., Hempleman, H.V., Paton, W.D.M. and Walder, D.N. (1960) Decompression sickness during construction of the Dartford Tunnel. *Br. J. Ind. Med.*, **17**, pp 167-80

Goodman, M.W. and Workman, R.D. (1965) Minimum Recompression Oxygen Breathing. *Approach to the Treatment of Decompression Sickness in Divers and Aviators*, 5-655 U.S. Navy Experimental Diving Unit Washington D.C.

Lam, T.H. and Yau, K.P. Manifestations and treatment of 793 cases of decompression sickness in a compressed air tunnelling project in Hong Kong. *Undersea Biomedical Research*, 1988, **15**, pp 314-88

CIRIA (1972) *Medical Code of Practice for Work in Compressed Air.* Construction Industry Research and Information Association, London, Report 44, 3rd edn.

Paton, W.D.M., Walder, D.N. (1954) *Compressed Air Illness - An Investigation during the Construction of the Tyne Tunnel, 1948-1950.* Spec. Report, Ser. No. 281, Medical Research Council, London

Republic of Singapore Factories Act 1973 (Building operations and work of engineering construction 28 June 1985) - excavation and tunnelling work

Triger, E. (1841) Memoire sur un appareil a air comprimé, pour le pencement des puits de mines et autre travaux, sous les eaux et dans les sables submergés. *C R Acad Sc.*, Paris, **13**, pp 884-896

20 Thermal stress in compressed air work

K.C. Parsons
Human Modelling Group, Department of Human Sciences, Loughborough University, UK

Abstract
There are many methods and thermal indices which can be used in the assessment of hot environments but none is directly applicable to work in compressed air environments. The International Standards Organisation (ISO) has recently published standards for the assessment of hot, moderate and cold environments, and ISO 7933:1989 provides an analytical approach to assessment of hot environments. The sweat rate required to maintain a worker in thermal equilibrium is calculated from values of air temperature, humidity, mean radiant temperature and air velocity in the environment and estimates of clothing insulation and metabolic heat production of the worker. This (SW_{req}) value is compared with sweat rates which can be achieved by workers to provide a basis for determining working practices in the environment. In compressed air environments, equations for calculating the heat exchange between a worker and the environment and also the physiological capabilities of the workers will differ from those presented in ISO 7933.

ISO 7933 has been proposed as a European standard which would then become adopted in the United Kingdom. This paper considers the appropriateness of ISO 7933 for the assessment of hot conditions in compressed air work.
Keywords: Heat stress, Standards and limits, Thermal indices, Heat balance equation, Compressed air work.

1 Introduction

The purpose of this paper is to consider whether traditionally used methods for assessing and evaluating workers in thermal environments are appropriate for use in compressed air environments such as those found in some tunnelling operations. Mathematical descriptions of the heat transfer between a worker and his or her thermal environment are being increasingly used for assessing thermal environments (e.g. Nishi and Gagge, 1977; Haslam and Parsons, 1988; ISO 7933:1989). These rational approaches, however, often assume in the calculation of heat transfer, that the prevailing barometric pressure (P_B) is one atmosphere (1 Atm = 101.3 kPa =

Engineering and Health in Compressed Air Work Edited by F.M. Jardine and R.I. McCallum. Published in 1994 by E & FN Spon. ISBN 0 419 18460 0

1.0013 bar). In compressed air environments, such as in tunnelling, this assumption is not reasonable. In such environments, heat transfer by convection and evaporation will be significantly different from that found under normal atmospheric pressure.

Almost all rational methods for assessing hot environments are based on a calculation of evaporation required (hence sweating required) to preserve workers' health and performance (e.g. the Heat Stress Index (HSI) – Belding and Hatch 1955; the Index of Thermal Strain (ITS) – Givoni 1976; the Required Sweat Rate (SW_{req}) – ISO 7933:1989). If equations for calculating evaporative heat loss are inappropriate or in error, the interpretation of the analysis will be in error and decisions (about allowable exposure times for example) could be made which would endanger workers. The principle of assessment of a thermal environment and of how methods could be modified to assess compressed air environments are discussed below.

2 The heat balance equation for the human body

2.1 The principle

That the internal temperature of a worker should be maintained at around 37°C dictates that there is a heat balance between the body and its environment. Therefore, on average, heat transfer into the body and heat generation within the body must be balanced by heat outputs from the body. This does not mean that a steady state occurs. A steady state involves unchanging temperatures. Temperatures within the body and avenues of heat exchange will vary, the point being that for a constant temperature there will be a dynamic balance. If heat generation and inputs were greater than heat outputs, the body temperature would rise; and if heat outputs were greater the body temperature would fall.

The heat balance equation for the human body can be represented in many forms. However, all equations have the same underlying concept and involve three types of term; those for heat generation in the body; heat transfer; and heat storage. The metabolic rate (M) provides energy to enable the body to do mechanical work (W) and the remainder is released as heat (i.e. $M - W$). Heat transfer can be by conduction (K), convection (C), radiation (R) and evaporation (E). When combined together, all of the rates of heat production and loss provide a rate of heat storage (S). For the body to be in heat balance (i.e. constant temperature), the rate of heat storage is zero ($S = 0$). If there is a net heat gain, storage will be positive and body temperature will rise. If there is a net heat loss, storage will be negative and body temperature will fall. That the body temperature remains within acceptable limits for health provides one of the criteria for assessing thermal environments.

2.2 The conceptual equation

The conceptual heat balance equation is

$$M - W = E + R + C + K + S$$

i.e. for heat balance ($S = 0$)

$$(M - W) - E - R - C - K = 0$$

Where (M - W) is always positive, E, R, C and K are rates of heat loss from the body (i.e. positive values for heat loss, negative values for heat gain).

2.3 Units

It is important that all of the above terms can be expressed as rates of heat production or loss. It is traditional (and useful) to 'standardise' over persons of different sizes by using units of Watts per square metre of the total body surface area. The units are then W/m^2.

2.4 Body surface area

Total body surface area is traditionally estimated from the simplified equation of Dubois and Dubois (1915), i.e.

$$A_D = 0.202\, W^{0.425}.H^{0.725}$$

where

A_D = Dubois surface area (m^2)
W = Weight of body (kg)
H = Height of body (m).

A standard value of 1.8 m^2 is sometimes used for a 70 kg man of height 1.73 m.

Although A_D provides only an estimate of body surface area and there are more accurate methods, A_D is traditionally used; the nature of the heat balance equation means that any error is systematic and unimportant.

3 The practical heat balance equation

For an analysis of heat exchange between the body and the environment, and hence to quantify components of the equation to allow overall calculations (e.g. to make it useful in practice), there are two objectives:

1. The specific avenues of heat production and exchange for the human body should be identified.
2. Equations must be determined for the calculation of (estimate of) heat production and exchange. It is important for practical applications that terms in the equations are those which can be measured or estimated (e.g. the six basic parameters – air temperature (t_a), mean radiant temperature (t_r), relative humidity ($ø$), air velocity (v), clothing insulation, and activity level).

Note that for 'normal' conditions, heat exchange by conduction is often assumed to be negligible. ASHRAE (1989) give the following equation of heat balance:

$$M - W = Q_{sk} + Q_{res} = (C + R + E_{sk}) + (C_{res} + E_{res})$$

where

M	=	rate of metabolic energy production
W	=	rate of mechanical work
Q_{sk}	=	total rate of heat loss from the skin
Q_{res}	=	total rate of heat loss through respiration
C	=	rate of convective heat loss from the skin
R	=	rate of radiative heat loss from the skin
C_{res}	=	rate of convective heat loss from respiration
E_{res}	=	rate of evaporative heat loss from respiration
E_{sk}	=	rate of total evaporative heat loss from the skin given by:

$$E_{sk} = E_{rsw} + E_{dif}$$

where

E_{rsw}	=	rate of evaporative heat loss from the skin through sweating
E_{dif}	=	rate of evaporative heat loss from the skin through moisture diffusion

and each term is in units of W/m^2.

A practical approach is therefore to consider heat production within the body (M-W), heat loss at the skin ($C+R+E_{sk}$) and heat loss by respiration ($C_{res}+E_{res}$).

The next objective is to quantify components of the heat balance equation in terms of parameters which can be determined (measured or estimated).

3.1 Heat production within the body (M - W)

Heat production within the body is related to the activity of the person. In general, oxygen is taken into the body (i.e. breathing air) and is transported by the blood to the cells of the body where it is used to burn food. Most of the energy released is in terms of heat. Depending upon the activity, some external work will be performed. Energy for mechanical work will vary from about zero (for many activities) to no more than 25% of total metabolic rate. Methods for the determination and estimation of metabolic rate, which is one of the six basic factors for defining human thermal environments, are described in ISO 8996:1990. It should be noted that for the same activity more work will be required, and hence there will be greater heat production, as pressure increases. For example the work involved in breathing will increase in pressurised air.

3.2 Heat loss at the skin (C + R + E_{sk})

3.2.1 Sensible heat loss (R + C)

ASHRAE (1989) gives the following derivation:

$$C = f_{cl}h_c(t_{cl} - t_a)$$

$$R = f_{cl}h_r(t_{cl} - t_r)$$

$$(C + R) = f_{cl}h_c(t_{cl} - t_o) \qquad (1)$$

where

$$t_0 = \frac{(h_r t_r + h_c t_a)}{h_r + h_c}$$

and

$$h = h_r + h_c$$

The actual transfer of heat through clothing (conduction, convection, radiation) are combined into a single thermal resistance value R_{cl} so:

$$C + R = \frac{(t_{sk} - t_{cl})}{R_{cl}} \qquad (2)$$

combining (1) and (2) to remove t_{cl} the final form is:

$$(C + R) = \frac{(t_{sk} - t_0)}{R_{cl} + \frac{1}{(f_{cl}h)}} \qquad (3)$$

Where *C* and *R* are as defined above and

f_{cl} = clothing area factor. The surface area of the clothed body A_{cl}, divided by the surface area of the nude body A_D (ND)
h_c = convective heat transfer coefficient ($Wm^{-2}K^{-1}$)
h_r = linear radiative heat transfer coefficient ($Wm^{-2}K^{-1}$)
t_o = operative temperature (°C)
t_r = mean radiant temperature (°C)
t_a = air temperature (°C)
h = combined heat transfer coefficient ($Wm^{-2}K^{-1}$)
R_{cl} = thermal resistance of clothing (m^2KW^{-1})
t_{cl} = mean temperature over the clothed body (°C)
t_{sk} = mean skin temperature (°C).

Air temperature t_a, mean radiant temperature t_r and the thermal resistance of clothing R_{cl}, are all basic parameters which must be measured or estimated to define the environment. Mean skin temperature can be estimated as a constant value (e.g. around 33°C for comfort and 36°C under heat stress), or predicted from a dynamic model of human thermoregulation (see Nishi and Gagge, 1977). All other values above can be calculated from values of basic parameters. However, for a seated person Mitchell (1974) gives:

$$h_c = 8.3v^{0.6} \quad \textit{for} \quad 0.2 < v < 4.0$$

and

$$h_c = 3.1 \quad \textit{for} \quad 0 < v < 0.2$$

where v is the air velocity in ms^{-1} (i.e. a basic parameter).

The radiative heat transfer coefficient h_r can be given by:

$$h_r = 4\epsilon\sigma\left(\frac{A_r}{A_D}\right)\left[273.2 + \frac{(t_{cl} + t_r)}{2}\right]^3$$

where

ϵ	=	the area weighted average emissivity of the clothing/body surface (ND)
σ	=	Stefan-Boltzmann constant, 5.67 x 10^{-8} ($Wm^{-2}K^{-1}$)
A_r	=	Effective radiative area of the body (m^2).

ϵ is often assumed to be between 0.95 to 1.0. A_r/A_D can be estimated as 0.70 for a sitting person and 0.73 for a standing person (Fanger, 1967). t_r is a basic parameter and t_{cl} must be calculated using iteration techniques. ASHRAE (1989) suggests that a value of h_r = 4.7 $Wm^{-2}K^{-1}$ is a reasonable approximation for 'most typical indoor conditions'.

3.2.2 Evaporative heat loss from the skin (E_{sk})

ASHRAE (1989) give the following equation:

$$E_{sk} + \frac{w(P_{sk,s} - P_a)}{\left(R_{e,cl} + \dfrac{1}{f_{cl}h_e}\right)} \tag{4}$$

where

P_a	=	water vapour pressure in the ambient air (kPa)
$P_{sk,s}$	=	water vapour pressure at the skin, normally assumed to be that of saturated water vapour at skin temperature, t_{sk} (kPa)
$R_{e,cl}$	=	evaporative heat transfer resistance of the clothing surface (m^2kPaW^{-1})

h_e = convection evaporative heat transfer resistance at the clothing surface (W/m^2kPa)
w = skin wettedness. The fraction of wetted skin area (ND).

h_e is calculated using the Lewis relation, $h_e = (LR).h_c$, i.e. the ratio of the evaporative heat transfer coefficient to the convective heat transfer coefficient is a constant (LR). This is an important development in the establishment and practical method of calculation of the body heat balance equation allowing comparison and combination of dry and evaporative heat transfer.

The values P_a and $R_{e,cl}$ are basic parameters. $P_{sk,s}$ is calculated from Antoine's equation (ASHRAE, 1989) using the value for mean skin temperature t_{sk} (assumed constant value or from a dynamic model of human thermoregulation). Skin wettedness varies from a value of 0.06 when only natural diffusion of water through the skin occurs (E_{dif}) to 1.0 when skin is completely wet and maximum evaporation occurs (E_{max}). E_{max} can be calculated by setting $w = 1.0$ in equation (4). Skin wettedness can be calculated from:

$$w = 0.06 + 0.94\left(\frac{E_{rsw}}{E_{max}}\right) \qquad (5)$$

where

$$E_{rsw} = M_{rsw}.h_{fg} \qquad (6)$$

and
h_{fg} = heat of vaporisation of water = 2430 $kJ.kg^{-1}$ at 30°C
M_{rsw} = rate at which sweat is secreted ($kg/s^{-1}.m^{-2}$).

E_{max} can be determined from the basic parameters and equation (4). Skin evaporation and skin wettedness are often calculated from heat balance equation in terms of wettedness required (w_{req}) or evaporation required (E_{req}), to maintain the body in heat balance. They can be calculated by determining E_{rsw} using equation (6) and

$$M_{rsw} = 4.7\times10^{-5}.WSIG_b.\exp(WSIG_{sk}/10.7)$$

where

$$WSIG_b = t_b - t_{b,n} \quad for \quad t_b > t_{b,n}$$

i.e. the deviation of mean body temperature from that which would occur with the body in thermal neutrality and

$$WSIG_{sk} = t_{sk} - t_{sk,n} \quad for \quad t_{sk} > t_{sk,n}$$

i.e. the deviation of mean skin temperature from that which would occur with the body in thermal neutrality

and mean body temperature

$$t_b = \alpha t_{sk} + (1 - \alpha)t_{cr}$$

i.e. a weighted average of skin and core temperature, weighting (α) depending upon degree of vasodilation.

$$t_{sk,n} = 33.7°C$$

i.e. mean skin temperature for a person in thermal neutrality.

$$t_{cr,n} = 36.8°C$$

i.e. core temperature for a person in thermal neutrality.

Typical values for α are 0.2 for thermal equilibrium while sedentary, 0.1 in vasodilation and 0.33 for vasoconstriction.

The mean body temperature in neutrality (neutral bulk temperature) $t_{b,n}$ is therefore

$$t_{b,n} = 0.2 \times 33.7 + 0.8 \times 36.8 = 36.18°C$$

3.2.3 Heat loss from respiration ($C_{res} + E_{res}$)

Heat loss from respiration is by 'dry' convective heat transfer (C_{res}) due to cool air being inhaled, heated to core temperature in the lungs and heat transferred in exhaled air to the environment. In addition, inhaled air is moistened (to saturation) by the lungs. When exhaled, therefore, there is a mass (evaporative heat) transfer from the body core to the outside environment (E_{res}). ASHRAE (1989) give the following empirical equations for total respiratory heat loss:

$$C_{res} + E_{res} = [0.0014M(34 - t_a) + 0.173M(5.87 - P_a)]/A_D$$

A_D is calculated as described above. All other values are basic parameters (P_a in kPa).

The above presents the principles and practice of the human heat balance equation. It provides the basis for the rational method for assessing thermal environments such as that used in ISO 7933:1989 for the analytical assessment of hot environments.

4 ISO 7933; an analytical method for the assessment of hot environments

4.1 Principles of the method

ISO 7933:1989 specifies a method for the analytical evaluation and interpretation of the thermal stress experienced by a worker in a hot environment. It was derived from the work of Vogt *et al.* (1981) and has been validated in a programme of studies sponsored by the European Coal and Steel Industries (Metz *et al.*, 1988). An equation is used to calculate evaporation required by the 'average' worker (E_{req}) to maintain heat balance:

$$E_{req} = M - W - C_{res} - E_{res} - C - R$$

The following are then calculated:

$$\textit{Required skin wettedness} \quad w_{req} = \frac{E_{req}}{E_{max}}$$

$$\textit{Required sweat rate} \quad SW_{req} = \frac{E_{req}}{r_{req}}$$

where r_{req} is the efficiency of sweating at the required sweat rate, used to account for sweat loss by dripping and not evaporation.

The analysis of a workstation consists of determining predicted values (w_p, E_p, SW_p) taking account of required values (w_{req}, E_{req}, SW_{req}) and limit values (w_{max}, SW_{max}) based on the maximum a worker can physiologically achieve. If the required values are achievable by the subject within limits then predicted values are the required values. But when

$$w_{req} > w_{max} \quad then \quad w_p = w_{max}$$

$$\therefore \quad E_p = w_p.E_{max} \quad and \quad SW_p = E_p/r_p$$

where r_p is the efficiency of sweating at the predicted sweat rate.
If SW_{req} (or SW_p above) > SW_{max}, then w_p and r_p are derived from:

$$w_p.E_{max} = SW_{max}.r_p$$

$$E_p + w_p.E_{max} \quad and \quad SW_p = SW_{max}$$

4.2 Allowable exposure times

When $E_p = E_{req}$ and $SW_p < D_{max}/8$ then no limit has to be suggested for an eight-hour work shift (D_{max} is the maximum dehydration allowed over an 8-hour shift, hence division by 8 allows comparison with hourly rates). If this is not the case, allowable exposure times are calculated taking account of heat storage (rise in body temperature) and dehydration. Physiological data and hence limits are provided to allow calculation (see Table 1). The equations used to calculate heat transfer between the clothed worker and the environment are presented in Table 2.

Table 1. Reference values for determining criteria and limits with which to evaluate a hot environment. ISO 7933:1989

Criteria		Non-acclimatised workers		Acclimatised workers	
		Alarm	**Danger**	**Alarm**	**Danger**
Maximum skin wettedness W_{max}		0.85	0.85	1.0	1.0
Maximum sweat rate					
Rest:					
$M < 65$ W/m^2					
SW_{max}	W/m^2	100	150	200	300
	g/h	260	390	520	780
Work:					
$M > 65$ W/m^2					
SW_{max}	W/m^2	200	250	300	400
	g/h	520	650	780	1,040
Maximum heat storage					
Q_{max}	W.h/m^2	50	60	50	60
Maximum water loss					
D_{max}	W.h/m^2	1,000	1,250	1,500	2,000
	g	2,600	3,250	3,900	5,200

Table 2. Formulae used in the calculation of the heat balance equation: $E_{req} = M - W - C_{res} - E_{res} - C - R$, ISO 7933:1989, and possible modifications to account for barometric pressure

ISO 7933:1989 At atmospheric pressure (PB=1)	**Changes made to account for barometric pressure**
$C_{res} = 0.0014M(35 - t_a)$	$C_{res} = 0.0014M(35 - t_a).PB$
$E_{res} = 0.0173M(5.624 - P_a)$	
$C = h_c.F_{cl}.(t_{sk} - t_a)$	
$h_c = 8.7v^{0.6}$	$h_c = 8.7(v.PB)^{0.6}$
if $v < 1,\ h_c = 3.5 + 5.2v$	*if* $v < 1,\ h_c = 3.5 + 5.2v.PB$
if $v < 0.15,\ h_c = 2.38\lvert t_{sk} - t_a\rvert^{0.25}$	*if* $v < 0.15,\ h_c = 2.38(\lvert t_{sk} - t_a\rvert.PB)^{0.25}$
$R = h_r.F_{cl}.(t_{sk} - t_r)$	
$E_{max} = h_e.F_{pcl}.(P_{sk,s} - P_a)$	
$h_e = 16.7h_c$	$h_e = 16.7h_c/PB$

Note: F_{pcl} is an index of the vapour permeation properties of clothing (see ISO 7933:1989).

It can be seen in Table 2, that the principles used by ASHRAE (1989) and by ISO are as described in Section 3 above. Although the principles are the same, there are slight differences between the equations used by ASHRAE (1989) and by ISO 7933:1989, depending upon particular studies used to determine coefficients of heat transfer.

5 Compressed air environments

If ISO 7933:1989, or any other analytical method for assessing thermal environments, is accepted as a European Standard, it will be particularly important to consider its appropriateness for application over a wide range of conditions. The effect of barometric pressure on mechanisms of heat transfer will be significant and ISO 7933 will not be applicable in its present form to hypo- or hyperbaric environments. Equations of heat transfer by convection and evaporation will vary from those presented and based on the assumption of normal atmospheric pressure.

Kerslake (1972) and Nishi and Gagge (1977) describe how heat transfer by convection and evaporation depend upon barometric pressure. In summary, the main effects on the human heart balance equation are on the convective and evaporative heat transfer coefficients (h_c and h_e respectively) and are related to the change in density of the air with pressure. Kerslake (1972) notes that 'The opposite effects of pressure on convection and evaporation make sense since extra air molecules might be expected to promote sensible heat exchange while hindering the diffusion of molecules'. Nishi and Gagge (1977) and Nishi (1988) provide an analytical approach. For both hypo- and hyperbaric environments they demonstrate the importance of pressure and its effects on the body within the comfort zone and on body cooling in hypobaric environments (e.g. up mountains). Table 2 presents changes that may be applicable to the heat transfer equations used in ISO 7933 in order to account for the effects of barometric pressure. The apparent reduction in heat loss from evaporation may present a serious limitation to workers who would otherwise essentially maintain heat balance and health by losing heat by the evaporation of sweat.

6 Illustrative example

Consider the example of tunnel workers 'working hard', in light clothing and in lightly moving air. An analysis can be performed for a range of air temperatures and barometric pressures. Suppose the following measurements or assumptions are made about the working conditions: air temperature = mean radiant temperature, air velocity = 0.5 m/s, metabolic heat production = 200 W/m^2, partial vapour pressure = 2 kPa and insulation of clothing worn = 1.0 Clo (e.g. estimate for a working overall with undergarments - 0.155 m^{2}°C/W).

6.1 Analysis using ISO 7933:1989

If the above work were carried out in a barometric pressure of 1 atm (i.e. normal atmospheric pressure) then ISO 7933:1989 would be applicable. The results of this analysis are provided in Tables 3 and 4 and can also be seen in Figures 1, 2, 3 and 4 (i.e. for a barometric pressure of 1 atm). The allowable exposure times are calculated for alarm and danger levels for both workers acclimatised to heat and also for workers unacclimatised to heat.

It can be seen that for the conditions described above, allowable exposure times range from 45 to 83 minutes for alarm criteria for non-acclimatised workers to 80 to 210 minutes for danger criteria for acclimatised workers for air temperatures of 35°C and 25°C respectively.

6.2 Analysis in compressed air work

As the barometric pressure varies from normal the equations used in ISO 7933 must be modified. The equations for convective and evaporative heat transfer were replaced with those presented in Table 2 so that barometric pressure could be considered. The results of the 'modified' analysis are presented in Table 3 and in Figures 1 to 3 for absolute barometric pressures from one (normal) to four atmospheres (atm).

Table 3. Convective heat loss, evaporation required and maximum evaporation possible from the body (W/m^2) for a range of environmental conditions. ISO 7933:1989 and modified to account for barometric pressure (see Table 2). It is important to note that the values presented apply only to the illustrative example provided in Section 6.

Absolute barometric pressure (atm)	**Convective loss at air temperatures (°c) of:**			E_{req}			E_{max}		
	25	**30**	**35**	**25**	**30**	**35**	**25**	**30**	**35**
1	27	13	-1	145	167	189	127	135	144
2	33	16	-1	139	164	189	68	72	77
3	36	17	-1	135	162	189	47	50	53
4	38	18	-1	131	160	189	35	38	40

6.3 Discussion of the results of the example

Figures 1, 2 and 3 present the evaporation required (E_{req}) for workers to maintain heat balance in the environment compared with the maximum evaporation 'possible' in that environment (E_{max}). If E_{req} is greater than E_{max}, a rise in internal body temperature can be expected and working practices used to control exposure times to within those 'allowable'. For all conditions considered it can be seen that $E_{req} > E_{max}$. Allowable exposure times are provided in Table 4.

$t_a = t_r = 25°C$; $v = 0.5$ m/s; $P_a = 2$ kPa
$M = 200$ W/m^2; $W = 0$; $Clo = 1.0$

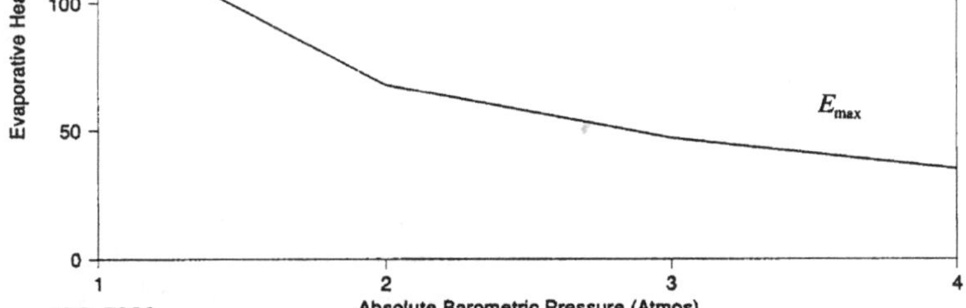

ISO 7933

Fig.1. Effects of barometric pressure on E_{req} and E_{max} (25°C)

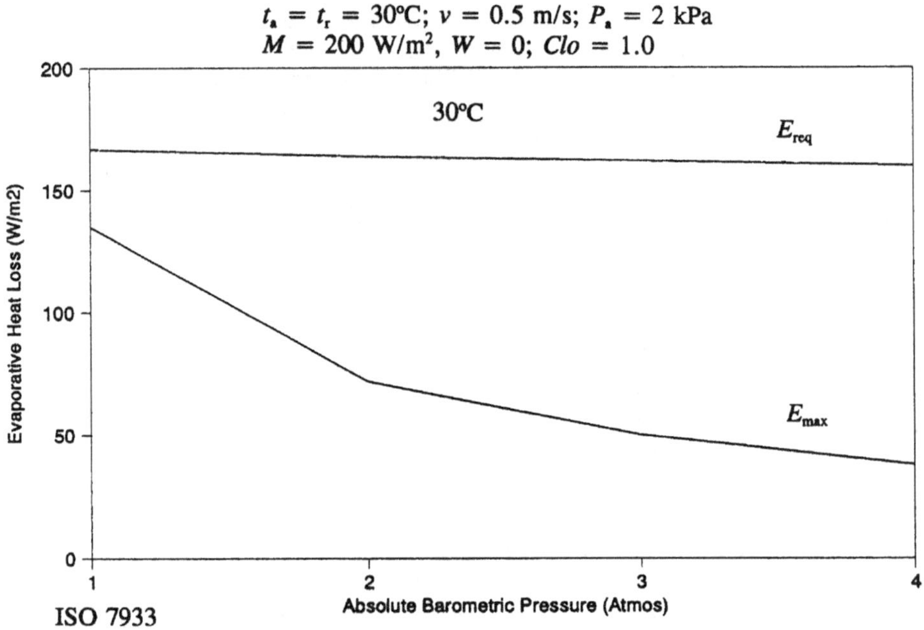

Fig.2. Effects of barometric pressure on E_{req} and E_{max} (30°C)

Table 4. Comparison of allowable exposure times for illustrative example described in Section 6

	Absolute barometric pressure	Allowable exposure times (min)*					
		Alarm limit			Danger limit		
		25°C	30°C	35°C	25°C	30°C	35°C
Workers not acclimatised to heat	1	83	58	45	99	70	54
	2	37	29	24	44	35	29
	3	32	25	21	38	30	25
	4	30	23	19	36	28	23
Workers acclimatised to heat	1	175	97	67	210	116	80
	2	42	33	27	51	39	32
	3	34	27	22	41	32	26
	4	32	25	20	38	30	24

*Based on ISO 7933:1989 modified to take account of barometric pressure

Figure 1 shows that at normal barometric pressures, at an air temperature of 25°C, workers would require 145 W/m² evaporative heat loss to maintain heat balance. However, the maximum evaporation possible (E_{max}) is 127 W/m². An allowable exposure time of 210 min (e.g. for an acclimatised subject and danger criteria) is calculated using ISO 7933:1989. As barometric pressure increases, E_{req}

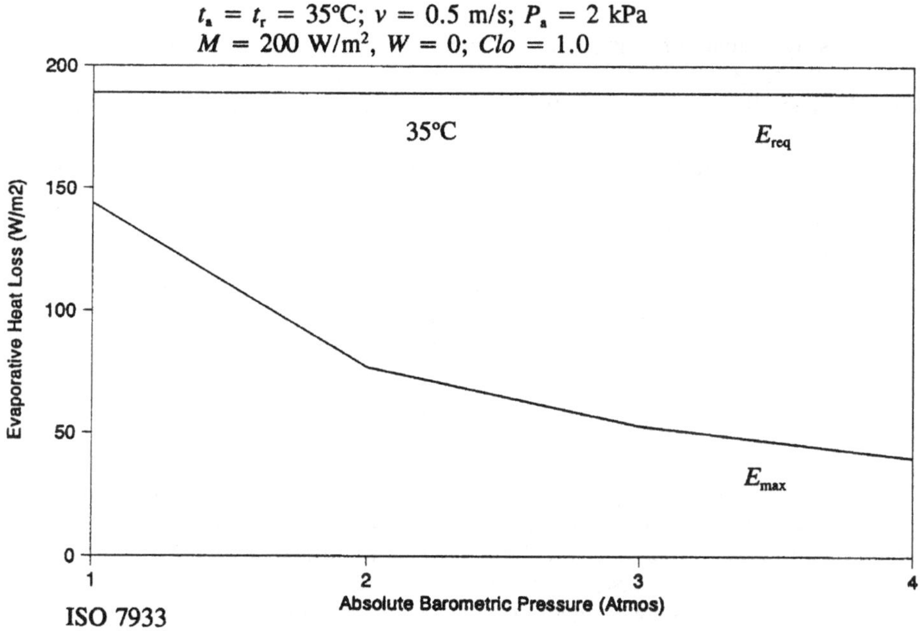

Fig.3. Effects of barometric pressure on E_{req} and E_{max} (35°C)

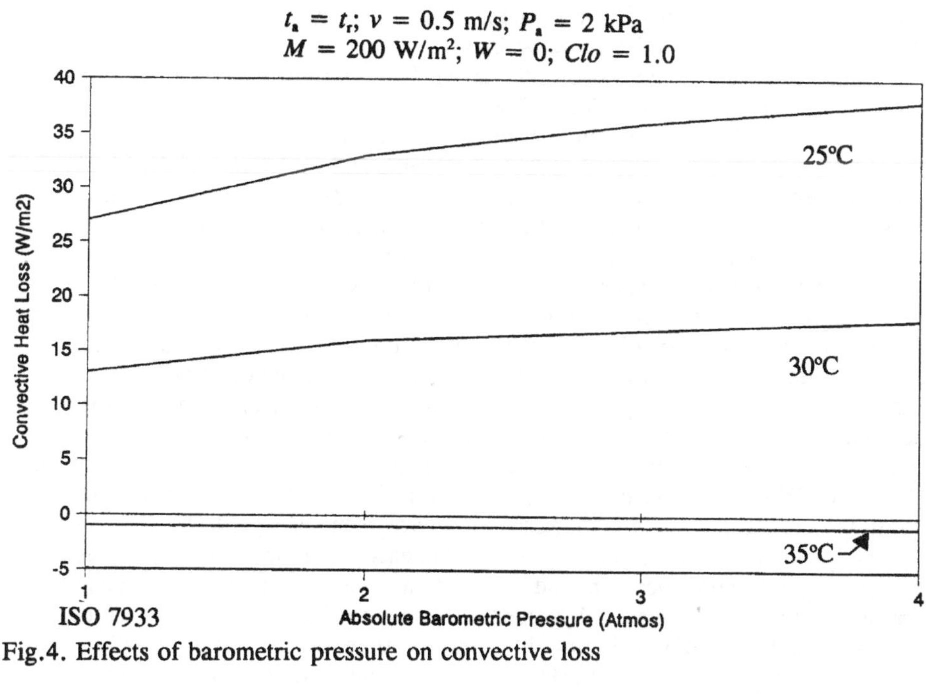

Fig.4. Effects of barometric pressure on convective loss

decreases by a small amount because of increased convective heat loss (Figure 4) but E_{max} greatly decreases. If ISO 7933:1989 were modified to account for these changes then allowable exposure times would be reduced from 210 min in 1 atm to 38 min in 4 atm.

As air temperature increases the small increase in convective heat loss with pressure will reduce (Figure 4) so E_{req} remains fairly constant. At 35°C air temperature, for example, the heat loss deficit (E_{req}-E_{max}) is 45 W/m^2 at 1 atm and 149 W/m^2 at 4 atm. Hence there is a very much greater predicted rate of increase in body 'core' temperature at 4 atm. The important practical point is that if ISO 7933:1989 is used in a non-modified form, in compressed air environments, workers may be in danger and suffer unacceptable thermal strain.

7 Discussion

The wider adoption of rational methods for assessing thermal environments, makes it particularly important that it is recognised that barometric pressure will greatly affect results. It should be clear that ISO 7933:1989 is not directly applicable to environments other than those of normal atmospheric pressure. The scope of the standard, however, cannot explicitly exclude all factors for which it is not applicable, and pressure is one of these. Even at barometric pressures of 1 atm, ISO 7933:1989 has limitations. For example, protective clothing is not considered, the method for estimating mean skin temperature is over-simplified and methods for determining work-rest cycles could be improved.

For compressed air environments the above analysis shows possible dangers and how some simple modifications could be made. The possible physiological responses of workers require further study however; and an evaluation comparing responses with predictions would aid in developing and analytical standard. The measurement of physiological responses such as of body core temperature, sweat loss and heart rate on workers can be used in working practices (ISO 9886:1992). However these may be impractical and unacceptable.

A simple direct index value used as a control measure could also be used. For example the temperature of a wet globe (WGT) or the Wet Bulb Globe Temperature instrument (WBGT – ISO 7243:1989) could be used and experience gained to establish limits for compressed air environments. A principle of using such instruments is that they respond to physical factors similar to those to which workers respond. In compressed air environments, therefore, evaporation from the instrument would be restricted in a way similar to that for evaporation of sweat from the body. Whether the instruments used in normal atmospheric conditions are those most appropriate for use in compressed air environments has yet to be determined.

This paper has not considered cold environments, where the barometric pressure is also of great importance. In cold environments the increase in convective heat transfer can potentially lead to hypothermia. Required evaporation will then be negligible and not useful and allowable exposure times will be based on fall in body skin and core temperatures. Working practices should therefore take account of the effects of pressure, e.g. in the calculation of the clothing insulation required for heat balance to be maintained in the environment (e.g. see ISO TR 11709:1992).

The overriding point of this paper is that evaporation of sweat from the body, the principal mechanism for maintaining workers' health and which is therefore the physiological parameter used for evaluating thermal strain is greatly decreased in compressed air environments. If this is not taken into account then workers could be exposed to unacceptable thermal strain.

8 References

ASHRAE (1989), *Fundamentals Handbook*, Chapter 8 Physiological Principles, Comfort and Health, Atlanta, USA: American Society of Heating, Refrigerating and Air-Conditioning Engineers

Belding, H.S. and Hatch, T.F. (1955) Index for evaluating heat stress in terms of resulting physiological strain. *Heating, Piping and Air conditioning*, Vol. 27, pp 129-136

Dubois, D. and Dubois, E.F. (1915) The measurement of the surface area of man. *Archs. Intern. Med.*, Vol. 15, pp 868-881

Fanger, P.O. (1967) Calculation of thermal comfort: Introduction of a basic comfort equation, *ASHRAE Trans.* 73, II

Fanger, P.O. (1970) *Thermal Comfort*, Copenhagen: Danish Technical Press

Givoni, B. (1976) *Man, Climate and Architecture*, 2nd Edition, Applied Science, London

Haslam, R.A. and Parsons, K.C. (1988) Quantifying the effects of clothing for models of human response to the thermal environment, *Ergonomics*, Vol. 31, pp 1787-1806

ISO 7243:1989 *Hot environments – Estimation of the heat stress on working man, based on the WBGT-index* (wet bulb globe temperature). International Standards Organisation, Geneva

ISO 8996:1990 *Ergonomics – Determination of metabolic heat production*, International Standards Organisation, Geneva

ISO 9886:1992 *Evaluation of thermal strain by physiological measurements.* International Standards Organisation, Geneva

ISO 7933:1989 *Hot environments – Analytical determination and interpretation of thermal stress using calculation of required sweat rate*, International Standards Organisation, Geneva

ISO 11079 (TR):1992 Technical Report. *Evaluation of cold environments. Determination of required clothing insulation, IREQ.* International Standards organisation, Geneva

Kerslake, D. McK. (1972) *The stress of hot environment*, Cambridge

McIntyre, D.A. (1980) *Indoor Climate*, London: Applied Science Publishers Ltd

Metz, B. *et al.* (1988) *Heat Stress Indices*, Proceedings of seminar of Commission of the European Communities, Health and Safety Directorate, Luxembourg, October 1988

Mitchell, D. (1974) Convective heat loss from man and other animals, in: Monteith, J.L. and Mount, L.E. (Eds.), *Heat loss from animals and man*, London: Butterworths

Nishi, T. and Gagge, A.P. (1977) Effective temperature scale useful for hypo- and hyperbaric environments. *Aviation Space and Environmental Medicine*, Vol. 48, pp 97-107

Nishi, Y. (1988) Prediction of severe body cooling in hypobaric environments. In: *Environmental Ergonomics*. I.B. Mekjavic, E.W. Banister and J.B. Morrison (Eds.), Taylor and Francis, London

Vogt, J.J., Candas, V., Libert, J.P. and Daull, F. (1981) Required sweat rate as an index of thermal strain in industry. In: *Bioengineering, thermal physiology and comfort*, K. Cena and J.A. Clark (Eds.), pp 99-110. Elsevier, Amsterdam

21 Oxygen decompression for tunnel workers in Brazil – The São Paulo Subway construction experience

I.J. Ribeiro
Centro Medico Hiperbarico de São Paulo, Brazil

Abstract
The use of oxygen decompression of compressed air workers is still regarded as too dangerous by many people involved with tunnelling, although it is acknowledged by some authors as a helpful method of decompression because of faster nitrogen elimination and a lower incidence of decompression sickness (DCS). In this paper we present our experience of this subject during the construction of the São Paulo Subway Tunnels, from October 14 to November 16, 1976, involving a total of 3,693 man decompressions from 1.5 bar using either air or oxygen, with no fires or explosions and no evidence of oxygen toxicity. A comparative study is made of the DCS incidence between the two groups. We discuss the method of decompression used (Behnke's oxygen decompression tables), system implementation, personnel training, safety measures and final results. It is considered that oxygen decompression for compressed air workers is a comprehensive and safe method to be used in working pressures of up to 1.5 bar for working shifts of 6 hours or more, provided its introduction to a working site is preceded by intensive and careful preparation of the personnel and materials involved and followed by close control and monitoring of the procedure.
Keywords: Oxygen Decompression, Decompression Sickness, Tunnelling, Compressed Air Work.

1 Introduction

It is well known, since the studies of Paul Bert in 1878, that inhalation of oxygen during decompression has beneficial effects in the prevention of caisson disease and osteonecrosis.

These days, its use is widespread among divers but not in caisson workers, mainly because of the potential hazards of fire and oxygen toxicity.

Back in 1976, during the construction of tunnels for the São Paulo Subways, we had to consider oxygen decompression for the workers because of the expected high working pressure (up to 1.5 bar) and long duration shifts (8 h) in a specific section

Engineering and Health in Compressed Air Work Edited by F.M. Jardine and R.I. McCallum. Published in 1994 by E & FN Spon. ISBN 0 419 18460 0

of the tunnels.

At that time we were using newly adopted Washington State Air Decompression Tables for Caisson Workers (now the OSHA Tables) which established 129 min of safe decompression for such an exposure to compressed air. Our intention was to reduce decompression time considerably yet maintaining a low incidence of decompression sickness and osteonecrosis. To achieve this we made a preliminary study comparing the German Oxygen Tables and Behnke's Oxygen Tables, finally adopting the latter.

Behnke's method recommends the inhalation of oxygen during the final stages of decompression, based on OSHA Tables, stating that 'decompression time in relation to the air tables can be reduced to half with a reduction in the incidence of bends'.

During the BART Project (San Francisco Metro) between 1967-1969, Behnke (1974) used oxygen inhalation with closed circuit breathing apparatus, observing that it was well tolerated in pressures of 40 psig for up to 2 h. The tests were performed in the worst possible conditions, the individual being 65 years old, overweight and with incipient diabetes mellitus, undergoing moderate exercise during oxygen decompression in the hyperbaric chamber with an ambient high degree of humidity and little ventilation. Behnke concluded that the decompression was highly safe and conservative.

Taking as a reference the OSHA tables, Behnke recommends the following method of decompression, using oxygen in the final stages.

'The total time of oxygen inhalation is the arithmetical mean of the last two stages of decompression using air, allowing 9 min for the decompression from 9 psig to the surface'.

Example	working pressure	40 psig
	working shift	4 h

OSHA Air tables		BEHNKE'S Oxygen tables	
pressure (psig)	decompression (min)	pressure (psig)	decompression (min)
40 to 24	3 - air	40 to 24	3 - air
24 to 8	45 - air	24 to 9	45 - air
8 to 4	35 - air	at 9	74 - oxygen
4 to 0	130 - air	9 to 0	9 - oxygen

In the example above, we have: $\frac{35 + 130}{2} = 82.5$ min

Total time for oxygen decompression is 83 min, being a 74 min-stage oxygen decompression at 9 psig, and 9 min bleeding oxygen decompression from 9 psig to the surface.

2 The planning of oxygen decompression

Oxygen under pressure constitutes, by itself, a serious fire hazard. Even at atmospheric pressure a 7% increase in oxygen concentration doubles the risk of fire. In the hands of unskilled persons or in locations where oil, grease or other combustible materials are present its use represents a great danger. During its use in a hyperbaric environment it is essential to monitor the atmosphere constantly in order to detect any early elevation in oxygen concentration in the ambient air. Also present is the risk of oxygen toxicity.

In order to avoid such risks we adopted the following measures:

1. The man lock was maintained clean and in excellent conditions of hygiene, free from oil, grease or other combustibles. Its interior was fire-proof painted and all electrical installations of low voltage were perfectly isolated. A sprinkler system was installed in connection with a pressurised water tank.
2. Oxygen inhalation was conducted using suitable breathing equipment with overboard discharge.
3. An oxymeter was installed inside the man lock for the early detection of any rise in oxygen concentration in the chamber air. It should never exceed the limit of 30%. By ventilation, when necessary, oxygen concentration was always maintained at a maximum level of 23% in ambient air.
4. All the workers involved in the procedure were previously trained in the use of the equipment, to recognise early signs and symptoms of oxygen toxicity, and were strongly advised to be aware of the risks of fire. They were told not to smoke nor to bring matches or other combustible materials inside the chamber.
5. During decompression an experienced supervisor was present inside the chamber in order that all the tasks were accomplished properly.

3 Introduction of oxygen decompression and results

The introduction of oxygen decompression for tunnel workers in São Paulo was accomplished in three phases:

Phase I (Testing Phase) - In this phase a group of 14 to 20 workers were submitted to oxygen decompression following intensive pre-operational instruction by medical and engineering staff. It extended for five working days, from October 14 to October 18, 1976, involving a total of 85 man-decompressions using oxygen (Behnke's Table). Working pressure was 1.5 bar, working period 6 h, total decompression time 58 min. There were no cases of decompression sickness in this phase. All decompression procedures were performed under the direct supervision of medical and engineering staff.

Phase II (Global Implementation Phase) - From October 20 to November 3, 1976, all of the workers in compressed air were submitted to oxygen decompression giving a total of 2,642 man decompressions from a working pressure of 1.5 bar, working shifts of 6 to 8 h, total decompression times 58 and 66 min. There were 52 cases of DCS among workers, an incidence of 2.98%. Such an incidence was very high by our standards, since the previous highest incidence of DCS during this contract had been 0.468% using air decompression from the same working pressure (21 cases in 4,479 man decompressions).

After a study of the situation, we concluded that the high incidence of DCS might have been due to such factors as follows: premature introduction of a new procedure without a sufficient explanation to all of the workers, lack of good supervision during all of the decompressions, misuse of the oxygen inhalation masks, lack of acclimatisation to the working pressure, and possible simulation of DCS symptoms by some workers.

Phase III (Reintroduction Phase) - The bad results found in Phase II, led us to adopt the same method used in Phase I, the testing phase: to perform oxygen decompression in a group of workers under direct supervision and after talks and training in the use of inhalation equipment. Between November 4 and 16, 1976, such a group of men was divided into two shifts, one of 30 men working 7 h at 1.5 bar with decompression time of 60 min and another shift of 15 men working more than 8 h at 1.5 bar with decompression time of 78 min.

Oxygen decompression was performed according to Behnke's method with a modification in the first stage (13 min air decompression from 22 to 9 psig, instead of 3 min). The remaining workers in the contract were decompressed in air from a pressure of 1.5 bar and a 7 h shift with a total decompression time of 132 min.

In this phase we were able to compare the incidence of DCS between oxygen and air decompression, as follows:

a) Oxygen decompression - one case of DCS in 902 man-decompressions (an incidence of 0.11%).
b) Air decompression - 9 cases of DCS in 2,048 man-decompressions (an incidence of 0.439%).

4 Conclusions

Although still experimental in Brazil, we consider oxygen decompression for tunnel workers has good prospects for lowering the incidence of decompression sickness.

The experiment we conducted during the São Paulo Subway Construction was fully satisfactory also in regard to fire safety, for in 3,700 decompressions, there was no fire incident.

From the standpoint of medical complications related to oxygen inhalation (respiratory or neurological damage) there were no cases of oxygen toxicity among the workers.

Provided its introduction in the working site is preceded by an intensive and cautious preparation of personnel, equipment and materials, and carried out with comprehensive supervision, we consider the use of oxygen decompression with Behnke's Tables for compressed air tunnel workers a useful and safe procedure to be employed at pressures of up to 1.5 bar for working shifts of up to 8 h.

5 References

Behnke, A.R. (1974), Personal Communication

Behnke, A.R. and Jones Jr., J.P. (1974) Preliminary BART Tunnel Results, in: *Dysbarism - Related Osteonecrosis* (eds Beckman and Elliott), U.S. Department of Health, Education and Welfare, Washington

Behnke, A.R. (1968) *Medical Aspects of Work in Pressurised Tunnel Operations*. San Francisco: Transit Insurance Administrators.

Behnke, A.R. (1974) *The Isobaric (Oxygen Window) Principle of Decompression*, abstract

Bennett, P.B. and Elliott, D.H. (1969) *The Physiology and Medicine of Diving and Compressed Air Work*. London: Baillière Tindall and Cassel

Bert, P. (1878) *La Pression Barometrique*. Paris: G. Masson

State of California, Department of Industrial Relations. (July 23, 1966) *Compressed Air Safety Orders*. Revised.

Department of Labour Occupational Safety and Health Administration - Safety and Health Regulations for Construction, Washington. (Saturday, December 16, 1972) *Federal Register*, Vol. 37, No. 243: 27559

Fire Journal. (November, 1965) Fire in a Hyperbaric Chamber

Fust, H.D. (September 6, 1971) Experiences with Oxygen Respiration Systems, in: *Caisson and Tunnel Airlocks, Diving Techniques*, Drager Tauch-technik

Hill, L. (1912) *Caisson Sickness and Physiology of Work in Compressed Air Control*. London: Edward Arnold.

Kindwall, E.P. (1974) Medical Aspects of Commercial Diving and and Compressed Air Work, in: *Occupational Medicine - Principles and Practical Applications* (eds for Carl Zenz)

Liebisch, H. (1976) Sicherheitsmabnahmen bei Schilvortriben unter Druckluft in Wien, *Der Aufban* No. 3

Matos, A. and Ribeiro, I.J. (1972) Doença Descompressiva, *Journal Brasileiro de Mediçina*, Vol. XXIII, No. 4: 54-74

Miner, A.C. (1968) Fire in the Pressure Chamber, *National Safety News*

Nashimoto, I. (1967) The Use of Oxygen during Decompression of Caisson Workers, in: *Decompression of Compressed Air Workers in Civil Engineering* (ed. McCallum, R.I.) London, Oriel Press Limited

Ribeiro, I.J. (1975) Osteonecrose Asseptica am Trabalhadores de Caixões Pneumáticos - Apresentação de um Caso, *Anais do XIV-CONPAT*, Rio de Janerio (to be edited)

Ribeiro, I.J. (to be edited) Variações de Pressão Atmosférica, in: *Perturbações causadas por Agentes Fisico*, Manual Brasileiro de Clinica. (eds Editora Guanabara-Koogan, Rio de Janerio)

Riberio, I.J., Chigo, G. and Robazzi, A.S.C. (1975) Aspectos Médicos do Trabalho sob Ar Comprimido na Construgão do Metrô de São Paulo - Comunicação Preliminar; *SOS Saude Ocupacional e Segurança*, Vol. 2

Riley, R.L. (1965) Gas Exchange and Transportation, in: *Philosophy and Biophysics* (eds. Ruch and Paton), Philadelphia and London

Turner, H.L. and Segal, L. (1965) Fire Behaviour and Protection in Hyperbaric Chambers, *Fire Journal*, November
Navy Department (1970) *U.S. Navy Diving Manual*, Washington
Bundesgesetzblatt (Oktober 14, 1972) Verordnung Über Arbeiten in Druckluft (*Druckluftverordnung*) Ausgegeben Zu Bonn

22 Decompression in tunnel construction using the hydroshield process

Experience in Switzerland with decompression tables adapted for construction sites in accordance with the ZH-L16 model

M. Rüegger, A.A. Bühlmann* and E. Völlm**
Occupational Medicine Department, SUVA, Lucerne, Switzerland
* *Erlenbach, Switzerland*
** *Dynatron AG, Zurich, Switzerland*

Abstract
Tunnelling by the hydroshield technique requires workers to be exposed to overpressure for the inspection and repair of cutting tools. For decompression purposes, we used the Bühlmann model ZH-L16 which was specially adapted to meet the needs of the construction site. In order to examine the validity of the model we retrospectively evaluated 1,961 overpressure exposures (OEs) registered during the first phase of construction, which ranged from 1 to 3.4 bar. Exposure times varied widely from 5 to 455 min. and decompression times from 3 to 271 min. The overall rate of decompression sickness was 0.56% (11 cases: 4 knee pains, 4 shoulder pains, 1 shoulder-chest pain, 2 skin bends. There were no central nervous lesions). Four of the cases happened within the highest pressure range of 3.0 - 3.4 bar (101 OEs), 3 within the next lower range of 2.5 - 2.9 bar (180 OEs) and 3 within the remaining pressure ranges (1,762 OEs). Therapeutic recompression with pure oxygen proved successful in all cases. In conclusion, we consider the Bühlmann decompression model to be a satisfactorily appropriate procedure for construction sites. One important risk factor for decompression incidents seems to be an uncomfortable body position resulting from a cramped air lock. A more spacious one should therefore be mandatory.

Keywords: Decompression for construction sites, Bühlmann model ZH-L16, Overpressure exposures, Decompression sickness, Therapeutic recompression, Risk factors.

1 Introduction

The demand for efficient traffic routes and environmental requirements resulted in the start of construction in 1990 of a new, 6.3 km-long rail tunnel at an altitude of 530 m above sea level in the Grauholz area (Berne). Construction was implemented using a tunnelling machine of the Herrenknecht type incorporating a hydroshield to cope with the water-bearing ground. This technique requires workers to expose

Engineering and Health in Compressed Air Work Edited by F.M. Jardine and R.I. McCallum. Published in 1994 by E & FN Spon. ISBN 0 419 18460 0

themselves to overpressure when inspecting and repairing the 11.6 m cutting wheel. To meet the problems of decompression, the construction consortium opted initially for the German decompression tables (1972). However in our capacity as the institution responsible for working safety in our country, we at SUVA rejected these as being too risky. We therefore had to provide other decompression tables which would also permit working times of more than 2 h at excess pressures of 3 bar and more.

2 The ZH-L16 model

In Switzerland, new decompression tables for air diving from three different altitudes (0-700 m, 701-2500 m, 2501-4500 m above sea-level) were introduced in 1986. These ZH-86 tables were calculated using the ZH-L16 model by Bühlmann (1986, 1988 and 1990). A detailed explanation of the background to this would go beyond the scope of this paper but, in summary, it represents an empirically derived mathematical algorithm expressing the almost linear relationship between absolute ambient pressure and tolerated inert gas overpressure in the tissue, represented by 16 half-value times. This model was considered a suitable basis for the construction-site decompression tables. DYNATRON, an engineering and consulting company, had the required software at their disposal and subsequently produced the tables we needed. They are structured in steps of 0.1 bar and permit decompression with air, as well as with pure oxygen, from 0.9 bar (Table 1). Preceding exposures automatically extend the decompression times of the following one, where the amount of excess time is indicated by 'repeat groups', marked by letters.

Table 1. Examples of decompression times

Overpressure (bar)	1.2		1.8		2.4	
Working time (min)	120	180	120	180	120	180
Decompression time (1) Druckluft-verordnung (1972)	4	4	27	46	67	112
(2) Bühlmann tables ZH-L16 (Grauholz/1989)	2	4	35	79	94	207

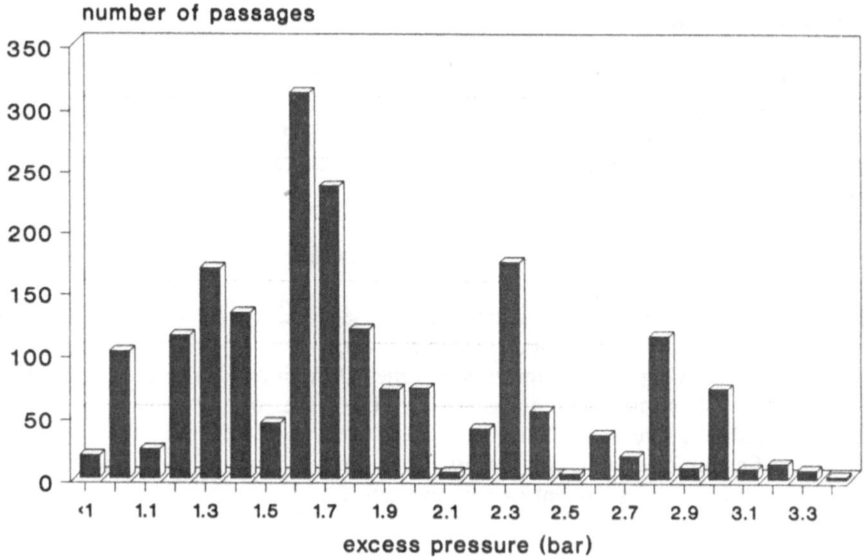

Fig. 1. Passages through air-lock/pressure stage

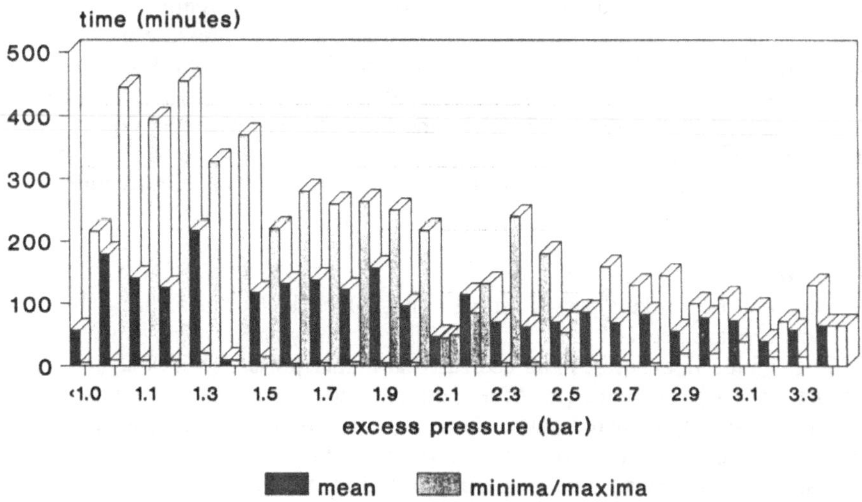

Fig. 2. Mean working times over individual pressure stages

3 Retrospective evaluation

The following items were at our disposal for evaluation purposes: the log books for the air-lock, the log book for the sickness air-lock on site and the records of each worker's preventive medical examination filed with us. What follows is therefore a retrospective evaluation of available data, which – we hoped – would provide the following information:

- Characteristics of the workforce concerned in terms of excess pressure medicine.
- Overall number of individual air-lock entry and exit movements and their distribution over the individual pressure stages.
- Average working and decompression times at the individual pressure stages.
- Characteristics, number, course and possible causes of any incidents attributable to excess pressure.
- Identification of certain risk factors deducible from the incidents observed.
- Interim, overall assessment of the new tables.

4 Results

Of 86 workers, 63 passed regularly through the air-lock, while 23 passed through once or only a few times. As a result, the latter group only accounted for 82 out of a total of 2,043 compressed air exposures. As we have no preventive medical records on them and none of them was affected by an excess pressure incident, they are not included in the following. The 63 workers from 9 different European countries in the first group were 34.6 years old on average (range 21 to 54) and had a body weight of 98.2% (76 to 134%) by the Broca Index (= weight (kg)/height (cm) - 100). Almost two thirds of them, 63.3% to be exact, were working for the very first time under the excess pressure conditions in the tunnel construction described here.

4.1 Compressed air exposures

As mentioned, 63 workers passed through the air-lock 1,961 times during the first construction phase between 15/1/90 and 16/7/91. Entry and exit movement peaked at 1.6 bar with 307 exposures and decreased at higher pressure with only 2 exposures at 3.4 bar (Figure 1). Access through the air-lock was for the purposes of inspection, maintenance and repairs. Such activities, by their nature, were not distributed regularly over the period mentioned and their duration was subject to wide variations. Working times therefore ranged from 3 to 455 min and decompression times consequently from 3 to 271 min. On average, working times gradually shortened from about 200 min in the lower pressure range to about 50 min in the highest one, and decompression times from 2 to about 120 minutes (Figures 2 and 3). For one worker there was a maximum of 2 to 3 exposures within 24 h.

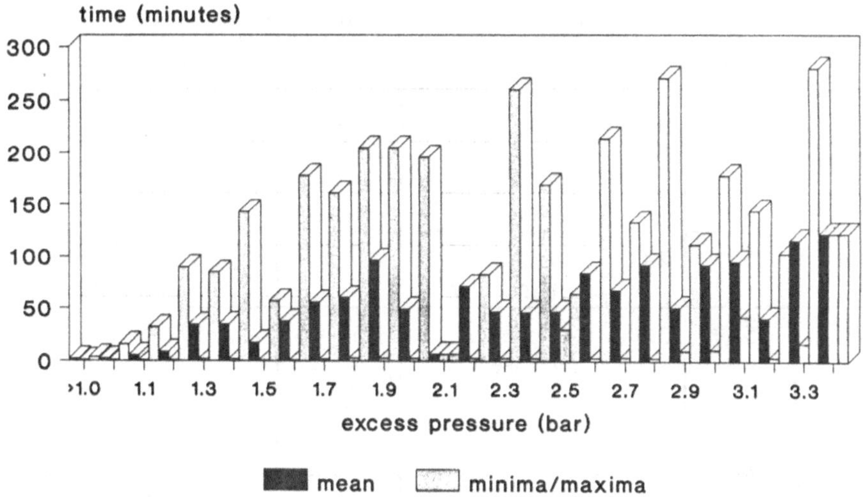

Fig. 3. Mean decompression times over individual pressure stages

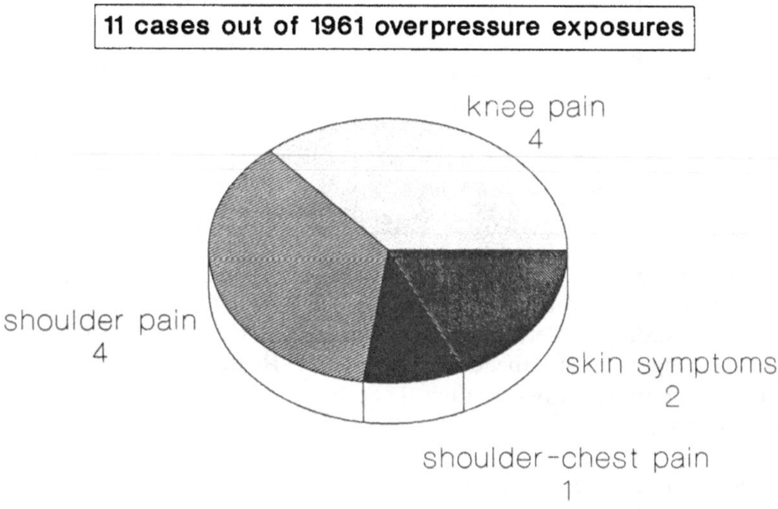

Fig. 4. Incidents attributable to excess pressure

Table 2. Decompression incidents occurring after exposures above 2.5 bar

Patient	Exp. time (min.)	Over-pressure (bar)	Symptoms	Latency after decom-pression (hours)	Therapy & duration (hours)	Remarks
F.F.	95	2.6	pain r. shoulder	4	* (4.5)	cramped posture
M.B.	75	3.0	pain r. knee	1.5	A (2.5)	change of rollers, cramped posture
5 days later	95	2.8	pain l. shoulder	6	* (4.5)	ditto
M.P.	75	3.0	pain r. shoulder	7.5	A *Hosp.	ditto
O.B.	94	3.0	pain l. arm, thorax	4	A (3.0) Hosp.	ditto
D.A.	90	3.0	pain l. knee	5	* (3.5)	Control and cleaning
K.J.	175	3.0	red spots on skin	3	* (3.0)	

* Treatment with pure oxygen and gradually decreasing overpressure-intervals of 55 min. from 1.5 bar to 0.6 bar

A Treatment according to German recommendation (Merkblatt für die Behandlung von Erkrankungen durch Arbeiten in Ueberdruck, Hauptverband der gewerblichen Berufsgenossenschaften, D-St. Augustin, April 1989)

( ) Length of therapy in hours

4.2 Incidents caused by excess pressure

During the period of observation, 10 workers were involved in 12 incidents. As Figure 4 shows, 11 cases were of decompression sickness and one case of barotrauma of the middle ear. Among the 11 cases of decompression sickness, nine concerned arthralgia and myalgia, with the knees and shoulders or upper arms being affected four times each plus one case of the left upper arm in combination with chest pains and headache. Skin symptoms were also observed twice (red spots,

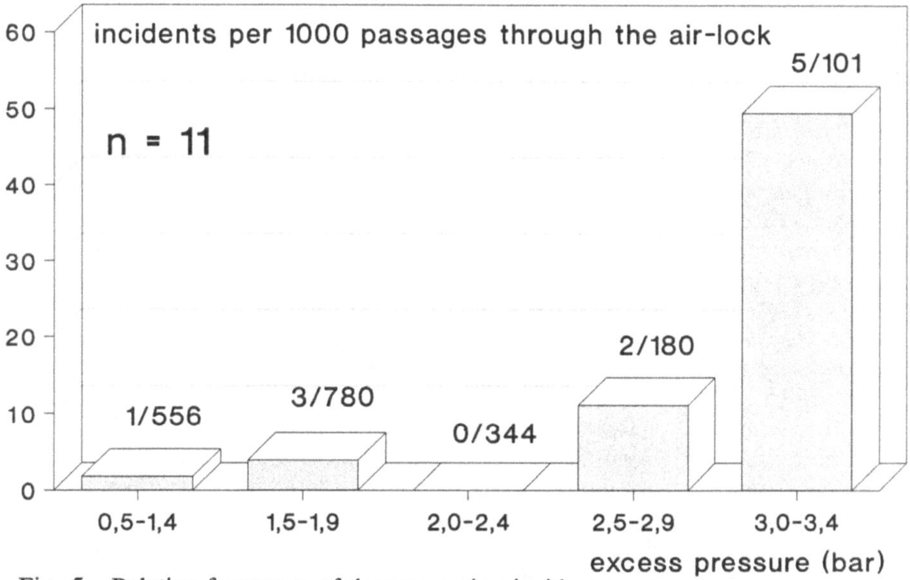

Fig. 5. Relative frequency of decompression incidents

itching, swelling). Details of the cases occurring after exposures over 2.5 bar are summarised in Table 2. The average latency between exiting the air-lock and the start of the complaint amounted to 5 h and 40 min (range 1.5 - 15 hours). Seven of these 11 incidents caused by excess pressure were concentrated on the 2.5-bar excess pressure stage, the remaining four, however, were distributed evenly over the lower pressure ranges in absolute terms. Four of the incidents took place on three consecutive days. As can be seen in Figure 5 – and this will come as no surprise – the risk of incidents increased as pressure rose.

4.3 The treatment and course of decompression incidents
With the exception of one case with skin symptoms, all the patients with decompression incidents were recompressed therapeutically following consultation with the doctor involved and, initially, on the construction site. This proved successful in 9 of the 11 cases. Two different styles of treatment were used, both of which, however, largely involved pure oxygen (Table 2). One patient felt no relief after the first course of treatment and another patient experienced no relief after the two combined methods. As a result, both had to be transferred for further treatment to the University Clinic in Zurich. However, after further HBO therapy, both were allowed to go home within 24 - 48 hours. It can be stated that all patients recovered fully.

5 Conclusions

5.1 Some thoughts on the causes of incidents resulting from excess pressure

Excess pressures and lengthy working times can lead to decompression problems as, due to the much greater volume of inert gas dissolved in the tissue, the limited predictability of individual and workplace-related factors have a greater effect, with the result that existing safety margins in the tables are more rapidly exceeded. Workplace factors of particular importance are, in our case, not only the heavy, physical work at times when changing the rollers and cutting tools, but also the deleterious environmental conditions, which occasionally reached a temperature of 30°C and 100% air humidity. It must also be pointed out that the workers were forced to adopt uncomfortable positions during their compulsory decompression phases when cramped in the 5.3 m^3 air-lock and leaning their shoulders against the corners, which could have had a contributory effect to the discomfort experienced in the upper extremities. Well-known risk factors such as being overweight or inexperience with compressed air are not considered to have played a role among the patients, as those concerned did not differ significantly from the entire group where these two factors are involved. The after-effect of previous entry and exit movements through the air-lock can largely be excluded as none of the patients had been exposed to excess pressure in the 24 h prior to the incident.

5.2 Overall assessment

The 0.56% incident rate of decompression sickness resulting from excess pressure in our group of patients is an expression of inadequate decompression (11 cases out of 1,961 overpressure exposures). Comparison with other construction sites is difficult, as hardly any relevant figures are available. In Munich (FRG) for instance, the 0.36% incident rate occurred within a much lower pressure range of 0.4-1.5 bar (Kessel, 1986; Mauermayer *et al.*, 1988) and in Kiel (FRG) with comparable pressures it was about 2% (personal communication). We believe therefore, that our results do not look too bad. With regard to the use of the tables, in particular the allowance for additional times in the form of repeat groups, those in charge of the construction site coped very well by trying to form consistent groups and by planning work assignments well in advance. As a result of the relatively low number of incidents, the additional costs caused by decompression times which were longer than anticipated were accepted by the construction consortium and the client. However, improvements are scheduled for the next construction phase and these will be implemented as far as possible. These mainly concern enlarging the working lock, which will improve seating comfort and prevent the adoption of a cramped posture. The possibility of decompression using oxygen is still a matter for discussion although the construction consortium has adopted a cautious attitude to this.

Acknowledgments
The authors thank Mr. W. Aebersold from the construction consortium and Mr. A. Zimmermann from Occupational Safety SUVA, for their help in providing data and support.

References

Bühlmann, A.A. (1986) The validity of a multi-tissue model in sport diving decompression. *Diving Officers' Conference*, November 1986, London, pp 25 - 27.

Bühlmann, A.A. (1988) Die Berechnung der risikoarmen Dekompression. Rechenmodell und Ergebnis der experimentellen Dekompressionsforschung. *Schweiz. med. Wschr.*, **118**, pp 185 - 197

Bühlmann, A.A. (1990) *Tauchmedizin*, Springer Verlag, Berlin, Heidelberg, London, pp 84 - 98

Kessel, R. (1986) Arbeitsmedizinische Aspekte zu Tätigkeiten im Ueberdruck, In: *Die sichere Anwendung der Spritzbetonbauweise unter Druckluft*. Tiefbau-Berufsgenossenschaft München, pp 15 -24.

Mauermayer, R., Kessel, R. and Praml G. (1985) Untersuchungen zur Festlegung neuer Ausschleuszeiten aus Druckbereichen um 1 bar. In: Bolt, H.M., Piekarski, C. and Rutenfranz, J. (Ed.) *Deutsche Gesellschaft für Arbeitsmedizin, 25. Jahrestagung*, Gentner, Stuttgart, pp 67 - 71

Verordnung über Arbeit in Druckluft (Druckluftverordnung), 4. Okt. 1972 Tiefbau-Berufsgenossenschaft, München

23 The effect of pressure on portable gas monitoring equipment used in tunnelling operations

D.W. Dabill and P.T. Walsh
Occupational Medicine and Hygiene Laboratory, Health and Safety Executive, Broad Lane, Sheffield, UK

Abstract
Portable workplace multi-gas monitors are used by the construction industry during compressed and free air tunnelling. This paper details a study of the effects of constant and transient pressure on the performance of two commercially available portable multi-gas monitors, and on the sensors used in these monitors. The monitors were subjected to gradually increasing and constant pressure (0 - 3.5 bar-gauge) and to pressure pulses up to 0.3 bar. Both types of monitor are affected by pressure and the reasons for this are discussed. Calibration of the monitors at the working pressure, under constant pressure, returns their performance to within specification. Transient pressure changes always affect the performance of the monitor, particularly the oxygen response, increasing the risk of false alarms. The merits of measuring volume fraction over partial pressure oxygen concentration in the workplace are argued. Finally, guidance is provided in the form of a limited number of recommendations on the use of portable gas monitoring equipment in hyperbaric atmospheres.
Keywords: Gas monitoring, Hyperbaric atmospheres, Sensors, Calibration, Flammable gases, Toxic gases, Oxygen

1 Introduction

While carrying out tunnelling operations in water-bearing ground, the construction industry sometimes pressurise a tunnel as a means of minimising water ingress at the tunnel face (BSI, 1990). Working pressures of up to 2 bar-gauge are typical, depending on the depth of the tunnel and the corresponding hydrostatic balance. However, legally, pressures up to 3.4 bar-gauge are permissible (CIRIA, 1982). Also, superimposed on this-steady state pressure are short-lived pressure pulses of up to 0.2 bar, which are generated by the opening and closing of pressure doors.

To monitor the pressurised (hyperbaric) atmosphere the industry uses the '3-in-1' or '4-in-1' type portable multi-gas monitors of which there are a number on the market. The monitors measure flammable gases, oxygen and one or two toxic gases

Engineering and Health in Compressed Air Work Edited by F.M. Jardine and R.I. McCallum. Published in 1994 by E & FN Spon. ISBN 0 419 18460 0

chosen from carbon monoxide (CO), sulphur dioxide (SO2), hydrogen sulphide (H_2S) and, more recently, carbon dioxide (CO_2). For tunnelling work the monitors are typically configured for CO and H_2S.

Portable multi-gas monitors of this type are used by many industries in a wide range of environments. However, there is no readily available information on their performance in hyperbaric atmospheres; consequently there is a need for guidance on their use under these conditions. Practical experience has reportedly shown wide fluctuations in readings when equipment is pressurised or depressurised, usually in the range 0.5 - 2.5 bar-gauge, and spurious alarms can occur. Research is therefore needed to address matters such as the range of pressures within which the instruments can operate and the accuracy of such instrumentation.

This paper reports on an investigation on the effect of constant and transient positive pressure on both portable multi-gas monitors, and the sensors utilised in monitors of this type. The objectives were to understand how the instruments are affected by pressure and to outline the steps needed to minimise these effects.

2 Experimental

2.1 Sensors

The monitors use three different types of sensor to measure flammable gases, toxic gases and oxygen. A brief description of the sensors is given below. The types of commercially available sensors used in the study are shown in Table 1.

Catalytic sensor (pellistor) – flammable gases

Pellistors are used to detect a wide range of flammable gases. They operate by catalytically oxidising the flammable gas on the surface of a heated element (approximately 520°C) (Walsh and Jones, 1991). The heat of oxidation is then detected by a platinum resistance thermometer embedded within the element. Normally the catalytically active element will be paired with an inactive element of the same construction and the two elements then form one arm of a bridge network. The role of the inactive element is to compensate for changes in ambient conditions, such as temperature and humidity.

Electrochemical fuel cell – carbon monoxide

Carbon monoxide (CO) is detected using an electrochemical fuel cell gas sensor. With these sensors the carbon monoxide diffuses into the cell *via* a diffusion barrier and is oxidised on a sensing electrode (anode). This generates a voltage between the sensing electrode and a counter electrode (cathode) and causes a current, proportional to the CO concentration to flow. The current is sustained by the electrochemical oxidation of CO at the sensing electrode matched by an equivalent amount of oxygen reaction at the counter electrode (Hobbs *et al.*, 1991).

Electrochemical metal-air battery – oxygen

Oxygen is detected using a diffusion limited metal-air battery, which in principle is very similar to the CO cell above, but the oxygen sensor is self powered (Hobbs *et al.*, 1991). In these sensors oxygen is reduced at the cathode to form hydroxyl ions and these in turn oxidise the anode (lead). Because the anode is constantly being consumed by oxidation to PbO, this type of cell has a finite lifetime (approximately 9 months in air).

Table 1. Sensors used in the study

Sensor	Type	Manufacturer
Catalytic (pellistor)	VQ2, VQ24	English Electric (EEV)[(1)]
Fuel cell	3ME/FC	City Technology[(2)]
Metal-air battery	C/S	City Technology[(2)]

(1) EEV (1992) *Combustible gas detector elements*
(2) City Technology Limited, (1990) *Product data handbook*

2.2 Diffusion barriers

In liquid electrolyte electrochemical sensors the response is dependent on the rate of reaction of the gas (O_2, CO, etc.) at the electrode surface. The design of the sensor is such that the rate of reaction is governed by the rate of diffusion of the gas across a diffusion barrier to the electrode, where reaction occurs. The type of diffusion barrier employed in electrochemical sensors has a profound influence on their characteristics (Hobbs *et al.*, 1991); this is discussed below.

Volume fraction sensors

Sensors which employ either a single-hole capillary barrier or a porous barrier, which can be in the form of porous plastic, metal or ceramic, measure gas concentration as volume fraction rather than partial pressure (Hobbs *et al.*, 1991). Bulk diffusion is the dominant process resulting in a sensor output which has a low temperature coefficient and provides a volume fraction measurement of concentration. However, different sensor designs result in devices that vary between those which have a very low pressure coefficient (100% volume fraction sensors), and those, which although not true partial pressure devices, exhibit a significant pressure coefficient. Single-hole capillary sensors fit the former category while sensors incorporating porous sinters fall into the latter category. All the sensors used in this study were of these two types.

Partial pressure sensors

Sensors incorporating non-porous solid-state barriers, where the reactant gas diffuses by solid-state diffusion through a polymer matrix, and sensors incorporating barriers based on Knudsen diffusion, provide a partial pressure measurement of the concentration of the reactant species (Hobbs *et al.*, 1991). The advantage of the Knudsen type is that it is far more robust than those incorporating solid-state barriers and has a much lower temperature coefficient.

2.3 Monitors

Two commercially available portable multi-gas monitors, commonly used by the construction industry when carrying out tunnelling operations, were used in the evaluation. They were the *Triple Plus* manufactured by Crowcon Ltd and the *Exotox 50* manufactured by Neotronics Ltd; the instruments were kindly loaned to us by the manufacturers. Each monitor was configured to measure flammable gas, carbon monoxide and oxygen. This is the usual configuration for these monitors when used in compressed air tunnelling operations. The extra channel on the '4 in 1' type *Exotox 50* measured hydrogen sulphide, although hydrogen sulphide was not considered in this study. Both types of monitor use very similar sensor technology. However, Crowcon utilise City Technology electrochemical sensors to monitor carbon monoxide and oxygen, while Neotronics manufacture their own sensors based on the same principles. Both manufacturers use EEV sensors for monitoring flammable gas: VQ24 (*Exotox 50*) and VQ32 (*Triple Plus*).

The monitors use a liquid crystal display (LCD) to display the gas concentrations and experimental results were recorded visually during the experiments. On the *Triple Plus*, all three gases are displayed simultaneously but on the *Exotox 50* only one gas can be displayed at a time. This complicated the evaluation because the monitor is designed automatically to switch itself from the channel being monitored to the appropriate channel if an alarm level has been exceeded. Therefore, it was necessary with the *Exotox 50*, when evaluating the carbon monoxide and flammable gas channels, to remove the oxygen sensor and replace it with a battery and potential divider to simulate the presence of the sensor. Otherwise, on compression, the monitor would immediately latch on to the oxygen channel.

2.4 Experimental set-up

Two pressure vessels, each tested to 7 bar, were used in the study. The smaller vessel (12 x 8 x 5 cm) was used for the sensor evaluation and a larger vessel (15 x 27 x 15 cm) was used for the monitor work. Each vessel was equipped with two 6-mm pipe fittings, one at each end, for connection into the gas flow system. The smaller vessel also had four gas-tight electrical feed-throughs which enabled power and signal connections to be made with the sensor. The larger vessel had a viewing panel on the top surface enabling the monitor display to be seen.

Each pressure vessel was incorporated into a gas flow system and the same basic system was used for both the sensor and the monitor evaluation. For gradually increasing and static pressure experiments it was important that a constant gas flow was maintained through the pressure vessel during compression. This was particularly important for sensors, such as catalytic sensors, which consume gas during operation and would therefore deplete the atmosphere if there was no gas replenishment. This was achieved using a mass flow controller upstream of the pressure vessel, a needle valve downstream and cylinder gas supply to pressurise the vessel. All the gas mixtures were derived from compressed gas cylinders.

A constant flow was set up with the needle valve fully open (minimal back pressure). The needle valve was then closed down until the outflow from the vessel, measured on a rotameter, was reduced to a set value and, by adjustment of the

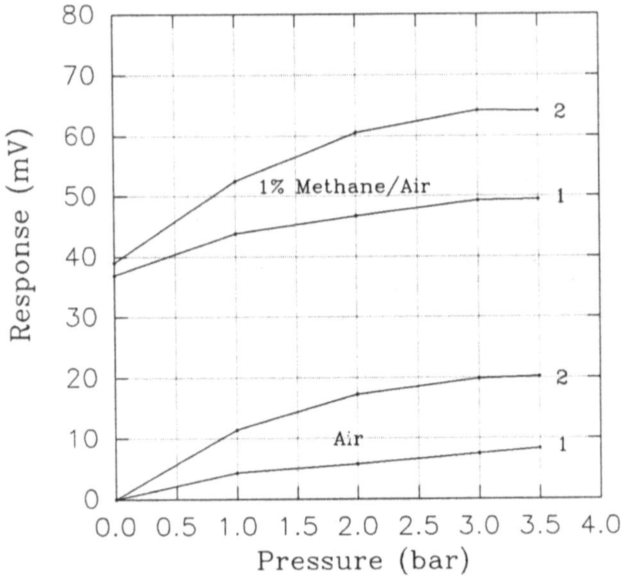

Fig. 1. The effect of constant pressure on the response of VQ24 pellistors to air and 1 % methane in air

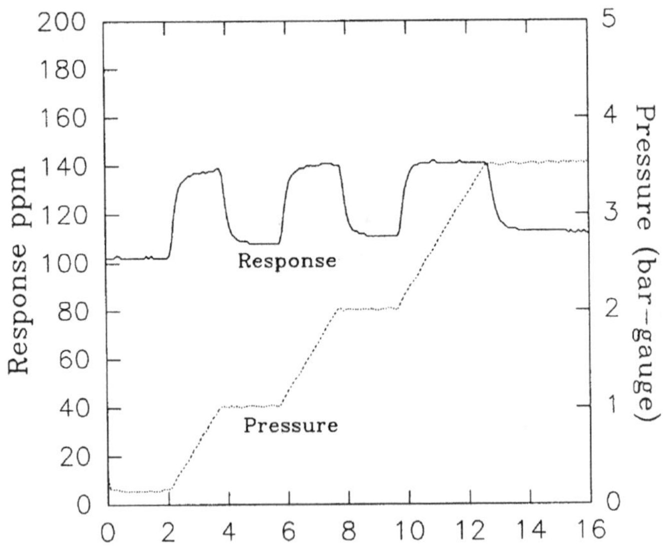

Fig. 2. The effect of increasing pressure on the response of an electrochemical CO sensor to 100 ppm CO in air

needle valve, was kept at this value during compression. Consequently, with the mass flow controller still providing a constant inflow, the pressure in the vessel increased at a constant rate. For the majority of the experiments the outflow was set to give a compression rate of approximately 0.5 bar/min. During compression the pressure in the vessel was monitored visually on a pressure gauge and also by a 0 - 7 bar pressure transducer.

For transient pressure experiments the arrangement was similar to above. However, the mass flow controller and the needle valve were replaced with manual gas flow valves, valves 1 and 2 respectively. The procedure was initially to close both valves 1 and 2. Valve 1 was then opened momentarily to allow a pulse of gas into the vessel. Valve 1 was then closed as valve 2 was opened to allow the pressure to return to normal. This technique allowed pulses of varying magnitude and duration to be produced.

A PC (computer) fitted with a differential analogue input board was used to collect data from both the sensors and from the pressure transducer. A BASIC programme controlled the input board, provided a real-time display of the sensor response and pressure profile, and allowed all the data to be saved to disc for further processing.

3 Results

3.1 Sensors

All three types of sensor were evaluated at pressures in the range 0 - 3.5 bar-gauge and, unless stated otherwise, at a compression rate of 0.5 bar/min. The effect of pressure pulses in the range 0.1 - 0.3 bar and approximately 5-s duration were also studied.

Pellistors

Experiments on three pairs of VQ24 elements were carried out. The results obtained (see Figure 1 where for clarity only the two extreme results are shown) show that the magnitude of the pressure effect is very variable, with measured percentage increases in the signal in 1% methane at 2 bar-gauge of 20, 40 and 52% (relative) for the three pairs. It is also clear that the absolute magnitude of the effect in methane is similar to, although slightly larger than, the effect of pressure on the air zero: absolute signal changes for the two sensors at 2 bar-gauge in air are 7,18 mV and for 1% CH_4/air 8,21 mV. The same absolute changes in signal were observed for 2% CH_4/air. These experiments indicate that pressure effects on the signal are caused by differential heat loss between detector and compensator rather than a pressure dependence of the rate of methane reaction.

Humidity (80% RH) was found to have no effect on the pressurised air or methane response.

Pressure pulses of approximately 5-s duration induce similar transients on to both the air zero and the flammable signal. For pulses in the range 0.1 - 0.3 bar then the effects are small, producing short-lived transients (< 10 s), equivalent to approximately 0.2% methane.

Electrochemical carbon monoxide fuel cell
Pressurising in air had no effect on the carbon monoxide fuel cell response. However, pressurising in 100 ppm carbon monoxide in air, resulted in a 40% increase in the response (see Figure 2). At constant pressure the response quickly (<1 min) stabilised and at 2 bar-gauge the sensor read approximately 11% high, e.g. 111 ppm for 100 ppm CO in air. Figure 3 shows that the observed pressure coefficient is variable over the range studied and decreases as the pressure increases. Averaged over the range 0 - 3.5 bar-gauge, the pressure coefficient is 3.7% relative/bar. The large increase in response of the sensor during compression is directly related to the compression rate. The reason for this is that during compression the cell response is no longer rate limited by the capillary diffusion barrier but is governed by bulk flow caused by the pressure differential set up across the capillary. In decompression, the reverse is true, and with no net inflow of CO into the cell, responses lower than the ambient gas concentration are observed as the sensor becomes depleted of CO.

A relative humidity of 80% had no effect on the response to 100 ppm CO in air.

Figure 4 shows the effect of pressure pulses on the air and 100 ppm CO response. Pulses of approximately 0.1 and 0.2 bar and less than 5-s duration, cause transient increases in the 100 ppm CO response of 20% and 80% respectively. There are no effects from pressure pulses in air.

Metal-Air battery oxygen sensor
The metal-air battery oxygen sensor is very similar in operation to the CO sensor and experiments here show that it is similarly affected by pressure. The main difference between the two is that the oxygen sensor is always actively measuring gas.

Although affected by constant pressure, Figure 5 demonstrates that the *Citicel Type C/S*, which is a capillary type oxygen sensor, maintains an approximately linear calibration over the pressure range 0 - 1 bar-gauge. Therefore calibrating the cell at the operating pressure would restore accuracy.

Figure 6 shows that on compression (0.25 bar/min) there is a measured increase of 25% in the response. Again, as with the CO sensor, the magnitude of the increase is directly proportional to the rate of compression and at constant pressure the sensor quickly recovers (within 1 min). Over the range 0 - 2 bar-gauge, there is a pressure coefficient of approximately 8 % relative/bar (see Figure 7).

It was found necessary with new sensors to 'condition' them at the working pressure (or, if possible at a pressure somewhat above the working pressure) for a period of 1-2 h before use. Figure 8 shows the first pressure run carried out on a new sensor and it can be observed that at approximately 1 bar-gauge a rapid rise in the response occurs. The reason for this is not clear (possibly mechanical effect) and has not been investigated in detail. However, if the sensor is left operating at that pressure for 1 to 2 h the signal will slowly return to normal and the sensor will then be conditioned to that pressure. Subsequent cycling to that pressure will show no rapid rise but if the sensor is taken above the conditioned pressure, then again at some point a rapid rise in the signal will occur. Therefore it is necessary to

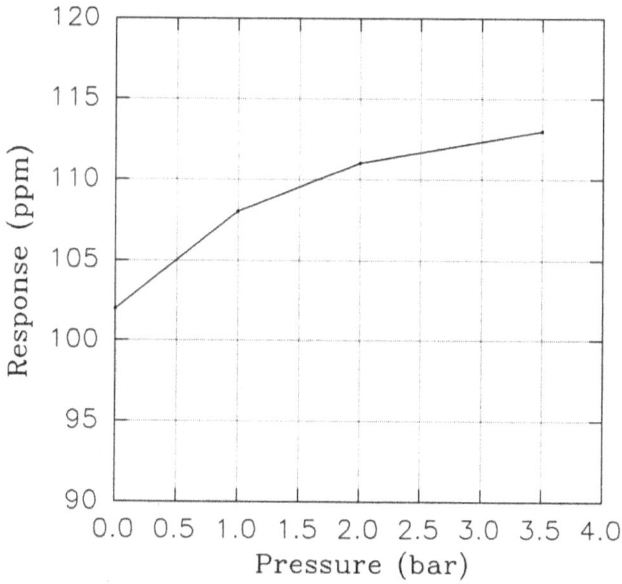

Fig.3. The effect of constant pressure on the response of an electrochemical CO sensor to 100 ppm CO in air

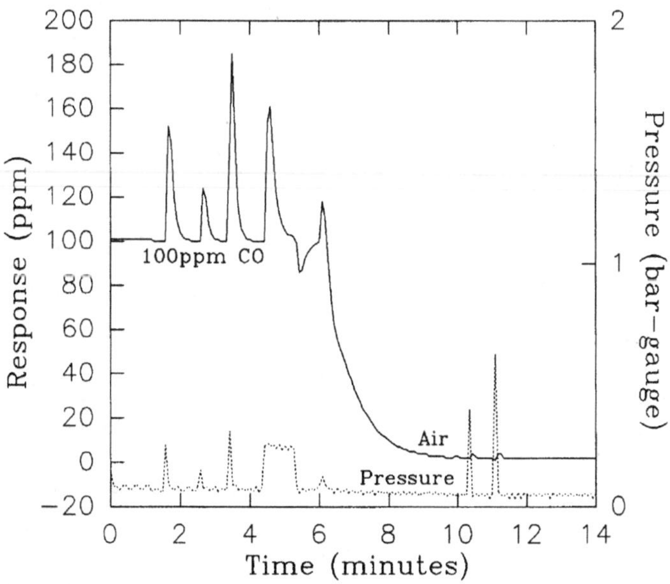

Fig.4. The effect of pressure pulses on the response of an electrochemical CO sensor in air and 100 ppm CO in air

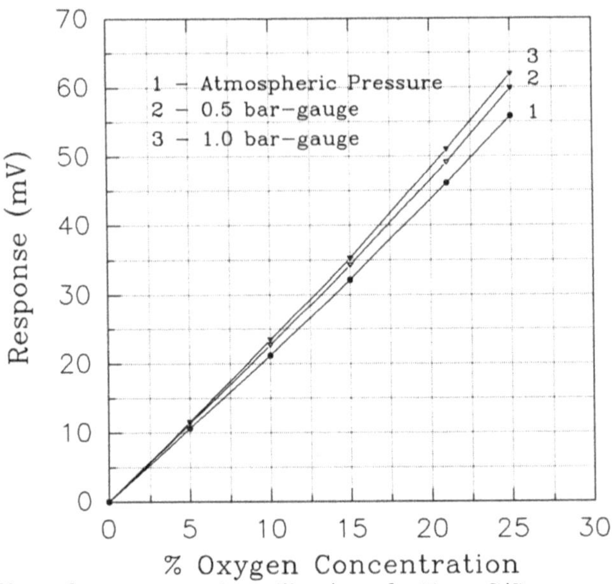

Fig.5. The effect of pressure on the calibration of a Type C/S oxygen sensor

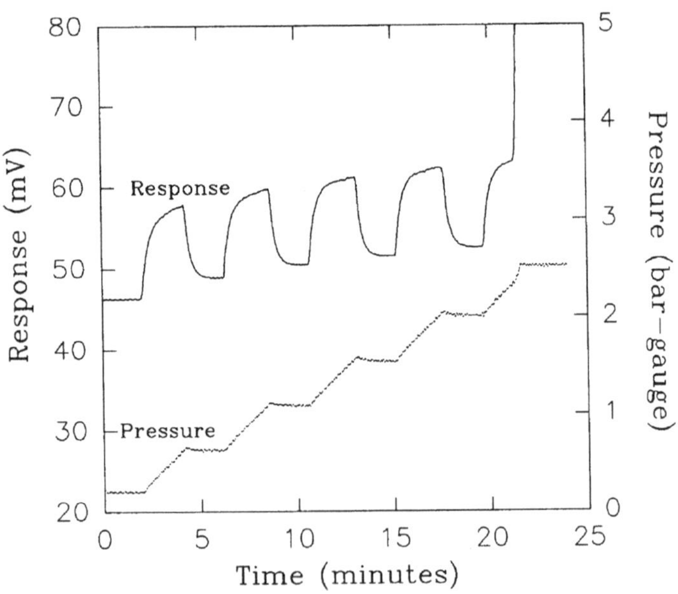

Fig.6. The effect of increasing pressure on the response of a Type C/S oxygen sensor

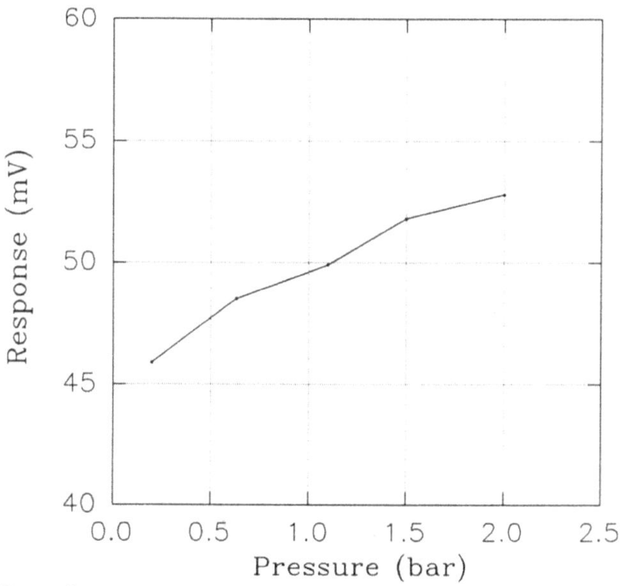

Fig.7. The effect of constant pressure on the response of a Type C/S oxygen sensor

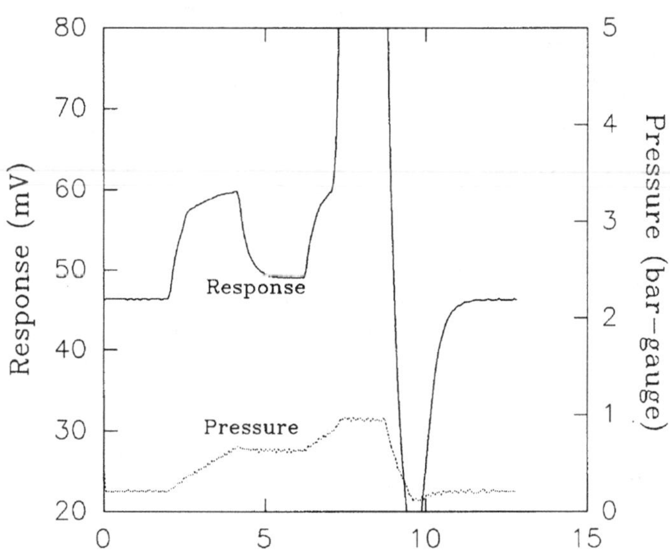

Fig.8. The first compression cycle performed on a new Type C/S oxygen sensor

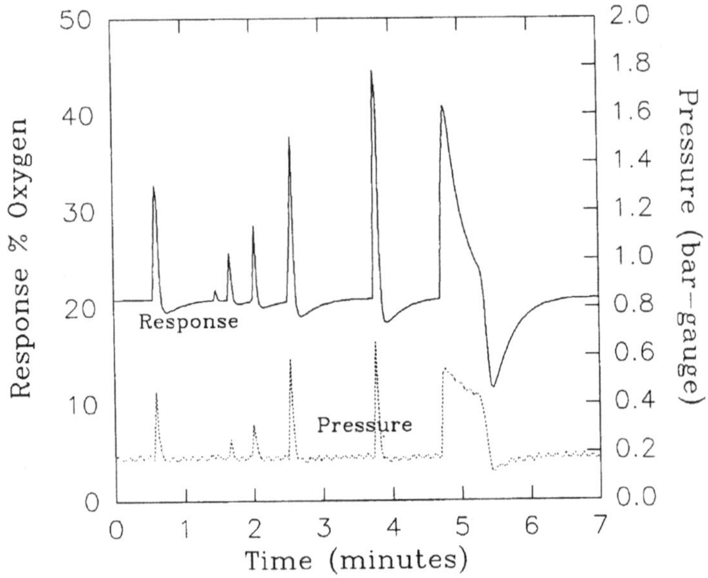

Fig.9. The effect of pressure pulses on the response of a Type C/S oxygen sensor

condition a new sensor at the working pressure for 1 to 2 h before use. Once this treatment has been performed the sensor remains conditioned.

No effects of humidity (80% RH) on the sensor over the pressure range were observed.

Oxygen sensors are very sensitive to pressure pulses. Figure 9 shows the effect of pressure pulses on the oxygen response and demonstrates that a pressure pulse of 0.1 bar would be sufficient to trip the high oxygen alarm (typically 24% oxygen).

Table 2. Response to air

Pressure bar (gauge)	**Crowcon triple plus Response**			**Neotronics Exotox 50 Response**			
	Oxy %	**CO ppm**	**Flam % LEL**	**Oxy %**	**CO ppm**	**Flam % LEL**	
0.2	20.7	0	0	22.2	1	0	-1
0.5					4	1	-1
1.0	22.1	-3	2	26.2	5	4	0
1.5					7	5	0
2.0	22.3	-3	5	28.9	9	6	0
2.5	22.9	-3	6		11	8	1

3.2 Monitors

The monitors were evaluated under laboratory conditions in air, carbon monoxide in air, and methane in air.

Air

The results are shown in Table 2. The *Triple Plus* monitor showed an 8% increase in the oxygen response at 2 bar-gauge; this was lower than that recorded in the sensor study which showed a 12% increase. Air had little effect on the CO sensor but there was a small increase in the flammable signal, equivalent to 5% LEL (lower explosive limit) at 2 bar-gauge.

The *Exotox* monitor, in contrast, showed a large oxygen pressure effect with a 30% increase at 2 bar-gauge. However, the oxygen sensors fitted in the monitors loaned were based on a sinter type of diffusion barrier and not a single-hole capillary design (see section 2.2). This type of diffusion barrier has a larger pressure coefficient than those based on a capillary design; however, they do have a smaller temperature coefficient. The effect of pressure on the flammable response shows some variability (data from two monitors shown in Table 2). In one case, pressure had very little effect on the air zero but the other showed an increase equivalent to 7% LEL at 2 bar-gauge. This variability was commented on in the sensor study (section 3.1). The carbon monoxide response was also affected and a reading of 9 ppm is recorded at 2 bar-gauge.

Table 3. Response to 100 ppm carbon monoxide in air

Pressure bar (gauge)	Crowcon Triple Plus Response			Neotronics Exotox 50 Response
	Oxy %	CO ppm	Flam % LEL	CO ppm
0.2	20.9	98	-1	131
0.5	21.4	105	1	143
1.0	21.5	110	2	163
1.5	21.7	116	3	179
2.0	22.3	118	6	190
2.5	22.2	121	6	196

100 ppm carbon monoxide in air

The results are shown in Table 3. Although the *Triple Plus* uses the same CO sensor as that investigated in the sensor study, the effect of pressure here is larger, showing a 20% increase in response at 2 bar-gauge, as opposed to 8% in the sensor work (section 3.1). The *Exotox* monitor showed a large CO pressure effect with a 45% increase in response at 2 bar-gauge. Again this is due to the different design of diffusion barrier used.

40% LEL methane in air

The results are shown in Table 4. In 40% LEL methane in air the *Triple Plus* showed an 11% increase at 2 bar-gauge. The absolute change was virtually the same value as that observed in air (i.e. 5 and 4 mV) and agrees with the findings in the sensor study (see section 3.1).

To demonstrate the variability of the effect of pressure on different flammable sensor pairs, the data from two of the *Exotox* monitors are shown. At 2 bar-gauge the percentage changes in response for the two are 39% and 28% respectively. The absolute changes here are much larger than those observed when pressurising in air; this is contrary to that found in the sensor work.

Table 4. Response to 40% LEL methane in air

Pressure bar (gauge)	**Crowcon Triple Plus Response**			**Neotronics Exotox 50 Response**	
	Oxy %	**CO ppm**	**Flam % LEL**	**Flam % LEL**	
0.2	20.4	3	36	41	39
0.5	20.9	0	38	44	40
1.0	21.5	0	39	48	43
1.5	21.5	0	40	53	46
2.0	21.8	0	40	57	50
2.5	22.0	0	41	62	54

Transient air pressure

The effects of gradually increasing and pulsed positive pressure in air were studied. It was found that pressure pulses of the order of 0.1 bar and of less than 5 seconds duration were sufficient to activate the high oxygen alarm on both monitors. Also, compression rates of approximately 0.07 bar/min will send the oxygen channel into a maximum or overload condition. Pulsed pressure has no effect on the carbon monoxide or the flammable gas signal but gradually increasing pressure shows a positive effect on the flammable gas air zero.

4 Discussion

4.1 Sensors and monitors

From the experimental work it is clear that pressure affects all the sensors to some degree. With the electrochemical sensors the same measured effect is observed between individual sensors of the same type. With catalytic sensors the effect is variable and three similar pairs of sensors studied at 2 bar-gauge showed measured increases in the 1% methane response of between 20% and 52%. Similar effects were observed with the monitors and, for the *Exotox 50* at 2 bar-gauge, increases of

28% and 39% were measured. This variability can be eliminated in the field by calibrating the monitor at the operating pressure.

The effects of pressure pulses on catalytic sensors are small, a pulse of 0.2 bar and less than 5-s duration produces an increase in response of approximately 0.1% methane.

In pressurised air containing no carbon monoxide, the electrochemical fuel cells are unaffected by pressure (constant or pulsed). In the presence of CO there is an observed increase in the response which is dependent on the compression rate: the higher the compression rate the larger the effect. At constant pressure the effects are much smaller and at 2 bar-gauge there is an approximate increase of 10% in the response to 100 ppm CO in air. Similar results were found with the monitors although the effects were larger. The *Triple Plus* monitor, which uses identical sensors to those used in the sensor evaluation, showed a 20% increase at 2 bar-gauge and the *Exotox 50* which utilises *Neotronics*' own sinter barrier sensors showed a 45% increase in response.

In the presence of CO, pressure pulses can induce large response transients and these can be significant and under certain conditions could lead to false alarms.

Unlike the other two sensors, the electrochemical oxygen sensor is always in the presence of the gas required to be measured and, like the similar CO fuel cell, these sensors are also affected by pressure in the presence of the gas. Therefore both constant positive and pulsed pressure will always cause increases in the sensor response. Results show that at a constant pressure of 2 bar-gauge there is an observed increase in the sensor response of approximately 12% (equivalent to 2.6% oxygen i.e. reading 23.6% in air). However pressure transients have a much larger effect, and transients of the order of 0.1 bar in air would be sufficient to trip the high oxygen alarm if set at 24% oxygen.

Again, with the monitors, the influence of the type of diffusion barrier used is highlighted by the much larger pressure effect exhibited by the oxygen sensors used in the *Exotox* monitor (30% increase in response at 2 bar-gauge) compared with the *Triple Plus* oxygen sensor (8% increase at that pressure). The *Triple Plus* uses oxygen sensors which incorporate a single hole capillary type of diffusion barrier as opposed to a porous sinter design used in the *Exotox 50*.

Both types of monitor are very sensitive to pressure pulses and pulses of the order of 0.1 bar and less than 5-s duration are sufficient to send the monitor into high oxygen alarm. Similarly, compression rates of 0.1 bar/min send the oxygen sensors into maximum or overload condition.

Calibrating the sensors at the working pressure will restore accuracy and reduce the number of false alarms when operating at constant pressure. Also it may be possible for manufacturers to dampen the effects of pressure pulses by fitting an auxiliary diffusion barrier in front of the sensor. However, slow increases in pressure of the order of 0.1 bar/min will always activate the alarm.

4.2 Partial pressure versus volume fraction monitoring in compressed air tunnelling operations

In the U.K. virtually all portable atmospheric gas monitoring equipment utilise sensors that are volume fraction devices and the two monitors evaluated here are of this type. However, there is an argument that monitors for use in hyperbaric atmospheres should incorporate sensors that are partial pressure devices, particularly for oxygen, as, physiologically it is partial pressure oxygen that is the important parameter in respiration and not the volume fraction. The risk is that an excess or a deficiency of oxygen may have harmful effects on the body.

However, over the pressure range encountered in tunnelling operations, typically 0 - 2.0 bar-gauge (but legally it can be as high as 3.4 bar-gauge), then the risks of both oxygen excess (hyperoxia) and oxygen deficiency (hypoxia) are very slight. For example, hypoxia is less of a problem in a compressed air tunnel than it would be in a tunnel being driven at atmospheric pressure because the partial pressure of the oxygen is higher in the compressed air tunnel than that required by the body for normal respiration. At the other extreme, the upper legal limit of 3.4 bar-gauge has been set such that the risks from hyperoxia over a normal working day are minimal and, in any case, compressed air tunnelling above 2.0 bar-gauge is uncommon.

Consequently, for background monitoring of the atmosphere during tunnelling operations, the authors contend that there are no real benefits in measuring the partial pressure of the oxygen. It is more useful to measure the volume fraction of the gas as this gives a direct measure of any changes in the atmospheric gas composition (e.g. O_2 deficiency caused by build-up of CO_2, fire) without the need to correct for any ambient pressure fluctuations that may have occurred.

5 Recommendations on the use of portable gas monitors in pressurised atmospheres

In all cases the effect of increasing pressure on the signal is positive and so the sensors under compression show a higher concentration than is actually present. However, on decompression the oxygen signal will go into low oxygen alarm and, if there is CO present, then the CO signal will show a *lower* concentration than the ambient.

Our recommendations as regards guidance for the use of portable gas monitoring equipment for general workplace monitoring under pressurised conditions are as follows:

1. Users must ensure when obtaining monitors, that the manufacturer is aware that they are for use in pressurised areas. Also users should try to make sure that capillary-type oxygen sensors are used in the monitor.
2. Sensor manufacturers and users must calibrate the monitor at the pressure at which it is to be operated. Monitors are supplied with gas manifolds for calibration purposes and zero and span controls are easily accessible. As the linearity of the sensors is unaffected by pressure, two gas mixtures will be sufficient for calibration; these could be:
   (a) A clean air mixture to set the zero and the oxygen level.

(b) A mixture to set the span levels. As there is no evidence for any cross interference between CO and methane at the concentrations used, one mixture of CO and methane will suffice. Suitable concentrations are 20% LEL for flammable gases and the occupational exposure limit (8-h long-term exposure limit: see HSE, 1992) for toxic gases, e.g. 1% CH4 and 50 ppm CO in air respectively. If other flammable and toxic gases are to be monitored, any potential interferences should be considered (e.g. CO and H_2S detection with electrochemical cells). In this case it may not be possible to use a single mixture to calibrate the span levels.

3. Oxygen electrochemical sensors are particularly sensitive to pressure pulses. This effect may be reduced by siting the monitor away from pressure doors, etc. It may be possible, however, for manufacturers to produce sensors that are less sensitive to pressure pulses by incorporating an auxiliary diffusion barrier forward of the active diffusion barrier.
4. Sensor manufacturers and users should be aware of the conditioning process that some oxygen sensors undergo. It may be necessary with a new oxygen sensor, initially to run the monitor at the operating pressure for some time (1 to 2 h), before a stable oxygen response is obtained.

If the recommendations are followed then, at constant pressure, portable multi-gas monitors provide useful warning of flammable and toxic gas hazards. However, if the pressure is fluctuating, the performance of the monitors will be affected, especially that of oxygen, and they will be of limited use.

Finally, it must be remembered, that this work examined the response of the monitors under hyperbaric conditions and that, as such, they were operating well outside their specification. Therefore, the results of this study should in no way be taken to represent their behaviour under the normal atmospheric conditions in which they were designed to operate.

References

British Standards Institution (1990) *Code of practice for safety in tunnelling in the construction industry*, BS 6164: 1990

Construction Industry Research and Information Association (CIRIA) Report No.44, *Medical code of practice for work in compressed air*, 3rd edition, (1982)

Health and Safety Executive (1992) EH 40/92, *Occupational exposure limits 1992*, HMSO, London

P T Walsh and T A Jones, Chapter 11: Calorimetric Chemical Sensors. In: *Sensors Vol. 2. Chemical and Biochemical Sensors Part 1*, (eds. W Göpel, T A Jones, M Kleitz, J Lundström & T Seiyama), VCH, Weinheim (1991)

B S Hobbs, A D S Tantram & R Chan-Henry, Chapter 6: Liquid Electrolyte Fuel Cells. In: *Techniques and Mechanisms in Gas Sensing*, (eds. P T Moseley, J O W Norris & D E Williams), Adam Hilger, Bristol (1991)

Discussion on Session 5: Occupational health and safety

Mr D R Lamont, the session chairman, introduced the speakers who were to present their papers. Surgeon Commander **J How** explained the arrangements for medical supervision of the Singapore Mass Rapid Transit construction by staff of the Republic of Singapore Navy and presented some of the results given in the full paper. **Dr K C Parsons** discussed an analytical approach to estimate thermal stress on workers in compressed air. **Dr I J Ribeiro**, after showing a video of its use on the site, reported the use of oxygen in decompression on the São Paulo subway tunnels. **Dr M Rüegger** explained the development and use of new decompression tables for use with a Herrenknecht Hydroshield at relatively high air pressures. **Dr D W Dabill** introduced the subject of the effect of pressure on portable gas-monitoring equipment used in tunnels. The discussion that followed these presentations is summarised as follows.

Dr E P Kindwall. I would like to comment on the difficulty of using electronic monitors in compressed air. We were faced with a possible methane problem in a tunnel at 15 psig (1 bar) pressure. We tried using electronic 4-gas analysers of more than one manufacturer, both American and German, but could not get them to function at pressure. The methane readings were spurious because of the high partial pressure of oxygen (42% effective) and the analysers burned out with exposure to compressed air. I finally put a standard analyser on the surface and fed tunnel air, from the face, to it *via* a long thin plastic tube (approx. 3 mm internal diameter). This worked well with less than one minute lag time over a 1/4 mile distance. We need a 4-gas monitor that works in compressed air but none is available.

For CO_2 I prefer to rely on hand-held gas sample glass tubes. When the suction pump is pulled the requisite number of times the reading is read directly, as the 'percentage' read is equal to the normal toxicity or physiologic effect of the gas. The same is true for CO.

For oxygen determinations, which are usually superfluous or unnecessary in a compressed air tunnel because of the high partial pressure of oxygen, any of the standard portable electronic sensors of O_2 partial pressure work well under pressure. They all use some sort of Clark electrode and are oxygen 'molecule containers'. Thus when calibrated to 21% outside the tunnel, they will read 42% at 2 atmosphere absolute (1 kg overpressure), 63% at 3 ATA, etc.). There are a number of manufacturers, including Beckman, Hudson, Biomarine, etc. It must be remembered, however, that the electrode has a service life of only about 3 months and a shelf life of about 6 months.

Fortunately methane problems in tunnels under compressed air are rare, as the air forces the methane away from the tunnel. Only when one takes the air off, does the methane return and then it may present an extreme danger. If one takes air off a tunnel that has penetrated ground containing methane, the length of the tunnel must be surveyed on foot for methane, before electronic locomotives are reintroduced. On one tunnel job in the USA, 6 men were killed in a methane explosion when they rode a muck train back into a tunnel that had had the pressure taken off 2 days before.

Mr J C Le Péchon. In Mr Dabill's studies, the test procedure is unclear, was the pressurisation with pure gas or with gas containing the contaminants?

If your analyser is reading concentration as you indicate, which is surprising for hyperbaric users, your system should not be recommended for use in compressed air since the parameters needed to assess gas quality are the partial pressures. A partial pressure sensor is needed inside the pressure chamber.

To control fire risk during oxygen breathing or oxyarc burning, oxygen concentration is to be monitored (oxygen fraction <0.25). An outside oxygen analyser is perfectly suitable provided the sampling system has an acceptable response time and its intake is properly located. In addition this analyser should be fitted with a high alarm set at 24%. Note that most of oxygen analysers used to control working atmospheres are fitted with a low alarm set at 16% to prevent the risk of hypoxia which is irrelevant in compressed air situation.

Dr Dabill. (Written response to question and comments submitted by Mr Le Péchon)

Test procedure for compression experiments. The compression experiments on both the sensors and the monitors were carried out using compressed gas cylinders containing the relevant gas mixture, i.e. 20 & 40% LEL methane in air and 100 ppm carbon monoxide in air. Regulators were used to reduce the pressure down to the required working level of 5 bar-gauge.

Volume fraction versus partial pressure monitoring of the working environment during compressed air tunnelling. Firstly, it should be stated that the monitors evaluated are specifically to monitor the ambient working atmosphere in the tunnel; their use in pressure chambers has not been considered.

Secondly, to our knowledge in the U.K. there are no portable multi-gas monitors that are designed to measure partial pressure. Indeed, probably the largest electrochemical sensor manufacturer in the U.K. does not now produce a partial pressure oxygen sensor. So to purchase a partial pressure monitor would be very difficult.

Finally, over the pressure range encountered in the majority of compressed air tunnelling projects in the U.K., i.e. 0-2 bar gauge, the risks from hypoxia and hyperoxia are very low. Therefore we feel there is little benefit from a health point of view in monitoring partial pressure oxygen, but from a safety stand-point there are advantages in monitoring the volume fraction concentration. The main advantage is that the monitor response will be independent of pressure. Thus, any observed changes in the oxygen concentration will be due to a change in the composition of the tunnel atmosphere and not because of a change in the tunnel pressure. Hence, the information is non-ambiguous: it quickly allows identification of problems such as the possible build-up of an asphyxiant gas and enables action to be taken swiftly. For this reason we believe volume fraction monitoring is advantageous over partial pressure which, unless pressure is monitored on the same instrument or at least in the vicinity, provides information that can be misinterpreted.

Fire risk. We agree that the oxygen concentration needs to be monitored and alarms set to provide warning of the increased risk of fire, etc. As we have stated above, volume fraction monitoring achieves this without any ambiguity.

Dr W Sterk. During this session as well as yesterday's sessions I heard greatly varying figures for the incidence of decompression illness (DCI). Some of the overall figures, presenting the percentage of DCI over the whole pressure range of a project, were rather low. However, this does not tell us very much about the DCI risk for the caisson workers. As an extreme example, we could take the DCI incidence for the whole population of Great Britain, since they are all exposed to barometric pressure changes, but only a minority is involved in diving or compressed air work, so the figure would be very low. From the projects presented here, it is clear that the incidence of DCI rises to around 2% or more if the working pressure is over 25 psig. Undoubtedly the probability of inducing DCI for work at these higher pressures can be decreased by more conservative decompression tables, the use of oxygen during decompression, reducing working time, etc. No matter what measures are taken, there will still be a risk of DCI, unless we stop exposing men to high pressure. Therefore, the crucial question is what incidence of DCI is considered acceptable? Only if this question is answered, can the decompression table designers can do their job properly. Could the authors answer this?

Dr Ribeiro. An acceptable level of DCS is difficult to establish and it has changed with time. It is a function of what is possible within the constraints of the project, its equipment and the national regulations. For the São Paulo subway, an exemption was needed from government to their using the decompression procedures that he had described, rather than the existing but out of date regulations.

Dr How. The question of how much DCS is acceptable is an important one, but if it cannot be eliminated, then we have to have a rapid reaction time for giving treatment, especially for Type II sickness.

Dr Rüegger. What is or is not acceptable is not a question for the doctor or the engineer to discover but for the public or the politicians to determine.

Dr H V Hempleman. Our problem is not knowing the *sequelae.* We have tried to reduce DCS to be as low as possible, but all approaches seem to be failing above 2 bar.

A major change, reflected in Dr Rüegger's paper, is the very much smaller size of the workforce – fewer people, fewer entries, when what would in number terms be few could be a relatively high percentage.

Dr R H McCaig. The CIRIA Code sets a limit of 27°C wet bulb temperature. Given the number of factors which determine thermal stress on the human, could I ask Dr Parsons whether this figure should be varied in any way to take account of different working conditions – for example different air flows, or should we leave well alone?

Dr Parsons. The type of wet bulb thermometer should be specified. Thermal stress does depend on the activity levels and the type of clothing being worn. It appears that more work is needed to establish better codes of practice.

(Reporters: J Buchanan, F M Jardine)

Part Six
Decompression Illness

24 Statistical investigation of working pressure and DCS for compressed-air workers

Y. Mano, M. Shibayama, N. Yamami, T. Nakayama and M. Kawashima
Tokyo Medical and Dental University, Japan

Abstract
Over recent years in Japan, compressed-air engineering has mainly been for pneumatic caissons; in tunnelling, higher working pressures are a result of mechanisation. The authors compiled the results of five years of surveys of decompression sickness (DCS) occurrence and compared the results with previous analyses. For this comparison, the control data were chosen from among those which had been collected by the same procedures as ours (*Japanese Journal of Industrial Health*, Vol. 29, 1987).

It is commonly said that DCS could not be developed under conditions with working pressures less than 1.0 kg/cm^2. Although we had previously encountered seven patients with DCS development under 1.0 kg/cm^2, no case was observed in the present survey which involved a total of 12,503 compressed-air exposures. For those compressed-air works which were at pressures over 1.0 kg/cm^2, there were: 3,623 more exposures in the period after 1986 than in the period before 1986; 5,827 more exposures were used in 1985 and earlier in the range of working pressures of 1.0-2.0 kg/cm^2 and there was a considerable increase in the exposures at pressures over 2.0 kg/cm^2 from 1986 onwards. This means that deep excavation works, needing higher working pressures than before, are increasingly used for today's compressed-air works. In other words, working conditions have become more severe in recent years and, therefore, more careful safety supervision was needed. When decompression control is carried out on the basis of the current decompression table, the development rate of DCS increases in line with the rise in working pressure, as was again observed in the present study.

Comparing the onset rates of DCS during these two periods, the results from the five years (1986-90) show lower rates (for working pressures intervals of 1.0 kg/cm^2) than those seen in the previous seven years. This, it is supposed, indicates that there is stricter decompression control today than previously.
Keywords: Compressed-air working, DCS rates, Bone lesions, Working pressures, Comparison with divers.

Engineering and Health in Compressed Air Work Edited by F.M. Jardine and R.I. McCallum. Published in 1994 by E & FN Spon. ISBN 0 419 18460 0

1 Introduction

In Japan, compressed air was first used for constructing a bulkhead and quay wall in Dejima, Nagasaki Prefecture, during 1789-1800. Foundation work by the caisson method started in earnest for constructing bridges, e.g. at Eitai, Kiyosu and Kototoi over the Sumida river, as part of the metropolitan reconstruction works following the Great Earthquake of 1923 in the Kanto area, when an American engineer, A.H. Hughes, with others, directed in these works. From 1935 on, the pneumatic method was actively employed in civil engineering works, reaching peak utilisation in construction works related to the Tokyo Olympics in 1964. Today, the pneumatic method is employed as a technically established method (Mano, 1992).

The pneumatic method, although accompanied by possible hazards, such as DCS, because of the compressed air, also has the following benefits: it is a reliable method for foundation construction; it is a particularly effective countermeasure to cope with groundwater and earth pressures; and there is little disturbance to the road traffic. These are essential considerations for engineering works in urban places.

Under these circumstances, the pneumatic method has frequently been used in tunnelling works for water supply as well as wastewater systems, electric power, telegraph, telephone and subway constructions. From about 1970 to 1984, compressed-air shield tunnelling, together with pneumatic caissons, were the principal types of pneumatic work.

But over the years, mechanisation has progressed for coping with poor ground by the development of slurry and earth-pressure balance tunnelling machines. From about 1985 on, the compressed-air shield method has been used only occasionally. As tunnelling has increasingly been at even greater depths, so vertical shafts also became deeper. In the same way as the innovations in shield methods, the caisson method was reappraised in order to avoid the influence of compressed air on human bodies, and to provide an alternative pneumatic method for deep foundation works. Although from 1985, the pneumatic method also seemed no longer to be needed, it was re-evaluated as both an economic as well as a reliable method from about 1989 on. Consequently, extremely deep excavation works, which rely on the pneumatic method, although requiring fewer entries, involve higher working pressures for their excavation operations.

Based on the above, the authors investigated the development rate of DCS in line with total numbers of worked exposures in each 0.2 kg/cm^2 pressure interval, treating the obtained results statistically; and they conducted a dynamic survey on the DCS incidence rate as well as yearly changes of working pressures. The authors carried out these investigations in conjunction with their other activities and involvement with the works such as; visiting the sites, conducting checks on safety and sanitary controls, researching actual conditions of the whole process of the work, and asking questions of those concerned.

2 Methods

The authors investigated the actual field conditions of compressed-air sites over the 12 years, 1979-1990. Dividing the 12 years into two groups, the first seven and the later five years, and considering working pressures greater than 1.0 kg/cm^2 (which could develop DCS) in intervals of 0.2 kg/cm^2, we surveyed the total numbers of both workers and exposures and the onset of cases of DCS.

It is specified in the law that compressed-air workers should undergo a medical examination every six months. We analysed the records of the medical examinations particularly in relation to bone lesions in the results.

The university has provided hyperbaric treatment facilities for DCS since 1966, and we have experience of the medical care of over 200 patients with Type II DCS. The authors examined the results from this experience, and analysed proportional changes between compressed-air workers and divers divided into three groups, i.e. those treated in the periods 1970-1980, 1981-1985, and 1986-1990.

3 Results

3.1 DCS incidence

Table 1 shows the incidence of DCS in compressed-air workers and for different working pressures for two periods. The total number of work exposures is 51,097, of which 38,594 were at pressures of 1.0 kg/cm^2 or over. DCS was observed in 264 of these, corresponding to a 0.68% (264/38, 594) incidence rate of DCS.

DCS incidence rates in both time periods increased as working pressures became higher. Comparing working pressures employed in the first seven years with those of the later five years, there was more work under higher pressures in the latter period - but, on the other hand, DCS incidence was shown to be significantly lower ($p<0.001$).

3.2 Bone lesions

Table 2 shows results of medical examinations in respect of bone lesions. Compressed-air workers, 2,696 in number, were examined by x-ray at A-P (anterior-posterior) positions on each of the shoulder, hip and knee joints, and the findings were analysed on the basis of the classification by Walder *et al.* The total number of cases with any findings of bone lesions was 298; of which 6 were Type A, i.e. of higher malignancy, corresponding to 2.0%; and 59 were Type B cases (19.8%). The rest, 233 cases, were Type C (78.2%). But, considering that the Type C cases had little correlation with compressed-air works, the actual cases with bone lesion, Types A and B together, 65 cases in total, corresponds to an incidence rate of 1.83% (65/2696). Results by site of the lesion, as seen in Table 3, show that most onsets were at the hip joints, corresponding to 73.3%.

Table 1. Comparison of DCS incidence for different working pressure

	1979-1985			1986-1990		
Working pressure (kg/cm^2)	Exposures (no.)	DCS events (no.)	Incidence (%)	Exposures (no.)	DCS events (no.)	Incidence (%)
~ 1.0	10,208	0		2,295	0	
~ 1.2	2,032	2		539	0	
~ 1.4	2,768	2	0.16	590	0	0.10
~ 1.6	2,152	5	(15/9,	935	0	(4/3,
~ 1.8	1,714	4	649)	800	1	822)
~ 2.0	983	2		958	3	
~ 2.2	743	9		2,479	11	
~ 2.4	489	16	1.37	1,485	9	0.86
~ 2.6	665	9	(49/3,	1,493	19	(191/20,
~ 2.8	1,287	11	586)	3,827	6	945)
~ 3.0	402	4		11,661	136	
~ 3.2	212	9		124	0	
~ 3.4	57	0	3.06	16	1	2.26
~ 3.6	25	0	(9/294)	36	2	(5/221)
~ 3.8	0	0		11	0	
~ 4.0	0	0		34	2	
4.1 ~	0	0		77	1	1.30 (1/77)
Total	23,737	73	0.31	27,360	191	0.70
> $1kg/cm^2$	13,529	73	0.54	25,065	191	0.76

Table 2. Results of medical examination on (CA) compressed air workers and (D) divers

	Persons examined (no.)	Mean age (years)	Persons x-rayed (no.)	Mean age (years)	Rate (%)	Persons with abnormal bones (no.)	Mean age (years)	Incidence (%)
C.A. Staff	2,747	34.3	2,384	34.5	86.8	263	34.8	11.0
C.A. Labourers	807	36.6	312	35.2	38.7	35	36.2	11.2
D. Fishermen	1,222	32.9	1,062	32.4	86.9	130	31.5	12.2
D. Engineering	356	29.5	328	29.7	92.1	34	29.8	10.4
D. Rescue	124	27.1	111	27.1	89.5	2	27.5	1.8
D. Sport	113	20.2	2	24.5	1.8	1	30.0	50.0
D. Instructor	6	37.5	6	37.5	100.0	0	-	0.0
Totals	5,375	33.6	4,205	33.4	78.2	465	33.6	11.1

Table 3. Sites of bone lesions

	Shoulder joints (no.)			Hip joints (no.)			Knee joints (no.)			Totals	
	L	R	B	L	R	B	L	R	B	(no.)	(%)
Type A	0	1	1	3	2	2	0	0	0	12	(2.2%)
Type B	8	8	7	30	23	6	12	3	2	114	(20.9%)
Type C	20	40	2	107	167	26	11	17	1	420	(76.9%)
Totals	28	49	20	140	192	68	23	20	6	548	(100%)
Rate	97/548	(17.7%)		400/548	(73.3%)		49/548	(9.0%)			

Table 4. Comparison of DCS and AGE events for compressed air workers and divers

	1970 ~ '80 Type of DCS			1981 ~ '85 Type of DCS			1986 ~ '90 Type of DCS			Totals	
	I	II	AGE	I	II	AGE	I	II	AGE	(no.)	(%)
Caisson workers	37	40	0	4	10	0	5	8	0	104	(27.4%)
Shield workers	27	19	1	15	5	0	3	3	0	70	(18.5%)
Total number of compressed air workers	61	59		19	15		8	11		174	(45.9%)
	120		1	34		0	19		0		
	(78.9%)			(37.4%)			(14.7%)				
Fishermen divers	2	8	0	16	14	1	42	17	0	100	(26.4%)
Other professional divers	8	8	2	2	8	0	8	10	0	46	(12.1%)
Leisure divers	3	3	0	9	8	2	19	14	1	59	(15.6%)
	(4.0%)			(18.7%)			(25.6%)				
Total number of divers	13	19		27	30		69	41		205	(54.1%)
	32		2	57		3	110		1		
	(21.1%)			(62.6%)			(85.3%)				
Totals	74	78		46	45		77	52		379	(100%)
	152		3	91		3	129		1		
	(100%)			(100%)			(100%)				

*AGE arterial gas embolism

3.3 Occupations

Table 4 shows a classification of the occupations of patients with DCS, who were treated in the University hospital during 20 years from 1970 to 1990, for two groups, compressed-air workers and divers. Ratios of compressed-air workers corresponds to 78.9% (120/152 patients) during 1970-1980, 37.4% (34/91 patients) during 1981-1985 and, very much less, 14.7% (19/129 patients) during 1986-1990.

4 Changing work patterns and DCS

Compressed-air work proceeds in different ways depending on whether it is a pneumatic caisson or shield tunnelling.

Caissons works go vertically downwards into the ground, e.g. for the foundations of a bridge, whereas shield tunnelling progresses more or less horizontally to excavate below ground, e.g. for subway construction. Thus, hyperbaric exposures to the workers increase in pressure as caisson works proceed, while, in shield works, the exposure pressures stay approximately constant from the beginning to the completion of the works.

As seen in Table 1, the peak-use working pressure from 1979 to 1985 corresponds to 1.2-1.4 kg/cm^2, and most of the works were carried out at pressures of 1.8 kg/cm^2 or less. This means that shield works were the main type of operation during these years (although there were also caisson and shafts using pneumatic works methods). Another peak in the working pressures is seen at 2.6-2.8 kg/cm^2 in Table 1; it is said that this was because sites try to keep below pressures of 3.0 kg/cm^2. The Safety and Hygiene Law for Labour in Japan identifies works at over 3.0 kg/cm^2 as being especially dangerous and requires that applications to do such works in any construction have to be made in advance to the Minister of Labour. Thus companies undertaking compressed air works use their best efforts to keep the maximum working pressure at 3.0 kg/cm^2 or less. For example, they try to lower groundwater levels by installing deep wells, or avoid increased working pressures by improving the ground with chemical treatment (grouting).

Since the compressed-air shield method was improved technically, resulting in the earth pressure balance method, with pressure exposure only at the cutting faces, work within the tunnels is at atmospheric pressure. Under these circumstances, compressed-air tunnelling rapidly decreased in number in the 1980s, and, from approximately 1986 on, the method has scarcely been used. Consequently, data on working pressures from 1986 to 1990, which we investigated in the present study, corresponded to those observed under pneumatic caisson works. In these data, working pressures of 2.8-3.0 kg/cm^2 are frequently seen in the numbers of hyperbaric pressure exposures. This, it is supposed, is because the companies concerned probably refrained from making the application to the Ministry of Labour.

Apart from the above, the general tendency is for the DCS incidence rates to increase with working pressures. This is primarily because of the standard decompression table itself.

The current decompression table has been compiled to take into consideration: (1) working efficiency and the risk of suffering from DCS, matters which have essentially inharmonious characteristics; (2) a favourable working efficiency with a minimum risk identified; (3) method improvement based on experience; and (4) the establishment of the currently used table of decompression as a standard, which was approved by the Ministry of Labour. It has long been widely known that as working pressures increase the time required for decompression is prolonged, resulting in a reduction of working efficiency. When compressed-air works are carried out in accordance with the current decompression table whether of Japan, European countries or U.S.A., the higher working pressures go, the more cases there are of decompression disorders, resulting in increased onset rates of DCS in line with the rise of working pressures. This, conceivably, is a natural consequence.

As seen in Table 1, however, the incidence of DCS was significantly lower, at similar pressures, in the later five years. This suggests that the work is being more strictly supervised with safer control of decompression.

5 Medical examinations and bone lesions

Compressed-air workers in Japan have to undergo a specific medical examination every six months. This is required by law, and the items of the examination are those shown in Table 5. In order to check the actual situation regarding medical examinations of compressed-air workers, the authors obtained copies of the completed forms showing the test items of examinations carried out for each working site and analysed them to show the results in Table 5.

At each work-site, the medical examinations were carried out by the particular institutes concerned. But, when we checked the items examined by the medical institutes, we found that only 68.4% of the test items specified by the compressed-air rules (Para. 1 of Art. 38) were satisfied.

In other words, measurement of blood pressure was exceptional in that it, alone, was conducted in medical examinations for all work-sites, but other test items were not necessarily carried out at each site. This possibly means that the institutes concerned do not acknowledge that the test items are essential for medical examination of compressed-air workers.

Under these circumstances, in which every work-site obtained medical examination for compressed-air works, it may be said that there remains a large task left for further study.

The number of medical examinations conducted so far is indicated in Table 2, which includes 3,554 compressed-air construction workers (mean age, 34.9 years). Of these, 2,696 workers had X-ray examinations at the major bone joints (shoulder, hip and knee).

The results of the X-ray examinations were classified by the criteria of Walder-Kawashima *et al.* The results showed 6 cases of Type A bone lesion, a type of articular surface disturbance, four of whom underwent artificial capital (head of the bone) replacement which had been authorised by the Labour Accident Department.

Table 5. Test items of medical examination for compressed air workers

	Test at employment	Regulation 38		Pneumatic caisson (12 sites)		Compressed air shield (7 sites)		Total (19 sites)	
		Item 1	Item 2	Number of sites	Rate of taking tests (%)	Number of sites	Rate of taking tests (%)	Number of sites	Rate of taking tests (%)
Height	•			7	58.3	1	14.3	8	42.1
Weight	•			7	58.3	1	14.3	8	42.1
Eye sight	•			6	50.0	1	14.3	7	36.8
Colour blindness	•			3	25.0	1	14.3	4	21.1
Disease hist.	•	•		9	75.0	5	71.4	14	73.7
Working hist.	•	•		6	50.0	2	28.6	8	42.1
Self-knowledge symptom	•	•		9	75.0	7	100.0	16	84.2
Symptom Motion func.	•	•		9	75.0	6	85.7	15	79.0
Breast radiog.	•			8	66.7	3	42.9	11	57.9
Bone radiog.			•	0	-	0	-	0	-
Medical exam				4	33.3	1	14.3	5	26.3
Eardrum		•		4	33.3	6	85.7	10	52.6
Hearing abil.	•	•		10	83.3	6	85.7	16	84.2
Blood press.	•	•		12	100.0	7	100.0	19	100.0
ECG			•	5	41.7	3	42.9	8	42.1
Blood chem.				1	8.3	0	-	1	5.3
Pulmonary volume		•		11	91.7	6	85.7	16	84.2
Pulmonary function			•	2	16.7	4	57.1	6	31.6
Urine Sugar	•	•		11	91.7	7	100.0	17	89.5
Urine Protein	•	•		11	91.7	7	100.0	17	89.5
Urine Blood				0	-	0	-	0	-
Urine pH				0	-	0	-	0	-
BSG				2	16.7	0	-	2	10.5
Other test				3	25.0	2	28.6	5	26.3
Working condition		•		1	8.3	2	28.6	3	15.8
Notice		•		3	25.0	2	28.6	5	26.3
Rate of sufficiency (%)	64.8	68.4	24.6						

Type B lesions were observed in 59 cases out of the 2,696 examined. Although Type C observations were recorded in 233 cases, this type is not currently included in the categories of bone lesions of compressed-air workers. So, it becomes necessary to consider Types A and B.

Bone lesion developments were mostly found at knee joints, i.e. 73% of the lesions were seen at knee joints on the articular surfaces, capitals and necks of the femoral bones. According to Kawashima *et al.*, bone lesions were observed at an approximately similar incidence at sites such as shoulder, hip and knee joints. This is supposedly due to the conditions and type of labour in pneumatic caissons. It has been separately reported that bone lesions were developed by over 50% of shell-fish helmet divers. Our own survey on divers working for fishery, salvage works and engineering showed a tendency of almost the same incidence of those with bone lesions (i.e. about 11%) as was observed with compressed-air workers (Table 2). On the basis of these findings, it can be concluded that the very high incidence of bone lesions appears to be limited to the group of shell-fish helmet divers.

6 Recent trends and future needs

The university has been treating DCS cases since 1966 and has experience with over 200 cases of severe DCS, i.e. Type II. The numbers of all the patients, who visited our hospital for the treatment of DCS, are given in Table 4, which shows the proportion who were compressed-air workers. Of the patients with this disorder visiting the hospital, the trends are of a steep increase in leisure divers, but a large decrease of compressed-air workers. These findings, one assumes, result from a synergistic effect between the following: scuba diving boomed in the 1980s in Japan (just as there has been an explosion in the ownership of credit cards in Japan, growing in geometric progression from only 10,000 persons in 1980 to 500,000 today) whereas, (1) the number of compressed-air shield tunnelling works decreased sharply in the 1980s and were hardly seen from 1988 on; (2) the number of pneumatic caisson works also reduced; and (3) when pneumatic caisson works were used, much stricter safety controls were required. These changes to compressed-air engineering works in Japan occurred rapidly. Although pneumatic caisson works are fewer in number, they tend to involve deeper excavation. Consequently, much stricter countermeasures for DCS are required. At the same time, the intention is carry out caisson excavation by remotely controlled means, but where human intervention is necessary, to use helium as an artificial air. We are coming to the stage, when we should presume that pneumatic caisson works will be at pressures up to 7.0 kg/cm^2, for which we should now be researching the safety controls that will be required.

References

Mano, Y. (1992), Decompression sickness (DCS) and its countermeasures (In Japanese), Kiso Koh (Foundation works), 30-37.

Mano, Y. (1992), *Sensui Igaku* (Diving Medicine, in Japanese), Asakurashoten, Tokyo, 1-320.

Mano, Y. and Shibayama, M. (1985), Decompression management and preventive counter plan of compressed-air works based on measuring bubbles (in Japanese, English abstract), *Journal of Japan Accident Medical Association*, **33**, 784-791.

Mano, Y. and Shibayama, M. (1987), Problems in compressed-air engineering fields in a view point of working environments (in Japanese), *Japanese Journal of Industrial Health*, **29**, 76-81.

Mano, Y., Shibayama, M., Ishiyama, A., Takahashi, S., and Maeda, M. (1984), An actual approach to prevent decompression sickness in compressed-air workers, In: *Underwater physiology VIII* (eds. A.J. Bachrach and M.M. Matzen), Undersea Medical Society, I. Bethesda, 191-190.

Walder, D.N. (1975), The prevention of decompression sickness. In *The physiology and medicine of diving and compressed-air work*, (eds. P.B. Bennett and D.H. Elliott), 456-471. Bailliére Tindall, London, second edition.

25 Analysis of decompression sickness and bubble appearance rates on a caisson project – Effects of working shift pattern and post-decompression oxygen breathing

I. Nashimoto, Y. Gotoh, K. Kobayashi, M. Nodera, R. Araki and F. Eda
Department of Hygiene, Saitama Medical School, Saitama, Japan

Abstract
Early in a pneumatic-caisson project, which lasted for 36 working days with working pressures ranging from 3.2 bar at the beginning to 4.2 bar in the end, the average bends rate was as high as 3.24% in a split-shift working pattern, i.e. two 2-hour compressed air working periods separated by a one-hour surface interval per day. To solve the high rate of bends in response to a request from the site manager when the working pressure had reached 3.6 bar, the working pattern was switched to a single shift of 3 hour's working time per day, using a modified Blackpool decompression schedule. The bends rate was reduced to 1.83%. Then, following our recommendation, oxygen breathing after surfacing in the caisson works further decreased the rate of bends to 0%. Bubble detection using the ultrasound Doppler method revealed that bubble appearance was also less in the single shift than the split shift. When the working pressure exceeded 4.1 bar towards the end of the work, however, the bends rate increased to 4.35%. After rescheduling the decompression procedure, the rate of bends was reduced to 2.11%, suggesting the necessity of further shortening the work time and/or redesigning the decompression schedule.
Keywords: Caisson Work, Bends Rate, Decompression Sickness, Oxygen Breathing, Single Shift, Split Shift.

1 Introduction

In general in pneumatic (compressed air) caisson work, the risk of decompression sickness (DCS) increases as the working pressure increases and working time is prolonged, because of increased nitrogen loading in the body. It has been recognised that a split-shift working pattern which consists of two separate working periods per day has greater risks of DCS than a single-shift working pattern, i.e. only one working period per day.

Upon a request for advice from a site manager who was seeking a countermeasure to reduce DCS occurring in a pneumatic caisson operation at which ambient (gauge) pressure was more than 3.2 bar, we recommended changing from the split-shift

Engineering and Health in Compressed Air Work Edited by F.M. Jardine and R.I. McCallum. Published in 1994 by E & FN Spon. ISBN 0 419 18460 0

working pattern (of which the total working period per day was four hours) to a single shift of three hour's working time. As a result, bends rate and bubble appearance rate in the blood stream detected by the ultrasound Doppler method were decreased. Oxygen breathing after decompression further decreased the rates of bends.

2 Description of the compressed air work and DCS prevention procedures

This pneumatic caisson was for the construction of a vertical shaft for a sewerage tunnel. As the caisson body sank, the ground became unstable and earth came into the working chamber, because of the higher than expected ground pressures. In spite of efforts to prevent it, the soil continued to flood into the working space. Excavation of the caisson was halted, sheet piles were driven around the caisson body to cut off the unstable ground, then excavation was resumed. The working pressure was 3.2 bar when work restarted. The working pressure was gradually increased as the caisson body was sunk, and became 4.2 bar after 41 calendar days when the caisson reached the planned depth. During this period, two to four teams consisting of 3 to 6 workers per team were engaged daily in the compressed air work for a total of 36 working days out of 41 calendar days (i.e. not working on Sundays). For the first 13 working days, the split-shift pattern was followed, where caisson workers were engaged in two hours of compressed air work followed by a two-hour decompression period twice a day and separated by a one-hour surface interval. After the first 13 days, the pressure was increased to 3.6 bar and limb bends which required recompression treatment occurred in 6 out of the total of 185 exposures (3.24%). The site manager was worried by this situation and called for our advice to reduce the high rate of limb bends.

We advised him to switch the split-shift working pattern to a single-shift system and to use a decompression schedule based on the Blackpool tables, modified for 3 hour's exposure to the working pressure of 3.6 to 3.9 bar. Subsequently, the rate of bends reduced to 4 cases out of a total of 218 exposures (1.83%) for the succeeding ten days. After installation of an oxygen breathing system according to our recommendation, workers were instructed to breathe oxygen for 20 minutes immediately after surfacing. No bends cases occurred in a total of 154 exposures during the following 7 days when the maximum pressure was 4.0 bar. The working pressure then was raised to 4.1 bar for the next two days during which 2 workers out of 46 exposures (4.35%) suffered from bends. In the face of this high DCS rate, we had to improve the decompression schedule for 3-hour exposure to 4.1 to 4.3 bar by lengthening the total decompression time by 30 minutes, i.e. a more conservative procedure. As a result, the rate of DCS was reduced to two events in a total of 95 exposures (2.11%) for the last 4 days when the working pressure was kept at 4.2 bar.

3 Preventive measures of DCS

3.1 Hyperbaric exposure and decompression procedure

Split-shift working pattern

Caisson workers worked for two hours in compressed air twice a day separated by a one-hour surface interval. Decompression followed an empirical stepwise method, based on experience, which lasted two hours in all.

Single-shift working pattern

Since the maximum pressure was expected to be 3.9 bar when the site manager called for our advice, we generated a decompression schedule of 3 hour's working time under 3.9 bar of pressure which was derived from the mathematical model of Blackpool tables (CIRIA, 1982), called Schedule One (S-1). To reduce the risk of DCS, the total working period was shortened to 3 hours rather than the 4-hour total in the split shift. The working pressure, however, eventually exceeded 3.9 bar as a consequence of the unexpected sinking of the caisson body, and finally reached 4.2 bar. Because of this, we had to regenerate a decompression schedule for 3 hour's exposure to 4.1 to 4.3 bar, called Schedule Two (S-2). These schedules, S-1 and S-2, are shown in Table 1.

3.2 Post-decompression oxygen breathing, PDOB

It is well known that oxygen breathing during decompression can safely shorten the decompression time and that it is effective in preventing DCS in divers and compressed air workers (Sterk, 1987; and Kindwall, 1989). The use of oxygen under hyperbaric conditions, however, is not always an easy task because of the possibility of oxygen toxicity and the hazard of fire. In compressed air operations not only is there concern about the discipline of workers regarding oxygen usage from the viewpoint of securing safety, but there is also a requirement for a system to exhaust oxygen-rich expired gas outside a pressurised man-lock. Taking these points into account, we chose oxygen breathing after surfacing as a preventive measure against DCS, which can be simply adopted without an expensive system.

After completing decompression in the man-lock, caisson workers moved to a resting room where an oxygen breathing system was furnished, and they breathed oxygen for 20 minutes *via* a demand-type regulator. Exhaled gas was exhausted outside the room through large-diameter piping *via* a one-way dump valve.

4 Evaluation of risks of DCS

Evaluation of the risks of DCS on this particular pneumatic caisson project was based upon the rate of DCS expressed as the ratio of the number of workers suffering from decompression sickness and who underwent recompression treatment divided by the total number of man-decompressions during each period when a specific set of preventive measure was taken. These periods were roughly divided into two periods, namely the split shift and single-shift working periods. The latter

Table 1. Modified Blackpool schedules, S-1 and S-2 for three-hour exposure times

Schedule	Working pressure (bar)	Decompression stops (min) at pressures (bar) of							Decomp. time (min)
		2.1	1.8	1.5	1.2	0.9	0.6	0.3	
S-1	3.6 to 3.9	-	10	20	35	40	50	85	240
S-2	4.1 to 4.3	5	10	20	40	45	60	90	270

was further divided into several periods depending upon whether or not there was oxygen breathing and the different decompression schedules. Detailed explanation on each period is given in Section 5 and Figure 1. We performed bubble detection in the blood stream after surfacing by the ultrasound Doppler method using Model 1032G made by Institute of Applied Physiology & Medicine (Seattle, WA, USA). The bubble detection measurements were taken three times as shown in Figure 1: the first measurement was performed on the last day of the split-shift period, immediately after each of the decompressions on 6 workers of 2 teams (3 workers per team) who were exposed twice to the compressed air on that day; the second measurement during the single-shift period with S-1 and without PDOB on 11 workers of 2 teams (one of 6 and the other of 5 workers); and the third during the single shift period with S-2 and PDOB on 12 workers of 2 teams (6 workers per team).

5 Results

The number of caisson workers who were engaged in the compressed air work was 28, whose ages ranged from 32 to 55 years old with an average age of 44 years. Their compressed air work experience ranged from 3 to 30 years.

The number of DCS cases, all of which were Type I (lower limb bends) and required recompression treatment, was 14 out of total 698 exposures (2.01%). In these 14 cases, two workers suffered from DCS twice. There was no tendency that DCS occurred more frequently on particular days, such as the beginning or towards the end of a week. We have divided the entire period into five working phases corresponding to the different measures to reduce DCS (as shown in Figure 1), and evaluated the effect of each measure based upon the rate of DCS.

Change in the working pressure throughout the caisson work, the rate of DCS in each period, and the times of bubble detection measurement are also shown in Figure 1. The rate of DCS was 6 out of total 185 exposures or 3.24%, 4/218 (1.83%), 0/154 (0%), 2/46 (4.35%), and 2/95 (2.11%) for the five periods in sequence. The ultrasound Doppler measurements detected bubbles in 10 of 12 cases or 83.3%, 2/11 (18.2%), and 6/12 (50%), at the three times of measurement, respectively.

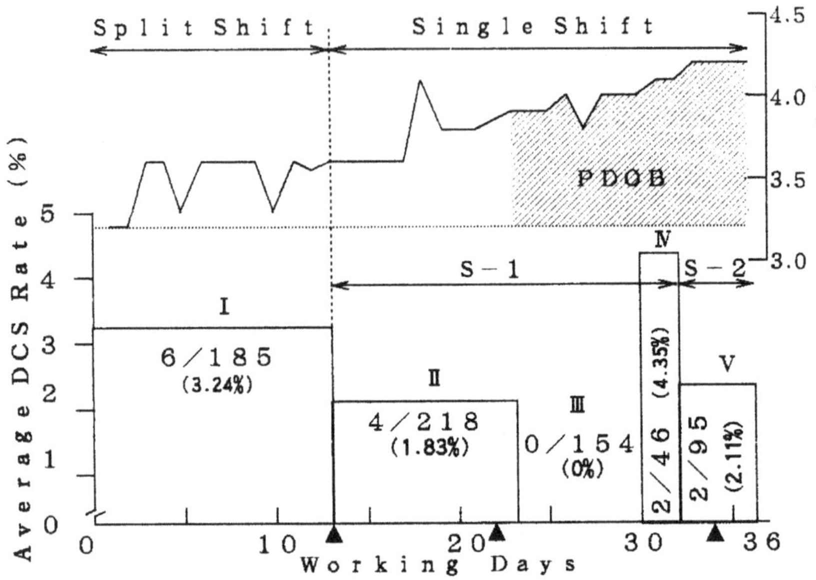

Fig.1. Change in working pressure (above) and average DCS rate (below).

Notes: PDOB indicates post-decompression oxygen breathing.
S-1, decompression Schedule One.
S-2, decompression Schedule Two.
Roman numbers divide working periods according to the DCS preventive measures.
Fractional numbers: the numerator is the number of workers who suffered from DCS and required recompression treatment, and the denominator is the total number of exposures in each period.
Numbers in parentheses are DCS rates in percent.
Triangles underneath the *x*-axis are days when bubble detection measurement was performed.

6 Discussion

Reduced rates both of DCS and detected bubbles in the second period in comparison with the first one indicated that changing the working pattern from the split to a single shift and shortening exposure time apparently decreased DCS risk. In the third period, no DCS occurred, suggesting that oxygen breathing after surfacing further decreased DCS risks by facilitating wash-out of dissolved excess nitrogen in the body. On the other hand there were two cases of DCS in the fourth period (4.35%), when the decompression schedule was designed for 3.9 bar exposure but the actual pressure was 4.1 bar. When the working pressure was elevated to 4.2 bar in the fifth period, and the decompression schedule was modified to suit, the DCS rate was halved (2/95).

The high bubble detection rate of 50% in this period, indicates a potentially high risk of DCS, even with the redesigned decompression schedule. It might be necessary to consider either shortening the exposure time or lengthening the decompression time, or a combination of the two, when the working pressure is extremely elevated. Further study is needed on this.

7 References

Kindwall, E.P. (1989) Caisson decompression, in: *The Physiological Basis of Decompression*, Thirty-Eighth Undersea and Hyperbaric Medical Society Workshop (ed. R.D. Vann) Undersea and Hyperbaric Medical Society, Inc., Bethesda, pp.375-395

Sterk, W. (1987) The use of oxygen in decompression, in: *Decompression in Surface Based Diving*, Thirty-Sixth Undersea and Hyperbaric Medical Society Workshop (eds. I. Nashimoto and E.H. Lanphier), Undersea and Hyperbaric Medical Society, Inc., Bethesda, pp.13-19

26 Review of the neuropathological effects of decompression sickness in animals and man

A.C. Palmer
Department of Clinical Veterinary Medicine, University of Cambridge, UK

Abstract
Compression/decompression procedures can lead to pathological changes in a number of organs in the body, both in animals and man. This review concentrates on observed pathological changes in the brain and spinal cord. Those in the spinal cord stem from deprivation of blood supply to the white matter, for a critical length of time. Changes in the brain, as recorded in man, are not so obvious and consist of small regions of oedema of white matter and necrosis of nerve cells. The walls of arteries in the white matter sometimes show evidence of thickening and may become surrounded by an abnormal space. Degenerative changes in the brain may arise in similar fashion to those in the spinal cord but are apparently not so severe because of a better collateral blood supply.
Keywords: Decompression Sickness, Animals, Man, Pathological Changes, Brain, Spinal Cord, Infarction.

1 Introduction

In the study of the effects of compression/decompression on man and animals, the roles of the physicist, physiologist, biochemist and clinician are well recognised. That of the pathologist is more problematical and usually less welcome. He comes on to the scene when the situation has gone catastrophically wrong. However, as Hughlings Hackson said 'The study of things caused must precede the study of the causes of things' and it is therefore surprising that pathological investigations have not been carried out more often especially in the field of experimental decompression in animals. Here a variety of exposures can be undertaken and the animals killed at different intervals, thus establishing the pattern and progress of pathological change. This knowledge can then be applied to man.

Unlike the marine mammals, man has not adapted his anatomy, physiology or biochemistry to cope with extremes of increased environmental pressure and subsequent decompression and it is small wonder that on occasions structural damage is done to internal organs.

Engineering and Health in Compressed Air Work Edited by F.M. Jardine and R.I. McCallum. Published in 1994 by E & FN Spon. ISBN 0 419 18460 0

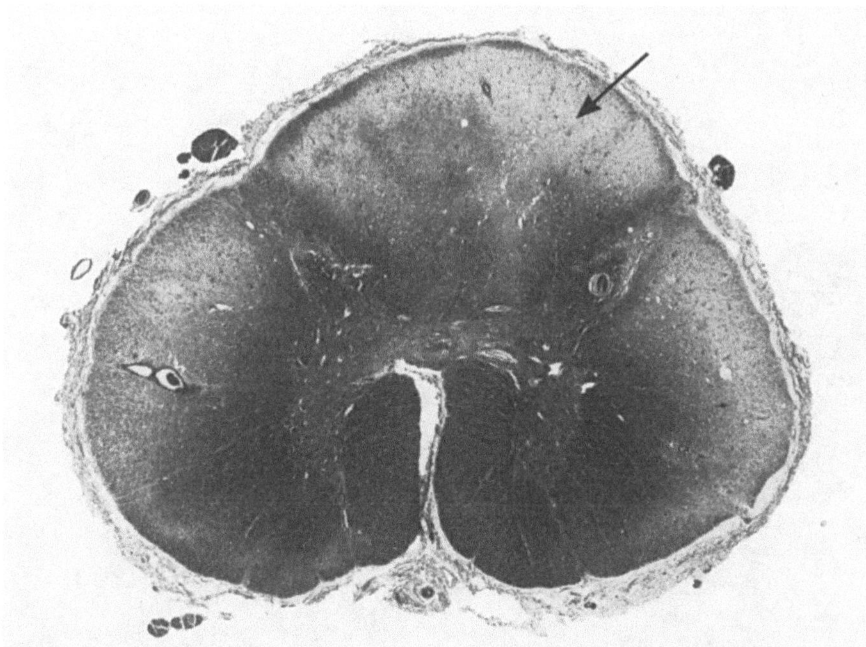

Fig.1. Spinal cord from a man who survived for 4 years as a paraplegic after an incident of spinal decompression sickness. Pale areas in the white matter indicate severe degeneration of myelin (arrow). Note the preservation of the myelin around the circumference of the cord, X10

Those organs in which clinical and pathological changes have been documented include the skin, lymph nodes, joints, eye, inner ear, bone, blood, spinal cord, brain and liver. Local erythema and oedema are often reported in the skin; pain is felt in joints; narrowing of retinal vessels has been reported in the eye (Polkinghorne *et al.*, 1988); deafness is associated with degeneration of the cochlea (Wilkes *et al.*, 1988); aseptic necrosis affects long bones; haemoconcentration and platelet changes are associated with thrombus formation; membrane-bound proteinaceous bodies occur in the space of Disse in the liver (Calder and Palmer, 1989). It is possible that all these effects share an aetiology common with those to be described in the spinal cord and brain and may be related to the evolution of gas bubbles in the blood or in other tissue fluids.

2 The spinal cord

'Spinal bends' or spinal paralysis is not an uncommon consequence of

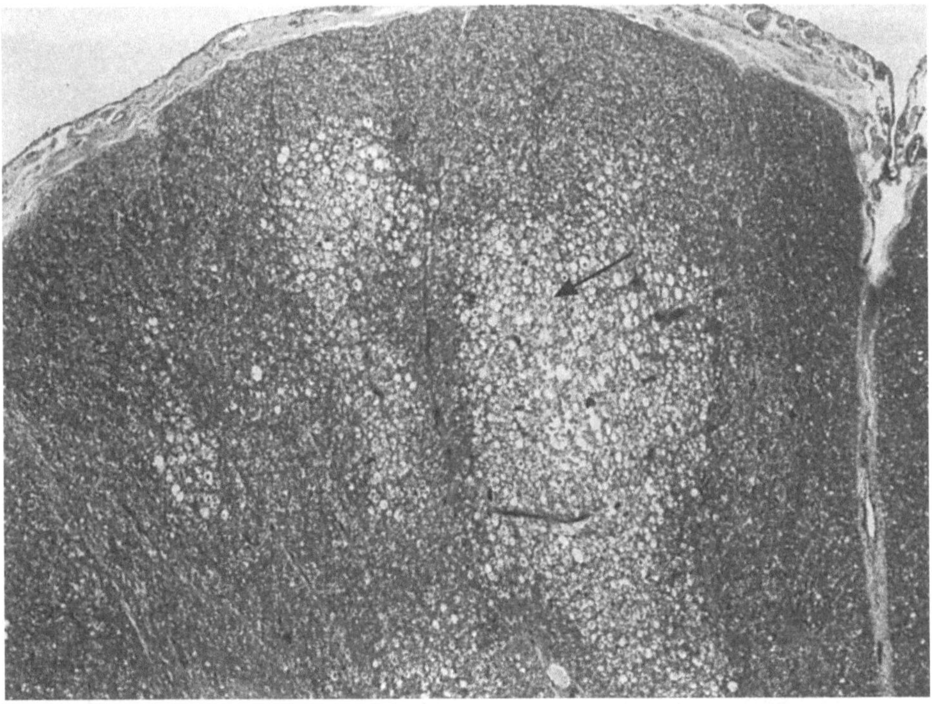

Fig.2. Infarcts in the dorsal columns of the spinal cord of a goat, 24 hours after decompression. Areas of necrosis are clearly separated from surrounding normal myelin. There is preservation of myelin at the circumference of the cord, X50

decompression. It occurs in both experimental animals and man and is related to the duration and depth of exposure and rapidity of decompression. The minimum depth of dive in which it has been reported is 4 m (sea water) (Elliott, 1985). The paralysis can affect the legs, arms and urinary bladder. Sometimes only one limb is affected which suggests that the lesion (damage) is confined to one side of the spinal cord so that the pathological process is not bilaterally distributed. After a period of time a certain degree of recovery can occur. In the goat, this recovery can appear to be complete clinically after 6 months, but this may be a reflection of the difficulty in detecting slight neurological abnormalities in animals (Palmer *et al.*, 1976). Almost total recovery has been reported in man, but even in these circumstances there may be residual damage in the cord (Palmer *et al.*, 1976, 1981 and 1987).

The nature of the damage in the spinal cord is seen in the white matter (Figure 1): the central grey matter, which consists of nerve cells and their processes, is usually unaffected, apart from sensory cells in the dorsal horn. The white matter,

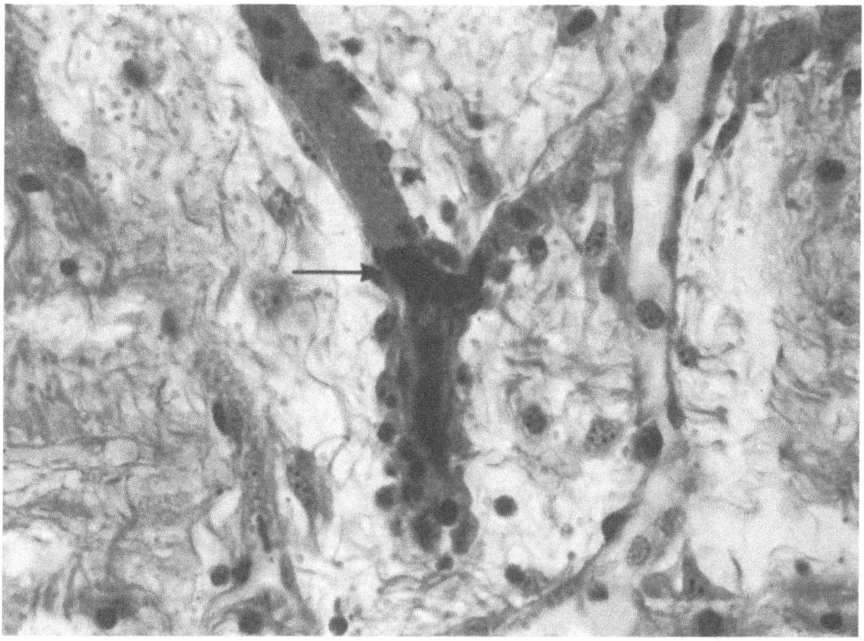

Fig.3. Microthrombus (a clot) blocking a blood vessel in the spinal cord of a goat, 48 hours after spinal decompression sickness, X360

which is made up of nerve fibres (many surrounded by an insulating fatty covering or sheath) carry messages up and down the spinal cord. Collections or bundles of fibres often subserve one specific function, such as conveying a particular sensation or directing a particular limb to move; these bundles of fibres are also known as tracts and they occupy a recognised anatomical position within the cord. For instance, many motor fibres occupy the pyramidal tract which is found in the anterior (man) or ventral (animal) part of the cord.

In decompression sickness, damage to the white matter consists of focal areas of acute degeneration, necrosis or destruction; the tissue is dead. The damage arises because the particular area has been deprived of blood (with failure of transport of oxygen and nutrients and removal of metabolic products) and the lesion is known as an infarct. An infarct arises when the supply of blood is interrupted. This can be brought about in a number of ways.

Interruption to this supply may be caused by cardiac arrest, by a catastrophic haemorrhage or by a narrowing of the lumen of a blood vessel. This narrowing can be caused by an external factor (e.g. by a tourniquet or theoretically in the case of decompression illness, by an expanding bubble of gas), mechanical constriction by contraction of the intrinsic muscles within the wall of a vessel (vasospasm), by

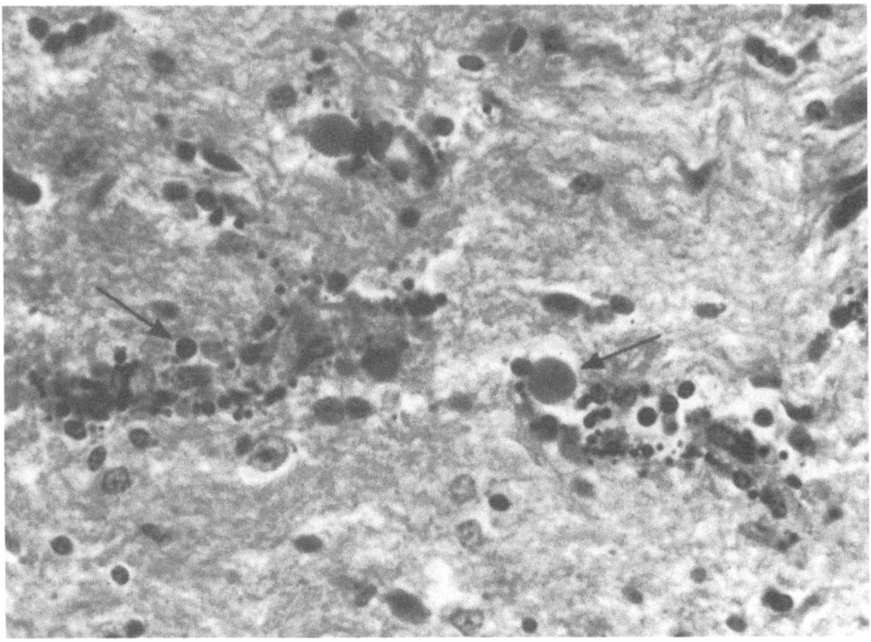

Fig.4. Leakage of proteinaceous globules from a spinal blood vessel of a goat, 48 hours after decompression sickness, X360

blocking the lumen of the vessel by a foreign object (embolism by a blood clot or gas bubble) or finally by obstructing the patency of the vessel by the deposit on its inner wall of a clot or thrombus which with time may increase in size.

In the case of decompression sickness, the nature of the process can be inferred from the experimental model in the goat. Evidence shows that affected regions of white matter in the spinal cord occur in so-called watershed areas; these are areas which lie between the territory supplied by two different arteries. Such 'end of the line' regions are the first to suffer deprivation of blood if the local supply is compromised.

One watershed area of particular interest is in the region of the dorsal horns in the grey matter of the spinal cord. The dorsal horn contains nerve cells concerned with sensation; if they suffer from ischaemia (loss of blood supply), the cells might generate impulses which are interpreted by the brain as sensations of pins and needles or of pain. This might account for the pain appearing to arise from joints, a pain which sometimes appears to flit from side to side.

Bundles of nerve fibres or tracts in the spinal cord which appear to be especially vulnerable are those in the dorsal and lateral columns (Figure 2); these convey impulses concerned with sensory and motor activity. Fibre tracts lying just within

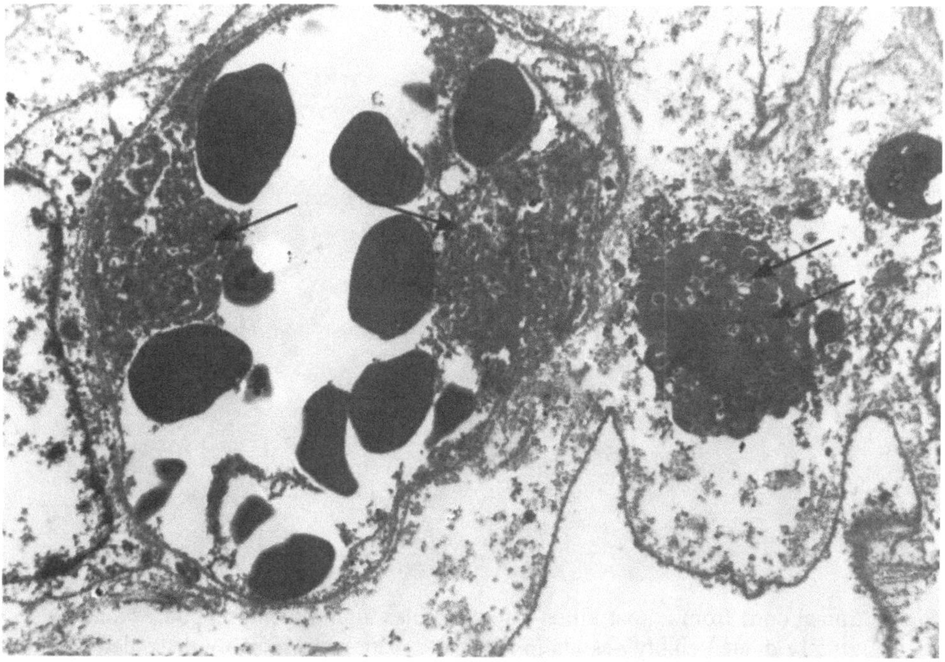

Fig.5. Electron microscope picture of a blood vessel from the spinal cord of a goat 48 hours after decompression sickness. Note the presence of 2 intravascular microthrombi (clots) – single arrows – and extravascular proteinaceous globule – double arrows. The dark bodies in the vessel are red blood cells. The endothelial lining of the vessel has been destroyed

the outer circumference of the cord are usually unaffected. Perhaps these tracts derive their nutrition from vessels in the adjacent meningeal coverings, or by diffusion from the surrounding cerebrospinal fluid. Preservation of this outer myelin in decompression sickness myelopathy is a feature in both animals and man.

As mentioned earlier, observations on experimentally-induced decompression illness in animals can help elucidate the nature of the genesis ('pathogenesis') of the lesions in man. In goats killed between 24 and 48 hours after a critical exposure, clot or microthrombus formation has been reported on the inner surface of small arteries (Figure 3) (Palmer and Blakemore, 1980). Around such vessels there can be an accumulation of proteinaceous material, evidence that the vessels are actually leaking (Figures 4 and 5) (Palmer *et al.*, 1978). Haemorrhage also occurs.

These phenomena suggest that the primary pathological event occurs within the vascular system, perhaps associated with the presence of intravascular bubbles (Figure 6). Hallenbeck and Andersen (1982) showed that conditions at the

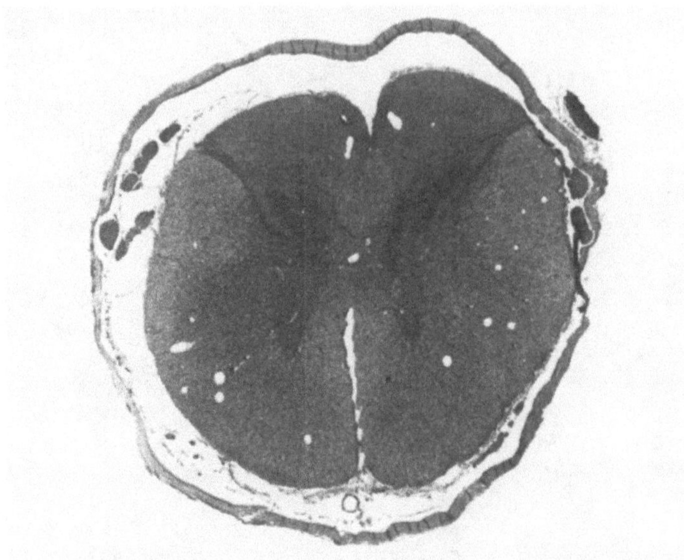

Fig.6. Spinal cord from a goat killed a few minutes after decompression. Note the grossly dilated empty vessels in the white matter, presumably distended by gas, X12

blood/bubble interface may lead to denaturation of protein molecules with resultant formation of protein gels in areas of stagnant flow, and with sludging of red cells and aggregation - events which precede clotting and thrombus formation.

Such a theory cannot by itself preclude a concomitant local vasoconstriction brought about by pressure induced on vessels by the formation of expanding gas bubbles in adjacent myelin, so-called autochthonous bubbles. Such a theory has been suggested by a number of earlier workers and has also been advanced by Francis *et al.*, (1990). However in my experience I have yet to see convincing morphological evidence of the presence of gas bubbles in the tissue of the central nervous system. This is not to deny the presence of grossly dilated empty vessels in both white and grey matter after rapid decompression in both animals and man.

When this dilatation is extreme the endothelial linings of these vessels are difficult to distinguish from surrounding tissue: special histological stains are needed. Moreover, the processing technique required to prepare tissue for microscopical examination involves dehydration in ascending concentrations of alcohol as well as temperatures of up to 56°C, all of which can lead to shrinkage of tissue. It would appear that under these circumstances, grossly distended vessels can disintegrate and collapse and in turn give the appearance of empty spaces or autochthonous bubbles within the tissue.

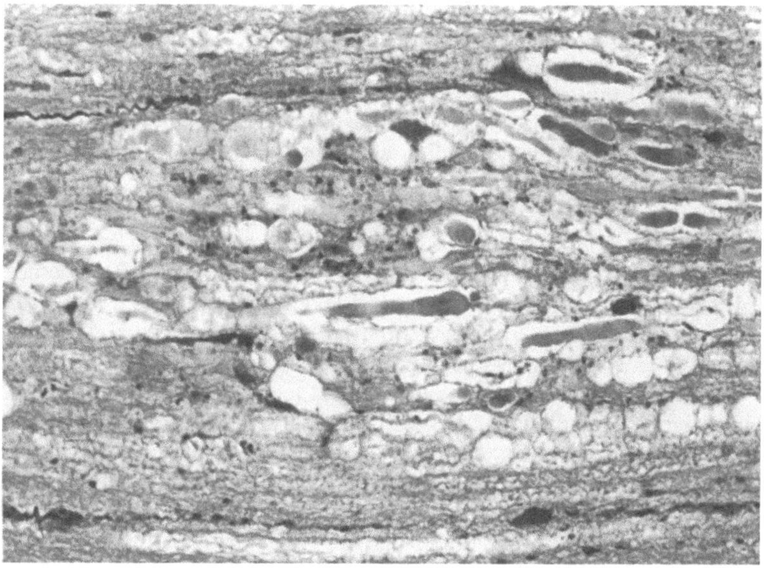

Fig.7. Longitudinal section through infarcted area in the spinal cord of a goat, 48 hours after decompression sickness. Note the swollen axons, indicating an arrest of flow of axoplasm, X140

The special vulnerability of the white matter of the spinal cord to injury in decompression sickness is surprising. In other situations, microembolisation can lead to infarction of the kidney but this is rarely observed in decompression sickness. A poverty of blood supply is a particular anatomical feature of the white matter of the spinal cord. Unlike other parts of the central nervous system, the spinal white matter possesses little collateral blood supply. In other organs of the body, a collateral supply provides another anatomical arterial route to a specific region so that in an emergency, an alternative supply can be provided. Thus in decompression illness, the white matter of the spinal cord is especially vulnerable.

The pathological process (infarction) in the spinal cord is focal and leads to an arrest of the flow of axoplasm within the affected nerve fibres (Figure 7). The flow depends upon the activity of local metabolism. If this metabolism can be reactivated within a period of perhaps 1.5 hours by an enhanced supply of oxygen, it is possible that the flow of axoplasm can be restored (Leone and Ochs, 1978; Blakemore and Palmer, 1982). This validates the treatment of cases of decompression illness with hyperbaric oxygen. If treatment is delayed, the degeneration will proceed to such a stage from which recovery cannot occur.

3 Brain

If the spinal cord can be damaged by compression/decompression procedures, it is reasonable to suppose that brain can also be affected. It too is composed of similar nerve cells, fibre tracts and supporting glial tissue, and one might therefore expect damage to be manifested both clinically and pathologically.

A number of authors have reported clinical signs related to a disability involving the cranial nerves (Elliott, Hallenbeck and Bove, 1974; Caruso *et al.*, 1977; Palmer *et al.*, 1981). Todnem *et al.* (1990) also described visual disturbances, vertigo, and transient episodes of cerebral dysfunction which included seizures and amnesia. Reports concerning long-term neuropsychiatric changes in both caisson workers and divers have included such conditions as loss of memory, retarded reactions, antisocial behaviour and difficulties in articulation and communication (Rozsahegyi and Roth, 1966; Peters, Levin and Kelly, 1977; Nix and Hopf, 1980; Vaernes and Eidsvik, 1982; Baddeley, 1983; Edmonds and Hayward, 1987 and Smyth, 1985).

The eyes are embryologically and anatomically closely associated with the brain. Involvement of the eyes in divers has also been observed by Polkinghorne *et al.* (1988). Using fluorescein angiography, a clinical technique which provides a visible indication of the patency of retinal vessels, they reported a lowered capillary density at the fovea, microaneurysms as well as small areas of capillary nonperfusion. The severity of the lesions was directly related to the duration of diving experience. The authors considered that the changes were consistent with obstruction of the retinal and choroidal circulations by bubbles or perhaps other intravascular phenomena. Visual defects were not reported.

Despite these clinical indications, there have been few investigations into the possibility of damage in the brain with which they may be related. The reason may be associated with the urgency with which the pathologist is required to establish the immediate cause of death so that he is not necessarily concerned with morphological abnormalities which might indicate evidence of long-term damage. Such reports as there are include that from Haymaker (1957) who described perivascular oedema and Sillery (1958) who reported intracranial haemorrhage and deposition of calcium in the walls of cerebral vessels. Akimov (1969) demonstrated fibrous thickening of the walls of blood vessels and enlarged perivascular spaces occupied by mononuclear cells. All these reports concerned individual cases.

A brief report on results of autopsies on 40 divers (Mork, 1988) indicated that in the cerebral white matter there was an increase in corpora amylacea, an excess of lipofuscin and hyalinisation of blood vessels. This author also described a focal loss of ependymal cells with degenerative and regenerative changes of the brain tissue adjacent to the ventricles, gliosis and eventually involvement of nerve cell processes and myelin. Kitano *et al.* (1990) examined the brains of 4 divers and found engorgement of meningeal vessels, cerebral oedema, haemorrhage and foci of pallor in the cerebral cortex and white matter: in one case there were also foci of necrosis in the cerebral white matter.

In 1982, Dr Ian Calder and I began to examine brains from professional and amateur divers.

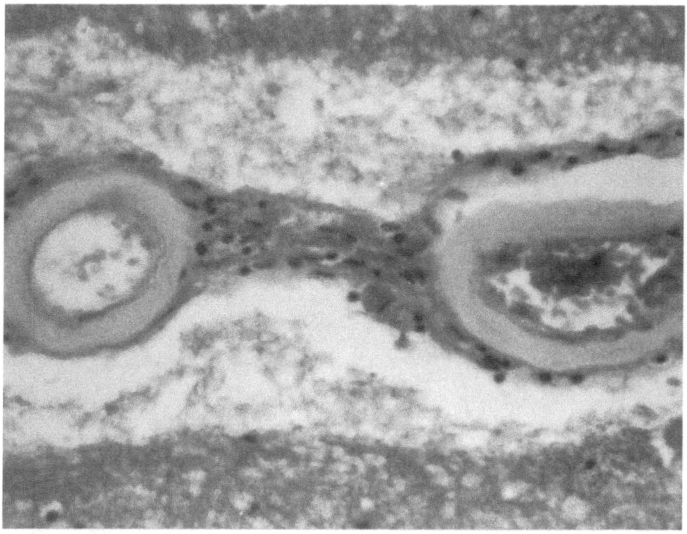

Fig.8. Hyalinised small artery in a lacuna found in the cerebral white matter from the brain of a diver, X370

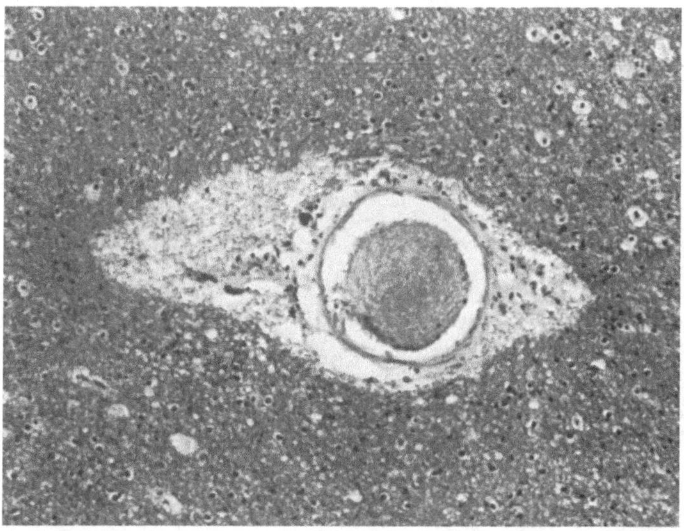

Fig.9. Lacuna, or space, around a vessel in cerebral white matter in a diver, X180

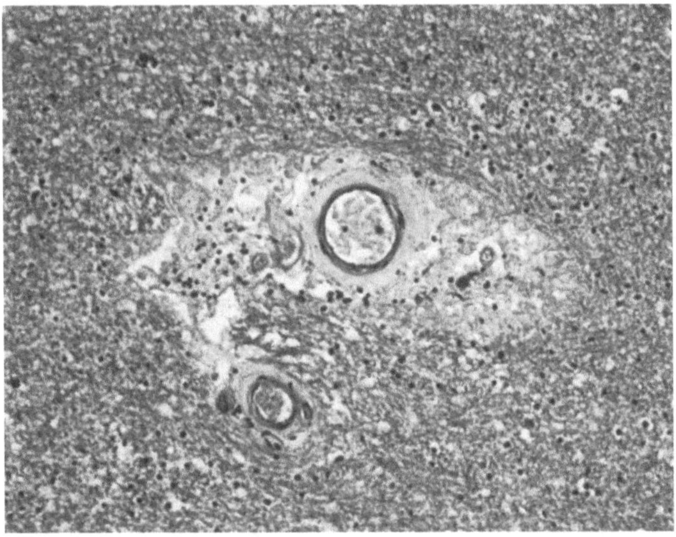

Fig. 10. Small artery in a lacuna surrounded by proteinaceous debris and mononuclear cells, X140

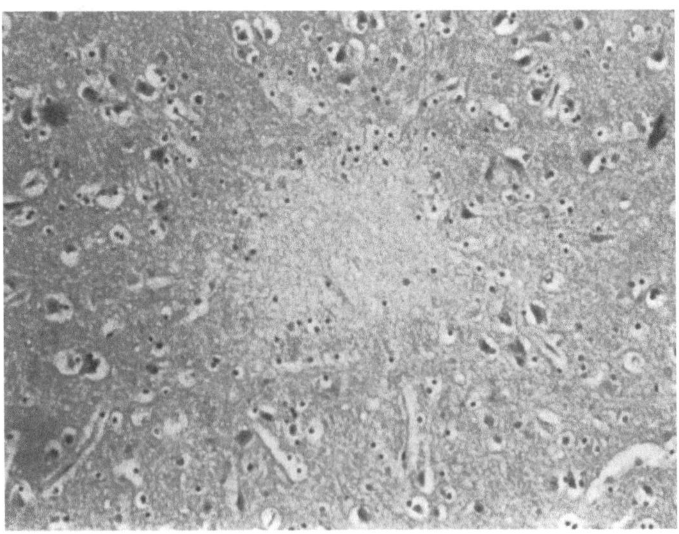

Fig. 11 Focal loss of nerve cells in the cerebral cortex of a diver, X140

Abnormalities we have found are subtle. In the brains of a number of amateur and professional divers degenerative changes occur in the walls of blood vessels (especially arteries) which are scattered mainly in cerebral white matter (Palmer, Calder and Yates, 1990 and 1992). The change concerns a hyalinisation of the wall of the vessel (Figure 8) which involved replacement of the normal muscle coat with fibrous tissue. Abnormal spaces (lacunae) (Figure 9) can occur also around the vessel within which are collections of small round cells, corpora amylacea and macrophages (Figure 10) which indicate a cellular reaction to injury.

Sometimes there is a spongy state in the local white matter and changes to the nerve fibres. In the cerebral cortex one can occasionally find small areas from which the nerve cells have disappeared or which are undergoing degeneration (Figure 11). The nature of this degeneration suggests that these areas have suffered from ischaemia. These abnormalities in white and grey matter are diffusely distributed and difficult to find. They may be sufficient to account for the clinical disturbances.

The nature of the morphological abnormalities in the brain are of a similar type to those in the spinal cord, but are not as severe. In both brain and cord it would appear that the primary insult occurs to the blood supply of white matter. Perhaps the difference in severity can be ascribed to the better collateral circulation in the brain.

4 Conclusions

In man, compression/decompression procedures can lead to structural damage to the spinal cord and brain. Spinal cord degeneration has also been recorded in experimental decompression illness in animals but so far, intracerebral damage has been restricted to haemorrhage (Palmer *et al.*, 1978). Long-term studies on animals have been negative (Palmer *et al.*, 1976).

The generation of the injury to the nervous system is likely to arise from ischaemia, perhaps due to the presence of intravascular bubbles of gas. A similar hypothesis can be suggested for the production of lesions in other organs of the body (skin, bone, eye and liver) although in the inner ear, cochlear damage may arise from bubble formation in the endolymph, disturbing metabolism of the Organ of Corti (Wilkes *et al.*, 1988).

If one considers dispassionately the likely effects of compression/decompression in man it is not too surprising that the production of gas bubbles in the blood can lead to pathological processes. After all, gas bubbles are foreign bodies (albeit sterile) in the blood and constitute an unwelcome disturbance in a medium which has to be maintained to a fine standard of cellular, biochemical and electrolyte purity.

Acknowledgements

I wish to thank the Editor and Publishers (Blackwell Scientific Publications Ltd) of *Neuropathology and Applied Neurobiology* for permission to reproduce Figures 2, 9, 10 and 11. Dr J Grimstad, Editor of the Proceedings of the 5th Scientific Meeting of the Euorpean Undersea Biomedical Society has kindly permitted reproduction of Figure 5. Permission has kinly also been granted by the Editor of *Paraplegia* to use Figure 1, by the Editor of *Undersea Biomedical Research* to use Figures 3 and 7 and by the Editor of *Progress in Undersea Science* to reproduce Figures 4 and 6.

References

Akimov, G.A., Elinskii, M.P. and Lovovskii, A.M. (1969) Changes in the nervous system with decompression sickness. *Zhurnal Nevropatologii i Psikhiatrii S.S. Korsakova* **69**, 979-984

Baddeley, A. (1983) In: *Long term neurological consequences of deep diving* (eds T.G. Shields, B. Minsaas, D.H. Elliott and R.I. McCallum). Proceedings of workshop organised by the European Undersea Biomedical Society and the Norwegian Petroleum Directorate, Stavanger, Norway, p. 165

Blakemore, W.F. and Palmer, A.C. (1982) Delayed infraction of spinal cord white matter following x-irradiation. *J. Pathology.* **137**, 273-280

Calder, I.M. and Palmer, A.C. (1982) Histological changes in liver of man and animals subjected to compression and decompression. *Undersea Biomed. Res.* **16** Suppl. 88

Caruso, V.G., Winklemann, P.E., Correia, M.J., Miltenberger, G.E. and Love, J.T. (1977) Otologic and otoneurologic injuries in divers: clinical studies on nine commercial and two sport divers. *Larygoscope* **87**, 508-521

Edmonds, C. and Hayward, J. (1987) Intellectual impairment with diving: a review. *Proceedings of the 9th International Symposium on Underwater and Hyperbaric Physiology.* Undersea and Hyperbaric Medical Society: Bethesdà, Maryland. pp 877-886

Elliott, D.H. (1985). Diving Medical Advisory Committee case history workshop. *European Undersea Biomed. Soc. Newsletter* **4**, 8.

Elliott, D.H., Hallenbeck, J.M. and Bove, A.A. (1974) Acute decompression sickness. *Lancet* ii: 1193-1199

Francis, T.J.R., Griffin, J.L., Homer, L.D., Pezeshkpour, G.H., Dutka, A.J. and Flynn, E.T. (1990) Bubble-induced dysfunction in acute spinal cord decompression sickness. *J. Appl. Physiol.* **68**, 1368-1375

Hallenbeck, J.M. and Andersen, J.C. (1982) Pathogenesis of the decompression disorders. In: *The physiology and medicine of diving*, 3rd edn. (eds P.B. Bennett and D.H. Elliott), Baillère Tindall, London pp. 435-459

Haymaker, W. (1957) Decompression sickness. In: *Handbuch der Speziellen Pathologischen Anatomie und Histologie. Nervensystem.*, eds O. Lubarsch, F. Henke and R. Rossle, Springer-Verlag, Berlin, pp. 1600-1612

Kitano, M., Urago, A., Hayashi, K., Kawashima, M., Funakoski, K., Yamada, K. and Tokufuji, S. (1990) A pathological study on cerebral lesions in diver's decompression sickness (DCS). *South Pacific Study.* **10**, 275-285

Leone, J. and Ochs, S. (1978) Anoxic block and recovery of axoplasmic transport and electrical excitability of nerve. *J. Neurobiology.* **9**, 229-245

Mork, S.J. (1988) In: *Report of M.R.C. Decompression Sickness Panel, Long Term Health Effects Working Group*, (ed M.J. Halsey), at Second International Symposium on Man in the Sea, University of Hawaii, p. 5

Nix, W.A. and Hopf, H.C. (1980) Schadigungen des zentralen nervensystems nach dekompressionszwischenfallen. *Deutsche Medizinische Wochenschrift* **105**, 302-306

Palmer, A.C. and Blakemore, W.F. (1980) Neuropathology of experimental decompression (goats). In: *Proceedings of the 5th Annual Scientific Meeting of the European Undersea Biomedical Society*, Bergen, (ed. A. Grimstad) pp 153-156

Palmer, A.C., Blakemore, W.F. and Greenwood, A.G. (1976) Neuropathology of experimental decompression (dysbarism) in the goat. *Neuropathol. Appl. Neurobiol.* **2**, 145-156

Palmer, A.C., Blakemore, W.F., Payne, J.E. and Sillence, A. (1978) Decompression sickness in the goat: nature of brain and spinal cord lesions at 48 hours. *Undersea Biomed. Research* **5**, 275-286

Palmer, A.C., Calder, I.M. and Hughes, J.T. (1987) Spinal cord degeneration in divers. *Lancet* 2, 1365-1366

Palmer, A.C., Calder, I.M., McCallum, R.I. and Mastaglia, F.L. (1981) Spinal cord degeneration in a case of 'recovered' spinal decompression sickness. *Br. Med. J.* **283**, 888

Palmer, A.C., Calder, I.M. and Yates, P.O. (1990) Cerebral vasculopathy in divers. *Proceedings of the International Congress on Hyperbaric Medicine, Amsterdam 1990*, 137-144

Palmer, A.C., Calder, I.M. and Yates, P.O. (1992) Cerebral vasculopathy in divers. *Neuropath. Appl. Neurobiol.* **18**, 113-124

Peters, B.H., Levin, H.S. and Kelly, P.J. (1977) Neurologic and psychologic manifestations of decompression illness in divers. *Neurology* **27**, 125-127

Polkinghorne, P.J., Sehmi, K., Cross, M.R., Minassian, D. and Bird, A.C. (1988) Ocular fundus lesions in divers. *Lancet* ii, 1381-1383

Rozsahegyi, I. and Roth, B. (1966) Participation of the central nervous system in decompression. *Industr. Med. and Surg.* **35**, 101-102

Sillery, R.J. (1958) Decompression sickness. A review of the literature and previously unreported histological observations. **66**, 241-246 *Archs. Path.*

Smyth, E. (1985) Deep-sea diving may cause loss of memory. *New Scientist* No. 1439, 8

Todnem, K., Nyland, H., Skeidsvoll, H., Svihus, R., Rinck, P., Kambestad, B.K. and Aarli, J.A. (1990) Nervous system involvement from occupational diving. An epidemiological study. *Undersea Biomed. Res.* **17**, suppl. 88-89

Vaernes, R.D. and Eidsvik, S. (1982) Central nervous dysfunctions after near-miss accidents in diving. *Aviation Space and Environ. Med.* **53**, 803-807

Wilkes, M.K., Palmer, A.C., Pearce, P., Luff, N.P. and Halsey, M.J. (1988) Cochlear degeneration in minipigs after repeated hyperbaric exposures using standard decompression schedules. *Undersea Biomed. Res.* **16**, 139-152

27 A descriptive approach to the decompression illnesses

J.J.W. Sykes, D.J. Smith and T.J.R. Francis
Institute of Naval Medicine, Hampshire, UK

Abstract
The currently accepted classification of decompression illness distinguishes between Cerebral Arterial Gas Embolism (CAGE) and Decompression Sickness (DCS) and further subdivides DCS into two categories, Type I or minor, and Type II or serious manifestations. The differentiation assumes certain mechanisms are responsible and requires an interpretation by a physician or supervisor to categorise the condition for appropriate treatment. The accuracy of any diagnosis is therefore highly dependent on the philosophical, intellectual and cultural background of the individual performing the classification. This is a far from ideal basis for data collection for national or international computer based databases or for the epidemiological studies which are derived from them.

The solution requires a reclassification of dysbaric illness which is not dependent on presumed mechanisms or interpretation of symptoms and signs. Reversion to the classic scientific method of description allows the almost infinite presentations of the decompression illnesses to be conveyed accurately. The proposed classification describes the decompression illness in terms of its evolution and presentation or manifestation.

Use of this classification removes the inherent inaccuracies and difficulties associated with the present convention permitting the acquisition of sound database information on an international basis which is independent of presumed pathological mechanisms. The Institute of Naval Medicine is fully committed to its adoption.
Keywords: Decompression illnesses, Classification, Diagnosis, Evolution, Manifestation, Time of onset, Gas burden, Barotrauma

1 Introduction

The first scientific description of the symptoms and signs seen in men returning to atmospheric pressure following work at raised pressure was written by Pol and Watelle (1854). Since then the disease complex known as caisson disease or decompression sickness has been described comprehensively and its pathological

Engineering and Health in Compressed Air Work Edited by F.M. Jardine and R.I. McCallum. Published in 1994 by E & FN Spon. ISBN 0 419 18460 0

consequences recorded in humans and animals. The current classification of the disease is based on the observations of Golding *et al.* (1960) who reported on cases arising during the construction of the Dartford Tunnel. This classification was further refined by Elliott and Kindwall (1982).

The diving diseases are separated into the barotraumata, caused by changes in volume of existing pockets of gas within the body, and decompression sickness which is generally thought to be caused by bubbles of gas forming from solution within the body tissues. Based on the perceived severity of the symptoms decompression sickness has been divided into Type I, consisting principally of joint and skin involvement, and Type II which gives rise to neurological, audiovestibular and cardio-pulmonary symptoms.

There are a number of theoretical problems associated with the current classification. Firstly the terms Type I and Type II have no inherent meaning. The disease symptoms have to be learned and then categorised. The Types are not mutually exclusive as one may progress to the other and symptoms from each frequently occur concurrently. Nor is the dynamic nature of the disease process conveyed. The diagnostician is required to make a decision on the likely location of the disease process in the nervous system in a disease known to be multifocal and multisystem – with complex biochemical consequences. Similarly in neurological presentations a mechanism has to be deduced to classify the condition either as a consequence of barotrauma (cerebral arterial gas embolism) or as Type II decompression sickness. Involvement of the middle and inner ear may have a similar aetiology also requiring an intuitive guess as to which mechanism is involved.

We believe this results in the inconsistent application of diagnosis by physicians. As a consequence therapeutic trials are difficult to undertake. Epidemiological studies are similarly compromised as pooling of data from different centres is virtually impossible. As the current system is based on assumptive diagnoses it is difficult to teach. Furthermore, misleading priorities for treatment may be decided on the basis of a diagnosis of Type I disease which may progress to Type II.

The Institute of Naval Medicine (INM), in attempting to define the decompression disorders for application to its own database on diving and decompression accidents, held an International Workshop in 1990 whose objectives were to define diagnostic criteria for the existing classification (Francis and Smith, 1990). It was soon clear that this could not be achieved. However it was agreed that cerebral arterial gas embolism cannot be distinguished from cerebral decompression sickness except under isolated and specific circumstances and that the pathogenetic terms for decompression sickness are misleading and should be abandoned in favour of Acute Decompression Illness.

2 The new descriptive approach

Based on basic scientific principles of observation and description a new protocol has been developed which circumvents the problems identified previously. It requires five essential items of information to describe a case. These are:

a. Evolution
b. Manifestation
c. Time of onset
d. Gas burden
e. Evidence of barotrauma.

With respect to *evolution*, a number of suitable terms have been identified which may describe the development of the disease in a diver or caisson worker. These are: progressive, static, spontaneously improving or relapsing.

The *manifestation* may be conveyed by the use of one or more terms describing the symptoms: limb pain, neurological, cutaneous, pulmonary and constitutional (to describe the vaguer symptoms often associated with the disease process). A patient may thus be described as a person with acute progressive limb pain and neurological decompression illness. When seen later the symptoms may have stabilised resulting in the use of the term, static, to describe the presentation. By recording the *time of onset* of various manifestations and by recording the evolution of the disease a much better description of the natural history of the disease will emerge, which in turn will lead to a better understanding of its aetiology and epidemiology.

An index of *gas burden* will ultimately give an indication of the possible mechanisms involved. Possible methods include $P\sqrt{t}$ (pressure x square root of time) or a measure of probability of developing decompression sickness. Similarly the existence of *barotrauma* should be recorded. This should be proven rather than assumed.

In providing a description of a complete history for clinical purposes it is also important to record the response to treatment and the results of any investigations.

3 Diagnostic concordance

In seeking to demonstrate the perception that diagnoses are applied inconsistently using the current terminology INM have recently conducted a study (Smith *et al.*, 1992) among experienced diving physicians. A total of 51 doctors from six continents took part in the study which required them to make diagnoses on 50 case records of diving accidents which summarised diving history, presentation, clinical course, treatment and outcome. Given the shortcomings of this summary each physician was asked to give an index of confidence on the diagnoses recorded which were permitted to include two or more ('a' and 'b') or alternatives suggesting indecision ('a' or 'b').

The following hypotheses were tested:

1. The distinction between AGE and DCS II cannot be made reliably.
2. Inner ear barotrauma and inner ear DCS cannot be distinguished consistently.
3. The terms DCS I and DCS II are not used consistently.

4. The anatomical subclassifications of neurological DCS II are not applied consistently.

The analysis of the study results supports each hypothesis and shows that the diagnoses of the diving disorders are not applied consistently either between or by individuals given identical clinical information. As treatment algorithms are chosen on the basis of diagnosis this suggests that these would be applied inconsistently. In turn this confirms the difficulties assumed in constructing multi-centre or even multinational trials. Similarly it is demonstrated that considerable variation exists in the ability to agree the location of lesions in neurological presentations.

It is concluded that if diving physicians cannot make diagnoses consistently despite possessing identical information, non-specialists cannot be expected to do better. The need for more precise and consistently applicable terminology is clearly identified.

References

Elliott, D.H., and Kindwall, E.P. (1982) Manifestations of the Decompression Disorders. In: Bennett, P.B. and Elliott, D.J., Eds. *The Physiology and Medicine of Diving*, 3rd Ed., London: Ballière Tindall, 461-472

Francis, T.J.R. and Smith, D.J. Eds. (1991) Describing Decompression Illness. *Proceedings of 42nd Workshop of Undersea and Hyperbaric Medical Society* 79 (DECO) 15/5/91. Bethesda MD. UHMS. May 1991

Golding, F., Griffiths, P., Hempleman, H.V., Paton, W.D.M. and Walder, D.N. (1960) Decompression sickness during construction of the Dartford Tunnel. *Br. J. Indust. Med.*, **17**: 167-180

Pol, B., and Watelle, T-J. (1854) Mémoire sur les effets de la compression de l'air appliqué au creusement des puits à houille. *Ann. Hyg. Publ.*, Paris. **2** (1): 241-279

Smith, D.J., Francis, T.J.R., Plethybridge, R.J., Wright, J.M. and Sykes, J.J.W. (1992) Concordance: A Problem with the Current Classification of Diving Disorders. *Undersea Biomed. Res.* **19** (suppl.): 40

Discussion on Session 6: Decompression illness

The session chairman, **Dr J D King**, who had introduced **Professor D H Elliott**, thanked him for giving the keynote lecture on the subject of the decompression illness. Professor Elliott identified a number of challenges for those concerned with managing and regulating hyperbaric work and for those wishing to understand its consequences for the human body.

Dr R W Hamilton asked about the future of the Decompression Sickness Registry at the University of Newcastle upon Tyne.
Professor Elliott, praising the foresight and effort that went into its creation and maintenance, regretted that the Registry could not have been funded to continue. There remains a need for data collection and analysis. In the latter respect the civil engineering records are more complete than the diving records in the Registry and have been, and are being, analysed. The diving world has a need for more hard data.
Dr King noted that there exists a considerable amount of data. His professional records, for example, relate to over 2800 individuals and include more than 2000 medical examinations of divers. The question is to find a mechanism for collating and compiling this type of data.
Mr F M Jardine said that as well as the initial difficulty of funding a new or revived Registry, the long-term commitment to run it has to be made. With the competition for research funding and the uncertainty that to some degree is inevitable with the data records from sites, perhaps it would be better to focus on one specific construction project monitoring that carefully.
Professor Elliott recognised that the open-endedness of the commitment of the Registry would have been a reason for HSE discontinuing its funding. He suggested that for both scientific reasons and to set finite boundaries to the research, the type of study should be to 'follow' individual workers from their recruitment through their hyperbaric experience and thereafter.

Dr King then introduced the speakers presenting their papers. **Professor Y Mano** presented the results of analysis of the incidence of illnesses from compressed air work comparing over two periods, 1979-85 and 1986-90. **Professor I Nashimoto** reported the changes made to working and decompression schedules in a rapid, interactive response to increasingly severe conditions in a caisson excavation and, in particular, the use of post-decompression oxygen breathing. **Dr A C Palmer** described the neuropathological effects that result from hyperbaric exposure, amplifying his important review paper with both explanations and photomicrographs of superb quality (some of which are reproduced in his paper). **Surgeon Commander J J W Sykes** argued the need for improved classification of the decompression illness, that it should be based on systematic description.

(Note: there was not time for a discussion from the floor after these presentations, but see the discussions of Sessions 7 and 10 in which some of these topics were raised).

(Reporters: G B Shutter and F M Jardine)

Part Seven
Decompression Illness and Clinical Practice

28 The technique of decanting in compressed air work

M.J. Catton
Appointed Doctor Services, London, UK

Abstract
Specific times for each of the three stages of decanting were measured after saturation exposures to compressed air at working pressures between 1.3 and 2.8 bar gauge at four contracts using different site design and equipment between 1963 and 1990.

Times were derived from 200 pressure recorder charts in the first study, by direct timing to 0.25 min blocks at the second study for 446 shifts over 13 months, and by precise timing to within 1 s at two minor contracts in 1989 over 25 and 118 shifts.

The men studied worked in compressed air for nominal 8-h shifts at all pressure levels and included men with and without previous compressed air and decanting experience. The study contracts employed 163, 130, 8 and 15 shift men respectively. All decompressions after decanting conformed to the decompression tables of the 1958 regulations in 1962/3 and to the Blackpool Tables from 1966. The results show that the decanting procedure fails to conform to the advisory statutory limitations of 5 min from working pressure to working pressure and significant delay to over 7 min occurred in 6-28% of procedures with the higher rates at working pressures over 2 bar g.

Analysis of the stage times contributing to the overall delay indicates specific areas that require alteration. Improvement during the progression of one contract is presented in detail.

Feasibility and practicality have been cited as reasons for prolonging times but only recompression delay is inevitable, the frequency can be reduced by improved pre-employment lock testing and equipment control.

The long-term effects of delay in decanting are not apparent. Further study of this group of men, small by comparison to other compressed air workers, is required.

Keywords: Decanting, Decompression, Recompression.

Engineering and Health in Compressed Air Work Edited by F.M. Jardine and R.I. McCallum. Published in 1994 by E & FN Spon. ISBN 0 419 18460 0

1 Introduction

During compressed air work at civil engineering projects in the UK, decanting has been shown to be practically feasible for working pressure (WP) up to 3.45 bar g. after exposure of 8 hours or longer.

The original experience from underwater salvage operations using a diving technique is the basis for the procedure formulated in the ICE Regulations Committee report 1936. Statutory regulations were promulgated in 1958 following preliminary drafts in 1945 and 1951, and the procedure adopted was simplified to a continuous rapid decompression, transfer and recompression to an equivalent pressure level. The time cycle of 5 min was retained but with the qualification of 'reasonably practicable' for compliance of both the total time and pressure level on recompression requirements.

There is no defined statutory duty to record the actual time taken and the official record form, *F 752 Lock Attendants Register*, issued from December 1957 has no columns relevant to the decanting procedure. As a consequence few records exist where the time taken during decanting can be allocated to specific phases of the procedure and even the total time recorded is not precise.

A study in 1963 at Tilbury (Catton, 1967) showed a wide variation in the times taken for both transfer and recompression and the total time exceeded 6 min in up to 34% of decantings in the early stage of the contract. These results were obtained at a site operating with locally trained medical lock-keepers supervising all decantings from surface level double blister locks at a working pressure of 2.4 to 2.8 bar g.

To achieve a rapid transfer and comfortable safe recompression requires a lock-keeper with authority, training, and practice. Observations at sites from 1958 confirmed the scant regard that compressed air workers had for safety procedures after decanting; shift men would visit the toilet block, canteen, and pay office, or even nearby shops if given the opportunity. Non-shift men regularly detoured to workshops and offices after decanting. The normal practice of completing one correct decompression after multiple short exposures persisted into the 1970s in my experience, but was not seen in more recent contracts.

Although it is clear that a significant percentage of decantings exceed the 5-min limit by a wide margin the significance of this observation has not been related to the occurrence of decompression sickness or the prevalence of bone changes. Men working before 1966 (MRC, 1974) show bone changes in up to 30% of those exposed to all varieties of decompression procedure, including decanting, using the 1958 Regulation tables.

Since the general adoption of the Blackpool tables by the publication of CIRIA Report 44 in 1973, the more reliable performance of the decompression offers hope of defining any effect of the decanting procedure. At present there is no clear evidence that decanting causes bone changes in excess of that of the decompression procedure.

2 Study method

Detailed times for each stage of the decanting procedure were recorded at four contract sites operating between 1963 and 1990 at working pressure of 1.1 to 2.5 bar g. (16-37 psig). The records were compiled by six medical lock-keepers supervised by one compressed air medical advisor.

All decompressions were recorded on 252-mm circular charts rotating in 2 or 4 h with centre zeroing of the pressure gradation 0 to 2.76 or 3.45 bar g. (40 or 50 psig).

The method of timing improved between contracts, from pressure record chart marking at Tilbury 1963/4 to direct timing from a wrist-watch second hand recorded to the highest 0.25 min unit at Blackpool 1966/7 and finally stop watch timing to $\pm$ 1 s at London 1989.

Consecutive records of shift exposures provided the most consistent information as the medical lock-keeper carried out regular shift decompressions on a planned predicted basis. At Blackpool, the site of the first trial study of a new decompression table, two medical lock-keepers produced records for 450 decantings with only 4 omissions or incomplete entries. At the two much smaller London sites, operating at a lower working pressure range for short periods only, there was no difficulty in obtaining complete records.

The method of pressure record marking used at the Tilbury site (Catton, 1967) required measurement of small-scale markings and although consecutive records were made throughout the contract the series reported was based on two sets of 100 fully readable consecutive charts, all assessed by one observer. This method was the least accurate and in the later period of the contract a high proportion of omissions or indistinct recordings occurred; the method was later modified using outer zeroing charts to obtain better definition on an experimental basis only.

3 Factors affecting decanting times

3.1 Working pressure range

At Tilbury and Blackpool, shaft sinking at pressures up to 1.5 bar g. was completed before the study. At Tilbury the 2 periods (a,b) included work at 2.8 reducing to 2.5 bar g. and 2.4 to 2.5 bar g. At Blackpool 3 consecutive periods (a,b,c,) started at 1.4 to 1.7 bar for 164 shifts, then 1.9 to 2.5 bar g. for 174 shifts with a final period at 1.3 to 1.4 bar g. for 108 shifts. The London contracts S and D were operated for 5 short periods of 8 to 35 days at 1.0 to 1.6 bar g.

3.2 Number of men in shift, shift frequency and duration

At both Tilbury and Blackpool, three nominal 8-h shifts of 4 to 7 men worked over 6 days for 16 shifts per week with an occasional Sunday shift. On occasions the day shift increased to 14 to 17 men.

The London contracts were organised on a strict 5 day week of 11 shifts nominally of 8h but reduced to 6 or 7h on Fridays involving 3 to 5 men only.

3.3 Previous experience of shift men

All compressed air contracts employ a proportion of men with previous experience of work at pressures below or above 1 bar, and during the contract a number of men without experience will be recruited. Men with experience of decanting tend to seek work in the higher working pressure ranges with financial reward as the important factor.

At Tilbury most of the 163 men were experienced and a number had worked at a similar contract in the area 2 years previously using the same decanting procedure.

At Blackpool, 62 experienced and 68 inexperienced men were employed, although the experienced men continued in work for a longer period.

In London, contract S used decanting for 3 months in 1986 and compressed air work continued for short periods to 1988. All the men exposed had recent compressed air experience and some had previous decanting experience.

Contract D employed men with no compressed air experience for the first phase but the 3 successive phases retained the majority of men. At both contracts the total number of men exposed was less than 30.

3.4 Medical lock-keepers

All 6 men were experienced in operating the lock equipment by the time the studies began; the 2 at Tilbury were trained on site and one transferred to Blackpool to assist the locally trained man already working during the shaft sinking over the previous 3 months.

The London contract was started by a research medical technician from Newcastle with support from a medical lock-keeper previously employed with the contractor at a similar decanting contract in 1986. The third man had experience of naval diving practice, and all three provided cover for the compressed air requirements of multiple sites in the area during a three-year period. This included supervision of conventional decompression in below ground locks, training lock-keepers and treating decompression sickness.

3.5 Site and equipment

(a) Decanting at surface level was from double blister locks fitted at manufacture with 0.75inch (20mm) level-operated valved outlets at all sites except one. The normal practice was for the outcoming shift to operate the controls but at the London sites only the medical lock keepers operated the blister lock controls using designated IN and OUT locks.

At Blackpool, decanting was from a large capacity (105m^3) below-ground combined muck/man-lock fitted with a 6-inch (152mm) outlet. This method required a mechanical hoist with a special man-carrying skip for transfer to the surface level. Contact between the lock-keeper at the base of the shaft and the medical lock-keeper at surface level was by visual and audible signals *via* the open vertical shaft.

(b) The surface recompression facility was sited within 20 m of the shaft area and at all but one site the medical lock was installed adjacent to or near the recompression lock. The recompression locks were of large capacity; two, constructed of boiler-lock sections 8.25 m long and 2.13 m high (capacity 44 m^3) divided with a smaller inner compartment (3.0 m x 2.13 m) with control valves mounted at the entrance area and viewing porthole in the door, were used at three contracts. London D contract used an integral welded medical lock of 5.2 x 2.0 m (27 m^3 capacity) with the larger compartment innermost and the controls mounted centrally with vision by viewing portholes mounted in the upper section of the lock wall.

(c) The air supply at the two large contracts was generous with 6,000 to 10,000 cfm (170-284 m^3/min) available at working pressure, but the small output of 1,000 cu.ft./min (28 m^3/min) available at the two London sites resulted in slower recompression rates taking more than 3 min/bar. When this problem was recognised the inner section of the recompression lock was pressurised, effectively reducing the 2-lock capability but maintaining the required rate of recompression over a short period. By adroit lock operation it was possible to utilise both lock sections simultaneously if required.

(d) Each contractor used similar pressure recording gauges from different manufacturers and these were found reliably accurate after installation and zero setting. Only at Blackpool was it found that a time error occurred during the period of rapid recompression, under reading by ½ to 1 min over a 5-min period. This fault appeared to be intermittent and fortunately direct timing of the recompression phase was also recorded independently. The charts remained valid to identify the pressure levels of any delay in recompression. At both London sites the recompression stage was not recorded when the inner compartment was pressurised.

(e) The rate of compression attained in the medical lock design was slower than that for the boiler-type decanting locks. To obtain a rapid recompression a combination of air supply at low pressure (LP), at WP level, and high pressure (HP) at 3.5-4.2 bar g. is required.

Figure 3 illustrates the recompression rate obtained with LP at HP air supply for a standard medical lock and can be compared with the actual recompression charts for boiler-lock design decanting locks shown in Figures 1 and 2. Without special cooling units HP air alone increases the heating effect during rapid compression to unacceptable levels.

4 Data

4.1 Table format

Table 1 gives data for the results of the studies at the three sites. Subdivisions (i.e. a, b or c) within each contract are given in chronological order with the range of working pressure in that period defined.

Table 1a. Decanting times recorded in seconds at 4 different contracts between 1963 and 1990 with percentage decants exceeding target times by more than 50% in time for each stage

Tilbury 1963/4 Shift size 4-17 men			Blackpool 1966/7 Shift size 4-10 men			London 1989/90 Shift size 2-5 men		
Stage 1, Decompression, Target 60-120 Seconds								
Range	Decants+	%	Range	Decants+	%	Range	Decants+	%
(a)60-180			(a) 90-120			S(a) 39-62		
Over 120	16	16	Over 120	0	0	Over 90	0	0
(b)60-180			(b) 60-120			D(a) 50-180		
Over 120	11	11	Over 120	0	0	Over 90	1	4
			(c) 75-120			D(b) 45-120		
			Over 120	0	0	Over 90	1	2

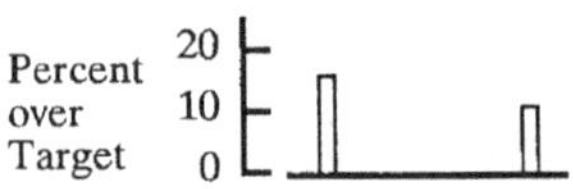

	Tilbury (a)	Tilbury (b)	Blackpool (a)	Blackpool (b)	Blackpool (c)	London S(a)	London D(a)	London D(b)
WP(bar g.)	2·5-2·8	2·4-2·6	1·4-1·7	1·9-2·5	1·3-1·4	1·1-1·4	1·4-1·5	1·5-1·7
Sample	100	100	164	184	108	25	25	50

Stage 2, Transfer, 60-120 Seconds								
Range	Decants+	%	Range	Decants+	%	Range	Decants+	%
(a) 45-210			(a) 90-300			S(a) 13-29		
Over 120	26	26	Over 180	6	4	(Over 60	0	0
Over 180	6	6	Over 240	3	2	Over 120	0	0
(b) 45-150			(b) 90-300			D(a) 25-100		
Over 120	4	4	Over 180	18	10	Over 90	1	4
			Over 240	4	2	Over 120	0	0
			(c) 90-240			D(b) 15-70		
			Over 180	23	21	(Over 60	1	2)
			Over 240	4	4	Over 120	0	0

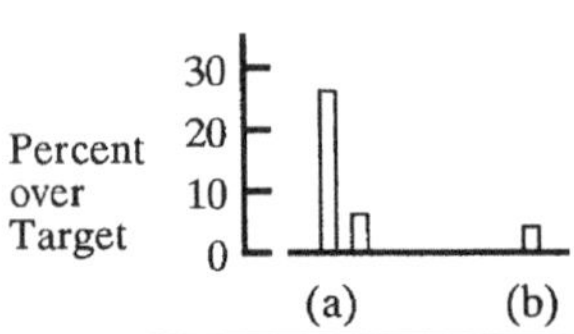

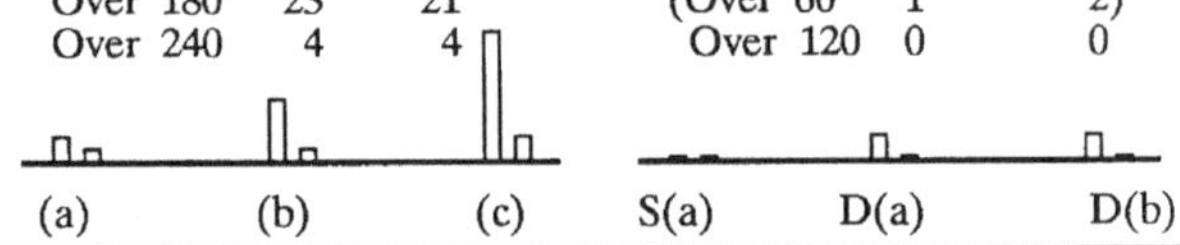

Stage 3, Recompression, Target 120-150 Seconds								
Range	Decants+	%	Range	Decants+	%	Range	Decants+	%
(a) 120-480			(a) 90-600+			S(a) 43-232		
Over 180	28	28	Over 180	19	12	(Over 150	2	8)
Over 240	7	7	Over 240	5	3	Over 180	1	4
(b) 120-600+			(b) 90-420			D(a) 85-504		
Over 180	27	27	Over 180	61	33	Over 180	7	28
Over 240	20	20	Over 240	24	13	Over 240	3	12
			(c) 90-300			D(b) 85-325		
			Over 180	17	16	Over 180	3	6
			Over 240	2	2	Over 240	1	2

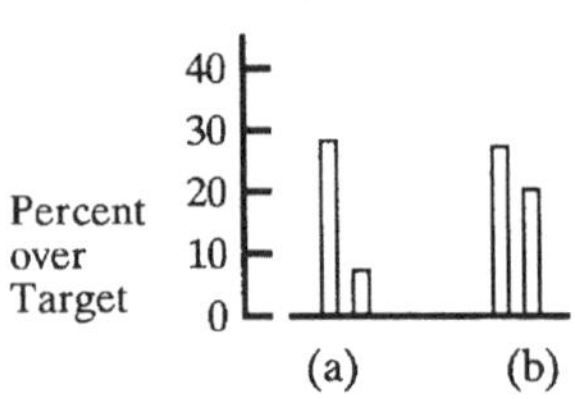

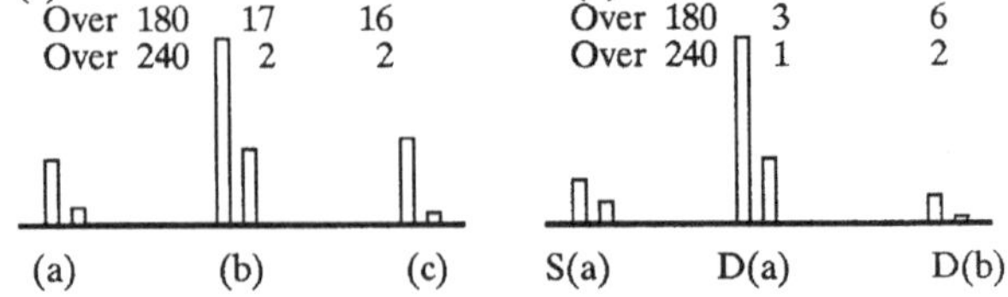

Table 1b. Percentage decantings exceeding 6 minutes and 7 minutes total time WP to WP for each phase of 4 contracts

Tilbury 1963/4 Shift size 4-17 men			Blackpool 1966/7 Shift size 4-10 men			London 1989/90 Shift size 2-5 men		
Total Time, Target 300 Seconds								
Range	Decants+	%	Range	Decants+	%	Range	Decants+	%
(a) 240-600+			(a) 270-600+			S(a) 103-285	0	0
Over 360	16	16	Over 360	23	14	Over 300	0	0
Over 420	16	16	Over 420	10	6			
						D(a) 170-600+		
(b) 240-600+			(b) 270-600+			Over 360	3	12
Over 360	23	23	Over 360	75	41	Over 420	4	16
Over 420	10	10	Over 420	48	26			
						D(b) 155-390		
			(c) 300-600			(Over 300	6	12
			Over 360	37	34	Over 360	1	2
			Over 420	11	10			

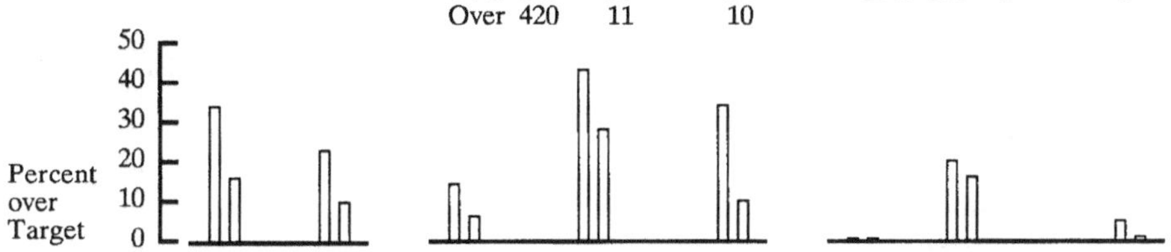

	(a)	(b)	(a)	(b)	(c)	S(a)	D(a)	D(b)
WP(bar.g.)	2·5-2·8	2·4-2·6	1·4-1·7	1·9-2·5	1·3-1·4	1·1-1·4	1·4-1·5	1·5-1·7
Sample	100	100	164	184	108	25	25	50

() < 10% over target time

The target times for decanting set to meet the 5-min rule at all sites were calculated as ¾ + 1 + 2½ min for surface decanting and 1 + 2 + 2 min for below-ground decanting at working pressures above 1.5 bar.

Decantings that failed to meet the target time for any stage by more than 50% and by 20% for the total time are expressed as percentages of the total in Tables 1a and 1b.

The third and fourth phases of contract D in London are not included as all decantings were completed within the target times.

4.2 Tables 1a and 1b

The range of times recorded in each stage of the decanting procedure are given with the number of decantings exceeding a given time expressed as percentages of the total sample.

Although the target times were attainable, the margin of achievement was wider in contracts S & D operating in a lower working-pressure range with fewer men. Improvement in overall performance is apparent at all contracts with more decantings within the 5-min limit, but transfer times at Blackpool phase (c) and recompression times at both Tilbury (b) and Blackpool (b) at WP above 2 bar g. were longer.

Table 2. Comparison of recompression and total (WP to WP) times at contracts S & D London 1989 for shifts of 3-5 men during surface decanting

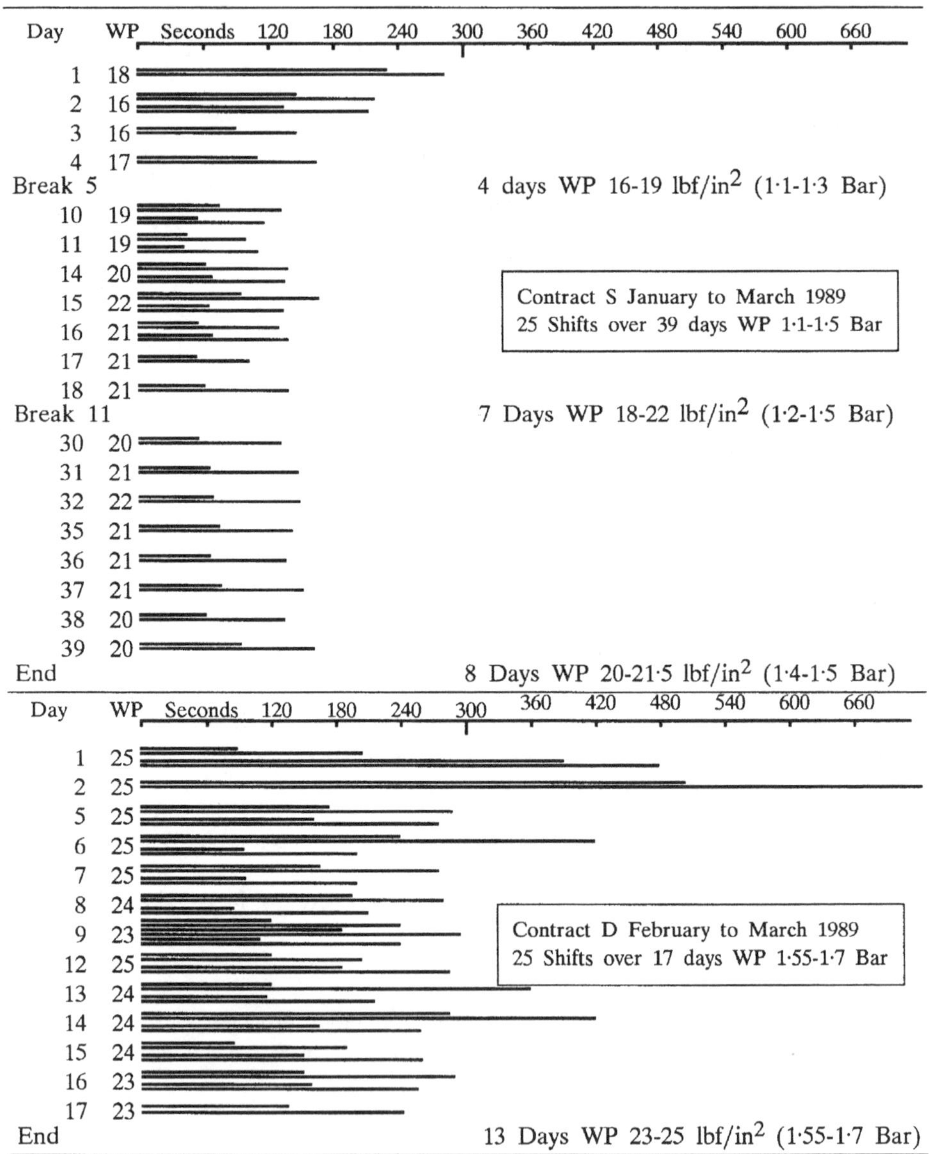

4.3 Table 2

Comparison of the recompression and total (WP-WP) times for the two London contracts show the problem encountered with slow recompression rates due to equipment design and low capacity of the low pressure air supply reserve.

After the first shift at site S, and for all shifts at site D, only the outer lock was used for recompression to achieve increased compression rates of 1-1.5 bar/min.

At Contract D the problem was compounded by 2 men each suffering from ear block on 2 occasions and delays also occurred during stages 1 and 2. In 8 of the 25 decants, the time to the recompression stage exceeded 2 min compared to less than 1.5 min at site S after the first 2 days.

A further 43 decantings at contract D were completed within the target times with one exception. The WP-WP times at site D(c), range 2 min 11s to 3 min 35s was not maintained at site D(d) range, 3 min 9s to 4 min 59s plus 1 at 6 min 48s due to air supply failure after the first shift.

Table 3a. Decanting times for stages 1, 2 and 3 for shift and non-shift men at Blackpool 1966-67 when climbing vertical latter system in shaft

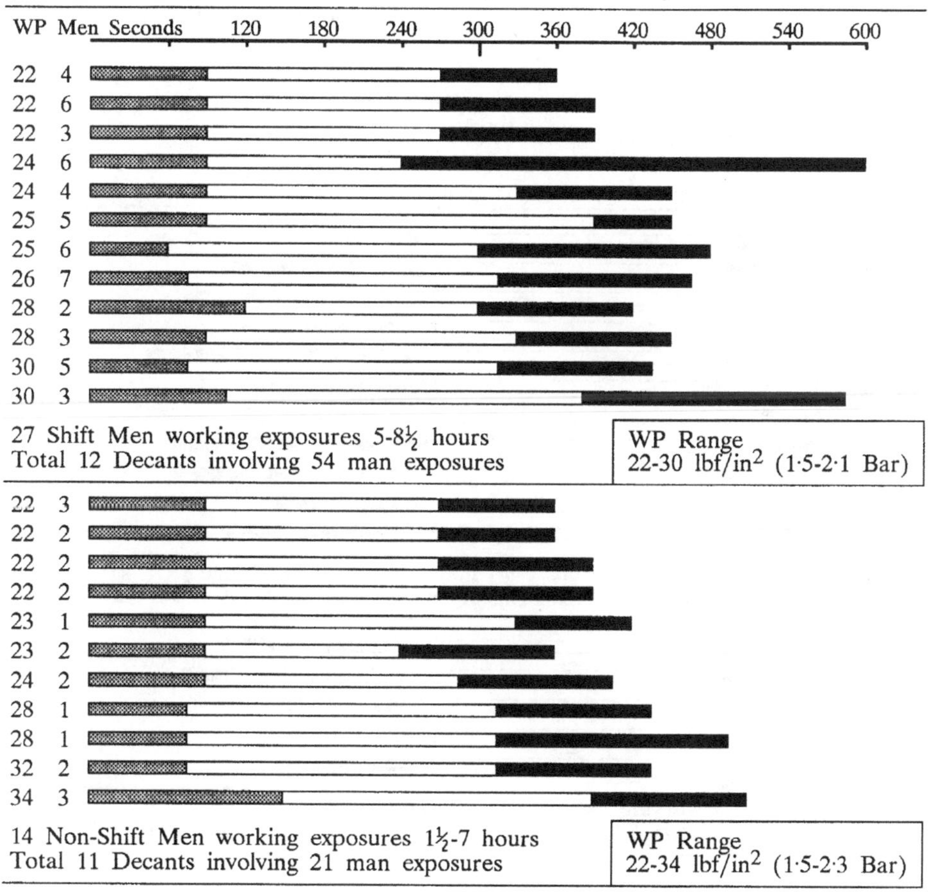

Contracts S & D were unusual with significant breaks in the continuity of compressed air work. At contract D the 4 separate sites operated on only 57 days over 5 months.

3 men during stages D(a) & (b) were each treated once for decompression sickness Type I from 348 exposures of over 4 h during all four decanting stages of this contract.

5 Blackpool data

5.1 Decanting times

At Blackpool the detailed time recording and annotated lock registers gives confidence in the accuracy of incidents noted. During the 2nd phase at working pressures of 1.2-2.5 bar only 29 (16%) of 174 decantings with 5-7 men in a group conformed strictly to the 5-minute limit, more than 27% exceeded the limit by 2 min.

Table 3b Decanting times for shift and non-shift men at Blackpool 1966-1967 when the normal transfer stage 2 time exceeds four minutes

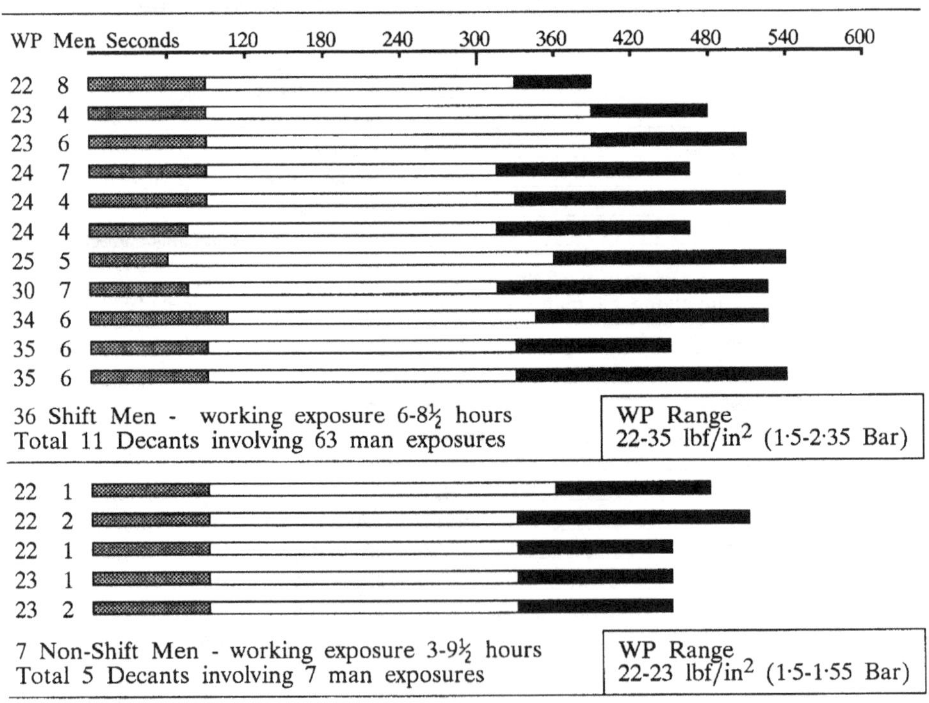

5.2 Physical exercise after decanting

When the hoist was not available men could climb the vertical ladder system in the shaft. 27 shift men after exposures of 4½-8½ h and 11 non-shift men after shorter exposures of 2½-4 h decanted and climbed up to the surface lock in 2½-4 min. Table 3a shows the times for shift and non-shift men when they had to climb the ladder, with Table 3b for comparison when the transfer was prolonged beyond 4 min with the hoist available.

5.3 Reverse decanting

A rare request is for men to leave the recompression lock during the stage decompression, either transferring to another lock or returning to the tunnel. Only 2 examples are recorded, both completed uneventfully within 5 min.

5.4 Delay in recompression stage

The overall picture (see Tables 1b, 2, 3a and 3b) confirms that this contract produced a population of compressed air workers with decanting experience outside the intended norm. Men employed for more than 3 months would be involved in one or more procedures in which the total time WP-to-WP was extended by more than 40% of the total time.

Men working during the winter months at WP above 2 bar could be involved in decantings taking more than 6 min 3-5 times during each month that they worked. Recompression times exceeded the target of 2½ min by 9%, 41% and 40% of decantings in the three periods defined in Section 3.1 with a peak of 49% during the period November 1966 to March 1967.

An obvious problem of one person with ear pain delaying the recompression in the early stage may justify transfer to the medical lock.

This occurred on 3 occasions, and recompression could be very slow taking up to 30 min. Table 3c shows the effect this can have on WP-WP times. Men with problems later in the recompression when pressures were higher than the stage-decompression level of about half the absolute WP were kept in the lock and completed the recompression in the front compartment with or without the other shift men. Examples of the different recompression curves are demonstrated in Figures 1 and 2.

It was unusual for men to miss the subsequent shift after ear symptoms, even with recurrence on one or more days and evidence of barotrauma of the eardrums was always minor and transient.

5.5 No recompression

A completely different episode, known from observations at other sites, is when men fail to attend for recompression after decanting. Single individuals in the supervisory or craft group used to make multiple tunnel visits in a normal working day may escape supervision after relatively short exposures but do later have a final planned decompression.

One Blackpool incident is well documented and involved 3 experienced shift men fully acclimatised over 4 weeks continuous daily exposures at 1.4-1.6 bar g. After a

Table 3c. Decanting times for shift men with delay in recompression due to ear pain at Blackpool 1966-1967 for stages 1, 2 and 3

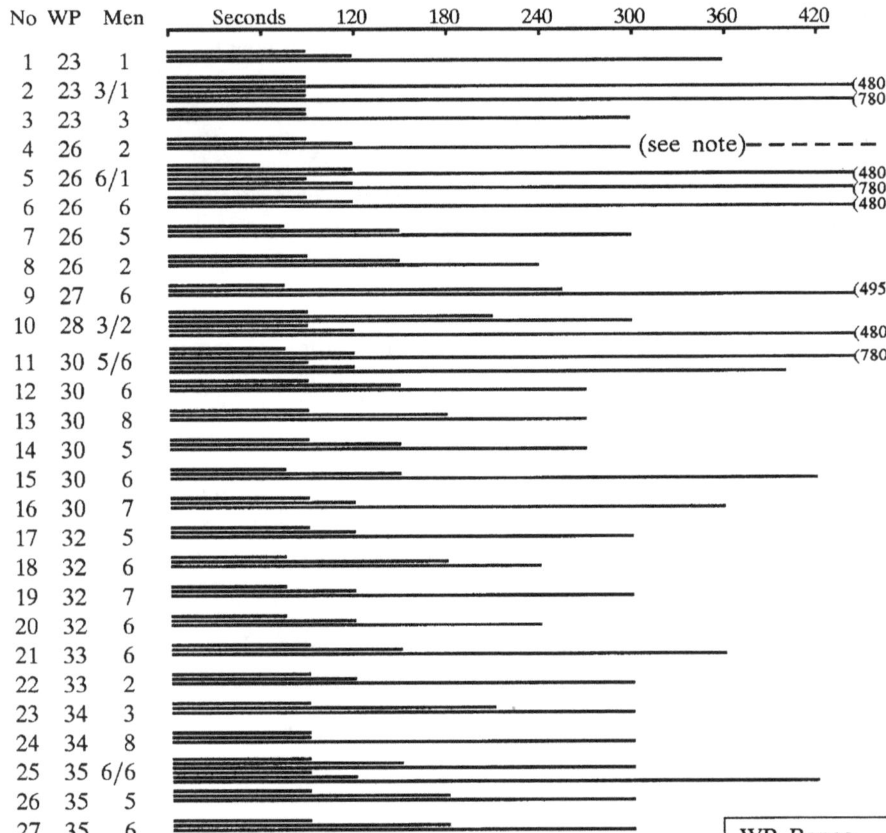

WP Range
23-35 lbf/in^2
(1·55-2·35 Bar)

36 Shift men 144 Man exposures
(No 4 - air supply failure, held at 1·1 bar 35 minutes
(No 2 - One man transferred to medical lock)
Range of times for Stage 1 Decompression 60-90 Seconds, Stage 2 Transfer 90-210 Seconds, Stage 3 Recompression 240-780 Seconds

Sunday shift of 6.75 h at 1.4 bar they climbed the vertical ladder system unannounced and found the recompression lock occupied by two previous groups; the men left the site and reappeared 32 h later apparently pain free.

These three men continued to work on the site for a further 10 months and 2 were treated for Type I decompression sickness 4-5 months after the incident. Bone x-rays taken 6-7 months after the incident showed no new changes and the men continued to work in the tunnelling industry.

An alternative explanation, that the men were recompressed later on the same day, is not supported by the lock-keeper's record and the recompression lock chart shows no further use for 16 h, until changed on Monday morning. The absence of symptoms was confirmed by the men when interviewed 10 days after the event.

5.6 Decompression sickness

All men at Blackpool were encouraged to report minor symptoms (e.g. niggles) and attend for advice and therapeutic recompression when required.

During the study period 28 shift men requested recompression on 37 occasions after 4,276 exposures of 7-8½ h plus 400 of 4-6½ h. The shift bends rate of 0.79% is acceptable for work in the pressure range above 2 bar for saturation exposures. The bends rate for each pressure phase is 0.4%, 0.88% at 0.93%. Phase c was of shorter duration with 642 exposures and all 3 men affected on 6 occasions were newly recruited to compressed air work.

21 men with no previous equivalent experience of compressed air work were treated on 30 occasions compared to the 7 men with high pressure experience treated on 7 occasions.

Reported symptoms and actual treatments rarely coincided with prolonged decanting times, and none of the shifts recorded in Tables 3 were related to treatment or symptoms on the day or after the subsequent 3 shifts.

5.7 Bone changes

Bone x-ray surveys were conducted at intervals between August 1966 and October 1967 with the support of the MRC Decompression Sickness Panel.

Due to the relatively short period that the contract operated only 4 non-shift and 8 shift men had radiographs taken more than 3 to 10 months after unusual decanting episodes were recorded. All had normal reports, or unchanged appearances, when previous bone radiographs were available, although the time scale is too short to completely exclude the possibility of undetected bone damage.

It is possible that some men will have been x-rayed later at other contracts as the same contractor operated a similar site in London 1967-8.

The need for consolidated records of compressed air workers was addressed by the MRC Decompression Sickness Panel in 1964 and all the Tilbury and Blackpool radiographs are deposited at Newcastle for future reference.

6 Discussion and recommendations

Although the number of men involved in the two London contracts was less than 28 the precise measurement assisted in clarifying aspects of decanting observed many years previously. These are discussed below.

6.1 Equipment specification

It is always difficult to set up a site and test the equipment before men are exposed, usually less than 24 h is available once the compressed air supply is fully operative.

Alteration to pressure-certified equipment on site is difficult and cumbersome to perform. The equipment specification must meet the operating times required for the specific layout and working pressure range. Minimum decompression and compression rates of 1.5 bar/min must be attainable. Figure 3 demonstrates the characteristics of low-pressure and high-pressure air supply using the standard inlets of a medical lock. The air supply must sustain the specified compression rate over more than 3 min and test pressure recordings should be obtained under simulated working conditions.

6.2 Target times

Target times for each of the 3 stages of decanting are set within the equipment specification to achieve the longest possible times for recompression. The ideal combination of 1+1+3 minutes is applicable to units in close proximity and can be extended to 1+2+2 minutes even in the higher pressure ranges, as rapid decompression and recompression are accepted by selected experienced compressed air workers. The feasibility of meeting the target times must be tested under realistically simulated site conditions.

Decompression is not recorded but is continuous without rapid or abrupt changes in rate; by theoretical calculation at least half the time is above half absolute pressure.

In practice, the time limitation of 5 min is exceeded at many sites, and prior to the statutory regulations many sites worked to a less severe limitation by measuring the transfer time only.

In diving practice different procedures are accepted. The USA divers use half absolute to half absolute pressure decanting, while UK divers use half absolute to half absolute + 1 bar, a moderate saving of time in the calculations, but neither method has been used in civil engineering in the UK since 1958. Study of the form of recompression curves show that delay usually occurs at pressures above half absolute, an effect that could extend the limitation to 6-7 min in a practical sense (see Figures 1 and 2).

If an increased total time WP-to-WP is accepted, the outcome at sites operating above 2 bar g. is considerably improved, leaving only the 2-14% of occasions when severe ear block occurred.

After full recompression to 2 psi (0.1 bar) above working pressure the pressure was held constant, nominally for 10 min, before completion of the staged decompression based on imperial units. When multiple groups decanted in sequence the period at constant pressure, 0.1 bar above WP, could vary between 8 and 20 min according to the interval between groups.

6.3 Medical selection and control of decanting

Medical selection for decanting is to the same standard for all compressed air workers, but specific lock testing of individuals is an extension of the medical examination. Failure during the lock test requires further medical assessment and may make an individual unacceptable for full shift exposures. Lock testing before decanting should include a second recompression at more than 1 bar/min, once the

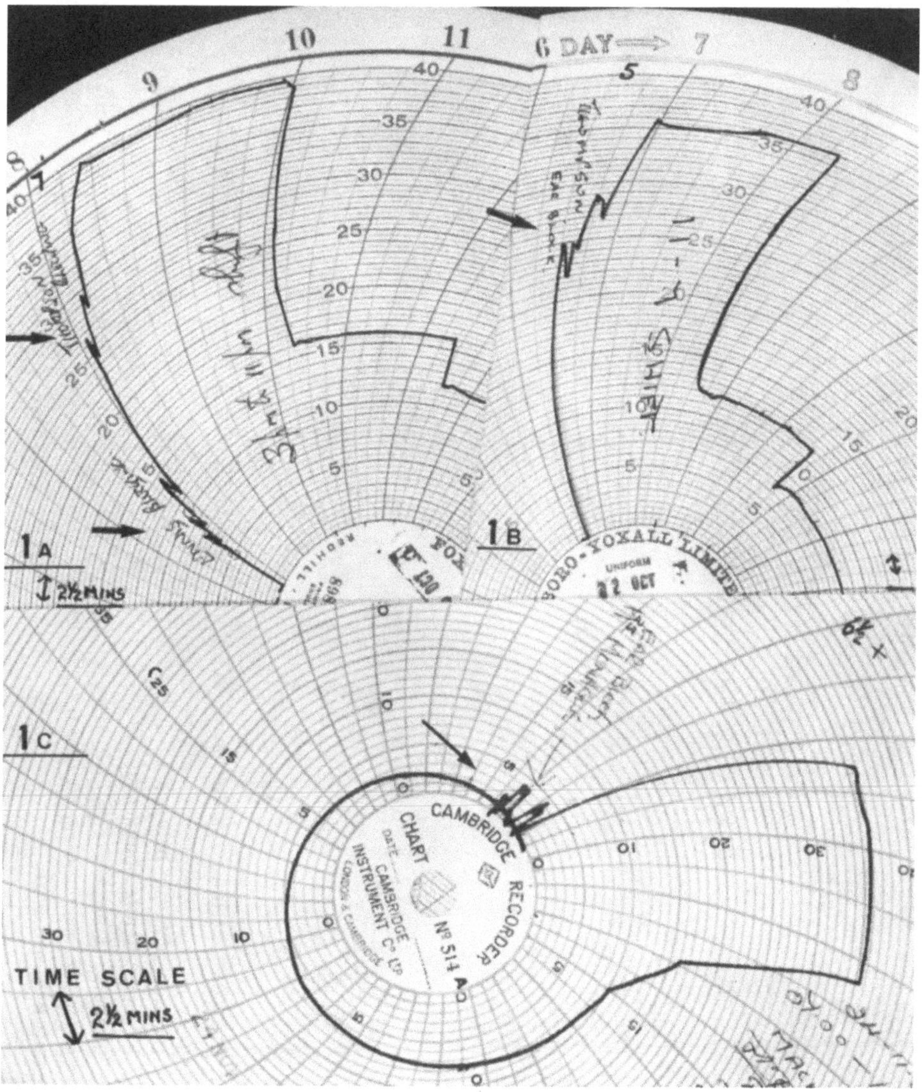

Fig.1. Decanting recompression charts showing delay due to ear symptoms

1a. 2 men affected causing 5 short stops above 7 psig (0.5 bar g.).
1b. Delay late in recompression above decompression stage level of 12 psig (0.85 bar g.) with 2 marked drops in pressure to relieve symptoms.
1c. Transfer of one man to medical lock after 2 attempts at recompression over 6 min. Shift then compressed at normal rate of 1.5 bar/min.

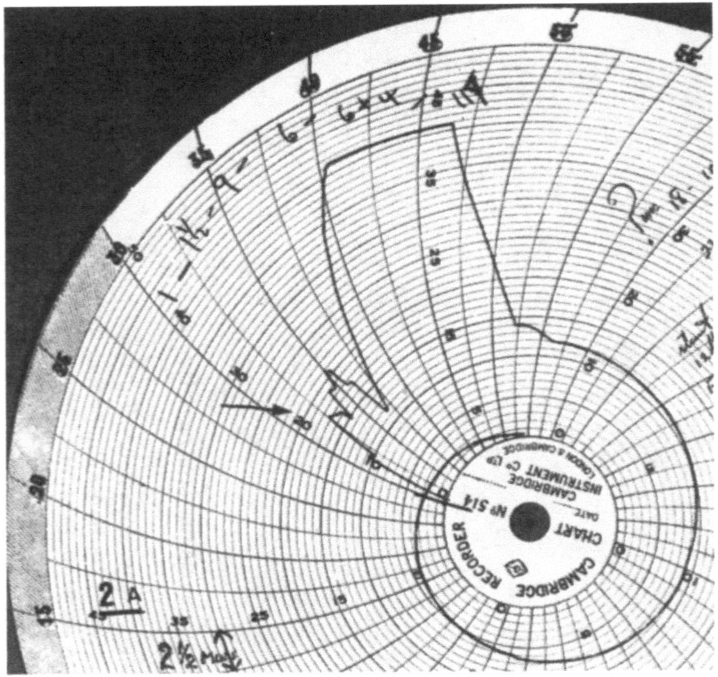

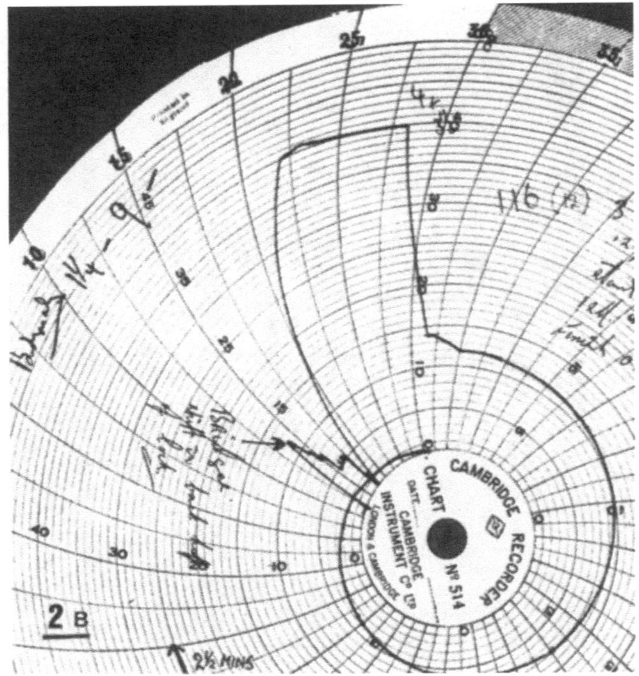

Fig.2. Prolonged delay in recompression with alternative solutions to reduce the risk to the rest of the shift

Fig.2a. Delay in recompression of 5 min between 12 and 17 psig (08.5 and 1.15 bar) before completing full procedure in 9 minutes. Time WP-WP 11½ min. Rate of compression after ear problem cleared 0.6 bar/min. All shift men involved.

Fig. 2b. Delay in recompression of 5 min under 12 psig (0.85 bar) with subsequent slow compression at 0.6 bar/min for affected man only. Shift compression in inner lock over 3½ minutes at 0.8 bar/min not recorded

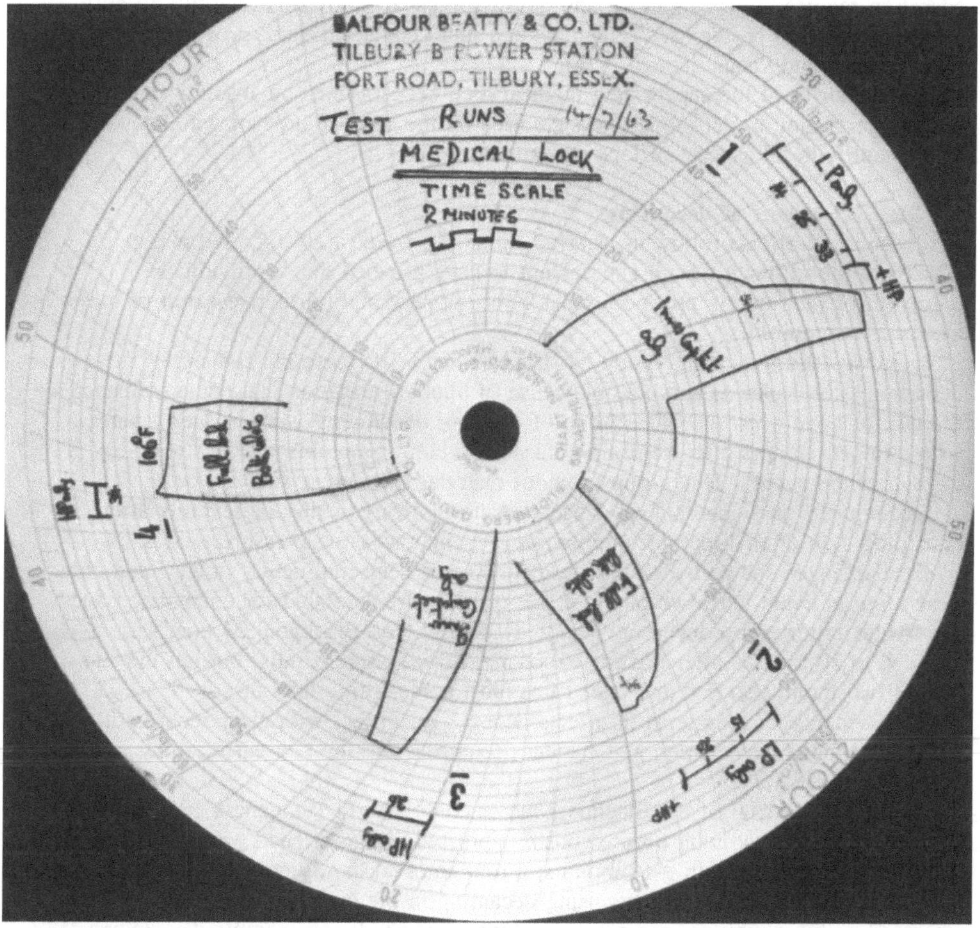

Fig.3. Test compressions of a standard medical lock

Tests 1 and 2 show compression rate of 0.4 to 0.5 bar/min using LP air supply at 36 psig (2.5 bar g.) with marked decrease in rate above 34 psig (2.4 bar g.) with one or both compartments in use.

Tests 3 and 4 using HP air supply at 50 psig (3.5 bar g.) gives a constant compression rate of 1.2-1.3 bar/min to above WP.

The size of the inlet in the inner compartment is a limiting factor during compression. HP increases the temperature during compression, in test 4 from 21° to 41°C in the summer season.

initial test at a slower rate as defined under regulations is satisfactory.

Organisation at the site is easier to control at the surface level and the co-operation required between above and below ground lock-keepers may fail because of communication problems. Telephone and radio contact has not proved reliable for site work and visual control is safer.

The authority of the supervising medical lock-keepers must be absolute in all matters pertaining to decanting and supported by the medical advisor who should be conversant with all aspects of lock operations.

6.4 Medical effects of decanting

Evidence of ill effects following delayed decanting were rarely reported and at Blackpool the ultimate failure appears not to have affected the 3 individuals involved, or men taking up to 30 min for recompression. Further research of these incidents is required.

Exercise during decompression has been shown to increase the risk of decompression sickness and possible cause of bubble formation that could influence the risk of bone necrosis. The group of 41 men, of which 9 worked over 2 bar g., who climbed ladders after decanting did not develop symptoms within a period of 5 days after the event. This group provides material for further study.

Decompression sickness Type I occurred no more frequently or severely after decanting than after direct decompression at similar working pressure ranges.

Bone changes shown by x-rays are difficult to assess confidently in less than one year after an event until two or more sets of films are available for comparison and the 4 sites reported produced only limited material. At Blackpool 24 men were x-rayed 4 to 10 months after starting exposure with negative results, but left the site within two months so that any later films must come from other sites.

The evidence of long-term ill effects is negative, in that with clear records of extended times for decanting and correct decompression the few early radiography results were negative.

At the second trial of the Blackpool tables completed at Dungeness 1967/8 (MRC, 1974) five of 59 men with no previous compressed air experience were shown to have bone changes within 9-12 months; a time interval hardly covered at Blackpool.

It is likely that most contracts using decanting in the early period before 1970 operated within a similar time pattern to those described; for example it is known that one London contract used the Blackpool equipment in 1968 with similar prolonged times during underground-to-surface decanting.

7 Conclusions

1. Experience over many years and several contracts shows that the 5-min rule of decanting is often exceeded.

2. Decompression sickness Type I occurred no more frequently or severely after decanting than after direct decompression over similar pressure ranges.

3. Decanting used with the ideal design of equipment and controlled by an experienced medical lock-keeper can regularly operate to close limits within the five-minute limit at working pressures up to 2 bar g.

4. The commonest major problem of delay during recompression is reduced by working with men selected and well versed in the art of equalising ear pressure, but even a slow recompression over 10-30 min appears to make no difference to the outcome.

5. The continuing acceptance of the qualification of reasonably practicable has never been tested, and the choice of 5 minutes is an empirical decision. The advantage of retaining the 5-minute limitation is to set a standard: without a guideline, time would extend indefinitely and the safety limit is not yet defined.

6. The material to establish the prevalence of bone changes after decanting may already exist but still needs to be collated.

Acknowledgements

The MRC Decompression Sickness Panel supported studies at Blackpool.

I thank the site agents and many medical lock-keepers who so enthusiastically supported my activities; Richard Liddle from Newcastle University was responsible for collecting and collating the London records.

References

Catton, M.J. (1967) Decanting Method of Decompression. In: *Decompression of Compressed Air Workers in Civil Engineering*, (McCallum, R.I. ed.). Proc. of International Working Party, CIBA, Foundation, London 1965, Oriel Press (Newcastle-upon-Tyne), 1967, p206-213

CIRIA (1973). *Medical Code of Practice for Work in Compressed Air*. Report 44, 1st edn., CIRIA, London

Institution of Civil Engineers (1936), *Report of the Committee on Regulations for the Guidance of Engineers and Contractors for work carried out under compressed air*. Institution of Civil Engineers, London

MRC Decompression Sickness Panel (1974). *Experience with a new decompression table for work in compressed air*. Technical Note No. 59, CIRIA, London

Work in Compressed Air Special Regulations 1958, Statutory Instrument No. 61, Factories. H.M.S.O., London

29 Observations of decompression sickness and effects of working conditions in two caissons in Bremen

K.H. Männche, K.H. Dreyer and D. Peusch-Dreyer
Arzt f Chirurgie, Bremen, Germany

Abstract
At two caisson construction sites, under the medical supervision of the authors, several cases of Type I decompression sickness occurred, notwithstanding the regular observance of working hours and decompression periods as established by the German employers' liability insurance associated table. Working pressures were up to nearly 4 bar absolute. 4% of all man decompressions on construction site 1 (15 cases out of 375 man decompressions), and 5.5 % of all man decompressions on site 2 (10 cases of 182 man decompressions) were affected. The possible causes of these problems are examined in this paper.
Keywords: Decompression sickness Type I, Working conditions, Caissons.

1 Construction site 1 (1989)

It took nine weeks to lower a concrete tank of a height of 30 m and a diameter of almost 30 m, whose purpose was to serve as a collector and a pumping pit for municipal drainage. The bottom itself was lined in form of a caisson working chamber. The structure was to be lowered entirely by means of flushing and pumping out the sand from underneath the working chamber. After passing through the lock at the upper rim of the tank, the working chamber was reached *via* a descent of 30 m under working pressure. Three caisson workers were here for each working shift. In the beginning, the work was performed in three shifts per 24 h. With increased depth of working, activities were carried out in four shifts per 24 h. As a rule, there was no work on weekends. On occasions, however, extra shifts were established on weekends when the progress of the work had fallen behind schedule.

Delays occurred and strenuous physical work effort was involved when rocky layers had to be penetrated. Smaller stones had to be collected by the miners while they were standing in the sump provided for the pumping, for subsequent outward hoisting in baskets. Boulders had to be broken by hydraulic breakers, and their fragments transported for subsequent outward removal.

Engineering and Health in Compressed Air Work Edited by F.M. Jardine and R.I. McCallum. Published in 1994 by E & FN Spon. ISBN 0 419 18460 0

The outward transfer of the miners on termination of their shift took place in the T-lock joined on to the chamber, so they stayed under pressure until the start of decompression. Decompression was in accordance with the German regulation relating to work activities under compressed air conditions – Table B : outward transfer assisted by the application of oxygen respiration (*Druckluftverordnung*, 1972). The same table also specifies the permissible maximum working time in a caisson, which was followed on this site.

We rendered medical assistance at the construction site from the moment that an absolute pressure of 1.5 bar had been reached in the caisson. In spite of strict observance of the exposure and decompression periods and even of the shortening of working times and the lengthening of the decompression times, a total of fifteen incidents of Type I decompression sickness was registered (see Table 1). This means that for almost 4% of all cases of admittance to the working chamber, we faced decompression illness needing treatment.

As humidity and low temperature were reported in the overwhelming majority of the sickness incidents, we repeatedly entered the working chamber in order to observe the working process personally. Investigation drilling had disclosed a mainly sandy soil with only a few stones in the subsoil. For this reason, the construction site had been set up in such a way as to permit the lowering of the tank by means of the flushing operations. Unexpectedly, however, a substantial rock layer had to be overcome. It was this circumstance that made it necessary for the miners, over extensive parts of the works, to collect the stones by hand while they were standing in the sump. This resulted in the 'effects of low temperature' on their lower as well their upper extremities (Rozsahegyi, 1989).

We attached temperature-measuring probes directly to the skin in the area of the knees of all workers active in two shifts and took many measurements. The temperature of the air of the caisson was 30°C (303 K). The temperature of the water in the sump was 26°C (289 K). The temperature values of the workers standing in the water averaged 29°C (302 K), while those standing on 'dry ground' showed an average of 33°C (306 K).

Despite a temperature of 37°C (310 K) in the decompression chamber, the skin temperature of the workers who had been standing in the pump sump could be established on termination of the decompression at only 33°C (306 K), although that of the other workers was at 35°C (308 K).

We had to do something in order to reduce the frequent occurrence of decompression sickness. In addition to the clothing used under the waterproof trousers, we supplied *Medima* knee-pads. In fact, this helped to reduce the number of decompression sickness events. We noted – at least as regards the aspect of decompression sickness pain of the lower limbs – that only a few cases occurred after the knee-pads had been supplied, and those actually presented themselves when the knee-pads were thoroughly soaked and their insulating effect thus had become nil.

A further finding of ours was based on the professional experience of somebody who has worked under excess pressure. Roughly half of the miners had been active

Table 1. Decompression sickness at construction site 1

No.	Patient[(1)] identification	Absolute pressure inside caisson	Possible causes of sickness, when exact compliance with maximum exposure and decompression times
1	JK	2.2 bar	First time under pressure, claustrophobia, hyperventilation, unrest
2	RK1	2.3 bar	Effects of low temperature, humidity
3	GP	2.5 bar	Effects of low temperature, humidity
4	MSCH	2.55 bar	Effects of low temperature, humidity
5	TF	2.6 bar	Effects of low temperature, humidity
6	SZ	2.6 bar	Effects of low temperature, humidity
7	RK2	2.6 bar	Effects of low temperature, humidity and short-time pressure loss inside caisson
8	AM	2.6 bar	Effects of low temperature, humidity
9	AK	3.3 bar	Effects of low temperature, humidity
10	AP	3.3 bar	Effects of low temperature, humidity
11	MG	3.3 bar	Effects of low temperature, humidity and cumulative exposure to pressure as had been working for 9 days[(2)]
12	MH	3.3 bar	Effects of low temperature, humidity and cumulative exposure to pressure as had been working for 9 days[(2)]
13	RK3	3.5 bar	Effects of low temperature, humidity
14	BW	3.6 bar	Strenuous physical efforts with dismounting work of firmly installed equipment
15	HB	3.6 bar	Strenuous physical efforts with dismounting work of firmly installed equipment

Notes: (1) Patient RK experienced decompression sickness three times
(2) Bennett and Elliott (1982)

in their trade for more than one year, the other half were beginners. The proportion of beginners in those affected by decompression sickness, however, was 60 per cent.

Once difficulties to meet the time schedule for lowering the caisson had arisen because of the unexpected obstacle of the stony layer in the subsoil, the managers of the construction works decided to have work carried out also during weekends in an effort to make up for the delay. Following nine days of uninterrupted shift operations, two cases of decompression sickness were registered (Bennett and Elliott, 1982).

Our findings lead to the conclusion that the working hours and decompression periods have not been sufficiently well assessed if they result in decompression illness, even for minor reasons (Buhlmann, 1989; Mauermayer *et al.*, 1985).

2 Construction site 2 (1990)

With the knowledge we had gained at the first construction site, we also looked after a second compressed air caisson. The purpose of this one, too, was the lowering of a storage basin and pumping pit for the municipal drainage. The one difference in the construction method was determined by the need for most of the subsoil to be dug by mechanical excavators and removed through the lock by means of suitable containers, while only a small amount of the subsoil could be removed by the flushing process. By working in this way there was no need for the strenuous physical work over long periods while standing in water.

Contrary to the first construction site, we registered practically no decompression sickness at the start. In this case, also, work was carried out in strict compliance with the official instruction tables; but from the start, great importance was attached to the use of thermal protection clothing. If under these circumstances, the decompression sickness rate had been 5.5%, it would have underlined our conjecture from site 1 about the inadequacy of the official tables.

In the course of the first 140 admittances to the chamber, we registered only two slight Type I cases. Then, all of a sudden and shortly before completion of the caisson work, there was a mechanical breakdown of the heating system in the decompression chamber, when the outdoor temperature was at freezing point. The miners were forced to spend the decompression period in a cold ambient temperature. All the woollen blankets and hot drinks brought inside proved to be insufficient to avoid the workers' supercooling. The consequence of this circumstance was that we registered so many Type I cases within a single day, that the decompression sickness rate jumped up to 5.5 per cent.

3 Type I decompression sickness pain sites

Under the two caisson construction works, there were 25 cases of Type I decompression sickness. The reported pain sites are listed in Table 2.

Table 2. Locations of limb pain in 25 cases of Type I decompression sickness

Site of Pain	Number
Knee-joints	20
Lower leg	4
Upper ankle-joint	2
Thigh	2
Forearms/wrist	3
Elbow joint	1
Upper arms	1
Shoulder-joints	2

4 Concluding remarks

Our main conclusion is that the working hours and decompression periods have not been sufficiently well assessed if they result in decompression illness, even for minor reasons – such as low temperature, humidity, or after strenuous physical work, and working every day in the caisson chamber for more than a week without a break.

References

Bennett, P.B. and Elliott, D.H. (1982) *The Physiology and Medicine of Diving.* Ballière Tindall, London, page 341

Bühlmann, A.A.; Zürich; (1989) Personal communication

Mauermeyer, Kessel, R. and Praml, G. (1985) *Untersuchungen zur Festlegung neuer Ausschleuszeiten aus Druckbereichen um 1 bar Überdruck.* Genter Verlag, Stauttgart, pages 67 - 71

Rozsahegyi, I.; Budapest; (1989) Personal communication

Verordnung über Arbeiten in Druckluft (*Druckluftverordnung*) (4.10.1972)

30 Memory and personality of caisson workers with a history of decompression illness

J. Leach, J.D. King*, P.E. Morris and J. Rawlins
Department of Psychology, University of Lancaster, UK
**London Hyperbaric Medical Service, London, UK*

Abstract
Memory (digit span) and clinical personality measures (EPQ) were recorded from 37 caisson workers. These measures were correlated with their length of working experience in caissons and the maximum pressure under which they worked. No significant relationship was found between memory performance and personality measure for either caisson experience or maximum pressure (all $p > 0.05$). Further analysis, however, revealed a significant impairment in memory performance (forward digit span) for those workers with a recorded history of decompression illness ($n=9$, $p<0.05$) compared with firstly, caisson workers without a history of decompression illness ($n=28$); secondly, a control group of non-caisson workers ($n=182$, $p<0.05$); and thirdly, a control group matched with the decompression illness group on age and level of formal education ($n=9$, $p<0.05$). There was no difference between the groups on personality measures. The non-decompression illness group and control groups showed no difference in memory performance. These results are consistent with recent findings by the authors of memory performance amongst commercial divers who have suffered decompression sickness.
Keywords: Caisson Workers, Memory, Digit span, Personality, EPQ, Decompression illness.

1 Introduction

Various studies have now accumulated to suggest that decompression sickness (DCS) or illness (DCI) may have long-term psychological consequences. In a now classic study Rozsahegyi (1959) originally reported the effects of decompression sickness on caisson workers involved in the building of the Budapest underground railway. All subjects, described by him as patients, who were unable to work for three or more days as a result of decompression sickness were clinically examined and 100 were followed up for a period ranging from two-and-a-half to five-and-a-half years after the decompression insult.

Engineering and Health in Compressed Air Work Edited by F.M. Jardine and R.I. McCallum. Published in 1994 by E & FN Spon. ISBN 0 419 18460 0

Rozsahegyi described four types of central nervous system (CNS) lesions occurring as the result of decompression sickness namely; (1) multiple focal injuries of the whole CNS, (2) multiple lesions to the cerebrum and upper brain stem, (3) injury to the medulla, pons and cerebellum, (4) spinal lesions. Of his follow-up group psychological changes were seen in one third of those with neurological DCS with particular emphasis being placed on personality changes. Men who were previously calm and self controlled before injury became impulsive and fell easily into rages. 'Pathological drunkenness' and alcohol intolerance were also frequent. Rozsahegyi reported that the psychotic syndrome persisted for several years only in those cases in which the upper segments of the CNS were affected particularly in group (1), those showing multiple focal lesions of the whole CNS. It was not found in group (4) i.e. those with only spinal cord lesions. More than half the cases in group (4) showed no other symptoms several years after the event, as did half of group (2); but in groups (1) and (3) only one fifth of the cases were without symptoms. The most frequent complaints were dizziness (49%), headache (46%) and increased irritability (40%). Rozsahegyi suggests a possible chronic cumulative encephalomyelopathy occurring as a result of repeated decompression procedures. Although this progressive disorder could occur in the absence of decompression sickness it was more common in individuals with a history of DCS. It should be noted, however, that many of Rozsahegyi's subjects continued with pressure exposure during the period between the initial study and his follow-up study. Consequently, the 'progressive' nature of the encephalopathy may be due to further decompression insults. In a separate investigation Rozsahegyi and Roth (1966) studied a group of 57 caisson workers, all of whom had suffered Type II decompression sickness. They reported complaints of anorexia, headache, sleep disorders and lassitude. EEG abnormalities were found in 66.7% of cases. Further chronic behavioural changes were reported among the group.

Similar behavioural changes in divers following diving operations have been reported (Baddeley, 1983; Leach, 1987) which include memory impairment, difficulty in concentration, retarded reactions, aggression and anti-social behaviour, loss of interest in appearance and reduced tolerance to alcohol. It is possible, however, that these may be more signs of situational maladjustment than any permanent clinical or psychological dysfunction, at least in those divers without a history of DCI. In fact, Leach (1987) found that most divers showed a functional psychological recovery within 3-5 days following diving operations with the exception of alcohol intolerance which did appear to be progressive. Similar psychological recovery was found by Curley (1988) following deep saturation dives with divers initially presenting with alterations in effect, visual focusing difficulties, increased lethargy and mild depression which resolved within 10 days. Furthermore, follow-up studies months and years after these dives revealed no substantial changes.

Levin (1975) claimed a high incidence of neurotic and psychosomatic symptoms among divers with a history of decompression sickness, while Peters *et al.* (1977), who studied 10 divers with a history of decompression sickness involving the CNS, reported that eight had neurological deficits implying spinal and supraspinal lesions.

The investigators concluded that cerebral disturbance following decompression sickness in divers is more common than had been previously thought. Kwiatkowski (1979) found a significantly higher proportion of pathological EEG records among professional divers coupled with a distinct increase in pathological records correlating with the number of years of diving. On the other hand Hallenbeck (1978) found no evidence for the development of a progressive encephalomyelopathy with DCI in the absence of further bubbling.

Vaernes and Eidsvik (1982) found impairment in psychological functioning following 'near miss' diving accidents. They argue that their results indicate impairment of complex spatial functioning, particularly spatial memory supporting the view that diving-related CNS decompression sickness includes lesions of the higher central nervous system and is not confined merely to the spinal cord. It should be emphasised, however, that of the nine subjects studied not all suffered DCI (two had hypercapnia and one had hypoxia). More recently Curley *et al.* (1988) reported the reappearance of neurologic and neuropsychologic impairment over a period of hours and days in five individuals who had sustained CNS decompression sickness (4) or arterial gas embolism (1).

Edmonds (1984) initially reported cases of decompression sickness followed by discernible psychometric impairment often with abnormal EEG patterns and occasionally with abnormalities in the brain scan. The following year Edmonds and Boughton (1985) reported a study in which they surveyed 30 compressed air abalone divers. They employed two separate psychometric tests, the ACER higher test WL and the Revised Visual Retention Test (Anastasi, 1982). The former is designed to measure general ability as revealed by performance on material of a verbal nature while the latter is designed to assess visual perception, visual memory and visuoconstructive abilities. Of the original 30 divers, six had to be excluded for technical reasons. From the remaining 24 subjects, Edmonds and Boughton found evidence of impairment in intellectual capacity in 11 (the evidence in four cases being suggestive and in seven cases strongly suggestive), and one subject was placed in the 'questionable' category. It was argued that the results obtained strongly suggest a general impairment of intellectual capacity (i.e. a dementia) and that such a finding would not be expected in normal subjects. Later, however, the view of a general dementia occurring in divers with a history of DCI was reversed, and Edmonds and Hayward (1987) reported further studies showing no difference in mental ability between divers with and without a history of DCI.

Morris *et al.* (1991) examined 282 offshore commercial divers on various psychological tests of whom 64 had a recorded history of DCI. They found that divers with a history of DCI performed significantly worse on working memory and complex non-verbal reasoning compared with divers without a history of DCS and compared with a matched control group of non-divers.

As well as cognitive dysfunction there have been a number of reports of possible personality and psychiatric changes following diving. Biersner and Ryman (1974) found a significantly higher rate of psychiatric admissions among 4,720 US Navy divers when compared to all Navy psychiatric hospital admissions. On the other hand Hoiberg and Blood (1985) compared 11,584 divers with 11,517 control subjects and found a significantly lower admission rate for psychiatric and stress-related disorders among divers than among the controls. Morris *et al.* (1991) found no difference in clinical personality traits (Eysenck Personality Questionnaires, EPQ) between commercial divers with a history of DCI (n=64) and divers without a history of DCI (n=218).

The evidence for and against psychological dysfunction occurring following DCI remains contradictory; nonetheless, anecdotal and some observational reports persist in suggesting the psychological impairment similar to that claimed for divers may be occurring in caisson workers following decompression illness. The aim of this pilot study was to investigate two areas of suspected dysfunction: memory span and personality.

2 Method

2.1 Subjects

(a) Caisson workers. The subjects were 37 caisson workers with a mean age of 34.19 years (sd=6.73) and mean caisson working experience of 3.63 years (sd=2.09). Their range of exposure varied between 10-40 psig (0.7-2.8 bar) with a mean of 23.78 psig (1.64 bar) (sd=7.77 psig). Nine subjects had a recorded history of decompression illness. Of these, eight had suffered one DCI incident each and one subject had suffered three incidents.

(b) Control subjects. The control subjects comprised 182 civil engineering workers who were mostly engaged on the same type of work as caisson workers but who had never worked under hyperbaric pressure. Their mean age was 32.34 years (sd=9.9).

The control group was matched with the caisson workers group on age and level of formal education. Taking the highest level of educational attainment is a broad but effective and practical measure of basic intellectual calibre. It has been found in previous research that educational achievement and intelligence levels are associated and that these may have an influence on test performance (Adams and Boake, 1981). Furthermore, it has been reported that educational attainment has an effect on test scores with lower levels of education being associated with lower-level test performance (Finlayson *et al.*, 1977).

2.2 Tests

(a) Digit span test. The digit span test is a measure of short-term or working memory capacity and is a component of the Wechsler Adult Intelligence Scale. In this test the subject is presented with a series of digits which he has to repeat in the given order. The series increases in number (from 3-9) until the subject fails to respond correctly. The digit span is given as the series above which he failed to respond correctly on two presentations. The backward digit test is presented to the subject in the same way as forward digit test but the subject must recall the digits in the reverse order to that presented.

(b) Eysenck Personality Questionnaire (EPQ). The EPQ is a personality test containing 90 self-assessment questions of the forced Yes/No type. It provides a personality profile along the scales of extroversion, neuroticism and psychoticism and includes a separate 'lie' scale to establish the tendency of subjects to 'fake good' (Eysenck and Eysenck, 1975).

2.3 Procedure

The caisson workers and non-hyperbaric control group were tested over a period of 12 months at the London Hyperbaric Medical Service. They were attending statutory medical examinations and were asked to complete the digit span test and EPQ. They were told that these additional tests were voluntary. Although voluntary, no subject refused to be tested. Additional data (e.g. age, education, work experience, etc.) were also obtained from each subject and clinical data (e.g. decompression illness history) were available to the medical officer (JDK). The EPQ and digit span tests were administered blind; that is, the testers were unaware at the time of testing whether or not the subject had a history of decompression illness.

3 Results

Memory and personality measurements were compared between three groups: (1) caisson workers with a recorded history of decompression illness, (2) caisson workers without a history of decompression illness, and (3) a control group of non-caisson civil engineering workers.

3.1 Test data and caisson experience

No significant correlation was found between experience of caisson work and either digit span performance or personality measures. There was a strong and significant correlation between experience and the age of caisson workers, which is to be expected ($r=0.44$, degrees of freedom (df)$=35$, $p<0.01$). Correlation data arc presented in Table 1.

3.2 Test data and maximum working pressure

No significant correlation was found between the maximum pressure under which a caisson worker had worked and his test scores. There was no correlation between either maximum pressure and age ($r=0.05$, df$=35$, $p>0.05$) or between maximum pressure and caisson experience ($r=0.19$, df$=35$, $p>0.05$). Correlation data are presented in Table 1.

Table 1. Correlations between caisson pressure, caisson experience and test scores

	Correlations with caisson pressure	Correlations with caisson experience	Test scores
Digit span:			
forward	-0.307	0.260	non-sig
backward	-0.224	-0.002	non-sig
mean	-0.317	0.161	non-sig
Personality:			
psychoticism	0.076	-0.151	non-sig
extroversion	-0.032	-0.162	non-sig
neuroticism	-0.143	-0.184	non-sig
'lie'	0.147	0.183	non-sig
Age	0.050	0.443	non-sig

3.3 Test data compared between caisson workers with and without a history of decompression illness

Analysis of the data for forward digit span revealed a difference in performance between the DCI and non-DCI groups with the workers with a history of DCI performing significantly worse than those without a history of DCI ($p=0.02$, df=35). The DCI group were also found to be significantly younger than the non-DCI group ($p=0.01$, df=35). There was no significant difference between the two groups on length of work experience, maximum working pressure or any other test or personality score (all $p>0.05$, df=35). Comparative data are presented in Table 2.

Table 2. DCI compared with non-DCI test scores

	Comparative scores (standard deviation)			
	non-DCI ($n=28$)	DCI ($n=9$)	t value (unpaired)	p
Digit span:				
forward	7.14 (1.08)	6.11 (1.05)	-2.51	0.02
backward	5.36 (1.06)	5.22 (0.97)	-0.34	non-sig
mean	6.25 (0.91)	5.67 (0.87)	-1.69	non-sig
Personality:				
psychoticism	4.00 (2.72)	5.44 (3.17)	1.40	non-sig
extroversion	11.82 (5.02)	12.67 (5.15)	0.44	non-sig
neuroticism	7.50 (4.17)	9.00 (5.94)	0.85	non-sig
'lie'	6.96 (4.29)	8.00 (6.52)	0.55	non-sig

3.4 Test data compared between caisson workers without a recorded history of decompression illness and the control group (unmatched and matched)

Analysis of the unmatched data revealed no significant difference in performance between the two groups on digit span (either forward, backward or mean). While the control group had a lower mean age (32.34 years, sd=9.9) than the caisson workers (35.75 years, sd=6.38) this difference was not statistically significant ($p>0.05$). Analysis of the personality measures revealed that the control group scored significantly higher than the caisson workers on extroversion ($p<0.05$), but there was no significant difference between the two groups on either psychoticism, neuroticism or 'lie' scale. The difference between the caisson workers and control group on extroversion may be due to the tendency for the control group to be younger than the caisson workers and the young do tend more towards extroversion. Furthermore, when the caisson workers and control groups are matched for age the statistical difference between them on extroversion disappears ($p>0.05$, df=27). There was found to be no difference between the matched caisson workers and controls on either psychoticism, neuroticism, or 'lie' nor on their performance on the digit span test. Comparative data for these two groups are presented in Table 3.

Table 3. Comparative scores for non-DCI compared with controls (matched and unmatched)

	Unmatched		**Matched**	
	***t* value (unpaired)**	***p***	***t* value (paired)**	***p***
Digit span:				
forward	0.24	non-sig	1.20	non-sig
backward	0.23	non-sig	1.06	non-sig
mean	0.26	non-sig	1.26	non-sig
Personality:				
psychoticism	0.54	non-sig	1.42	non-sig
extroversion	-2.43	0.02	-1.36	non-sig
neuroticism	-1.11	non-sig	-0.36	non-sig
'lie'	0.09	non-sig	-0.26	non-sig

3.5 Test data compared between caisson workers with a history of decompression illness and the control group (matched and unmatched)

Analysis of the unmatched data shows that caisson workers with a history of decompression illness performed worse on the forward digit span test than the control group ($p<0.05$, df=189). This difference increased statistically when the two groups were matched on age ($p=0.01$, df=8). There was no significant difference in performance between the two groups on either backward digit span or mean digit span for either matched or unmatched samples.

Analysis of the personality measures revealed no significant difference between the

DCI group and the control group on psychoticism, extroversion, neuroticism, or 'lie' for either matched or unmatched conditions. While the DCI group were slightly younger than the control group this difference was not statistically significant. Comparative data for the two groups are presented in Table 4.

Table 4. Comparative scores for DCI compared with controls (matched and unmatched)

	Unmatched		Matched	
	***t* value (unpaired)**	***p***	***t* value (paired)**	***p***
Digit span:				
forward	-2.29	0.02	-3.28	0.01
backward	-0.16	non-sig	-0.71	non-sig
mean	-1.34	non-sig	-1.90	non-sig
Personality:				
psychoticism	1.75	non-sig	-1.90	non-sig
extroversion	-0.90	non-sig	-0.71	non-sig
neuroticism	0.22	non-sig	0.48	non-sig
'lie'	0.88	non-sig	0.05	non-sig

4 Discussion

Anecdotal reports from tunnel workers supplemented by research into decompression sickness amongst divers suggested the possibility that some psychological impairment may be occurring in caisson workers following incidents of decompression illness. This study examined caisson workers with and without a history of DCI (n=9 and 28 respectively) and a control sample of construction workers (n=182) who had not worked in caissons. Analysis revealed no significant difference between controls and caisson workers on age. Each man completed the digit span test and the Eysenck Personality Questionnaire.

The only significant difference found was the poorer performance on the forward digit span test of those caisson workers with a recorded history of decompression illness compared with both caisson workers without a DCI history and a control group (both $p<0.05$). There was found to be no significant difference between any of the groups on performance of backward or mean digit span. These findings are consistent with studies of DCI in commercial divers for both general memory performance (e.g. Vaernes and Eidsvik, 1982; Curley *et al.*, 1988) and specific digit span capacity (Morris *et al.*, 1991).

Despite anecdotal reports there was found to be no significant difference in the personalities of caisson workers with a history of DCI (as measured by the EPQ) compared with either their non-DCI colleagues or a control group of construction workers. In fact the single notable difference was between the non-DCI caisson workers and the control group with the latter scoring higher than the former on extroversion ($p < 0.05$). Further analysis however revealed that extroversion was inversely correlated with age and that the control group tended to be younger than the caisson group. This suggests that the effect of decompression illness is cognitive in origin, that it affects the processing of information and that personality appears to remain stable over time. This is not to dismiss entirely anecdotal and observational reports of personality changes following DCI but that many of these may be due to situational maladjustment consequent upon any trauma and which is usually a temporary condition. Finally, and again contrary to some anecdotal reports, there was found to be no significant correlation between caisson experience and changes in either memory span or personality.

5 Conclusion

Our findings suggest that insult from decompression illness produces impairment in working memory capacity but that clinical personality remains stable over time and does not generally appear to be affected either by hyperbaric exposure (diving and compressed air tunnelling) or by decompression illness. Unfortunately this initial study was not able to define the scope of the problem, but further research in this area would be desirable, particularly research aimed at locating the range of cognitive abilities which may or may not be affected.

It is proposed, therefore, that firstly a short, easily administered neuropsychological test battery be developed to explore further the apparent susceptibility of working memory to decompression illness and secondly, that an information-processing approach (Norman and Lindsay, 1977) to cognition and neuropsychological function be adopted both in the research investigation and in the design of the test battery.

6 References

Adams, R. and Boake, C. (1981) *Bias in neuropsychological test classification related to education, I.Q. and ethnicity*. University of Oklahoma

Anastasi, A. (1982) *Psychological Testing*. 5th edn. Collier Macmillan, New York

Baddeley, A. (1983) In: Shields, T.E., Minsaas, B., Elliott, D.H. and McCallum, R.I. (eds) *Long-Term Neurological Consequences of Deep Diving*. Proceedings of the European Undersea Biomedical Society and the Norwegian Petroleum Directorate, Stavanger, Norway

Biersner, R.J. and Ryman, D.H. (1974) Psychiatric incidence among military divers. *Military Medicine*, **139** (8), pp 633-635

Curley, M.D. (1988) U.S. Navy saturation diving and diver neuropsychologic status. *Undersea Biomedical Research*, **15** (1), pp 39-50

Curley, M.D., Schwartz, H.J.C. and Zwingelberg, K.M. (1988) Neuropsychologic assessment of cerebral decompression sickness and gas embolism. *Undersea Biomedical Research*, **15**, (3), pp 223-36
Edmonds, C. (1984) Bubbles from down under. *Pressure*, April 1984, p6 & p11
Edmonds, C. and Boughton (1985) Intellectual deterioration with excessive diving (punch drunk divers). *Undersea Biomedical Research*, **12**, (3), pp 321-6
Edmonds, C. and Hayward, L. (1987) *Ninth International Symposium on Underwater and Hyperbaric Physiology*. 877-866 UHMS: Bethesda
Eysenck, H.J. and Eysenck, S.B.G. (1975). *Manual of the Eysenck Personality Questionnaire (Junior and Adult)*. Hodder and Stoughton, Kent
Finlayson, M., Johnson, A. and Reitan, R. (1977) Relationship of level of education to neuropsychological measures in brain-damaged and non-brain-damaged adults. *Journal of Consulting and Clinical Psychology*, **45**, pp 536-542
Hallenbeck, J.M. (1978) On long term health hazards of diving with regard to the central nervous system. *Mines Safety and Health Commission Workshop*. Commission of the European Communities, Luxembourg
Hoiberg, A. and Blood, C. (1985) Age-specific morbidity and mortality rates among U.S. Navy enlisted divers and controls. *Undersea Biomedical Research*, **12**, (2), pp 191-203
Kwiatkowski, S.R. (1979) Analysis of the EEG records among divers. *Bulletin of the Institute of Maritime and Tropical Medicine, Gdynia*, **30** (2), p 131
Leach, J. (1987) Post-dive syndrome. *Progress in Underwater Science*, **12**, pp 33-48
Levin, H.S. (1975) Neuropsychological sequelae of diving accidents. In: *International Symposium on Man in the Sea*. (S.K. Hong, ed.). Undersea Medical Society, Bethesda
Morris, P.E., Leach, J., King, J. and Rawlins, J. (1991) *Psychological and Neurological Impairment in Professional Divers*. Department of Energy Report, Project N 2050
Norman, D.A. and Lindsay, P.H. (1977) *Human Information Processing*. 2nd ed. Academic Press, New York
Peters, B.H., Levin, H.S. and Kelly, P.J. (1977) Neurologic and psychologic manifestations of decompression illness in divers. *Neurology*, **27**, pp 125-127
Rozsahegyi, I. (1959) Late consequences of the Neurological forms of decompression sickness. *British Journal of Industrial Medicine*, **16**, pp 311-7
Vaernes, R.J. and Eidsvik, S. (1982) Central nervous dysfunction after near-miss accidents in diving. *Aviation, Space and Environmental Medicine*, **53** (8), pp 803-7

31 Blood changes associated with decompression and their possible clinical significance

R.B. Philp
Department of Pharmacology and Toxicology, The University of Western Ontario, Canada

Abstract
We and others have reported that decompression of divers is accompanied by changes in cellular and chemical components of the blood. Some of these have been seen following surfacing from as little as 3 ATA. While some may be regarded as laboratory curiosities, others have the potential to influence the progression of decompression sickness (DCS) and perhaps dysbaric osteonecrosis (aseptic bone necrosis); namely, those changes which indicate an acceleration of the blood clotting process and which appear to be triggered by the presence of intravascular bubbles, even 'silent' ones that do not lead to DCS. They include: (1) a loss of blood platelets of about 25% over 48-h post decompression, (2) an increase in levels of fibrinogen/fibrin split products, (3) changes in complement levels, and (4) other trends suggesting a mild hypercoagulable state. These changes could contribute to the formation of tiny thrombi which, in conjunction with vascular bubbles, might occlude small blood vessels to delay nitrogen washout, and interrupt the blood supply to critical areas of the longbones, leading to bone death. Increases in circulating enzyme levels, indicative of tissue damage, have been seen in divers in the absence of overt DCS. Human studies to determine whether intervention with drugs can prevent these changes, have been equivocal. Most of these have been conducted in divers. Similar investigations in compressed air workers could be instructive.
Keywords: Blood Platelets, Clotting, DCS, Dysbaric osteonecrosis, Drug prophylaxis

1 Introduction

The possibility that the blood clotting system might play a role in the pathogenesis of decompression sickness (DCS) was noted very early on, even before theories emerged proposing that bubbles in the blood and tissues were responsible for the pain and other, more serious, manifestations of 'the bends'. These early suggestions stemmed from observations made by physicians like Andrew Smith and Alphonse

Engineering and Health in Compressed Air Work Edited by F.M. Jardine and R.I. McCallum. Published in 1994 by E & FN Spon. ISBN 0 419 18460 0

Jaminet, medical consultants for the construction of the Brooklyn Bridge and the Saint Louis Bridge respectively. While conducting autopsies on caisson workers who suffered fatal attacks of DCS, they saw areas of haemorrhage and blood clots in the spinal cord and became proponents of the 'congestion' theory of DCS. As knowledge developed regarding the role of bubble formation and the success of recompression therapy, interest in the possible contribution of haematological factors faded, even though some experiments continued to focus on this aspect, revealing changes associated with decompression (see Philp, 1974 for a review). The possibility remains that activation of the blood clotting system and of platelet aggregation by intravascular bubbles could result in the formation of small thrombi which could complicate the outcome of DCS and possibly of dysbaric osteonecrosis. Should this be the case, recent developments in the treatment and, more importantly, the prophylaxis of thrombotic diseases might be exploited to advantage in these pressure-related disorders.

2 Platelets and blood clotting

As a result of work done in this and other laboratories from the '60s on, and due to improved knowledge of the process of haemostasis, there has been a renewed interest in blood cellular and chemical changes associated with decompression.

During World War II, both Allied and German workers suggested that blood platelets might adhere to bubbles in the blood but our laboratory appears to have been the first to report a fall in circulating platelet count in both experimental animals (Philp *et al.*, 1967) and volunteers decompressed from elevated pressures (Philp *et al.*, 1972) whereas Martin and Nichols (1972) in a study of Royal Navy divers, first found that this loss progressed for up to 27 h post decompression. Others have observed these changes in association with DCS in experimental divers (Valeri *et al.*, 1974). Our electron microscopical examination of tissues of rats succumbing to massive intravascular bubble formation revealed that a layer of denatured protein, probably fibrinogen, formed at the bubble surface and probably constituted the thrombogenic attractant to which platelets adhered, forming a layer several cells thick around the bubbles (Philp, 1974). Others subsequently confirmed this observation (Letho and Laitinen, 1979; Letho *et al.*, 1981). There is no doubt that bubbles are capable of activating platelets and initiating aggregation and release of granular constituents, including procoagulant substances. More recent work has demonstrated that this occurs when platelets are exposed to nitrogen bubbles *in vitro* (Thorsen *et al.*, 1987).

Subsequent investigations in several laboratories, including ours, have detected changes in blood chemistry suggestive of increased activity in the clotting system, such as elevated levels of fibrin/fibrinogen split products and other evidence of increased plasminogen activity in divers with DCS (Philp, 1974). These changes are moderate, however, and only detectable when compared to pre compression control values. Their usefulness as a diagnostic aid is thus limited. There have, however, been suggestions that severe DCS might precipitate a mild form of disseminated intravascular coagulation (DIC) (Holland, 1969).

The possible consequences of intravascular platelet aggregation include the formation of microthrombi with consequent occlusion of fine blood vessels, delayed elimination of nitrogen (with increased bubble formation as a result), tissue hypoxia, the release of promoters of coagulation from the platelets themselves and the development of delayed or persistent sequelae such as organ damage if the microthrombi do not dissolve quickly. These potential hazards led to investigations of drugs which might prove useful by virtue of their ability to inhibit platelet aggregation or blood clotting. This coincided with the intensive search for such drugs for the treatment or prevention of thrombotic diseases such as coronary thrombosis and stroke. Both antiplatelet and anticoagulant drugs received attention. In human studies the latter, because of their attendant risks, have generally been reserved for the treatment of severe DCS while the former have more often been the subject of experimentation.

We have investigated a number of experimental antiplatelet drugs for their ability to prevent or reduce the post decompression fall in circulating platelet count and found that a number of these did so, both in hyperbaric chamber dives and in open-water saturation dives. In Figure 1, the results of one such study are shown. The drug (VK744) is a pyrimidine chemically and pharmacologically related to dipyridimole. It prevented the loss of platelets which was observed in the placebo-treated divers (Philp *et al.*, 1975). In other studies of divers, we have not been able to show consistently a similar effect with aspirin. In one study, five divers had ingested aspirin (325-975 mg) incidental to the study 24-96 h before a saturation dive to 50 feet of seawater (2.5 ATA, 22.4 psig) (Philp *et al.*, 1975). Their initial fall in platelet count was no different from that of 11 divers who did not take aspirin, but the counts rebounded to normal more quickly (Figure 2). In a hyperbaric chamber dive, divers who received 325 mg three times daily in a double-blind study, did not show a loss of platelets. The group receiving the same dose of aspirin in combination with dipyridamole lost platelets over 24-48 h post dive as did the control group (Philp *et al.*, 1979). Dipyridamole alone did not prevent the platelet loss.

Controlled clinical trials of drugs potentially useful for the treatment of DCS in divers are impractical because of the scattered nature of such occurrences and the fact that they are often remote from medical centres. The use of 'substituted endpoints' (parameters taken to reflect the phenomenon of interest) such as post decompression platelet loss are useful, but their predictive value remains unconfirmed, and it is difficult to acquire enough subjects to conduct meaningful statistical comparisons. In one double-blind study which we conducted on 24 diving students, half receiving aspirin, 325 mg daily, the findings were rendered inconclusive because the dive profile (1 h at 4 ATA or 45 psig with stage decompression) did not induce a significant loss of platelets in the control group.

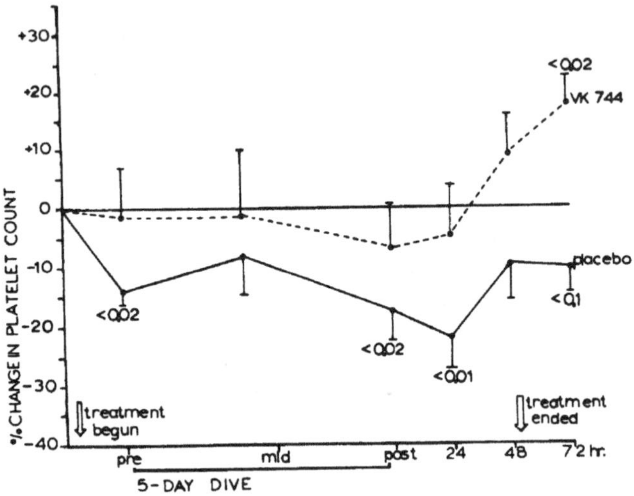

Fig.1. Percent fall in platelet count (mean value ± standard error) of placebo-treated vs drug (VK-744)-treated divers (from Philp *et al.*, 1975). Statistically-significant *p* values are shown adjacent to their data points

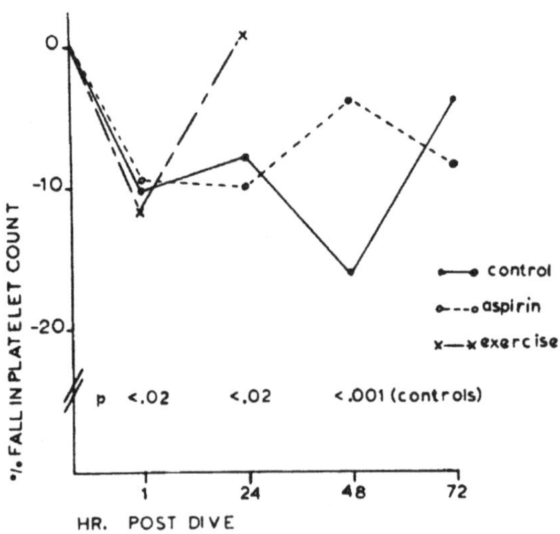

Fig.2. Comparison of post dive platelet counts (mean value ± standard error) of divers who had taken aspirin with those of divers who had not and an additional group had undertaken strenuous exercise (from Philp *et al.*, 1975)

3 Drugs in an animal model of DCS

Using a rat model of DCS developed in this laboratory we have screened a number of agents with actions involving the haemostatic system for effects on morbidity and morality (Philp, 1981). The anticoagulants heparin and bishydroxycoumarin had no detectable effect. Arvin, which accelerates the removal of fibrinogen from plasma, significantly reduced both when given 90 min prior to compression. Epsilon-amino-caprioc acid, which prevents the breakdown of clot (fibrinolysis), caused significant *increases* in morbidity and mortality as did trasylol, which has a similar action. This provided additional evidence that the acceleration of blood clotting could contribute to the pathology of DCS. The nonsteroidal anti-inflammatory drugs aspirin and indomethacin had no effect, whereas the antiplatelet agent VK774 (related to VK744 noted above) significantly reduced both morbidity and mortality.

More recently we completed a similar study in which a number of closely related antiplatelet agents (all inhibitors of phosphodiesterase) were tested in our rat model of DCS. It differed in that the dose was 10 mg/kg orally instead of 20 mg/kg used for VK774 in the above study, and drugs were given immediately, instead of 90 min, before compression. The results are shown in Table 1. No agent significantly reduced the incidence of DCS (bends) although one (RA433) did approach the $p=0.05$ level of significance. This drug was selected for further study in combination with aspirin. Aspirin was given at 50 mg/kg intraperitoneally (i.p.) which is considerably lower than the dose (200 mg/kg) used in the previous study. Neither aspirin nor aspirin in combination with VK433 significantly altered the incidence of bends (Table 2). Two experimental antiplatelet agents, a prostaglandin analogue and a thromboxane synthetase inhibitor, were investigated with the rat model. Neither significantly altered the bends incidence (Table 3).

Table 1. Effects of a group of phosphodiesterase inhibitors on the incidence and morality of DCS in rats

Drug	No DCS	DCS	Died	Chi^2	p
Control	8	17	5		
RA 233	15	12	3	2.54	NS
RA 433	16	14	0	3.40	$<<0.1$
RA 255	12	16	1	1.17	NS
VK 744	13	12	5	1.17	NS
VK 774	12	15	3	0.68	NS

All drugs were given at 10 mg/kg orally, just prior to compression. Each group had 30 animals weighing 385-415 g. NS = not significant.

Table 2. Effect of aspirin, alone and in combination with RA 433, on the incidence and mortality of DCS in rats

Drug	No DCS	DCS	Died	Chi²	*p*	Total
Control	21	9	14			44
Aspirin	12	3	8	0.1	NS	23
Aspirin + RA433	10	5	6	0.1	NS	21

Aspirin dose 50 mg/kg intraperitoneally, RA 233 10 mg/kg orally. Rats weighed 420-445 g. NS = not significant

Table 3. Effects of a prostaglandin analogue (BW 245C) and a thromboxane synthetase inhibitor (BW 149H) on the incidence and mortality of DCS in rats

Drug	No DCS	DCS	Died	Chi²	*p*	Total
Control	22	9	2			33
BW 245C	9	8	1	0.76	NS	18
BW 149H	10	2	3	0.11	NS	15

Drugs were given intraperitoneally. BS 245C at 1 mg/kg, BW 149H at 50 mg/kg according to the supplier's recommendation. Rats weighed 465-425 g. (See Tai and Yuan, 1978 and Town *et al.*, 1983 for drug references). NS = not significant.

In summary, there is considerable evidence from animal experiments that decompression activates the coagulation system and platelet adhesion/aggregation, probably as a result of intravascular bubbles acting as a foreign surface. Species differences in response to drugs may complicate the extrapolation to the human condition and such differences have been shown to exist for antiplatelet drugs (Philp and Bishop, 1970). There is, however, evidence from human studies that decompression dives are frequently associated with a post dive loss of circulating platelets and an increase in clotting activity (both remaining within normal limits) even in the absence of overt signs and symptoms. While the clinical significance of these changes remains unclear, some success in the ancillary treatment of severe decompression sickness has been claimed for anticoagulants such as heparin (Kindwall and Margolis, 1975; Elliott, 1974), and at least one study (Hart, 1976) has established a connection between elevated levels of fibrinogen-fibrin degradation products and severe DCS. The question of dosage may be critical to the use of aspirin in DCS. It is now well-established on both pharmacological and pharmacokinetic grounds that low doses (325 mg every second day) are more effective in preventing coronary and cerebral thrombosis than higher ones (Physician's Health Study, 1988). There has been no study of such a low dose in the context of DCS.

In order to understand the rationale for low-dose aspirin therapy it is necessary to understand something of its action. Aspirin inhibits an enzyme, cycloozygenase, which is responsible for the synthesis of a group of fatty substances known as prostaglandins and thromboxanes. In platelets, the predominant end-product is a thromboxane (thromboxane A_2) which promotes platelet aggregation and hence microthrombosis. In the cells lining the blood vessels, the end-product is a prostaglandin (prostacyclin) which protects the vessel against thrombotic deposits. At high doses of aspirin, the synthesis of both substances will be blocked and the effects will cancel out. At low doses, only the platelets are affected and an antithrombotic protection will occur.

4 Changes in blood chemistry

We and others have reported changes in serum enzyme levels associated with diving activity and especially with the occurrence of DCS. In 1976 we had the opportunity to measure levels of several serum enzymes, indicators of occult tissue and organ damage, in two divers who had severely violated decompression tables (Freeman and Philp, 1976). One of these divers (RW) suffered an attack of Type II DCS (neurological) immediately upon surfacing from his last dive. His total lactate dehydrogenase (LDH) was significantly elevated 6 h later, returning to normal 20 h post dive. His companion (MS), who was asymptomatic, showed the reverse, with elevated levels at 20 h. These increases were accompanied by increases in glutamic-pyruvic transaminase (GPT) and sorbitol dehydrogenase (SDH), the latter being specific for liver damage. These results suggested that significant tissue damage might occur in the absence of overt signs of DCS. Subsequently, we monitored serum enzyme levels in a group of 24 experimental divers who did a simulated 1 h saturation dive to the equivalent of 60 feet of seawater (2.8 ATA) (Philp *et al.*, 1979). Five of the divers incurred Type I DCS on surfacing. Many of the divers had modest elevations of glutamic-oxaloacetic transaminase (GOT) and GPT with the affected ones generally showing the highest levels. These peaked 72 h post dive. Creatine phosphokinase (CPK) levels also have been shown to be elevated in association with diving activity (Martin and Nichols, 1974). The delayed nature of the peak increase in serum enzyme levels, however, limits their usefulness as diagnostic aids but they could possibly be of use in estimating the occurrence of 'silent' bends.

More recently there has been considerable interest in the role of complement in the pathogenesis of DCS. Complement, a family of serum proteins, mediates the inflammatory response and interacts with blood clotting. Ward *et al.* (1990) have found that there is a high degree of correlation between the sensitivity of plasma to complement activation by exposure to bubbles *in vitro* and the susceptibility of the donor to DCS. This was true both for rabbits and for human volunteers. In the rabbit experiments, decompression of the sensitive animals resulted in loss of complement fraction C_3 and the authors theorise that this could account for the acclimatisation (reduced susceptibility to DCS) which has been observed in divers and pressure workers and they provided experimental evidence to support this. Many years ago we reported that decompression of a group of divers from a simulated dive resulted in a significant loss of circulating complement, but there was no difference in the levels of divers who developed Type I bends versus those who did not (Philp *et al.*, 1974).

5 Dysbaric osteonecrosis

A 1974 editorial in *The Lancet* states that 'Despite the production of dysbaric osteonecrosis in laboratory animals, there is little to support the various hypotheses and phenomena that may be cited to explain these lesions – osmotic effects of dissolved gas, alterations in the blood after normal dives, lipid emboli, and platelet thrombi' (Editorial, 1974). A survey of the literature since that time gives little reason to change this assessment. One fairly consistent observation seems to be, however, that the incidence of the disease is higher when decompression procedures are likely to be lax. Two studies have indicated very high incidence of dysbaric osteonecrosis in fishing divers in both Hawaii (Wade *et al.*, 1978) and Japan (Kawashima, 1976). Efforts at prevention have largely focused on improving decompression tables. A more recent review (Gregg and Walder, 1986) re-emphasised the fact that no direct evidence exists to support various etiological theories, including vessel occlusion by bubbles, but that existing evidence does support the hypothesis that prolonged ischemia resulting from impaired local vascular supply remains the most plausible explanation. Using radiolabeled microspheres injected into the external iliac artery of rabbit hind limbs, they found that the proximal distribution of the spheres matched the more frequent proximal occurrence of aseptic necrosis. They also found that exposure to a hyperoxic environment caused swelling of fat cells which could constrict blood vessels through the expansion of extravascular fat cells. They concluded that the etiology of dysbaric necrosis was probably multifactorial.

An American group recently reported on 15 caisson workers who were diagnosed radiographically in 1972 as having dysbaric osteonecrosis (Van Blarcom *et al.*, 1990). None had been exposed to hyperbaric conditions for at least 10 years (mean 14 years). Twenty-three heads (8 humeral, 15 femoral) remained unchanged (normal). One normal humeral head showed new lesions and 5 with pre-existing lesions showed progression to more advanced stages. The progression of lesions in the absence of further opportunities for bubble occlusion suggests that some permanent impairment of vascular supply has occurred. Histopathological studies of femoral heads excised from divers with dysbaric osteonecrosis by Kawashima *et al.* (1978) showed a dense fibrous wall with poor vascularity surrounding the osteonecrotic lesion. Post mortem examination of femoral heads of four divers who died of acute DCS revealed bubbles in the marrow of the femoral heads with associated blood sludging, platelet thrombi and haemorrhage in arterioles and venous sinusoids around the bubbles. They concluded that these pathological findings were important for the progression of osteonecrosis.

6 Concluding remarks

In summing up the current status of dysbaric osteonecrosis, Elliott (1989) recently reiterated the lack of answers regarding etiological factors but pointed out that procedures developed by the MRC panel regarding early diagnosis are largely responsible for the fact that there have been only 12 cases of sub-chondral collapse of bone in nearly 7,000 divers. He also pointed out that this success should not detract from the pursuit of research to identify the causative factors. If indeed the formation of platelet microthrombi contribute to the pathogenesis of the disease by prolonging the period of ischemia, could the application of new information regarding the antithrombotic drugs be exploited to minimise this effect? In addressing the question of which antiplatelet drug might be the subject of an investigation into the prophylaxis of DCS or dysbaric osteonecrosis, practical considerations must prevail. Unfortunately most of the agents which we have investigated either have been abandoned because of untoward effects (the phosphodiesterase inhibitors) or they are too hazardous for routine use (anticoagulants, fibrinolytic agents). Although aspirin has provided equivocal results in animal tests of DCS, the question of low dosage has never been adequately addressed and it is difficult to do so in small rodents because of the rapidity with drugs may be absorbed, metabolised and excreted. The success of a single aspirin tablet every two days in reducing the incidence of myocardial infarction in the Physician's Health Study in the USA at least confirms the efficacy of this prophylaxis in a condition in which thrombosis plays a critical role. Safety, low cost and acceptability to experimental subjects are also points in its favour.

7 Acknowledgements

The author is grateful to the Burroughs-Welcome and Karl Thomae Companies for providing experimental drugs and to the Department of National Defence (Canada) for some funding.

8 References

Downs, G.J. and Kindwall, E.P. (1986) Aseptic necrosis in caisson workers: a new set of decompression tables. *Aviat. Space Environ. Med.*, **57**, pp 569-74

Editorial (1974) Bone necrosis in divers. *The Lancet*, ii, 263-4

Elliott, D.H. (1974) Acute decompression sickness. *Lancet*, ii, 1193-200

Elliott, D.H. (1989) Long-term sequelae of diving. *J. Royal Soc. Med.*, **82**, 79-80

Freeman, D.J. and Philp, R.B. (1976) Changes in blood enzyme activity and haematology of rats with decompression sickness. *Aviat. Space Environ. Med.*, **47**, 945-9

Gregg, P.J. and Walder, D.N. (1986) Caisson disease of bone. *Clin. Orthop. Rel. Res.*, **210**, 43-54

Hart, G.B. (1976) Screening for decompression sickness. *Aviat. Space Environ. Med.*, **47**, 993-4

Holland, J.A. (1969) *Discussion of disseminated intravascular coagulation in decompression sickness*. Sub. Med. Res. Lab., Nav. Sub. Med. Centre. Report 585

Kawashima, M. (1976) Aseptic bone necrosis in Japanese divers. *Bull. Tokyo Med. Dent. Univ.*, **23**, 71

Kawashima, M., Torisu, T., Hayashi, K. and Kitano, M. (1978) Pathological review of osteonecrosis in divers. *Clin. Orthop. Rel. Res.*, **130**, 107-17

Kindwall, E.P. and Margolis, I. (1975) Management of severe decompression sickness with treatment ancillary to recompression: case report. *Aviat. Space Environ. Med.*, **46**, 1065-8

Letho, V.-P., Kantola, I., Tervo, T. and Laitinen, L.A. (1981) Ruthenium red staining of blood-bubble interface in acute decompression sickness in rat. *Undersea Biomed. Res.*, **8**, 101-11

Letho, V.-P. and Laitinen, L.A. (1979) Scanning and transmission electron microscopy of the blood-bubble interface in decompressed rats. *Aviat. Space Environ. Med.*, **50**, 803-7

Martin, K.J. and Nichols, G. (1972) Observations on platelet changes in man after simulated diving. *Aerospace Med.*, **43**, 827-30

Martin, K.J. and Nichols, G. (1974) Serum creatine phosphokinase in man during diving training. *Aerospace Med.*, **45**, 67-71

Philp, R.B. (1974) A review of blood changes associated with compression-decompression: relationship to decompression sickness. *Undersea Biomed. Res.*, **1**, 117-50

Philp, R.B. (1981) *Methods of Testing Proposed Antithrombotic Drugs*. CRC Press, Boca Raton, FL pp 183-7

Philp, R.B., Ackles, K.N., Inwood, M.J., Livingstone, S.D., Achimastos, A., Binns-Smith, M. and Radomski, M.W. (1972) Changes in the haemostatic system and in blood and urine chemistry of human subjects following decompression from a hyperbaric chamber. *Aerospace Med.*, **43**, 498-505

Philp, R.B., Bennett, P.B., Andersen, J.C., Fields, G.N., McIntyre, B.A., Francey, I. and Briner, W. (1979) Effects of aspirin and dipyridamole, alone and in combination, on platelet function, haematology, and blood chemistry of saturation divers. *Undersea Biomed. Res.*, **6**, 127-45

Philp, R.B. and Bishop, B. (1970) Species-dependent inhibition of platelet aggregation by adenosine and dipyridamole: paradoxical potentiation in the rat. *Thromb. Diath. Haemorrh.*, **23**, 129-39

Philp, R.B., Freeman, D., Francey, I. and Bishop, B. (1975) Hematology and blood chemistry in saturation diving: I. Antiplatelet drugs, aspirin and VK744. *Undersea Biomed. Res.*, **2**, 233-249

Philp, R.B., Gowdey, C.W. and Prasad, M. (1967) Changes in blood lipid and cell counts following decompression sickness in rats and the influence of dietary lipids. *Can. J. Physiol. Pharmacol.*, **45**, 1047-59

Physicians' Health Study (1988) The steering committee of the physicians' health study research group. Preliminary Report: Findings of the aspirin component of the ongoing physicians' health study. *New Eng. Med. J.*, **318**, 262-70

Tai, H.-H. and Yuan, B. (1978) On the inhibitory potency of imidazole and its derivatives on thromboxane synthetase. *Biochem. Biophys. Res. Comm.*, **80**, 236-42

Thorsen, T., Dalen, H., Berkvig, R. and Holmsen, H. (1987) Transmission and scanning electron microscopy of N_2 microbubble activated human platelets *in vitro*. *Undersea Biomed. Res.*, **14**, 45-59

Town, M.-H., Casals-Stenzel, J. and Schillinger, E. (1983) Pharmacological and cardiovascular properties of a hydantoin derivative, BW 245C, with a high affinity and selectivity for PGD_2 receptors. *Prostaglandins*, **25**, 13-28

Valeri, C.R., Feingold, H., Zaroulis, C.G. and Adams, G.M. (1974) Effects of hyperbaric exposure on human platelets. *Aerospace Med.*, **45**, 610-6

Van Blarcom, S.T., Czarnecki, D.J., Fueredi, G.A. and Wenzel, M.S. (1990) Does dysbaric osteonecrosis progress in the absence of further hyperbaric exposure? A 10-year radiologic follow-up of 15 patients. *Amer. J. Roentgen.*, **155**, 95-7

Wade, C.E., Hayashi, E.M., Cashman, T.M. Jr. and Beckman, E.L. (1978) Incidence of dysbaric osteonecrosis in Hawaii's diving fishermen. *Undersea Biomed. Res.*, **5** 137-47

Ward, C.A., McCullough, D., Yee, D., Stanga, D. and Fraser, W.D. (1990) Complement activation involvement in decompression sickness of rabbits. *Undersea Biomed. Res.*, **17**, 51-66

Note

In his presentation in Session 7 at the conference, Professor Philp showed a number of slides of bubbles in small blood vessels. He has kindly provided four of these micrographs and electronmicrographs for reproduction here.

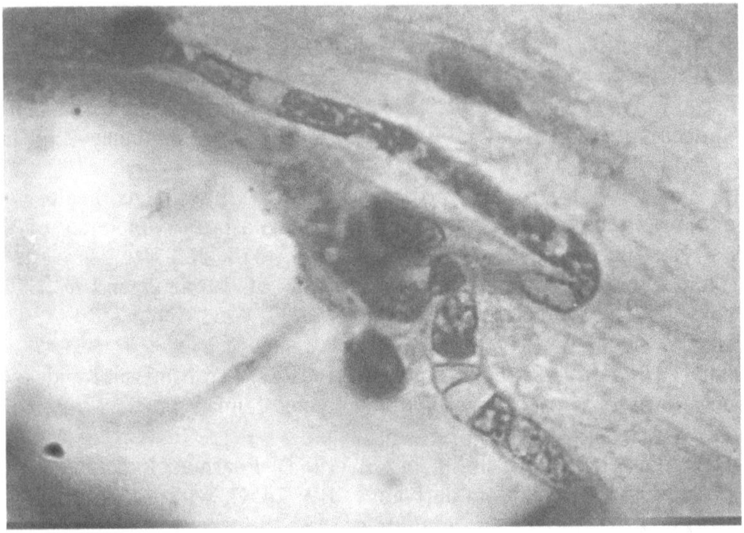

Fig.3. 'Trains' of bubbles in a small blood vessel of the mesentery of a decompressed, anaesthetised rat. The dense, oval mass in the upper left corner is the nucleus of an endothelial cell lining the vessel. Interruption of the oxygen supply to this nucleus for several minutes will result in cell death and sloughing (X250 microscope magnification).

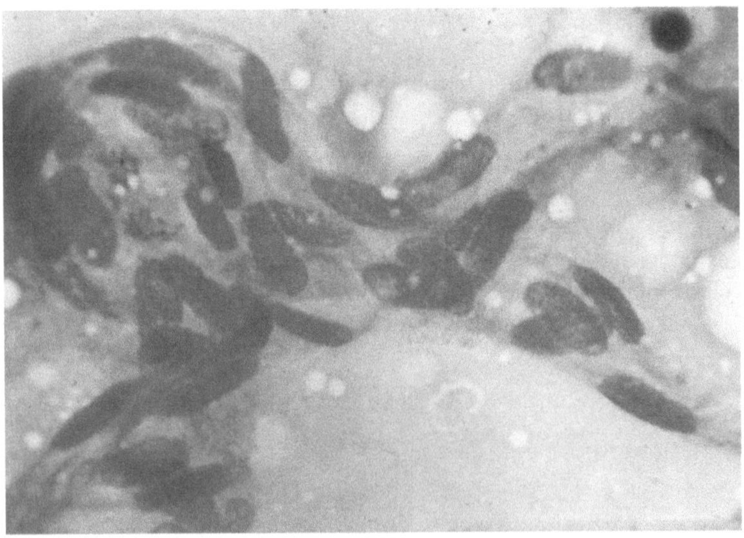

Fig.4. A cluster of endothelial nuclei sloughed from blood vessels of a rapidly-decompressed, anaesthetised rat (H-E stain, X900 microscope magnification)

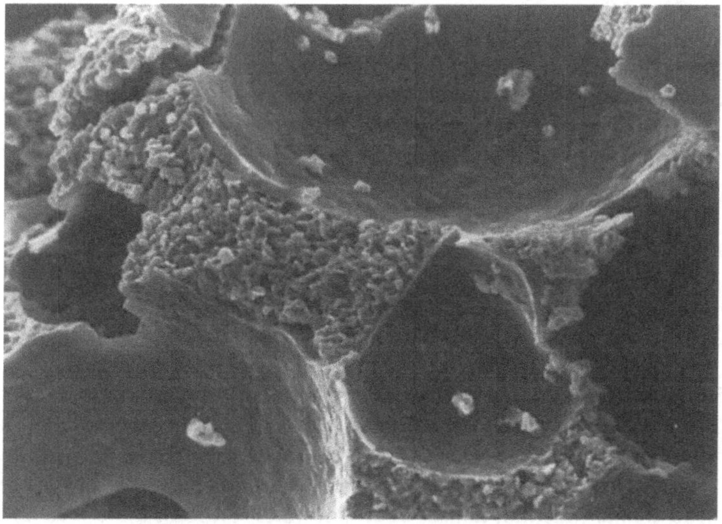

Fig.5. A scanning electron photomicrograph of intravascular bubbles in a vessel of an anaesthetised rat (fixation *in situ*). The view is into the interior of the bubbles, which appear to be lined or coated with an opaque substance. The granular material to the left of centre and at the bottom is circulating red blood cells. Each red cell is about 7μ in diameter (EM magnification X2,500)

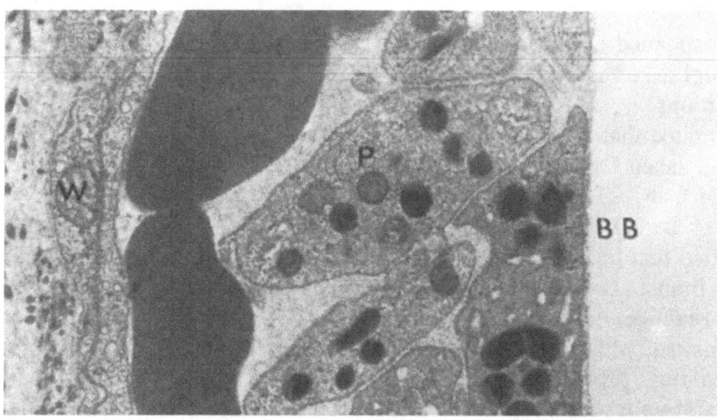

Fig.6. A transmission electron photomicrograph of an intravascular bubble. The bubble is to the right, separated from the blood by the vertical blood-bubble interface (BB). Platelets (P) coat the surface of the bubble. The vessel wall (W) is to the left. Dark masses are red blood cells (EM magnification X15,000)

Discussion on Session 7: Decompression illness and clinical practice

The chairman of the session, **Dr R H McCaig**, introduced the speakers presenting their papers. **Dr M J Catton** described the management and performance of decanting on compressed air projects. **Dr K H Männche** reported observations made during the medical supervision of two caisson projects in Bremen. **Dr J Leach** presented the results of a study of memory and clinical personality measurements on caisson workers with a history of decompression sickness. **Professor R B Philp** addressed the question of the possible clinical significance of blood changes associated with decompression.

The discussion that then followed covered both the above presentations and those in the previous session (decompression illness).

Dr E P Kindwall. I would like to add a comment to Dr Leach's excellent paper. In our MRI studies in Milwaukee we also carried out psychological tests on our sample of 19 compressed air workers and 11 controls, tunnellers who had never worked in compressed air. Their average age was almost identical, as was their history of head trauma, alcohol abuse, education, etc. At least there was no striking difference in those parameters that we recorded.

The compressed air group scored lower on every one of 10 sub tests which we administered, albeit, the differences were small, and in themselves not significant. The tests included trail making, the digit symbol test, crossing out repetitive letters, the peg board (each hand), etc.

Although the lower scores themselves were not significant, the chances of randomly scoring lower on *all* the tests appeared to be less than random, with a p value of 0.05.

Dr Leach responded by explaining the limitations to the sensitivity of the testing.

Dr G C Voigt asked if there were differences in, for example, alcohol use between the tested groups.

Dr Leach replied that no differences had been noted.

Dr Kindwall asked Dr Männche if the caisson work in Bremen had been carried out under the German 1972 decompression tables.

In replying to **Dr J D King's** question about his attitude to decanting, **Dr M J Catton** replied that he is in favour of decanting carried out properly – that is within the planned limits. **Dr King** then commented that based on his experience he was against this practice.

Mr L Lofthouse, who had been the medical lock keeper there, recorded his support for the 'decanting' procedure on the Blackpool outfall, voiced by Dr Catton. The miners, maximum 6 persons, were brought up the shaft in a bucket (man rider) to the surface less than 6 feet from the 'decant lock', and recompressed within 5 minutes. The Factory Inspector visiting the site often timed the procedure without the site personnel knowing that he was doing so.

Dr McCaig pointed out that part of the analysis of the compressed air records of the DCS Central Registry related to the relative incidence of DCS with decanted decompressions being 54% higher than with normal decompressions, although some of the lowest DCS incidence rates were on decant projects. The results of the analysis are given in CIRIA Project Report 6, *Results of the analysis of compressed air records in the Decompression Sickness Registry* (F.M. Jardine, ed.).

Mr J Buchanan, first admitting his involvement in one of the largest projects which used decanting – and that in terms of decompression sickness it was one with the lowest overall rate – re-emphasised the value of the information gained by the analysis of the Newcastle data and given in that CIRIA Project Report.

Mr D R Lamont posed the following question to Dr Catton. In the Compressed Air Working Group consideration was given to whether there was a limiting maximum pressure for decanting set by the rate at which men could be decompressed and recompressed. At what rate does Dr Catton consider such decompression and recompression can be carried out bearing in mind the guidance in CIRIA Report 44 on speed of decompression and compression?

Dr Catton replied that they always accepted that in the compression stages the normal rate for shift men was around 1 bar/min and expected them to accept 1.5 bar/min easily. Decompression is not a problem – rates of 1.5 to 2 bar/min were used – but there is no upper limit.

Mr Lofthouse appealed to manufacturers/suppliers of man or medical locks to continue using the circular barograph charts, as the strip charts do not give an overall picture. On the circular type it can be clearly seen at a glance if there is any change of pressure inside the lock. These records are also more convenient for calculating the stage times. The 'outside-zero' type is advised for medical decompression with a 12-hour revolution.

Dr A Evans put the following question to Professor Philp. How thick were the protein or lipid layers which he had described on the surface of air bubbles and did he consider that the coating would significantly reduce gas transport and lead to prolonged gas nuclei *in vivo*? If so how long might they last?

Professor Philp's reply was that the protein coating seen was about 500 Å thick and we have always suspected that it would stabilise bubbles, but we have no direct evidence. The fat coating seen by Dr Kawashima was much thicker (2-4 microns) and presumably would have a stabilising effect. We did not stain for fat so it may have been present on our bubbles as well. We have seen individual fat globules on bubbles.

Dr McCaig asked **Dr Männche** about the measurements of skin temperature, and received the reply that the thermocouple electrodes were simply taped to the skin.

Mr M H Olsen commented that there are no time delays in Boyle's Law. The total volume of the nitrogen bubbles depends only on the pressure. But the symptoms of DCS occur with a time delay, although the expansion of the bubbles takes place instantaneously. Based on the present knowledge about the harmful effects of the bubbles on the human body the question is: can we take the responsibility of allowing decanting of compressed workers to continue?

Dr Catton commented that there was no evidence of effects from decanting.

Dr Evans reminded the audience of the risk factor analysis he had carried out using the Newcastle data and, in particular that the 'decanting' risk factor for bone

necrosis was the last of 15 and that at $p = 0.044$, it was barely significant.
Professor D N Walder, likening the question of decanting to 'a bit of a closed book', said it is viewed as a suspect procedure. A bad decant followed by a bad decompression could cause trouble in comparison with a 'good' decompression which may avoid it.
No one answered **Mr R P Lewis's** question of what is a 'bad' decant.
Dr King commented that there had to be less confidence in a system with the additional complications of rapid decompression, transfer and recompression time. With the 3-chamber systems now available, he saw no need for it. Asked about the French experience of decanting, **Mr Le Péchon** replied that in France, there has been no experience of decanting at least since 1959.

Surface decompression was forbidden by the 1974 *Décret* on divers' safety. In 1978, he had carried out a careful statistical analysis of bubble grades after matched dry-chamber exposures followed with oxygen decompression (French Ministry of Labour 1974 Tables) and US Navy oxygen surface decompression. (Ref. *Médecine Aéronautique et Spatiale, Médecine Subaquatique et Hyperbare*, Tome XVII, n° 68 - 1978 pp 353-357). No bubbles were observed on any dive during surface interval. No statistically significant difference in bubble grade was observed. No symptoms of DCS were reported. Many bubbles (grade 3) were observed in some divers in both situations after completion of decompression.

A dispensation was obtained to use a modified US Navy table when in-water decompression was impracticable.

For compressed air, no decanting tables or procedures are proposed for the 1992 regulations, the professionals never having asked that they should be.

Mr Le Péchon does not like decanting, because of the probable high incidence of ear barotrauma, in addition to the unknown effects of surface interval, but what he certainly would never recommend is to decant on air breathing. If no other ways exist, decanting with oxygen recompression is probably acceptable within exposure time/pressure limits still to be determined.
Dr R W Hamilton. Surface decompression, the diving analogy to decanting, has a bad reputation, but it works well if done properly. It got a bad name in the North Sea because most of the long, deep, high-stress dives were done with surface decompression, and in-water decompression was used only for the less stressful dives.
Dr H V Hempleman mentioned the history of the adoption of surface decompression and choice of surface interval by the US Navy. In the UK, the same decompression table is used with decanting as for normal decompressions.
Dr Kindwall expressed his agreement with Mr Le Péchon's views against decanting on air breathing. He added that oxygen treatment provides scope for overcoming the problems of air tables, all the existing ones, in his view, being inadequate.
Dr Hempleman repeated the observation that there is no hard, statistical evidence that decanting is harmful.
Mr F M Jardine pointed out that the analysis by Dr Evans of the incidence of decompression sickness relates to data from engineering works that were many years ago. In that comparisons were made on the basis of decanted and normal

decompressions, there are many unknowns and many differences between the projects. One aspect, now difficult to isolate, but which could well be the most critical is the varying quality of the management of the compressed air procedures. Thus, for example, when compressed air is brought in at a late stage, perhaps not having been originally planned, or when the contract is a short one, there is greater risk of adverse effects on health.

Dr Hempleman asked if **Dr R D Vann** had statistical data on DCI risk and decanting.

Dr Vann replied that he had no data on decanting, nor knew if suitable data exist. For data to be suitable, accurate pressure-time profiles must be available with information on individual exposures and recompressions or symptom occurrence. Without such data, differences in exposure or decant interval are potentially confounding, and decanting might be blamed for DCI incidents which were actually a consequence of inadequate decompression. In the terms which Mr Anderson introduced in his keynote address, decanting is a hazard, but a risk assessment is required to determine if it causes unacceptable morbidity.

There followed a short discussion on the questions to be resolved about bottom-time and surface interval and the subsequent decompression procedure. The view was repeated that we still cannot objectively prove that surface decompression or decanting is inevitably more dangerous. The attitude of divers from their North Sea experience was that the combination of depth and duration was what creates the hazard rather than decompression.

Dr R L Marcus asked if there was evidence of spinal cord changes in compressed air workers, **Dr Palmer** replying that he knew of no spinal cord specimens being available for examination.

Mr Buchanan noted that we were still uncertain as to what is the cause of bone necrosis, of what is going wrong.

In response to an earlier question from **Dr W Sterk** about what in UK would be considered an acceptable incidence of decompression sickness, **Dr Hempleman** answered that there is no 'acceptable' incidence rate; the approach is to examine data for compressed air work, discover the decompression illness incidence, and then try to assess how it can be reduced.

(Reporters: G B Shutter, F M Jardine)

The following written contributions were received after the conference

From Dr C E Lehner

The importance of recompression therapy in cases of so-called 'pain-only' limb bends needs to be emphasised. Our initial studies with sheep indicate that hyperbaric osteonecrosis can be prevented with prompt recompression treatment. We found that recompression treatment of decompressed sheep affected by limb bends prevented the development of bone necrosis in the affected sheep. An identical hyperbaric exposure without recompression treatment of affected sheep produced bone necrosis in sheep long bones.

Also, the low prevalence of hyperbaric osteonecrosis among commercial divers in the Gulf of Mexico and among US Navy divers may partly reflect the beneficial effects of routine recompression treatment of those divers with even minor cases of limb bends. Prompt recompression treatment of limb bends cases would appear to be a cost-effective means for reducing the prevalence of hyperbaric osteonecrosis in compressed air workers.

From Dr M J Catton

Delay in presentation of decompression sickness Type I following an error in setting of the stage decompression pressure level repeated over 3-5 days

Episodes of decompression sickness in compressed air work are usually related to the decompression procedure with errors that occur sporadically and are non repetitive. At a small-bore tunnelling site with a working pressure of 18-20 psig, 5 of 10 shift men developed decompression sickness Type I on the same day after a consistent repetitive error in all the planned decompressions (to CIRIA Report 44) after exposures for 3-5 days.

Due to a faulty new manufactured gauge, all stage decompressions were 1.5 to 2 psig above the correct pressure setting but were correct in time length.

The error arose during the initial 2-day training period of the lockkeepers by a diving-trained medical lock keeper, who used the dial gauge mounted above the outlet controls of the lock. A recording pressure gauge at the other side of the lock was disregarded because it has minor errors of 0.5 psig in zero setting.

During the five days of operation, the pressures of the decompression stages varied slightly by up to 0.5 psig from a constant value, but the settings were 9.5-10 psig (rather than 8 psig) and 6.5 or 6 psig (instead of 4 psig), irrespective of the zero-setting variation.

The four shift-men affected had all completed full shifts of 3.5 to 8.25 h on either three or five of the days, and a foreman had been daily in the tunnel for two shifts and three short exposures (less than 2h) over the five days (see Table 1). No further cases occurred after correction of the error, three of the men returning to work after treatment.

Table 1 Exposures of the five men treated for Type I DCS on the fifth day

	Exposure time (h:min) on days:				
	1	2	3	4	5
Foreman	4:02	1:00	1:00	0:28	7:11
Man A	4:02	5:00	7:45	8:03	8:19
Man B	-	-	3:55	6:20	8:19
Man C	-	-	6:04	7:30	7:10
Man D	-	-	3:34	7:30	7:10
Barograph pressure (psig)	19?	17.5-19	18-19	17-18.5	18

The error was not recognised because of the inexperience of the lock keepers, and was compounded by the position of the gauges in relation to the operator's sitting position which, being low, induced a parallax error in the mid range of the gauge needle traverse. The supervising engineer who was monitoring the barograph record remained standing by it across the doorway from the lock gauge.

A correct setting-up-and-test procedure should have revealed the errors.

Part Eight
Codes and Regulations

32 CEN standardisation – Air locks for tunnelling work safety requirements

M. Bandmann
Tiefbau-Berufsgenossenschaft (TBG), Munich, Germany

Abstract
The current work on European standards includes the preparation of a draft standard on air locks in tunnelling. The paper examines the provisions of the draft in terms of general safety requirements, personnel locks, pressure bulkheads.
Keywords: Air locks, Compressed air, Equipment, Standards.

1 Introduction

On December 12, 1992 the trade barriers of the market within the European Community will fall. This also means that the legal provisions regulating safety in the individual countries will have to be converted into EEC law. Through EEC directives the basic requirements for the protection of health and safety need to be stipulated. These directives will be backed up by technical standards, which will be worked out by the independent European standardisation committees CEN and CENELEC, and according to which the European manufacturers can make their products. Three different types of harmonised European safety standards are to be provided, i.e.

A-type, i.e. basic standards
B-type, i.e. safety group standards
C-type, i.e. safety product standards.

The last type contain the specified safety requirements for particular machines or machine assemblies.

2 Draft CEN standard for air locks

The safety requirements governing air locks for tunnelling work will be developed to a C-type standard within the framework of Technical Committee TC151,

Engineering and Health in Compressed Air Work Edited by F.M. Jardine and R.I. McCallum. Published in 1994 by E & FN Spon. ISBN 0 419 18460 0

'Construction Equipment and Building Material Machines' by the working group on 'Tunnelling Machines'.

Delegations of the following countries have been involved in this drafting: Denmark, France, Germany, Great Britain, Sweden, Switzerland.

The scope of this standard covers the dangers to be regulated. Additionally, air locks must be in accordance with the rules of EN292 *Safety of Machinery - Basic Concepts, General Principles for Design*. Dangers existing in conjunction with all mechanical, electrical, hydraulic and pneumatic parts of machinery will be covered by higher level standards.

The present draft for air locks applies to the design, construction, equipment, marking and testing of air locks for personnel and material used in tunnelling work. This draft covers all specific risks relevant to air locks when used as directed and in accordance with the manufacturer's instructions, as well as measures for reducing such risks.

The standard will apply to tunnelling air locks built following its publication, i.e. from 31.12.1992 at the earliest.

This paper presents the factual contents of the safety requirements and measures of the draft, which contains the following sections:

General requirements
- Calculation, leak-tightness and performance testing
- Electrical equipment
- Control panel
- Communication
- Doors, closing and sealing
- Fire protection.

Personnel locks
- Number of chambers
- Dimensions
- Seats
- Doors, dimensions, locking
- Minimum equipment requirements
- Ventilation (air scavenging)
- Protection against line breaks.

Material locks

Combined locks

Pressure bulkheads
- Sealing
- Service ports
- Testing of bulkheads.

Servicing and maintenance
Operating, servicing and repair instructions
Marking.

3 General safety requirements

Some of these safety requirements are explained in more detail:

Air locks shall comply with the CEN standard for pressure vessels, EN286 Parts 1 and 2, and shall be calculated for 1.5 times the maximum operating pressure.

Electrical equipment in air locks shall comply with EN60204-1 and IEC64 as well as enclosure code IP55 to be suitable for humid and wet spaces and protected from dust deposits and spraying water. The electrical equipment shall be designed to minimise fire risk and shall withstand pressure changes to the maximum pressure of the air lock.

Control equipment and indicating instruments shall be assembled in a control panel. They shall be easily readable and so arranged and designed that function and switching direction can be clearly recognised.

Between the control panel and the various chambers of the air lock and the work chamber technical means for two-way voice communication shall be installed.

Doors of personnel and material locks shall be self-sealing and kept shut by differential pressure or be fitted with a mechanical interlocking system preventing opening of the door unless the pressures are equalised. On personnel locks the door interlocking system shall permit opening in both directions following pressure equalisation.

As regards fire protection, the draft standard requires personnel and material locks to consist exclusively of materials and components of low flammability.

4 Personnel locks

Personnel locks shall consist of at least two separate chambers having an overhead clearance of 1.5 m, whereby one chamber shall have a minimum volume of 3 m^3 with enough space for 3 persons and 1.0 m^3 for each additional person. The second chamber shall have a minimum volume of 2 m^3 and enough space for 2 persons.

The future aim in establishing the minimum dimensions of air locks is to achieve a minimum criterion as regards ergonomic and humane working conditions to be the same all over Europe, whilst also taking into account the need for escape and rescue measures in an emergency.

Seating shall also be designed in keeping with minimum ergonomic requirements by equipping seats in personnel locks with back-rests and affording each person a sitting width of a least 0.5 m and a sitting depth of a least 0.4 m. Seat and back-rest surfaces shall be insulated against cold contact.

The clear width of door openings shall not be less than 600 mm, i.e. more where possible. These door dimensions of air locks differ from the minimum dimensions of access openings of other tunnel-machinery (e.g. access openings to cutter heads or bulkheads).

Personnel lock compartments shall feature at least the following equipment:

- equipment for pressure control
- pressure monitoring devices with a scale which can be read to 0.05 bar
- pressure recorder to chart the pressure profile in the personnel lock
- thermometer
- clock, inside and out, self-contained with a display of seconds
- sampling line for gas analysis
- observation windows of safety glass for visual contact between control panel and lock compartment and between compartments
- dummy flanges
- electrical interior lighting with a minimum illumination of 200 lux, measured at head height above the seat
- emergency lighting, switched on automatically and independently of the main lighting system, with a minimum illumination of 15 lux for 1 hour
- heat-insulated seating
- voice communication (intercom) system
- sound suppressors for fresh and waste air; maximum sound level in the chambers shall not exceed 85 dB(A)
- regulatable heating system
- fire-fighting equipment.

Personnel locks shall be equipped with a means for supplying fresh air (scavenging air) at the rate of a least 50 litres per minute per person measured at chamber pressure.

Compressed gas lines shall be equipped with means for shutting off the supply (e.g. non-return valve) to prevent a collapse in pressure as a result of line rupture.

In material locks part of the minimum equipment requirement for personnel locks may be omitted, whilst combined locks shall comply with the same requirements as those of personnel locks.

Where combined locks are not used for personnel decompression some of the safety requirements for personnel locks such as seats and heating may be omitted.

5 Pressure bulkheads

The safety requirements on pressure bulkheads are given below: the first is that they ensure an effective seal between spaces having different working pressures.

Where bulkheads can be proof tested the regulations pertinent to pressure vessel testing shall apply (1.5 times maximum working pressure), otherwise they shall be designed for a pressure 2.0 times the maximum working pressure.

Air locks shall be designed and manufactured so that hazards during servicing and maintenance are avoided or reduced. Hazards which cannot be avoided shall be highlighted in the appropriate instruction manual or indicated by warning signs.

Any manufacturer of air locks and pressure bulkheads shall give all the necessary information for safe transportation, installation, commissioning, testing, operation, servicing and dismantling in the operating and maintenance manual, including the weights (in kg) of transportable parts of the air lock, as well as a complete record of warning signs, pictograms and their location and meaning. Basic operating and safety instructions for the air lock shall be clearly displayed at the lock attendant's control position in a waterproof format.

The repair manual shall also provide all information necessary for safe repair, assembly and dismantling off the air lock. Damaged air locks shall be suitably tested after repair before reuse.

In addition to the usual marking requirements of machines and equipment, the following data shall be durably and clearly indicated on air locks inside and out: maximum safe working pressure and number of persons permitted as a maximum in each compartment.

In future, the approach to be adopted for many of the aspects of machinery will be self-certification by the manufacturer: this would also apply to tunnelling machines. An exception, however, will be air locks for which type testing closely adapted from the testing of pressure vessels will continue to be valid.

6 Concluding remarks

In drafting this CEN standard for air locks, it was established that all the parties involved in Europe place high priority on safe engineering and thus on accident-prevention in tunnelling under air-lock conditions; this being all the more important, since this type of work is gaining increasing significance. The present draft standard for air locks will improve the safety level in this field throughout the whole of Europe. It contributes not only to improving accident-prevention, but it also signifies a step in the direction of achieving the European common market.

33 Are the German regulations for compressed air work safe?

I. Holthaus
GKSS Forschungszentrum Geethacht, Germany

Abstract
In one respect the German regulations for compressed air work are enlightened, in that they provide for the use of oxygen breathing during decompression; this is not yet common practice worldwide, despite being widely recognised as beneficial to the health of the workers. A review of the requirements of the *Druckluftverordnung*, issued 1972, in the light of developments over the last 20 years in hyperbaric physiology and occupational medicine, reveals some weaknesses in the regulations. They do not cover working pressures of more than 4 bar absolute; the need for which is shown by the tunnel recently dug under Kiel Harbour. While 'exceptional permission' can be granted for deeper work on the basis of safety equality with other methods, there are no provisions for dealing with the special aspects of such exposures. The regulations call for a doctor, but do not specify the special training needed to manage such high-pressure exposures; further, there is no requirement for either the workers or the site manager to be trained in the physiology of work at pressure. We propose a seminar for training of compressed air worksite supervisors. The German philosophy is for work regulations to cover all hazards; and this one fails to deal with noise protection, hygiene aspects, gas analysis, and protection against breathing concrete spray. Facility requirements do not include recording of all pressure exposures, redundancy in communications systems, or back-up supplies of gas. Perhaps the greatest need is in the field of decompression procedures.
Keywords: Compressed air work, Medical fitness, Oxygen decompression, Air saturation

1 Introduction

The German compressed air regulations in the currently valid edition were published in 1972. The comparable regulations for commercial air diving have been updated several times since 1972, the last edition being issued in 1985.

Why is this so? The reasons are: the higher risk for divers, their better training

Engineering and Health in Compressed Air Work Edited by F.M. Jardine and R.I. McCallum. Published in 1994 by E & FN Spon. ISBN 0 419 18460 0

and their keener lobby.

In order to become a commercial diver in Germany, the training period is two years, ending with an exam; in contrast, compressed air workers are only given a talk on the subject at 6-month intervals.

Since 1972 there has been considerable progress in baromedicine, especially to meet the needs for deep saturation diving in the exploration of oil and gas offshore. In commercial diving, depths of 200 to 300 m are standard, the world record being 686 m (Atlantis II, Duke University). The equivalent maximum depth for compressed air work is still 30 m. However new techniques for deep tunnelling in materials like glacial sands may require greater compressed air working pressures.

2 Status of safety technology

The standards for pressure testing of pipes, chambers and other hyperbaric equipment are much lower than in diving.

There is no stipulation for compressed air work, that there should be redundancy of equipment and safety systems. As the most important parameters of a hyperbaric exposure are time and pressure, dissimilar redundancy is required, especially to avoid same-mode failures. This means that, at least, there should be two separate sets of pressure gauges and time recorders. Pressure recording is not required below 1.3 bar, even though it is known that decompression sickness may occur at pressures as low as 0.6 bar. Unhappily it is no joke that the regulations specify the requirement for a mercury thermometer. The fresh air supply requirement is for 0.5 m^3 per worker per minute, regardless of pressure and work rate.

There are no regulations concerning noise, despite techniques being available to provide nearly noiseless conditioning of the atmosphere. Also lacking is any requirement for back-up communication systems. The regulations require the working pressure to be maintained, but no information is given about providing spare capacity of air volumes. In diving, enough breathing gas must be readily available to recompress the plant, should there be a rapid loss of pressure. One cannot find in the compressed air regulations any recommendations concerning fire safety precautions, even though fire hazards are most critical in hyperbaric conditions.

No gas analysis of the working atmosphere is even requested, although in compressed air work toxic gases like methane, carbon monoxide and dioxide, welding fumes and others may be present. Moreover there is no requirement for alternative or back-up oxygen analysis in the decompression chamber. This, like the above mistakes, does not represent the status of current technology.

The regulations yield no information regarding breathing in circumstances of sprayed concrete work (which now is standard in subway tunnelling) or when toxic gases are in the atmosphere, nor about built-in-breathing systems (BIBS).

There is no hint that equipment which might present hazards on changes of pressure must not be used.

A chemical toilet must be provided, while a sanitary installation in the chamber is much the better solution.

3 Baromedical status

There is no requirement to provide a chamber for therapeutic recompression when the pressure is less than 1 bar, although DCS begins at 0.6 bar.

Decompression time after 8 hours of work at 1 bar is only four minutes, which is much too short.

In addition to oxygen decompression, there are also tables for air decompression which, from 3 bar, require a decompression time of 250 min after a work time of 230 min. This is not only dangerous, but also totally ineffective. In the section of the regulations about air decompression, it ought to mention that air decompression is an emergency method should the oxygen supply fail, a situation which can be excluded by providing redundancy measures.

After the end of decompression, an observation period for bends of only 0.5 h is asked for. This certainly is much too short, especially at weekends, when workers might travel, perhaps by plane, far away from a clinical centre able to provide therapeutic compression.

Oxygen decompression at set pressures of 1.2 or 0.6 bar would save the industry much money, without additional installations, by allowing safe longer work-shifts with quicker decompressions. Up to now oxygen decompression is used at 0.9, 0.6 and 0.3 bar.

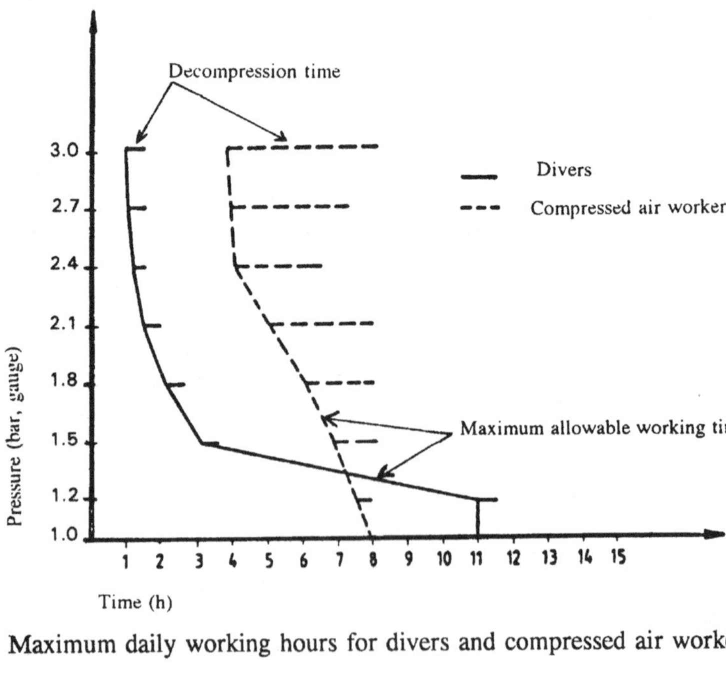

Fig.1. Maximum daily working hours for divers and compressed air workers

Table 1. Air saturation with repetitive air excursions

Depth (bar)	Stop time (min)	Clocktime (h:min)	Mixture	pO_2 (bar)	Comments	OTU
00	00	00:00	Air	0.21	All ascents and descents at 1 bar/min. Habitat atmosphere air. Breathe air on all excursions	
					Day 1	00
1.4	00	00:00	Air	0.21	Compress to 1.4 bar in habitat	00
4.5	480	08:00	Air	0.50	Begin excursion on air	00
2.3	240	12:03	Air	1.16		298
2.0	25	12:30	Air	0.69		311
1.7	25	12:55	Air	0.63		319
1.4	75	14:10	Air	0.57		331
	00	14:11	Air	0.50	Remain in habitat overnight	331
					Day 2	
4.5	1069	08:00	Air	0.50	Begin excursion on air	00
2.3	240	12:03	Air	1.16		298
2.0	25	12:30	Air	0.69		311
1.7	25	12:55	Air	0.63		319
1.4	75	14:11	Air	0.57		331
	00	14:11	Air	0.50	Remain in habitat overnight	331
					Day 3	
4.5	1069	08:00	Air	0.50	Stop at 1.4 bar for *** min	00
2.3	240	12:03	Air	1.16		298
2.0	25	12:30	Air	0.69		311
1.7	25	12:55	Air	0.63		319
1.4	75	14:11	Air	0.57		331
	00	14:11	Air	0.50	Remain in habitat overnight	331

If one compares oxygen decompression times from the literature, the oxygen load, measured in Oxygen Tolerance Units (OTU) is only 80 after 2 h work in a pressure of 3 bar, whereas the U.S. Navy Diving Manual which does not contain very conservative tables, sets a limit of 150 OTU. A safe limit for daily use is 300 OTU.

When industrial needs demand higher working pressures and useful work times, decompression times cannot be short. However there is a possibility, staying in air saturation at 1.4 bar, of performing air excursions to 5 bar without long decompression times. Figure 1 shows the maximum allowable work and decompression times for divers and compressed air workers. The divers' curve is approximately exponential; that of the compressed air workers shows a nonlinearity at 2.4 bar which indicates a discrepancy from physiological values. Exposure times are very different. Both time curves cannot be correct.

Table 1 shows a part of a saturation exposure at 1.4 bar in air with repetitive excursions. The schedule was produced by the decompression analysis and computation programme from R.W. Hamilton (Tarrytown Laboratory), in 1986. It demonstrates that work at, for example 4.5 bar, is feasible, repeatedly, for 4 h daily. Decompression time is acceptable at approximately 2 h. Oxygen decompression is not needed but, if used, it would further facilitate the procedure.

The regulations permit decompression to be performed according to the pressure at the end of the work-shift, but it ought to be mandatory to base it on the highest pressure during the shift.

When there is a case of decompression sickness, the rule in the regulations is not to decompress further, but to wait until pain subsides. Should this not happen, after a 'certain' time, therapy is to be given according to the recommendations of a medical doctor. In contrast, the state-of-the-art procedure is immediate recompression to 1.8 bar at which oxygen breathing is used. Further therapy would then be as advised by a medical doctor, specialised in compressed air work.

4 Conclusions

1. The regulations are out of date and take no account of the current state-of-the-art in technology or in baromedicine. Moreover they are incomplete, as they do not tackle subjects like toxicological aspects, breathing protection, or noise. They need updating urgently.

Fortunately this process has started. In October 1991 DLR (German Aerospace Research Association) presented medical recommendations, and the federal authorities (minister of labour) have now started to prepare new regulations.

2. There is no adequate text for either engineers or doctors which sets out the relevant aspects of safety technology, decompression schedules and therapeutic compression. There is also a need for structured training schemes for engineers managing sites which involve compressed air work.

Such a course should address:

- Hyperbaric work, technical and physical background, and its hazards
- Practical application of gas laws
- Relevant regulations in handling gases, especially oxygen
- Basic explanations of human anatomy and physiology, sufficient to understand the principles of compressed air medicine
- Medical aspects of compressed air work, hygiene, ear prophylaxis, noise, medical fitness
- Decompression procedures and tables, decompression illnesses, lung rupture, osteonecrosis, therapeutic recompression, oxygen breathing and therapy
- Air saturation with excursions to and from higher working pressures
- First aid, cardio-pulmonary resuscitation, transfer under pressure.

References

Bennett, P.B., and Elliott, D.H. (1982), *The Physiology and Medicine of Diving*. 3rd edn. Baillière Tindall, London

Hamilton, B., Holthaus, I., Vollbrand, J. and Heineke, M. (1986), GUSIS capability of developing new decompression procedures. In: *Proceedings of the 2nd International GUSI Conference*, Jg. 20/1-13, GKSS-Forschungszentrum, Geesthacht

Hauptverband der gewerblichen Berufsgenossenschaften (1977), *Berufsgenossenschaftliche Grundsätze für arbeitsmedizinische Vorsorgeuntersuchungen*. Gentner-Verlag, Stuttgart

Hauptverband der gewerblichen Berufsgenossenschaften, Zentralstelle für Unfallverhütung und Arbeitsmedizin (1979) *Spezifische Einwirkungsdefinitionen*, Entwurf Oktober 1979, Bonn

Holthaus, I. (1987), Längere Tauchzeiten in Tiefen bis 50 m. *Tiefbau Berufsgenossenschaft*, **99**: 30-32

Holthaus, I. (1988), Menschliches Arbeiten jenseits normaler Druckverhältnisse. *Arbeitsmed Sozialmed Präventivmed* **23**: 78-80

Holthaus, I. (1989), Stand der therapeutischen Kompression, 20/80-Heloxsaturation Mittel der Wahl bei Spätfallen. *Zbl Arbeitsmed* **39**: 30-36

Kessel, R. (1986) Gesundheitsschäden durch Arbeiten unter Druckluft im modernen Tunnelbau. Habilitationsschrift. *Arbeitsmed Sozialmed Präventivmed* **21**: 93-94

National Oceanic and Atmospheric Administration (1975), *NOAA Diving Manual*. Superintendent of Documents, United States Government Printing Offices, Washington

Tiefbau-Berufsgenossenschaft (1972), *Verordnung über Arbeiten in Druckluft vom. 4.10.1972*, München

Tiefbau-Berufsgenossenschaft (1985), *Unfallvorschriften Taucherarbeiten VGB 39*, April 1985

United States Navy Diving Manual 1986, Best Publishing Corp., San Pedro, California

34 Some medico–legal aspects of modern work in compressed air

F.B. Wright and L. Marks*
European Occupational Health and Safety Law Unit, University of Salford, UK
**Finers, Solicitors, London, UK*

Abstract
Some leading cases concerning breaches of the statutory requirements and of the common law are analysed with a discussion of the applicable regulatory legislation, and proposed new developments, where the activity has involved or will involve tunnel mining and the use of compressed air. Compressed air is widely used in tunnel mining because of its effectiveness and versatility as a tunnelling method notwithstanding its possible danger to health to tunnel miners and others who work alongside them such as engineers, fitters and electricians. Special hazards of compressed air work include the development of decompression sickness and dysbaric ostenecrosis. The evidence suggests firstly that a system of strict medical examination and surveillance and careful compliance with the relevant statutory regulations will normally ensure a relatively low incidence of morbidity and, secondly, that the statutory rules might usefully be strengthened or improved. The authors of this paper examine the legal consequences of this activity and make proposals for reform.
Keywords: Legal liabilities, Statutory duty, Proof, Limitation, Health and Safety legislation

1 Compressed air illnesses

Men who work in compressed air are liable to suffer from decompression sickness. There is more than one kind of this sickness. Type I usually consists of a pain in the joints varying in intensity from a mild form know as the niggles to an acute form known as the bends which can be agonising. Type II may affect the brain, heart, central nervous system or lungs. It is capable of causing paralysis and death. If either of these two conditions occurs it usually happens within hours of decompression. It is rare for it to occur after 12 hours have elapsed.

Workers in compressed air may also suffer from aseptic necrosis of the bones. The fact that this has been set in train is not always immediately apparent. It will not be visible by X-ray examination for at least three months and it may be a year or

Engineering and Health in Compressed Air Work Edited by F.M. Jardine and R.I. McCallum. Published in 1994 by E & FN Spon. ISBN 0 419 18460 0

even much longer before the symptoms are felt by a patient. Almost exclusively, symptoms are associated affecting the shoulder, knee and the hip joints. Similar bone changes can occur from other causes. Diagnosis is therefore partly circumstantial and partly by exclusion of other causes.

In Britain the 1958 statutory regulations provide a code to be observed by both employers and workmen. In particular they specify the procedure to be followed when a man is entering or leaving an area where work is being carried out in compressed air. Use is made of a man-lock which consists of a chamber in which the pressure can be increased or reduced as required. No significant long-term problem usually arises with regard to compression. Decompression is the important aspect. The rules specify that decompression shall be carried out in accordance with the decompression table, Table 1, annexed to the *Work in Compressed Air Special Regulations 1958*. The responsibility for operating a man-lock during a decompression rests on the man-lock attendant. He must personally perform all the decompressions. The purpose of the decompression is to prevent the formation of the nitrogen bubbles or at least to reduce the risk of them forming or becoming dangerously large.

If in fact bubbles form so as to give symptoms of Type I or Type II decompression sickness the procedure of recompression followed by slow decompression is carried out not in a man-lock to which we have referred and which is usually to be found at the pit bottom, but in a special medical lock under the supervision of trained persons skilled in medical decompressions. It has been asserted that the statutory decompression table is inadequate. It has been said by Mr Justice Cusack in *Ransom v Sir Robert McAlpine & Sons Ltd* [1971] 11 KIR 141 that the table came into existence as a compromise between medical opinion and the desire of contractors, which is readily understood, to economise on time because decompression is a process which can occupy a good deal of time. It wrongly assumes that there is no risk arising if the working pressure is below 18 lbs. per square inch. It wrongly assumes that full saturation occurs after four hours and the times allowed for decompression are now regarded as being too short. Investigation carried out over a number of years by members of the Medical Research Council Decompression Sickness Panel showed that the statutory decompression procedures laid down in the *Work in Compressed Air Special Regulations 1958* were unsatisfactory in that they did not reduce to an acceptable level either form of decompression sickness. Under the regulations there is power to approve of alternative rules as to decompression and it is for this reason that the Blackpool Decompression Tables were introduced in 1966 (See *Medical Code of Practice for Work in Compressed Air*, Report 44, third Edition, CIRIA 1982). From the point of view of the courts such a substitution is acceptable but it is, however, important that any substitute rules should be at least as effective for securing adequate decompression as the statutory rules. Use of the Blackpool Decompression Tables is permitted only when a Certificate of approval has been granted to a contractor under Regulation 10(3) of the *Work in Compressed Air Special Regulations 1958*. The Blackpool Decompression Tables are now in wide use.

2 Legal liability

Legal action may be founded on either the civil or the criminal law. Criminal liability is considered in section 3 below. Civil liability concerns the rights, duties and obligations of individual members of the community to each other and actions may be founded, in this connection, on the law of tort and more specifically on the law of negligence or on breach of statutory duty. Torts are civil wrongs. Generally speaking, torts are committed by wrongful behaviour, although there are exceptions to that broad rule, and that behaviour may infringe another human being's interest, which the law regards as being worthy of protection. Additionally, the plaintiff is entitled to obtain compensation from the defendants in the form of damages by taking action in a civil court. Negligence is a universal concept of legal systems but as a ground of liability for causing damage it is not so common. In 1932, the House of Lords in *Donoghue v. Stevenson*, had to decide whether a case of action in negligence existed under English and Scots law. Lord Atkin's judgement is of most importance in this respect. He said that whether there is a cause of action in negligence depends on whether it can be proved that one person has committed a breach of the duty of care binding upon himself and owed to another and has therefore caused injury to that other. Whether a duty of care is owed to another depends on whether that person is in law one's neighbour. Lord Atkin described a neighbour as a person 'so closely and directly affected by [my] act that I ought reasonably to have them in contemplation as being so affected when I am directing my mind to the acts or omissions which are called in question'. Negligence, is quantatively at least, the most important of torts in English law.

Torts may also be created by statute. It is particularly the case where industrial health and safety legislation has been passed for the protection of workers. This legislation is generally construed as conferring a right on, for example, an injured workman, to pursue a civil action for breach. In some cases, employers will be liable if they are found to be negligent whilst in other cases the statutory duty is absolute and even a non-negligent failure to comply will render an employer liable. Perhaps it would be appropriate at this point to consider the tort of breach of statutory duty in the contents of the *Work in Compressed Air Special Regulations 1958* (as amended).

2.1 Breach of statutory duty

The *Work in Compressed Air Special Regulations 1958* (as amended) provide:

> By regulation 4 that: '(1) It shall be the duty of every contractor and employer of workmen who is undertaking any work to which these Regulations apply to comply with such of the requirements of Regulations 5 to 21 as affect any workman employed by him...'
>
> By regulation 6: 'Every bulkhead, air lock or other structure used in connection with work in compressed air shall be of good construction, sound material and adequate strength and shall be properly maintained.'

By regulation 8: '(1) Every man-lock shall be of adequate internal dimensions and capacity for the purposes for which it is used and shall be suitably equipped. (2) .. there shall be provided in connection with every man-lock (a) pressure gauges which will readily indicate (i) to the man-lock attendant the pressure in the man-lock and the pressure in each working chamber to which the man-lock affords direct or indirect access, and (ii) to persons in the man-lock the pressure in the man-lock; ...'

By regulation 10: '(1) Every man-lock shall, whilst any person is in that man-lock, ... be in charge of a competent lock attendant who shall control the maximum rate of compressions and shall perform all decompressions in the man-lock .. (3) Compression of a person in a man-lock shall not be carried out otherwise than in accordance with the Rules set forth in Part I of the Schedule to these Regulations; and decompression of a person after being in a working chamber shall not be carried out otherwise than in accordance with the Rules set forth in Part II of the said Schedule: ...'

N.B. When considering the regulations it should be noted that although provision is made for the man-lock to be in the charge of a competent attendant the regulations do not specify how that attendant is to be trained or how his competence is to be tested. The regulations do, however, provide for medical supervision and certification of the workmen and for every workman to have an individual Compressed Air Health Register which should be handed to him on the termination of his employment.

2.2 Duty

So far as common law liability is concerned it is not sufficient to rest simply on the minimum requirements of the statutory regulations. In this connection one might refer to the opinion of Lord Hunter in *Paterson v. Charles Brand & Sons Ltd.* (unreported, January 14, 1965) which was heard in the Court of Session. That case also was concerned with decompression sickness and bone necrosis. Lord Hunter, in observations which must admittedly be regarded as *obiter* and which in any event are not binding on an English court, but may be persuasive, held that the defenders in this action could not escape liability at common law merely by showing that they had complied with the duties imposed upon them by the statutory regulations, but where (as in the case he was adjudicating upon) the regulations, that is to say the *Work in Compressed Air Special Regulations 1958*, had been very recently passed, they might have evidential value as to what safety procedures a reasonably careful employer should follow. (It is to be understood that the opinion of Lord Hunter has been paraphrased rather than quoted.) In the case with which he was dealing the regulations had become law less than 18 months before the pursuer entered the defenders' employment.

Moreover it would be quite wrong to seek to exclude the operation of the Regulations simply because the actual kind of illness suffered by the plaintiffs may not have been in the contemplation of the makers of the Regulations. One cannot in any event be sure what information they had about the varieties of decompression sickness or bone necrosis. The mischief against which the Regulations, and particular Regulation 10, is directed is the risk of physical injury caused by too rapid a transfer from compressed air to ordinary atmospheric pressure. In fact that risk is mainly the risk arising from bubbles. The Regulations are not confined in their application to any particular kind of injury. They are just as apt to cover bone necrosis as they are to cover decompression sickness Type I and Type II and indeed there may be still further kinds of physical injury to which they could extend.

2.3 Causation

The plaintiff must next show that the breach was the cause of the illness from which he has suffered. The medical evidence might be that even if the statutory decompression table was followed exactly illness could still occur. Moreover illness could still occur even when substitute and improved decompression tables such as the Blackpool Decompression Tables were used. No method of decompression has yet been devised which totally eliminates bends. It follows that no method has yet been devised which totally eliminates bubbles. It is not enough for the plaintiffs to prove that there might be a connection between the defendant's breaches of statutory duty and negligence and the illness from which they suffer. They must prove that those breaches of statutory duty and the negligence have caused or materially contributed to the illness, not merely that it was possible that the illness might have been so caused. See *Bonnington Castings Ltd. v. Wardlaw* [1956] A.C. 613, H.L. The effect of the medical evidence can probably be summarised in this way. Faulty decompression increases the risk of bubbles forming. The greater the number of faulty decompressions the greater the number of occasions upon which that risk arises.

The court must then ask if the defendant's breaches and did their negligence make a material contribution to the incidence of the plaintiff's necrosis? If the bubble theory is correct then any specific area of necrosis was set in train at some particular moment of time when a bubble cut off the blood supply. If there is more than one such area such as in bilateral necrosis, each would have been set in train at different moments of time. Nevertheless no such area arises from the cumulative effect of bubbles; each is due to a single particular incident. In *Nicholson v. Atlas Steel Foundry & Engineering Co. Ltd.* [1957] 1 W.L.R. 613 H.L., the House of Lords was dealing with a case of pneumoconiosis which arose from the cumulative effect of inhaling deleterious particles over a period of years. The workman in that case would have inhaled some particles in any event, but it was held that his employer's breach of statutory duty had exposed him to greater risk than he would ordinarily have incurred and so his widow succeeded in her claim for damages. In *Gardinery v. Motherwell Machinery & Scrap Co. Ltd.* [1961] 1 W.L.R. 1424 H.L. (Sc.), the plaintiff was in the defendant's employment only for about three months. He developed industrial dermatitis from working amongst dust, dirt, oil and grease.

This was a risk to which the nature of his work exposed him. It was suggested that he could have contracted the disease otherwise than from his work. It was held that where a workman, who had not previously suffered from the disease, contracted it after being subjected to conditions likely to cause it and also showed that it started in a way typical of such conditions, he had established a *prima facie* presumption that his disease was caused by those conditions.

2.4 Remoteness

As far as breach of statutory duty is concerned it is not relevant whether or not the defendant employers ought to have known the danger to which the plaintiffs were exposed. Foreseeability is, however, a relevant consideration where common law liability is concerned. An employer is at common law responsible for the foreseeable consequences of his negligence. It is clear from the case of *Hughes v. Lord Advocate* [1963] A.C. 837, H.L.(Sc.), that if an injury is caused by a known source of danger the defendant cannot escape simply because it was caused in a way which could not have been foreseen. An illustration of this is to be found in the case of *Bradford v. Robinson Rentals Ltd.* [1967] 1 W.L.R. 337; 1 K.I.R. 486, in which the plaintiff was exposed by the defendants to conditions of extreme cold and considerable fatigue. A risk of injury was foreseeable although it may not have been foreseen that the actual injury would be frostbite, which is unusual even in an English winter. Defendant employers should foresee that their negligence both in the operation of the system of decompression and in their failure to utilise a satisfactory decompression table would give rise to a risk of injury to the workforce due to bubbles.

2.5 Limitation

Until 1963 most common law disease claims remained wholly or partially barred by an absolute statutory period of limitation. At that date this stood at three years from the date the cause of action accrued which, in claims of negligence or breach of statutory duty, is when damage occurs. For common law claims there was no relief granted by a 'reasonable cause of delay provision' so that the fact that a claim was not brought within three years of accrual because of misdiagnosis, or because the doctor had given the patient a euphemistic diagnosis to minimise anxiety or because of the latency of symptoms, would not save it. Because of the relative rarity of bone necrosis it has been known for a diagnosis of arthritis or rheumatism to be confirmed.

In *Ellis v. Charles Brand* [1969] SLT 132 a workman who had been employed from May 1959 to December 1962 in driving a tunnel below the bed of the River Clyde claimed damages from his employers on the ground that he had contracted bone necrosis in that occupation. For most of his employment with the defenders the pursuer worked in the construction of that tunnel and he alleged in his pleadings that from May 1959 until about December 1962 that employment involved his working in compressed air at various pressures. The pursuer claimed that he contracted bone necrosis affecting his right hip joint by reason of certain failures in the duties owed by the defenders to him as an employee working under such conditions, and accordingly he sought reparation from the defenders. The summons was signeted on 3rd October 1966. The failure in duty which the pursuer alleged took place during a period which ended in December 1962. The defenders pleaded that the action was time-barred. Proof before answer directed to that plea-in-law was allowed. The proof established that the pursuer had no symptoms of the disease before December 1963. In November 1964 the pursuer complained of pain in the region of the right hip. Radiographs at the Southern General Hospital showed changes consistent with bone necrosis of the right hip. That was the first radiological indication of any change and was in any event minimal and difficult to see. Until that time he had sustained no real damage. The pursuer was then informed for the first time that he had sustained injury and that it would be reasonable for him to claim Industrial Injury Benefit.

The reform process was not precipitated in Britain until 1961 when substantial criticisms were made of the decision in *Cartledge v. Jopling & Sons Ltd.* [1961] 2 Lloyds Reps. 61;[1963] AC758. This decision held that the plaintiffs who were suffering from pneumoconiosis from the previous working conditions were statute barred from suing the defendants because their cause of action had accrued more than the relevant limitation period before the date of claim. The fact that the plaintiffs' symptoms had not become reasonably discoverable until the limitation period had expired was considered irrelevant. It was held that one could suffer damage and therefore have a cause of action in tort and yet be unaware of it and the limitation period could therefore elapse before one is aware of any damage and in a practical position to sue. The reasoning was highly unsatisfactory and led to a fundamental change in the law as it applies to claims for personal injuries.

The *Limitation Act 1963* began a series of statutory reforms. First, time would not begin to run against the victim until he was, or should reasonably have been, aware of certain material facts, for example that the victim had knowledge that he was suffering from the disease or that the victim knew that the disease arose because of a relevant man-made cause. Now time does not begin to run until the worker was, or should reasonably have been, aware of the material facts. Thus time would not run whilst the disease was either latent or misdiagnosed. Today, following *Firman v. Ellis* [1978] 2 All E.R. 851 the courts have a discretion – if it is equitable to do so – to permit claims brought out of time even if the delay was not due to the ignorance of a material fact.

As we have said with a disease such as aseptic necrosis of the bones the fact that this has been set in train is not immediately apparent. It will not be visible by X-ray examination for at least three months and it may be a year or much more before the symptoms are felt by a patient. In one recent extreme English case fourteen years elapsed before the symptoms were felt by the victim. It will only be when the articular surface of the joint is damaged and the joint begins to 'break down' that the tunnel miner becomes aware of pain. It may only be at that point that an X-ray examination will be sought. Alternatively the worker may be advised that he has contracted aseptic necrosis following a compulsory medical examination for a new assignment.

Similar difficulties have been experienced in the United States of America. In *McCann, Walsh Construction Company et al., Appellants, and Special Funds Conservation Committee, Respondent. Workmen's Compensation Board, Respondent,* [1954] 306 N.Y. 904; 119 N.E. 2d 596, the claimant had been employed by appellant employer as a cableman and labourer in the construction of the Queens Midtown Tunnel in the city of New York from July to September, 1938. His last exposure to compressed air was in 1938. His condition was first diagnosed as compressed air illness or caisson disease in 1950, and he first learned that he had that illness in December, 1950. The Workmen's Compensation Board determined that the date of claimant's disablement was December 11, 1950, and that the appellant employer was liable for the award of compensation. Section 40 of the Workmen's Compensation Law formerly provided that 'Neither the employee nor his dependents shall be entitled to compensation for disability or death resulting from disease unless the disease is due to the nature of his employment and contracted therein, or in a continuous employment similar to the one in which he was engaged at the time of his disablement, within the twelve months previous to the date of disablement, whether under one or more employers.' Section 28 of the Workmen's Compensation Law formerly provided that a claim for compensation due to an accident must be made within one year from the date of the accident, but that a claim could be filed, in the discretion of the Workmen's Compensation Board, within two years. In 1947, section 40 of the Workmen's Compensation Law was amended (L. 1947, ch. 624, eff. July 1, 1947) to provide that the right to compensation in cases of compressed air illness shall not be barred by the failure of the employee or his dependents to file a claim within the two-year period described by section 28 of the Workmen's Compensation Law, provided such claim was filed after such period of two years, within ninety days after disablement and after knowledge that the disease is or was due to the nature of the employment. Section 28 of the Workmen's Compensation Law was similarly amended (L. 1947, ch.624), and the time for filing claims was extended to two years (L. 1947, ch.77). Appellants argued that the award against the employee gave the amended sections of the Workmen's Compensation Law unwarranted retroactive effect and violated the Fourteenth Amendment to the United States Constitution, section 10 of article 1 of the United States Constitution and section 6 of article 1 of the New York Constitution. Respondents argued that the amended statutes had properly been applied since the disablement occurred after the date of such amendments.

2.6 Damages

Damages have to be assessed on the basis of disablement, pain and suffering, past, present and future, on the loss of amenity suffered by a particular man, his capacity to earn and his likely loss of earnings.

2.7 The burden of proof

The law decides a civil action on the balance of probabilities. Thus where there is a conflict of medical evidence the court must decide which of two opposite opinions is more likely to be true. It is therefore important to emphasise that although there are medical issues in such a case any judgement upon them only constitutes a decision made upon such evidence as has been presented in the court, and judgment is made in accordance with the ordinary rules applicable to civil litigation.

Ransom v. Sir Robert McAlpine & Sons Ltd

Probably the leading case in England is *Ransom v. Robert McAlpine & Sons Ltd.*, Q.B.D. [1971] 11 KIR 141. In this case the plaintiffs were employed as labourers by the defendants and from 1962 to 1964 worked for periods on the construction of a tunnel below the sea bed. The work involved the use of compressed air and necessitated the plaintiffs working in conditions to which the *Compressed Air Special Regulations 1958* applied. When entering or leaving work in the tunnel the plaintiffs underwent the procedures of compression or decompression in a man-lock. After the plaintiffs ceased to work in compressed air both contracted necrosis of the bones and as a result both had to have operations on their hips and became permanently partially crippled. The plaintiffs alleged that they had contracted the disease because the procedures and equipment used in the course of decompression were defective. Each plaintiff claimed damages on the grounds that his injury was due to the negligence and breach of statutory duty of the defendants.

On the plaintiffs' claims against their employers for breaches of statutory duty as set out in the *Compressed Air Special Regulations 1958*, and negligence, the court held, giving judgements for the plaintiffs:

(1) That, on the balance of probabilities it had been established that bone necrosis in compressed air workers was caused by bubbles of nitrogen forming in the blood as a result of exposure to excess atmospheric pressure.

(2) That, at the relevant times, the decompression procedures were not carried out in accordance with the statutory tables, that the door of the man-lock leaked and was ill-fitting, that there was overcrowding inside the man-lock, that the man-lock attendants were inadequately trained and accordingly that the employers were in breach of regulations 4(1), 6, 8(1), 10(1) and (3).

(3) That, as allegations of breaches of statutory duty had been established against the employers and as they had failed to ensure the periods and/or rates of decompression were adequate to reduce the risk of bone necrosis to a minimum, they were guilty of negligence and that they should have foreseen the risk of injury to the plaintiffs resulting from that negligence.

(4) That, the probability was that the plaintiffs contracted the disease because of the greater risk to which they were exposed during decompression as a result of their employer's breaches of statutory duty and their negligence and that those matters were the cause of the injuries suffered by the plaintiffs.

Practical problems of proof to substantiate a civil claim.

Tunnel miners wishing to prove a claim have arguably a more difficult task than other personal injury litigants for the following reasons.

1. The victim must be able to prove that he is suffering from aseptic necrosis of bone. In the light of what has been said, this requires specialist knowledge of a high order on the part of general practitioners, orthopaedic surgeons, hyperbaric specialists and radiologists. The disease can easily be wrongly diagnosed. Real specialist knowledge of the dysbaric diseases is possessed by perhaps no more than forty medical minds in the United Kingdom and what is required of general practitioners and orthopaedic consultants is a high level of suspicion and preparedness to seek further specialist advice.
2. The aseptic necrosis must be of a disabling character, be a cause of pain or restrict the person's ability to work as a tunnel miner. It may be that the presence of a lesion and its nature only prevents a man from working in compressed air or in tunnels or in heavy labour; that becomes a disability or a limitation in social if not in physical and medical terms.
3. The tunnel miner, and his legal adviser, has then the task of identifying and proving which employment the disabling feature, arising from aseptic bone necrosis, could be attributed. Sometimes more than one contract is involved.
4. The employer or employers must be proved to have been negligent.
5. The problem of proof is made more and more difficult as time progresses.

2.8 Compensation: social security provisions

By the *Social Security (Industrial Injuries) (Prescribed Diseases) Regulations 1985* (SI 1985 No 967) Sch 1, Part 1 lists both Prescribed Diseases and the occupations for which they are prescribed. Specifically this includes:

	Prescribed disease or injury	**Occupation**
A.	Conditions due to physical agents	Any occupation involving:
A3.	Dysbarism, including decompression sickness, barotrauma and osteonecrosis.	Subjection to compressed or rarefied air or other respirable gases or gaseous mixtures.

Statistics for the year ending March 1991 made available by the Department of Social Security show that the current awards number one hundred under the category of dysbarism. The average disability under this head is thirty per cent. No initial assessments are shown for 1990-1991. Additional problems occur with claims for this benefit.

Workers may be aware that their injury is work related but may not realise that such an injury gives a legal entitlement to compensation. This rule is particularly harsh for disease victims. This is because the industrial compensation system in Britain is focused on accidents and coverage is extended to accidents on a highly selective basis. The result is a much lower consciousness of the opportunity to make a benefit claim for disease as opposed to traumatic injury. As Commissioner Lazarus noted in R(1) 14/74,

> 'It is expecting a lot of people that they should suspect that a scheme mainly designed to deal with accidental injury and described as an injury scheme may also deal with diseases which develop otherwise than by accident and are not injuries in the sense in which that word is commonly used.'

A particularly pertinent feature of such a claim, and which distinguishes it significantly from a claim for civil liability against an employer, is that it does not require the claimant to prove a breach of the statutory regulations or the common law in order to establish his entitlement thereto.

3 Criminal liability

Criminal law deals with wrongful acts against the State. Criminal liability attaches for a deliberate and wilful breach of statutory duty, where recklessness is present, where there has been negligence and also, in some cases, for purely inadvertent conduct. The *Health and Safety at Work etc. Act 1974* and the *Factories Act 1961* are criminal statutes. Section 47(1) *Health and Safety at Work etc. Act 1974* provides that nothing in Part 1 of the Act shall be construed as conferring a right of action in any civil proceedings in respect of any failure to comply with any duty imposed by sections 2 to 7 or any contravention of section 8, but this shall not affect the actionability of any breach of duty imposed by any of the existing statutory provisions. The *Factories Act 1961* and the *Compressed Air Special Regulations 1958* are existing statutory provisions. The burden of proving the commission of an offence rests on the prosecution. The burden of proof is higher than the civil burden. In relation to these offences the prosecution must prove its case to a high degree of probability (see the speech of Denning J. *Miller v. Minister of Pensions* [1947] 2 All E.R. 372). Many of the general duties examined below require the accused to do something so far as is reasonably practicable. Section 40 makes it clear that it is for the accused to prove that the state of affairs required by the statute has been met once the prosecution have established a *prima facie* case. The standard of reasonable practicability in the criminal law broadly approximates to the standard of negligence in the civil law. The phrase is not defined in the Act but its meaning was judicially determined in the leading case of *Edwards v. National Coal Board* [1949] 1 K.B. 704 in the Court of Appeal and subsequently approved in *Marshall v. Gotham* [1954] A.C. 360 in the House of Lords. In *Edwards v. National Coal Board* the Court considered the phrase in relation to its meaning in section 102(8) of the *Coal Mines Act 1911* and Lord Justice Asquith in that case stated:

> '"Reasonably practicable" is a narrower term than "physically possible" and seems to me to imply that a computation must be made by the owner, in which the quantum of risk is placed on one scale, and the sacrifice involved in the measures necessary for averting the risk (whether in money, time or trouble) is placed in the other; and that if it be shown that there is a gross disproportion between them – the risk being insignificant in relation to the sacrifice – the

defendants discharge for the onus on them.. Moreover, this computation falls to be made by the owner at a point of time anterior to the accident.'

3.1 Health and Safety at Work etc., Act 1974 section 2

Section 2(1) provides that it shall be the duty of every employer to ensure so far as is reasonably practicable the health, safety and welfare at work of his employees. This wide duty is owed to all 'employees', namely those who work under contracts of employment or apprenticeship, provided that they are 'at work'. An employee is 'at work' throughout the time that he is in the course of employment but not otherwise.

Without prejudice to the generality of this all-embracing duty, section 2(2) makes specific that duty in five examples which are themselves of fairly general application and all of which use the test of reasonable practicability.

(a) *Plant and systems of work: section 2(2)(a)*

The employer is obliged to provide and maintain so far as is reasonably practicable plant and systems of work which are safe and without risks to health. The term 'plant' is defined in section 53(1) and 'includes any machinery, equipment, or appliance'. The central notion is of something which is used for the purposes of the employer's business or enterprise such as the 'fixtures, implements, machinery and apparatus used in carrying on any industrial process', (according to the Oxford English Dictionary). The term means something used in the industrial process and not the products of that process or the objects on which that process is carried out. The term is redolent of the duty placed upon the employer under his common law duty of care as stated in the leading case of *Wilsons and Clyde Coal Co v. English* [1938] A.C.79, in which the employer's obligation to exercise reasonable care for the safety of his employees was classified into sub-categories, namely: proper personnel and supervision, proper plant and equipment, and a safe system of work – a threefold classification which has been judicially criticised as providing no more than three instances of a single general duty of care.

The general duties in sections 2 to 9 of the Act 'are largely modelled on the common law duties of care' see *R. v. Swan Hunter Shipbuilders Ltd and Telemeter Installations Ltd.* Accordingly, the common law on 'safe systems of work' provides a basis in the present statutory context. Work which varies from one project to another such as construction work is included. The employer must take account of the inexperience, personal infirmities and known propensities of his employees, although all work to a varying extent involves judgement and discretion in the employee the exercise of which would fall outside the concept of a 'system' for which the employer is responsible. The concept of 'a system of work' breaks loose from the central notion in the codes that the statutory duties apply only within defined spatial limits, the factory, the shop, the office and so forth. An employer's obligation under section 2(2) (a) is not geographically limited in this way and comes to resemble the mobility of common law negligence in which employers have been held subject to a duty of care in respect of premises outside their control (*General Cleaning Contractors Ltd. v. Christmas* [1953] A.C.180).

(b) *Articles and substances: section 2(2)(b)*
The second application of the general duty of the employer relates to the arrangements which he must make to ensure, so far as is reasonably practicable, the safety and absence of risks to health in connection with the use, handling, storage and transport of articles and substances.

(c) *Information, instruction, training and supervision: section 2(2)(c)*
Unless employees have proper information, training and supervision the best-laid plans of management are apt to fail, something which section 2(2)(c) recognises when it requires the employer to provide such information, instruction, training and supervision as is necessary to ensure, so far as is reasonably practicable, the health and safety at work of his employees. No longer is it possible to place health and safety behind a shroud of secrecy; the worker has a 'right to know'.

The fourfold duty under section 2(2)(c) with regard to information, instruction, training and supervision is part of a battery of requirements which have as their objective the involvement of well-informed employees and others in activities at work. The information which is provided must be accurate and as comprehensive as possible. *Vacwell Engineering Ltd. v. BDH Chemicals Ltd.* was a case in which suppliers of a dangerous chemical were held liable in contract and for the tort of negligence; but it is submitted that had the defendants been employers they would, as a company dealing in chemical agents, have been obliged to conduct adequate research into the scientific literature on a particular chemical where that research would have shown a risk of explosion if the chemical were to be brought into contact with water. In other words it is not merely the duty of the employer to communicate the knowledge which he has but also the knowledge which as an employer in that position he ought to have possessed.

In the case of *Graham v. Co-operative Wholesale Society Ltd* [1957] 1 W.L.R., 511, it was held that it was the duty of an employer to take reasonable steps to keep up to date his knowledge of new developments and discoveries which might reveal hitherto unsuspected sources of danger to his workmen and that the test to be applied was whether he had failed to acquire information that was commonly known in the trade or had been guilty of neglect. By 1962 it was common knowledge among those concerned with and particularly those in charge of compressed air tunnelling, first, that bone necrosis was a hazard and, secondly, that the statutory decompression table was inadequate. Proper and clear instructions must be given as to what the employee is required to do or not to do bearing in mind that workers performing repetitious routine tasks are not infrequently careless, as is sometimes the case with young and inexperienced workers. The duties are *ad hominem* so that, for example, the information and instruction which must be given to employees whose command of English is weak must take account of their linguistic difficulty.

(d) *Access and egress to a place of work: section 2(2)(d)*
By virtue of section 2(2)(d), the general duty of the employer is specifically extended to include, so far as is reasonably practicable a place of work under his control which is safe and without risks to health and the provision and maintenance of means

of access and egress to and from such a place of work which are safe and without such risks.

(e) *The working environment: section 2(2)(e)*
There was a newly introduced requirement in section 2(2)(e) that the employer must provide and maintain 'a working environment' for his employees that is, so far as is reasonably practicable, safe, without risks to health, and that he furnishes adequate facilities and arrangements for their welfare at work. It recognises the entitlement of all work people to reasonable minimum standards of health and amenity in their working environment.

3.2 Work in Compressed Air Special Regulations 1958
The material requirements of the *Work in Compressed Air Special Regulations 1958* made under the *Factories Act 1937* and the *Factories Act 1948*, are listed above in Section 2.1.

Employing organisations and their staff, including directors, managers and indeed any member of the workforce, may be prosecuted either under the *Health and Safety at Work etc. Act 1974* or the *Work in Compressed Air Special Regulations 1958*. Section 7 provides that it is the duty of every employee while at work:

(a) to take reasonable care for the health and safety of himself and of other persons who may be affected by his acts or omissions at work; and
(b) as regards any duty or requirement imposed on his employer or any other person by or under any of the relevant statutory provisions, to co-operate with him so far as is necessary to enable that duty or requirement to be performed or complied with.

Directors and employees at all levels in companies have been prosecuted under *Health and Safety at Work etc. Act 1974* but the most common defendant is the company. Section 33, which creates the offences under the Act, specifies fifteen categories of offence. The offences under the Act differ according to their gravity.

The most serious offences are indictable, that is triable by judge and jury in the Crown Court which has wide powers of punishment. Fines of up to £250,000 have been imposed for these offences. Some offences are triable either way, that is they may be tried either before magistrates or by way of indictment. A small number of offences are triable only summarily. From October 1992 the maximum penalty for these offences will be £5,000 and further increases in the scale of fines and offences for which custodial sentences have been made available have been promulgated by Parliament in the *Offshore Safety Act 1992*. These provisions enable magistrates to impose fines of up to £20,000 for breaches of sections 2-6 *Health and Safety at Work etc. Act 1974*. Custodial sentences of up to six months imprisonment are available to Magistrates Courts where there has been a failure to comply with an improvement or prohibition notice. It has also been known for the Crown Prosecution Service to become involved in this area. The Crown Prosecution Service has recently initiated a successful prosecution for manslaughter following a fatal accident in the North West of England.

In addition to the power to prosecute, inspectors have the power to issue improvement notices and prohibition notices. If an inspector is of the opinion that a person is contravening one or more of the relevant statutory provisions or has contravened one or more of these provisions in circumstances that make it likely that the contravention will continue or be repeated, he may serve on him an improvement notice stating that he is of that opinion, giving particulars of the reasons why he is of that opinion, requiring that person to remedy the contravention or, as the case may be, the matters occasioning it, within such period (ending not earlier than the period within which an appeal against the notice can be brought) as may be specified in the notice.

Inspectors also have the power to require the discontinuance of an activity carried on or about to be carried on by or under the control of any person, being an activity to or in relation to which any of the relevant statutory provisions apply, or will, if the activity is so carried on, apply if as regarding such activity the inspector is of the opinion that it does or will involve a risk of serious personal injury. The inspector who decides to issue such a prohibition notice must state that he is of that opinion, must specify the matter giving rise to the risk of any contravention of the relevant statutory provisions (where relevant) and must direct the discontinuance of the activity in question until such time as the matter giving rise to the risk or any contravention specified above is remedied.

4 Proposals for reform

4.1 Proposed changes to health and safety law to foster improvements in the working environment: European Directives

The 1989 framework Directive (89/391/EEC), which is addressed to Member States, requires compliance by 31st December 1992, and was the first Directive to be proposed as part of the implementation of Article 118A of the EEC Treaty. The overall object of the framework Directive is stated as being, 'to introduce measures to encourage improvements in the safety and health of workers at work', and in its Preamble declares that:

(i) Member States have a responsibility to encourage improvements in the safety and health of workers on their territory and that
(ii) taking measures to protect the health and safety of workers at work also helps, in certain cases, to preserve the health and possibly the safety of persons residing with them.

It is further noted 'that the incidence of accidents at work and occupational diseases is still too high' and that 'preventive measures must be introduced without delay in order to safeguard the safety and health of workers and ensure a higher degree of protection'.

The Directive contains re-stated general principles concerning, in particular, the prevention of occupational risks, the protection of safety and health and the informing, consultation and training of workers and their representatives, as well as principles concerning the implementation of such measures. This measure constitutes a first attempt to provide an overall complement to the technical harmonisation directives designed to complete the internal market. A recent comparative analysis of existing occupational health law within the Member States indicates that all will need to amend their legislation to some extent in order properly to implement the 1989 Directive. The extent of such amendments at national level varies considerably between Member States, although even the most developed national systems of occupational health and safety regulation will face some need for change, at the very least in order to accommodate certain administrative and organisational aspects.

4.2 Scope of the 1989 Framework Directive

The scope of the 1989 Directive's provisions is widely drawn. Thus, *an employer* is defined as 'any natural or legal person who has an employment relationship with the worker and has responsibility for the undertaking and/or establishment', while the term *worker* means 'any person employed by an employer, including trainees and apprentices but excluding domestic servants'. Some obligations are also owed by employers to 'workers from any outside undertakings and/or establishments engaged in work in his undertaking and/or establishment'. This includes the duty to provide information and appropriate instruction. In situations where several undertakings share a workplace there are specific provisions requiring co-operation between these in relation to measures of protection, prevention and information for their respective workforces.

The overall objectives indicated in Article 1 of the Framework Directive may be summarised as being:

1. Humanisation of the working environment.
2. Accident prevention and health protection at the workplace.
3. To encourage information, dialogue and balanced participation on safety and health by means of procedures and instruments.
4. To promote throughout the Community, the harmonious development of economic activities, a continuous and balanced expansion and an accelerated rise in the standard of living.
5. To encourage the increasing participation of management and labour in decisions and initiatives.
6. To establish the same level of health protection for workers in all undertakings, including small and medium-sized enterprises and to fulfil the single market requirements of the Single European Act 1986.
7. The gradual replacement of national legislation by Community legislation.

As a general approach, the Directive (which applies to all sectors of activity, both in the public and the private sectors – although Member States are permitted to exclude certain public services, such as the Police and the Army – lays down minimum standards and requirements for health and safety matters. Although it is not concerned with the protection of the public from work activities, non-employees may, subject to certain provisions, be protected by virtue of the Directive.

4.3 General duties placed upon the employer

An approach adopted by the 1989 framework Directive which has given rise to much comment has been the establishment across the European Community of a number of 'general duties'. This particular method of introducing occupational health and safety protection draws upon successful experience in some of the more developed national systems for regulation in this field.

Various kinds of duty for employers emerge from the framework Directive, which may be enunciated as follows:

1. *Duties of awareness.* These are threefold, consisting, first, in a broad duty which follows from the Preamble to the Directive – for the employer to keep himself informed of the latest advances in technology and scientific findings concerning workplace design; second, in a duty to identify and to evaluate risks to the safety and health of workers in the undertaking or establishment, and third, a duty – closely linked to duties of instruction below – to be aware of the capabilities of workers at the undertaking or establishments as regards health and safety. Employers must designate competent personnel for health and safety purposes to aid them in their task.
2. *Duties to take direct action to ensure safety and health.* These firm duties constitute the 'heart' of the Directive. They comprise (a) a duty to eliminate avoidable risks; and (b) a duty to reduce the dangers posed by unavoidable risks.
3. *Duties of strategic planning to avoid risks to safety and health.* These two duties reflect the needs for both a general overview of safety and health needs and the implementation of a specific programme at the undertaking or establishment. The former is covered by the duty in Article 6 (2) (g) to develop a 'coherent overall prevention policy'. The latter is detailed in Article 7 and includes the duty to develop and implement a system for the protection and prevention of occupational risks.
4. *Duties to train and direct the workforce.* The general duties of training for the workforce in matters of safety and health are set out in Article 12. In addition, there is a broad requirement in Article 6 (2) (i) that employers, as part of their normal activity in managing and directing the work, should provide appropriate instructions to their workers, while Article 6 (3) (d) makes it clear that employers must take appropriate steps to ensure that 'only workers who have received adequate instructions' are to have access to areas where there is 'serious and specific danger'.
5. *Duties to inform, consult and involve the workforce.* These duties are extremely widely cast by the Directive. The details of the information which is to be given by employers to workers and/or their representatives are set

out in Article 10, while broad duties to provide for consultation and participation of workers are set out in Article 6 (3) (c) and, most importantly, in Article 11.

6. *Recording and notification duties.* A variety of obligations on the part of employers and undertakings are set out in Article 9. These include duties to list certain accidents and occupational illnesses at the undertaking or establishment and the preparation and provision of reports, for the benefit of relevant national authorities, on occupational accidents and occupational illnesses suffered by the workforce.

4.4 Information, consultation and participation

The development of rights to information, consultation and participation for workers and their representatives has been a long-standing ambition of the European Commission. A study conducted by the Health and Safety Commission in 1980 suggested that subcontractors were employed on some fifty six per cent of construction sites. Moreover that same body has calculated that some seventy per cent of fatal accidents could have been prevented by management action. The Health and Safety Executive has proposed that a competent project manager be appointed to take responsibility for health and safety. Shortcomings in this area in the future will be vigorously pursued. Both companies and their senior managers and directors will be prosecuted in appropriate cases.

4.5 Further directives

A further directive, on minimum safety and health requirements for the workplace, which is addressed to Member States, will form the basis of new regulations which will be laid before Parliament before 31st December 1992. This will be followed within the next twelve months by a further directive on health and safety requirements at temporary or mobile work sites.

4.6 Implications

A number of recent cases before the European Court of Justice have emphasised the importance of this new European legislation. Article 189 of the EEC Treaty provides:

> 'A directive shall be binding as to the result to be achieved, upon each Member State to which it is addressed, but shall leave to the national authorities the choice of form and methods'.

The Court of Justice held almost twenty years ago in *Van Duyn v Home Office* 41/74 [1974] E.C.R. 1337 that a directive might contain directly effective provisions. For this to occur two conditions must be fulfilled. Firstly, there must be a clear and precise obligation and, secondly, it must not require the intervention of any act on the part of the institutions of the Community or the Member States.

In two recent cases (joined), *Francovich v. Italian State* and *Bonifaci v. Italian State*, it was held that where a Member State fails to enact the legislation required in order to achieve the objective prescribed by a directive it may be possible for an aggrieved individual to take remedial action directly against his Member State government. However, it would appear that three conditions must be fulfilled before such liability can be created. In the first place, the objective sought by the directive must include the creation of rights for individuals. The second is that the content of those rights must be ascertainable from the provisions of the directive itself. The third condition is the existence of a causal link between violation by the State of its duty to implement the directive and the loss sustained by the individual. Where these three conditions are met, Community law directly confers on individuals the right to obtain compensation.

Francovitch implied that where damage is suffered through the acts or omissions of a private party which are incompatible with directly applicable provisions of a directive which has not been implemented by appropriate legislation, an action in damages will lie against the State.

Of more direct relevance to the *Work in Compressed Air Special Regulations 1958*, the European Court of Justice held in *Marleasing SA v. La Commercial Internacional de Alimentacion SA* (1990) that in applying national law, whether the provisions concerned pre-date or post-date the directive, the national court when asked to interpret national law is bound to do so in every way possible in the light of the text and the aim of the directive to achieve the results envisaged by it and thus to comply with Article 189 (3) of the Treaty. The European Court held that a national court is compelled to disapply a provision of national law (or ignore its application) in order to comply with the obligations of European law. The Health and Safety (General Provisions) Regulations and an accompanying Approved Code of Practice (ACOP), which give effect to the 1989 Framework Directive (89/391/EEC) on the introduction of measures to encourage improvements in the safety and health of workers at work and a 1991 Directive (91/383/EEC) supplementing the measures to encourage improvements in the safety and health of workers with a fixed-duration employment relationship or a temporary employment relationship are expected to become law in the Autumn of 1992. For the reasons outlined above these regulations and accompanying ACOP must be read in conjunction with the relevant directives.

5 Reform

To summarise then, and turning our attention back to the *Work in Compressed Air Special Regulations 1958*, there is wide agreement that the statutory rules currently in force are inadequate, outdated and should now be improved. In many respects these regulations have already been amended. As has been seen, the *Health and Safety at Work etc. Act 1974* itself brought about major changes.

Section 2 (1) *Health and Safety at Work etc. Act 1974* provides that 'It shall be the duty of every employer to ensure so far as is reasonably practicable, the health, safety and welfare at work of all his employees', whilst Section 3 (1) provides that 'It shall be the duty of every employer to conduct his undertaking in such a way as to ensure, so far as is reasonably practicable, that persons not in his employment who may be affected thereby are not thereby exposed to risks to their health and safety'.

The Management of Health and Safety Regulations 1992 and its accompanying Approved Code of Practice will come into force on or possibly a little before 31st December 1992. This will mean that from that date, *inter alia*, a risk assessment must be carried out and that the risks identified with that assessment must lead to the identification of the protective and preventive measures needed to ensure compliance with all the requirements of health and safety law. The measures to be introduced must be related to the level of risk. Both the assessment and the health and safety arrangements will need to be recorded where employees have five or more employees. Employers' risk assessments would have to take account of everyone who might be affected by their activities including non-employees. Employers would have to provide appropriate health and safety arrangements and where the risk assessment shows a need for it, appropriate health surveillance. Employers will also need to establish procedures, in the light of their risk assessments, to be followed in the event of serious and imminent danger to their employees. Finally, employers will need to ensure that their employees receive adequate information on health and safety risks and adequate health and safety training. Such training should adequately prepare and enable their employees to deal with the risks to which they are exposed. Employees will be empowered to stop work in the event of their being exposed to serious, imminent and unavoidable danger.

The provisions of the *Work in Compressed Air Special Regulations 1958, Health and Safety at Work etc. Act 1974* and *The Management of Health and Safety Regulations 1992* and following *Litster* the relevant EC directives must be read together. In the light of the changed state of knowledge since 1958 employers must be prepared to raise standards beyond the standard set out in the *Medical Code of Practice for Work in Compressed Air*, (CIRIA Report 44, Third Edition, 1982). Failure to recognise this could bring claims for negligence, prosecution or a *Francovich* claim against the British Government, in certain situations, for example where decanting takes place.

As has been said under the *Work in Compressed Air Special Regulations 1958*, there is a power to approve of alternative rules as to decompression. This has been done on many occasions when rules known as the Blackpool Decompression Tables have been substituted. Notwithstanding this amendment in practice it is submitted that the *Work in Compressed Air Special Regulations 1958* (as amended) should be brought up to date and the *Medical Code of Practice for Work in Compressed Air,* (CIRIA Report 44, Third Edition, 1982) should be revised and have the status of *either* an Approved Code of Practice *or* a voluntary code under *Health and Safety at Work etc. Act 1974*. The differences between these two different methods of regulatory reform are important. An Approved Code of Practice does not have direct effect but any provision in such a code which appears to a court to be relevant to an alleged contravention of a requirement or a prohibition is admissible in evidence. If it is proved that there was at any material time a failure to observe any provision of a code which the court considers to be relevant for the prosecution to prove in order to establish a contravention of a requirement or prohibition, that matter will be taken as proved unless the court is satisfied that the requirement or prohibition was in respect of that matter complied with otherwise than by way of observance of that provision of the code. On the other hand voluntary codes have no legal significance. Their aim is to be helpful and assist employers and others to comply with the law. Whichever method is chosen it ought to result in a modernisation and strengthening of the legal framework whilst at the same time it should allow employers the maximum flexibility to order their working arrangements.

In addition, the following changes should be made to the *Work in Compressed Air Special Regulations 1958*:

1. The person in charge of the compressed air equipment, design and facilities, the Compressed Air Operations Director, should be a qualified civil engineer and be required to have clearly defined duties, undergo approved training, and be certificated or licensed. This recommendation is most important since the presence on site of a suitably qualified person with the ability and the ultimate responsibility for carrying out these duties is very much needed.

2. Adequate compression and decompression facilities approved by the appropriate statutory body must be specified and priced for each project before the commencement of each job.

3. Decanting is an inherently dangerous procedure. It adds to the risk of decompression sickness and should only be permitted following specific official approval of the comprehensive arrangements proposed to be provided and the issue of a separate renewable certificate. Such arrangements should include, *inter alia*, the provision of adequate sanitary facilities within the decompression chamber and synchronisation of time recording at all relevant chambers.

4. Lock attendants should be required to have clearly defined duties, including the maintenance of accurate contemporaneous registers and records, undergo training, and be certificated or licensed by an appropriate statutory body.

5. Medical lock attendants should be required to have clearly defined duties, including the maintenance of accurate contemporaneous registers and records, undergo training and be certificated or licensed by an appropriate statutory body.

6. All contemporaneous compression and decompression registers and records must be required to be checked and countersigned also contemporaneously and not later than twenty four hours of the end of each shift by the Compressed Air Operations Director.

7. All tunnel workers should be physically examined prior to both the commencement of their contract and at the end of their contract by specially trained medical practitioners, and the results of these examinations lodged with the National Registry.

8. Rules should be laid down specifying rules for repeated compressions within a given period.

9. Adequate fire fighting equipment should be provided in the tunnel.

10. No person should work alone in a pressurised tunnel.

11. A permanent specialist X-ray screening facility should be available free of charge *via* the National Health Service or the Employment Medical Advisory Service for the use of tunnel miners and former tunnel miners. Radiographs of the shoulder, hip and knee should be repeated once a year up to two years after ceasing work in compressed air and at five-yearly intervals thereafter. This is an essential requirement not only for the reasons stated above, but also because a large number of tunnel miners, and the figure is estimated to run into many hundreds, may be suffering from dysbaric osteonecrosis which has hitherto been wrongly diagnosed or inadequately investigated. Radiographs taken by expert radiographers in a planned 'rolling' programme could do much to alleviate this problem and at the same time, provide valuable information for research and study, as well as extending the opportunity for compensation, where merited, to those who have suffered and who are currently unaware of the source of their symptoms.

In addition it would seem sensible, and in accord with the legislation outlined above, to provide for accurate records of the time taken for compression and decompression. These should be kept and verified by employers. Records used to be kept at the Decompression Sickness Central Registry in Newcastle upon Tyne, although this was on an informal basis and records were by no means complete. Because of the length of time for aseptic necrosis of the bones to manifest itself and because of the peripatetic nature of the working life of the average tunnel miner it would surely be beneficial if these records were to be housed in a national registry. Such a move would both facilitate medical treatment and provide a neutral scientific basis for the principal purposes of assisting further research and the monitoring of applicable legislation. Such a move would aid the progressive improvement of standards. These records could then be made available to the Health & Safety Executive and, following appropriate assurances, to legal and medical practitioners.*

The authors of this paper are of the view that in order to effect the necessary changes the suggested reforms should be seen as a package of measures to be adopted as a whole.*

---

* This paper draws on material in the introductory section of *The European Communities Health and Safety Legislation*, by Alan C. Neal and Frank B. Wright, Chapman & Hall, London, 1992, and on *The Law of Health and Safety at Work: The New Approach* by Charles D. Drake and Frank B. Wright, Sweet & Maxwell, London, 1983

Discussion on Session 8: Codes and regulations

The chairman of the session, **Dr H V Hempleman**, introduced the speakers presenting their papers. **Mr M Bandmann** explained the CEN standardisation proposals for the safety and other requirements for tunnelling air locks. **Professor I Holthaus** drew attention to inadequacies of the German and, indeed, other regulations for compressed air work and the need for much better education and training. **Mr F B Wright**, whose paper with **Mr L Marks**, examines the legislative and medico-legal aspects of work in compressed air, identified specific changes which, in their opinion, should be brought into new regulations. (These recommendations are included in the revised version of their paper printed in these proceedings).

The discussion that followed these presentations is given below.

Mr D R Lamont. As a principal inspector for HSE, I feel I must respond to the suggestions for changes to the *Work in Compressed Air Special Regulations 1958* made by Mr Wright and Mr Marks.

Regulation 5 currently requires the attendance on site of a person 'competent to take charge of compressed air operations' and that the said person is in charge of the operations when persons are working under pressure. The concept of the 'competent person' is well understood and has been established in law. Certification or licensing is unnecessary.

I believe that the current situation where the 1958 Regulations are supplemented by the CIRIA *Medical Code of Practice for Work in Compressed Air* and the forthcoming CEN standard is acceptable. HSE is involved in the work of CEN/TC151/WG4.

HSE has already expressed concern at the practice of decanting and is pursuing this approach in the review of the regulations for work in compressed air currently being undertaken. Certificates of Approval to use the Blackpool Tables in the UK include certain requirements in respect of the use of recording pressure gauges and sanitary facilities.

The duties of lock attendants are clearly defined in Regulation 10. The requirement in Regulation 10 for 'competent lock attendant' cannot be satisfied unless the attendant has received training. Regulation 18 applies to medical lock attendants.

I see little to be gained by the Wright and Marks proposals on record-keeping over the requirements in Regulation 10. The current arrangements for medical examinations and records meet the needs of HSE. I do not believe that there is a major problem with repeat compressions.

HSE Guidance note GS20 and BS 6164: 1990 set out good standards for fire precautions in compressed air tunnels. I agree that lone working in tunnels is wrong.

I have no comment on the proposals about X-ray facilities. HSE will be considering its requirements for records during the review of the regulations for work in compressed air.

Mr L Marks expressed a concern that there should be no complacency, for the current legislation had proved ineffective. It would also be desirable that a mechanism be found to keep a central registry of records.

Mr J C Le Péchon described the French legislation which provides a unified framework covering all types of hyperbaric work. (Subsequently, Mr Le Péchon submitted a written contribution which explains this legislation more fully and which is printed immediately after this report). These were being offered by France for consideration as the basis of a European Standard.

Dr H V Hempleman commented that in this age of prescriptive rules it is essential that they are understood as being 'reasonable'.

Mr A J Howard. Professor Holthaus said that divers have a very extended training, whereas tunnellers have little or no training. This I do not find surprising: divers have to manage their own decompression with considerable sophistication; tunnellers are largely passive. What would Professor Holthaus say is an appropriate training period for tunnel miners?

Professor Holthaus replied that the point was not the length of a training period, but that first each worker has a right to know what he is being asked to undergo and that what is needed are structured systems of training for different types of work and responsibility.

Mr Le Péchon mentioned the different training requirements for French hyperbaric workers which are linked to the exposure conditions.

Mr H R Davies asked what, in the absence of a national registry, is to be done with contract records.

Mr Marks re-emphasised the need for the record to be kept and protected, which a registry would do. Without this the claimants are effectively deprived of supporting evidence for their cases.

Mr V J Belman asked who is being proposed to pay for a registry.

Mr Marks pointed out that the tendency would be to save no records.

Mr Belman wondered if it would be appropriate to keep records of exposures above a particular pressure.

Dr E P Kindwall commented that hospitals keep records in effect 'for ever', but companies keep records that the law requires them to keep.

Mr Le Péchon noted that there is not a European (EC) requirement.

Dr Hempleman asked if there was any proposal for a European registry.

Mr Marks wondered about the attitude of insurance companies.

Mr Lamont pointed out that there is a requirement to keep records of exposures over 1 bar for 15 years: this also applies to successor companies.

Dr R D Vann commented on pressure/time records. Having worked with paper records of the depth-time profiles of divers for many years, he had found that one is soon overwhelmed by the volume of data, and it becomes unusable. Even data entry into computer-readable form by keyboard requires a substantial infrastructure and introduces errors which must be checked. A better solution may be digital pressure-time recorders in which basic information such as date, location, etc. are entered at the same time the pressure data are recorded. This becomes part of a 'paperless' record which can be accessed and analysed by standard software modules. Such a programme is under development by the Divers Alert Network for recording information on recreational diving.

Dr R W Hamilton asked what the prospects are for using alternative decompression procedures.

Dr Hempleman replied that there was no problem if the chosen alternative is more conservative, but the burden of justification is on the contactor if the procedures are different. (Dr Hamilton subsequently wrote to say that this in keeping with the UHMS Validation Workshop guidelines).

Mr Davies remarked that the adoption of decanting depends on the size of the tunnel and the pressure level. He thought too sweeping the proposal to cease from decanting. The three-lock system might not be seen invariably as beneficial, e.g. if the client finds it too costly or if alternative construction methods are preferred to compressed air.

Mr Marks expressed his view that decanting was wrong. In practical terms, it has often been carried out with huge breaches of the specified timing.

Professor Holthaus supported Mr Marks' view.

Mr Lamont reminded the UK members of the audience that all the CEN documents will be made available for public comment though the British Standards Institution.

Mr Belman made a formal presentation, first making the point that the industry urgently needs new, clear regulations. His presentation is summarised as follows.

There are three main groups of people who have to be considered in compressed air work:

1. Those who can be classed as 'non-workers', e.g. engineers, chainmen, foremen and supervisors, fitters and electricians, inspectors, visitors and medical personnel. These people are the least likely to suffer DCS or bone necrosis.
2. The 'hard core', i.e. experienced miners. Generally, those men at risk of being susceptible to ill-effects from the nature of the work have been 'weeded out'.
3. The new-start miners. These are the most vulnerable to DCS, and should be given the strictest medical examinations.

The purposes of compressed air in caissons and tunnels are to keep out water and stabilise the ground. The hazards associated with its use are:

1. Of a 'blow'. It is worth being reminded that the potential energy of the air in a 1000-m long, 2-m dia. tunnel at 2 bar-gauge is comparable with that of a 100-t train travelling at 80-90 km/h.
2. Of ill health. The commonest is damage to the ears. The others, as we know, are Types I and II DCS and osteonecrosis.

On the other hand, the use of compressed air has particular advantages:

- it is the least expensive of comparable options
- it drives out noxious gases
- it helps to reduce ground-surface settlement
- it reduces vibration, e.g. from blasting
- it is the best cure for a hangover!

At present in UK, the risks of compressed air are controlled through three documents:

- *Work in Compressed Air Special Regulations 1958*
- *Medical Code of Practice for work in compressed air* (3rd edn.)
- BS 6164: 1990 *Code of practice for safety in tunnelling in the construction industry.*

Currently, HSE is working towards new regulations to control, contain and reduce these risks. They will have to take into account and make provision for the following matters:

- that there are four times as many compressed air contracts below 1 bar than there are above this pressure
- simplification of format and procedures
- shift durations and decompression times – in order not to jeopardise the viability of the methods
- whether the medical lock with a full-time attendant is on site
- decanting
- smoking
- training
- multiple exposures
- the possibility of saturation-excursion techniques
- women in compressed air
- medical examinations and follow-ups.

(Reporters: J Buchanan and F M Jardine)

Editor's note. Two written contributions relevant to the subject of codes and regulations are presented in the following pages. While they have been put into the same format as the papers before conference, they have not been formally referred, as all the other papers were.

35 French regulations 1992 for hyperbaric works

J.C. Le Péchon and J.L. Pasquier*
JCLP Hyperbarie, France
**Ministère du Travail Bureau CT4, Paris, France*

This is a post-conference supplement to the discussion, not a refereed paper.

Abstract
This note explains the context and content of the hyperbaric regulations recently issued in France. These cover the health and safety requirements for all work at increased pressure with specific provisions for different hyperbaric situations.
Keywords: Hyperbaric regulations, Compressed air, Decompression tables.

1 Introduction

When tunnel boring machines (TBM) began to be used in France in the '80s, the *décret* (statutory order) July 9, 1974, concerning safety rules for compressed air workers, proved to be inappropriate to this new technology.

Very soon after the publication of the French *décret* July 11, 1974, which covers the rules applicable to the safety of commercial divers, it appeared that these rules would need improvement. The developing offshore industry asked for more detailed and better adapted legislation.

As previously reported (Le Péchon and Pasquier, 1989), new French regulations have been published (*décret* March 21, 1990), and supplemented by additional *arrêtés* (departmental orders) in 1991 and 1992; others are expected in 1993.

This new regulation is based on the *Code du Travail* and is in agreement with the law (December 31, 1991) which implements the European Directive dated June 12, 1989 on health and safety at work.

The general organisation of the Health and Safety at work legislation in France is shown on Figure 1.

Engineering and Health in Compressed Air Work Edited by F.M. Jardine and R.I. McCallum. Published in 1994 by E & FN Spon. ISBN 0 419 18460 0

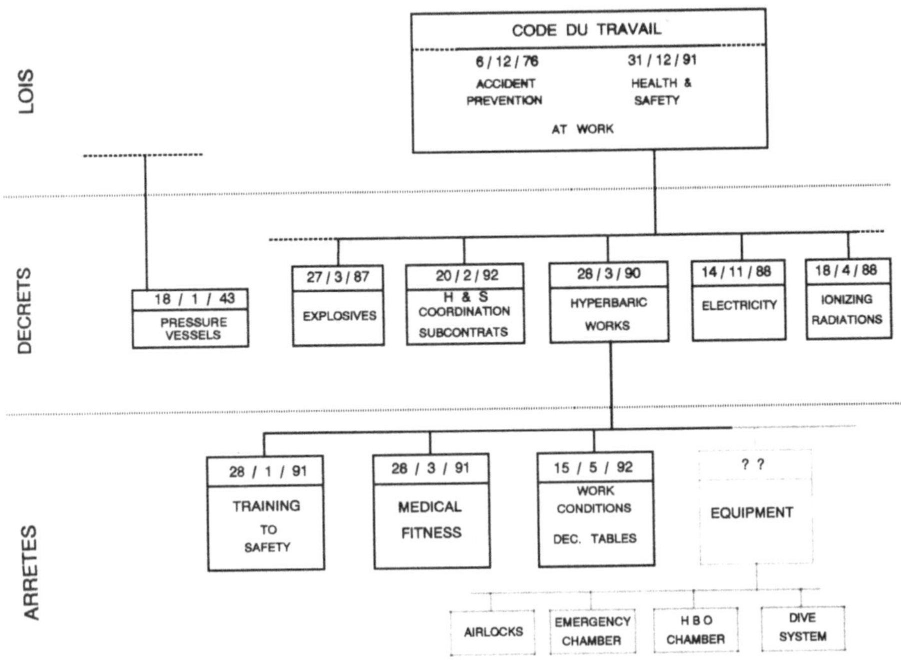

Fig. 1. French health and safety regulations for hyperbaric works

2 Décret February 20, 1992: Coordination of health and safety in subcontracts

For diving activities, the *décret* for the coordination of health and safety between a company and its subcontractors is of major importance since most diving works are carried out by specialist underwater contractors.

This *décret* stipulates in particular:

1. The company employing the subcontractor on its own premises or sites is responsible for the initiation and coordination of the health and safety measures that are to be applied.
2. Each firm remains responsible for its own personnel's health and safety.

3. A *Plan de Prévention* must be prepared and agreed by both parties that addresses co-ordination of safety between each other's activities, which could interfere, and contingency plans. This is to be understood widely. The *Plan de Prévention* also has to be approved by employees' representatives.

3 Décret March 28, 1990: Prevention measures for hyperbaric works

3.1 Who is concerned?

This *décret* is general to all hyperbaric works and makes no specific reference to the various professions or activities which expose personnel to hyperbaric situations, by immersion or in the dry; the field of application is very wide:

> Anybody exposed, at work, to an increased pressure above 0.1 bar relative to local pressure, needs to be protected against the possible harmful effects of pressure and, when applicable, immersion according to the requirements of this *décret*.

It therefore applies to professional divers of all sorts:

- commercial oilfield divers
- commercial onshore divers
- scientific divers (biologists, archaeologists..)
- film industry divers (TV reporters etc...)
- professional diving instructors
- divers involved in aquaculture
- divers on board fishing vessels, etc.

It equally applies to compressed air workers in situations such as:

- tunnel boring machines
- compressed air tunnelling
- hyperbaric testing of nuclear plant confinement structures
- attending patients during hyperbaric oxygen therapy exposures
- submerged caisson works
- experimental dry chamber exposures
- dry transferred hyperbaric subsea welding operations.

This *décret* is not about the different types of work but about the physical risks associated with being under pressure.

3.2 Content of the décret of March 28, 1990

Breathing gases quality standards

Compressed air is permitted up to 6 bar(g). The limits on partial pressures are given in Table 1.

Table 1. Limits on partial pressures

Carbon dioxide, CO_2	10 hPa = 10 mb = 1 kPa
Carbon monoxide, CO	0.05 hPa = 50 μb = 5 Pa
Water vapour, H_2O	In saturation, 60 to 80% relative humidity
Volatile hydrocarbons	0.5 mg/m^3 (m^3 measured at pressure)
Dust, other pollutants	Standard Threshold Values expressed in partial pressures
Nitrogen, N_2	5.6 bar - air at 6 bar(g) -
Oxygen O_2	> 0.160 bar i.e. >160 hPa
In water	< 1.6 bar and less according to the duration of exposure
Chamber decompression	< 1.2 bar
Saturation atmosphere	0.3 to 0.45 bar
Emergency recompression	< 2.8 bar
Fire prevention (FO_2)	< 25% by volume in atmosphere

Selection of proper gas is the Employer's responsibility. There are three stipulated responsibilities for quality control of the breathing gases:

- Gas supplier (analytical data to be supplied with delivery)
- Employer if gas mixed on site
- Oxygen control mandatory (except for air diving).

Personnel requirements
Training in safety for the various hyperbaric situations is mandatory for all concerned personnel undergoing pressure exposures, with various levels of training objectives (*arrêté* January 28, 1991). In addition there are further requirements:

1. A competent person nominated in writing by the employer must look after the operation and be ready to take any action required for the safety of the personnel (*Chef d'Opération Hyperbare*).
2. A competent person nominated in writing by the employer must, at all times, supervise the person(s) under pressure and make sure that safety rules are adhered to and that the procedure is followed safely (*Surveillant*).
3. A qualified and medically fit person must be ready at any time, on signal from the supervisor, to give assistance to the person(s) exposed to pressure

and to carry out a rescue operation if needed. This person must hold a first aid certificate.

Equipment and procedures
At all times equipment and procedures must allow for a possible rescue of the person(s) exposed to the hyperbaric situation (as a consequence a stand-by person, a suitable airlock configuration, a system for unconscious diver recovery, etc. are needed).

Documentation
In addition to the *Plan de Prévention* when work is subcontracted, the documentation must include:

- employer's hyperbaric safety manual
- site safety rules and procedures
- logging of exposures
- personnel certificates of classification and medical fitness
- personnel log book
- accident reporting
- equipment maintenance books.

Medical fitness
The medical fitness certificate is signed by an occupational medicine doctor, who may ask a specialist hyperbaric doctor to carry out specific examinations on his behalf in accordance with the recommendations of *arrêté* March 28, 1991.

Working conditions and decompression tables
Working conditions are addressed in the separate *arrêté* May 15, 1992. Decompression tables are published as five individual annexes to this *arrêté*.

4 Arrêté January 28, 1991: training and certificates

There are no training certificates for superintendents (*Chef d'Opération Hyperbare*) or supervisors (*Surveillant*); these persons are nominated by the employer. It is the employer's responsibility to make sure that they have received the proper training and are competent for the purpose.

Persons going under pressure must have a safety training certificate called *Certificat d'Aptitude à l'Hyperbarie* (CAH). There are various types of CAHs. A CAH includes a *Mention* describing the type of hyperbaric activity and a *Classe* showing the maximum pressure for which the training objectives have been achieved. Example: for a compressed air intervention at 2.4 bar, the workers must hold a CAH *Mention* D (eventually A or B) *Classe* I (eventually II or III). Table 2 shows the various combinations of CAHs.

Table 2. Training certificates (CAH)

Pressure (bar)		0 to 1.2	0 to 4	4 to 6	>6
Activities		***Sous-Classe IA***	***Classe I, II or III***	***Classe II or III***	***Classe III***
With immersion	Industrial divers	NA	A	A	A
	Other divers	B* or A	B* or A	B or A	B or A
Without immersion	Medical personnel	NA	C* or A, B	C* or A, B	C or A, B
	Compressed gas workers	D* or A,B	D* or A, B	D or A, B	D or A, B

NA = Not applicable
* = Authorised Employer
A, B, C = Corresponding necessary *Mention*

The various CAHs (with proper *Mention* and *Classe*) can be obtained from various sources:

1. Approved training centres. The approval procedure is described.
2. For certain categories (marked * in Table 2) from the employer if he has received formal authorisation from the Administration. Application conditions are given.
3. Until March 3, 1993 and only for those who were in that work before October 1990, an interim procedure is available for obtaining a 'classification from experience' through an *ad hoc* commission.
4. Some certificates or diplomas, issued by various recreational diving organisations, military diving schools or universities, lead to the issue of the CAH with no additional training.

All CAHs are issued by Institut National de la Plongée Professionnelle (INPP) on behalf of the French Ministry of Labour.

5 Arrêté May 15, 1992: working conditions, decompression tables

All decompression tables published have been established by COMEX. For the air diving tables and their evaluation, calculation and validation, COMEX received financial support from the Ministry of Industry. Validation is based on the COMEX dive-logging system (Imbert and Montbardon, 1990; Le Péchon *et al.*, in press).

5.1 List of official decompression tables

Diving operations

AIR AND NITROX MIXTURES:

- Minitable: Air
- Table: Air/Standard
- Table: Air/O_2/6 m
- Table: Air/O_2/12 m
- Table: Air/Surface decomp./O_2
- Nitrox diving procedure
- Multilevel diving procedure (ascending)
- Altitude diving procedure (up to 3000 m)
- Bentonite diving procedure
- Repetitive diving procedure.

HELIOX MIXTURES:

- Table: Heliox/O_2/6 m
- Table: Heliox/O_2/12 m
- Table: Heliox/O_2/Diving bell
- Saturation diving procedure to 200 msw.

Compressed gas works

H B O THERAPY ATTENDANTS:

- Table/*Mention* C/Standard
- Table/*Mention* C/O_2

COMPRESSED AIR WORKS:

- Table/*Mention* D/Standard
- Table/*Mention* C/O_2
- Tables for one extra bottom time.

Emergency recompression

- Decompression illness symptoms: Type I
- Decompression illness symptoms: Type II.

5.2 Immersed working

The *arrêté* reviews the various diving situations. The maximum daily immersion time is specified (3 hours for commercial diving *Mention* A and up to 6 hours for non-commercial divers *Mention* B) as well as the duration and frequency of saturations. It stipulates the minimum permitted team content for the various diving procedures and the items of equipment.

The decompression tables to be used in connection with the diving works and published are listed above. No decompression computer is allowed for professional diving.

5.3 Works in compressed gases

Maximum shift durations in compressed air are given as:

8 hours per day if pressure is lower than 0.75 bar
6 hours per day if pressure is higher than 0.75 bar.

Repetitive exposures are restricted to pressures lower than 2.1 bars.

The decompression tables to be used in connection with hyperbaric works in the dry are listed in Section 5.1.

5.4 Emergency recompression

Based on the principle that whatever the quality of a decompression procedure used, a case of decompression illness could arise, two emergency recompression tables (for so-called Type I and Type II decompression symptoms), to be used on site before the advice of an hyperbaric medical doctor is obtained, are published. The table for Type II cases requires the patient to breathe a heliox mixture 50/50 by mask from 3 bar(g) to 1.8 bar(g), and then pure oxygen. The provision for the decompression of the attendant on pure oxygen is included.

5.5 Remarks, new technology, alternative methods

No limitations are given for pressure exposures. When pressures or durations foreseen during a specific operation are higher or longer than those basically displayed in the decompression tables, authorisation is required. To obtain this, an application describing the project, the safety rules and the proposed decompression table must be submitted to the Ministry of Labour for approval.

When methods or equipment not described as standard in the *arrêté* are needed, an application for authorisation also has to be made to the Ministry of Labour.

As the decompression tables have been published as Annexes, it should mean that they will be easy to amend if so requested following field evaluation.

This legislation is framed so that it is open to improvement.

5.6 Emergency recompression chamber

Availability of a recompression chamber is necessary for all types of situations, but the permitted time to reach this chamber depends on the hyperbaric exposure contemplated.

Diving

(a) If the decompression time is less than 15 minutes:
  Mention A activities: less than 1 hour to the recompression chamber
  Mention B activities: less than 2 hours to the recompression chamber.
(b) If the decompression time is to exceed 15 minutes: recompression chamber on site.

Compressed gas works

(a) Decompression time less than 15 minutes:	recompression chamber available in less than 2 hours
(b) Decompression time more than 15 minutes:	recompression chamber available in less than 1 hour
(c) Maximum pressure more than 1.8 bar(g):	recompression chamber on site.

6 The future and conclusion

Still to be issued are *arrêtés* specifying the equipment for hyperbaric works. For practical reasons, there may have to be several publications covering, for example:

- airlocks for tunnelling
- emergency recompression chambers
- hospital walk-in HBO chambers
- diving systems and equipment.

These regulations are currently in preparation and under review by the groups of users.

These French rules, 1990-92, are based on a new approach (only Brazil has used it before), which takes into consideration the physical risks involved in staying under increased pressure. These rules cannot be taken out of the context of the *Code du Travail* which covers the other physical risks which might be associated in the hyperbaric exposures; e.g. noise, electricity, fire, ionising radiations and atmospheric pollution. These points are not addressed in the hyperbaric regulation except where specific additional precautions are needed because of the actual pressurised conditions, as in the cases of fire prevention and fire fighting or breathing-gas purity standards.

While the new regulation is probably not perfect, most of the problems identified in recent years as major issues have been addressed (Holthaus, 1994). When these rules are offered to other European countries as a draft for a future European Directive for hyperbaric health and safety, the chosen limits, restrictions, methods, tables and recommendations will have to be discussed. Amendments will be offered for corrections and improvements as needed.

References

Regulations

Journal Officiel de la République Française, Brochure 1636: *Travaux en Milieu Hyperbare, mesures particulières de prévention*

European Directive 89/391, dated June 12, 1989, relative à la mise en oeuvre de mesures visant à promouvoir l'amélioration de la sécurité et de la santé des travailleurs. (JOCE L 183 June 29, 1989

Brasilian regulations: *Trabalho sob condiciones hiperbaricas*, Diario Oficial do Brasil, 15 Sept. 1987, p. 16024-16028

Holthaus, I. (1994) Are the German regulations for compressed air work safe? In: *Engineering and Health in Compressed Air Work*, (Jardine, F.M. and McCallum, R.I. eds.), E&FN Spon, London

Imbert, J.P. and Montbardon, S. (1990) The Comex diving data base. In: *Operational dive and decompression data: collection of analysis.* (W Sterk and R.W. Hamilton eds.) EUBS Workshop publication (DATA) 17-8-90, p. 36-40

Le Péchon, J.Cl. and Pasquier, J.L. (1989) New French regulations for man working under pressure. *Proceedings EUBS Annual Meeting*, Eilat, 1989

Le Péchon, J.Cl., Imbert, J.P. and Pasquier, J.L. (in press) *Les nouvelles tables de décompression du Ministère du Travail*, 1992, MedSubHyp, in press

36 Comparison and prediction of decompression sickness rates

F.M. Jardine
Construction Industry Research and Information Association, London, UK

This is a post-conference supplement to the discussion, not a refereed paper.

Abstract
Data from past compressed air projects can be represented in terms of a single expression which combines pressure, working time at that pressure, the number of exposures, and the consequent treatments of decompression sickness. In this way, these data are used to define probable upper and lower bounds of the incidence of decompression sickness, either per exposure or per worked hour. Records from overseas compressed air projects are compared and shown to be consistent with the UK project database. This method of representing exposure records is used to compare the relative effectiveness in terms of the amount of decompression sickness of UK projects carried out under the 1958 regulations and under the Blackpool tables/CIRIA Report 44 procedures. The empirical relation between the exposure parameters is examined to see if it can give any insight into reducing the occurrence of decompression sickness.
Keywords: Compressed air projects, Decompression sickness, 1958 regulations, Blackpool tables.

1 Introduction

The requirement for health and safety risk assessments to be made in advance is one of the important provisions of *The Management of Health and Safety at Work Regulations 1992* (Anderson, 1994). To take one aspect of this, it would seem that it will be necessary to make a rational estimate of the probable incidence of decompression sickness (DCS) that could occur on a compressed-air tunnel or caisson project.

Published values of DCS incidence rates (in terms of recompression treatments per exposure) of past projects are a poor guide for making such an estimate. Quite apart from the difficulty of matching to a previous job, the range of incidence rates that has been reported is at least one order of magnitude, e.g. 0.2 to 2.0% or similar, as can be seen from the values quoted in papers to this conference.

Engineering and Health in Compressed Air Work Edited by F.M. Jardine and R.I. McCallum. Published in 1994 by E & FN Spon. ISBN 0 419 18460 0

A more rational method of estimation is needed, and it is suggested that the approach outlined in the appendix to Jardine *et al.* (1991), although empirical, could provide this. The approach is based on the numerous exposure records of the several UK compressed-air projects held in the Newcastle Central Registry, and was made possible by the data collation and formatting carried out by Dr A Evans in his further analysis of these records (see Evans, 1990, for examples of this).

Several of the papers to this conference provide project data, though in far less detail than the Newcastle records, but which can be seen to be quite consistent with the UK incidence of DCS when compared on the same basis.

With some confidence then, it is possible to use the approach as a robust method not only to estimate the probable amount of DCS on a new project, but also to identify where there appears to be scope for improvement in decompression procedures.

2 Background

2.1 Representation of empirical data

In an appendix to the paper, *Studies of the incidence of decompression sickness on UK compressed air projects,* (Jardine *et al.*, 1991), the writer presented in one chart (repeated here as Figure 1) the aggregated exposure records of the high-pressure (> 1 bar, 14 psig) compressed-air projects in the DCS Central Registry. This representation takes account of exposure (working) time, t, at pressure, p, total numbers of exposures (or man-decompressions), E, and the numbers of cases of

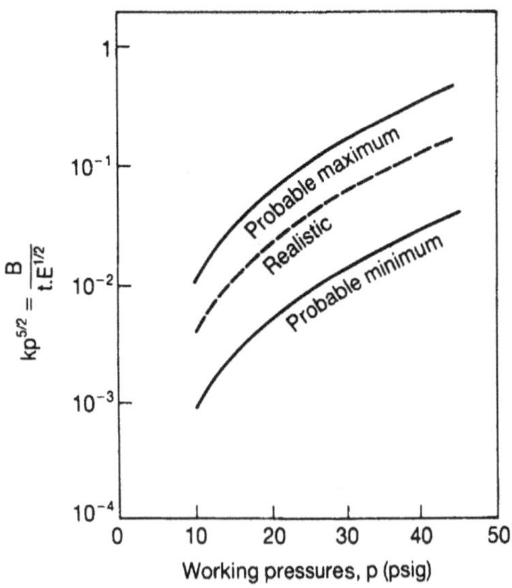

Fig.1. Envelope of the UK normal decompression dataset

DCS (irrespective of whether it is Type I or Type II) given recompression therapy, B. On a project-by-project basis, the aggregated values of B and E were known for pressure ranges of 4 psig (0.28 bar) and for working times in intervals of 2 h. The assumption was made that the mid point of the pressure range was the mean of the pressures actually used. Average working time, t, was calculated by dividing the known cumulative worked hours, H, by the total number of exposures for the same pressure-time intervals; i.e. $t = H/E$.

The approach developed to represent the incidence of DCS was that it was not only a function of pressure and working time but also an inverse function of the total number of exposures, i.e.

$$B = f\,(t^x, p^y, 1/E^z) \tag{1}$$

An empirical relation was put forward of the following form:

$$kp^{5/2} = \frac{B}{tE^{1/2}} \tag{2}$$

where k is a composite factor, defining either an upper or a lower bound to the UK exposure data, when plotted as points in terms of the right-hand expression, i.e. $B/t.E^{0.5}$. Note that if $B = 0$, $k = 0$; and that any point that can be plotted on Figure 1, because it has a log scale, means that there has been some DCS, i.e. $B \geq 1$. In amplification it is pointed out that the equation includes two implied relations: (1) the number of DCS treatments per manhour worked, i.e.

$$\frac{B}{tE} = \frac{B}{H} \tag{3}$$

and (2) a reduction factor inversely dependent on the square root of the number of exposures, $\sqrt{E}$, a measure of the size and duration of the project. The derivation of the reduction factor is as follows.

$$B_{predicted} = f_1(p^{5/2})H \tag{4}$$

Comparison with the actual number of DCS treatments gave:

$$\frac{B_{actual}}{B_{predicted}} = f_2\left(\frac{1}{E^{1/2}}\right) \tag{5}$$

Thus, as $B_{actual} = B$,

$$B = f_2\left[\frac{1}{E^{1/2}}\right] f_1(p^{5/2})H \tag{6}$$

$$or \quad B = f_1\, f_2\, p^{5/2} t\left[\frac{E}{E^{1/2}}\right] \tag{7}$$

transposing, $\dfrac{B}{tE^{1/2}} = kp^{5/2}$ *as equation* 2 *above.*

The upper curve of $kp^{5/2}$ in Figure 1 is a reasonable bound to the data - almost all of the projects incurred less DCS than it represents. The lower curve, which has the same general form, is not a true lower bound as there were many sets of exposures where there was no DCS (usually lower pressures, shorter times, and relatively few exposures), but it represents a probable minimum amount of DCS on all but small jobs or at short working times at low pressure.

For any set of exposures (p,t,B,E), the point plotted as $B/t.E^{0.5}$ can be considered as lying on a curve of the general form, $kp^{5/2}$, but defined by a particular value of k. A higher value of k means that there was more DCS for the same conditions of pressure, time and number of exposures. As k can be back-calculated knowing the values of B,t,E and p, it can be used as a comparator of DCS incidences.

The parameter, $B/t.E^{0.5}$, is the mean DCS rate per exposure-hour corrected for sample size, in effect normalised to a single exposure.

2.2 Sample size reduction factor

There are some interesting implications in the apparent dependence on the number of exposures (i.e.$\sqrt{E}$). First, on small projects, i.e. with few exposures, there tends to be relatively high DCS rates. Secondly, at large numbers of exposures, when it might be expected that several factors combine to reduce the incidence of DCS, the reduction with size of sample seems reasonable. The factors which could be expected to result in proportionately fewer DCS treatments include:

- inurement of men and 'weeding out' of the more susceptible
- the likelihood of better discipline with routine
- the overall incidence of DCS being less affected by isolated extremes.

By far the largest sets of exposures are for the longer working periods because of the traditional UK use of 8-h shifts. This bias in the data population presents difficulty in comparisons, but the reduction factor for size ($f_2\surd E$) goes some way to overcome this problem.

An important point which it allows us to address is that raised by Evans (1990). He postulated by extrapolation from the incidence of DCS per worked hour on the historical projects that there would be less DCS by using shift times of less than 3 h and of 8h and more, which he termed the 'excluded middle' type of restriction. His corollary was that regulations which shorten the working-time limits could increase the overall amount of DCS on a project, because the consequent increase in the number of exposures needed to complete the work would not be offset by the marginal reduction of the rate of DCS per exposure from longer to shorter shift times.

Such an argument took no account of the sample size reduction factor. If it had been applied, it would have reduced the DCS rate to be used in the calculation for the greater number of shorter shifts and correspondingly increased the DCS rate for the fewer longer shifts. The net result if recalculated in this way might well have been a prediction of less overall DCS with restricted (shorter) shift times. If so, it would reconcile the ethical difficulty in the 'excluded middle' approach - the greater risk in long exposure to an individual per exposure and the apparently lower risk overall for the whole project.

2.3 UK projects

The way the Newcastle exposure records have been sorted and aggregated by Evans allows comparison, project by project, of normal and decanted decompressions under either the 1958 tables (or modified versions of these) or the Blackpool tables. Strictly, the comparison is not of the tables but of the projects. What could well be a most useful method of comparison, the Standardised Bends Ratio (SBR), was developed (see Evans, 1992) but, because of lack of time, the first use of the SBR in comparing the overall DCS occurrence on projects under the two regimes was not followed up or examined in more detail. The result appeared to be that, overall, there was little difference in the DCS incidence between the two regimes (Jardine *et al.*,1991). This rather disappointing result should be re-examined, as Evans (1992) has suggested. Use of the normalised DCS rate per exposure-hour, described in Section 2.1, does show certain distinct improvements in the Blackpool tables/CIRIA Report 44 projects compared with the earlier ones.

When the normal decompression pressure-time cells of each of the projects are plotted, the 1958 points are generally higher, i.e. k is greater, than the Blackpool ones, as can be seen in Figures 2 (a) and (b). This is not the case for the decanted decompressions (Figure 3). Although Figure 3 does not distinguish the two sets of data, some of the highest points as well as those with the least DCS were on the decant decompression projects under the Blackpool tables regime. Overall the range of DCS with the normal decompression exposures compared with the decanted ones is much the same. On the basis of DCS alone, it is apparent that decanting can be managed well, but there were projects when it gave rise to higher DCS rates.

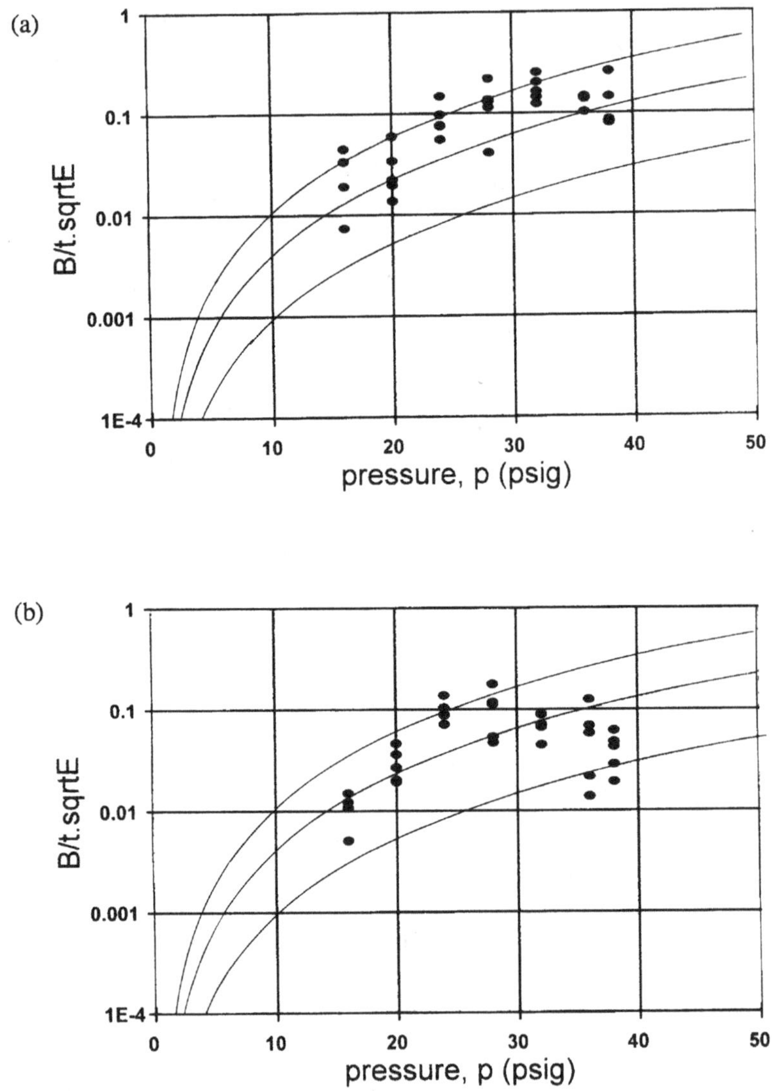

Fig.2. UK normal decompression dataset, (a) 1958 Tables projects, (b) Blackpool Tables projects (aggregated data)

This plotting method, unlike the only use so far made of the SBR approach (although that could be developed to the same effect) does show the spread of DCS occurrence. Many of the 1958 regulations projects were at least as successful as the Blackpool Tables/CIRIA Report 44 ones, but both sets of projects were uneven in their DCS performance. This aspect is discussed further in Section 4.

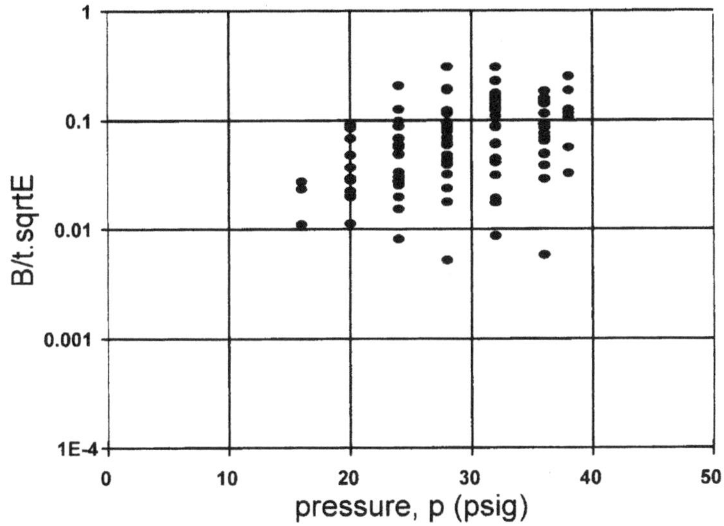

Fig.3. UK decanted decompression dataset (aggregated data)

3 Overseas projects

Several of the papers to this conference give the occurrence of DCS in sufficient detail to allow use of the above approach. The data in these papers are necessarily summarised into broader intervals of pressure and working time than available from the UK project dataset.

The first projects which are considered in this way are the large tunnel works of the Hong Kong MTR (Contract 107), the Singapore MRT works and the Cairo Wastewater contracts (Campbell and Gutteridge, 1994; How and Wong, 1994; and Court, 1994). The Hong Kong and Cairo results have been plotted in Figures 4 and 5. The Singapore results are given in very broad pressure bands, so they are not so amenable to plotting in this way. Typically they have values of $B/tE^{0.5}$ of between 1 and 6 x 10^{-2} for pressures of 15 to 40 psig. It is noteworthy that DCS was reported at less than 1 bar (14 psig). These overseas results are entirely consistent with the UK records and very similar to each other.

A point to be noted is that there appears, by this form of comparison, to be no difference between the DCS incidence in people of different races - a finding that is by no means clear when the comparison is by crude 'bends rates'.

Another important set of information is that given by Mano *et al.* (1994) in their Table 1, which is a summary of DCS incidence in Japan over two periods of about 5 years. Their data are sorted by pressure intervals of 0.2 kg/m^2, but without the breakdown of exposure times. With reasonable assumptions of what these would have been on average, it is possible to calculate the parameters, $B/t.E^{0.5}$, for the pressure steps. These are shown in Figure 6. The assumptions made for average

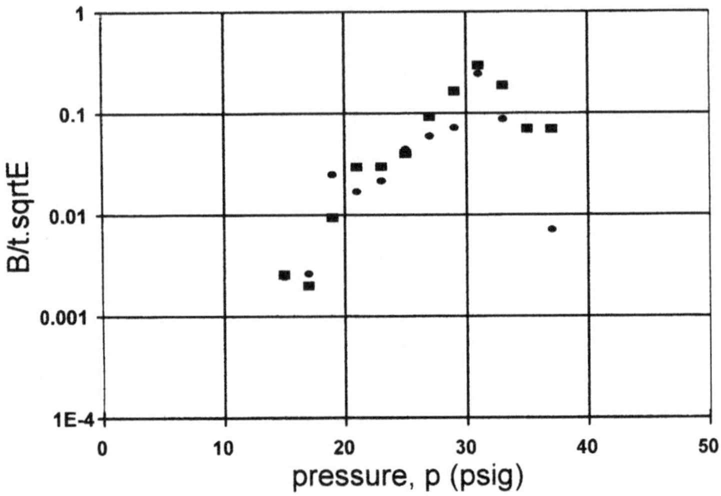

Fig.4. Contract 107, Hong Kong MTR (data from Campbell and Gutteridge, 1994)
Key: Europeans ● Chinese ■

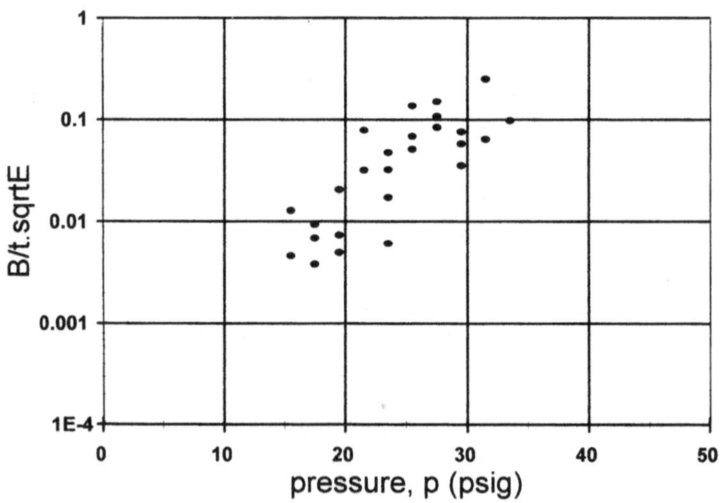

Fig.5. Contracts 3, 12 and 12/5A, Cairo Wastewater Project (data from Court, 1994)

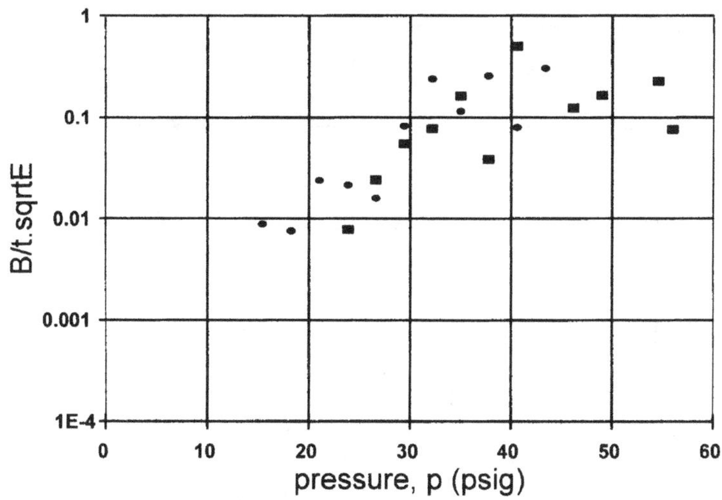

Fig.6. DCS incidence in Japan (data from Mano *et al.*, 1994)
Key 1979-85 ● 1986-90 ■

working times were that they were 7 h at the lowest pressures reducing steadily to 2.5 h at the highest and the same for both periods of years. The plots are not particularly sensitive even to gross errors of, say, a factor of two in the chosen value of working time, because of the log scale. It can be seen that the Japanese record on DCS is much the same as that in UK, i.e. within the same ranges.

The third type of information given in papers to this conference is about new decompression procedures. Of particular interest are the papers by Rüegger *et al.* (1994) and Nashimoto *et al.* (1994), because these relate to new and modified decompression procedures to cope with high pressures. These results are given in Figure 7. In terms of the crude 'bends' rate, they appear as high incidences of 2% and more, but plotting in terms of the normalised-exposure parameter shows that they were not untoward for the conditions. Indeed, the rapid response to changing conditions on the Japanese site by modifying the work pattern and decompression procedures was clearly of considerable benefit, achieving the lower practical bound of DCS in new and severe conditions.

4 Comparing DCS occurrence on pre- and post-Blackpool Tables projects

The factor, *k,* which defines the curve on which a particular set of exposure data lies, can be back-calculated. The higher the value of this factor, the greater is the amount of DCS for given values of working pressure, time, and number of exposures. Thus *k* can be used as a comparator in three ways:

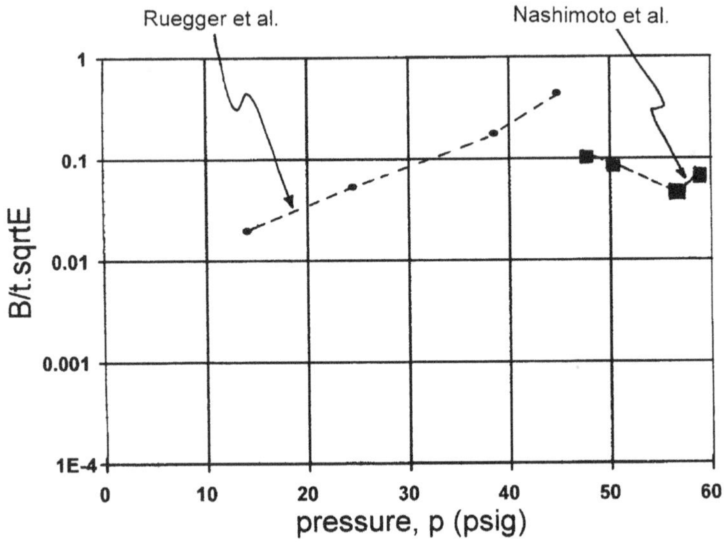

Fig.7. DCS incidence on modified decompression procedures (data from Rüegger *et al.* and Nashimoto *et al.*, 1993

1. On its own as an absolute value, remembering that the target of zero DCS means a k value of 0.
2. Relatively, by comparing the k value for one project or set of results against the equivalent value on another project or set.
3. Again relatively, but comparing the k value for one project or set of results against the 'probable minimum' value for the equivalent pressure, i.e. the lower bound value k_{LB}. This is perhaps a more pragmatic comparison than the first two, as it is comparing against what, overall, was the best that had been achieved rather than the ideal of no DCS.

All the Figures presented so far here have the purpose of allowing comparisons to be made on an equal basis, but it would be wrong to use those of any specific project to judge whether its DCS performance was better or worse than another project or the overall average, for that would require much more detailed exposure information than is available to the author. At the very least it would have to be as detailed as that compiled by Evans from the DCS Central Registry records.

Nevertheless, as comparable information is available for the UK projects carried out either under the 1958 regulations or under the procedures associated with the Blackpool Tables and CIRIA Report 44, this comparison is worth making. Specifically, the cumulative exposure data under the two regimes are compared, i.e. those in Figures 2(a) and (b), the normal decompressions. Again it is important to remember that it is the projects which are being compared not the decompression tables. All sorts of effects are hidden in the comparison, including such variables as: the quality and effectiveness of site and medical management, the selection of workers, and the degree to which DCS was reported by the workmen; not least is the implicit assumption that each exposure and each resulting case of DCS is independent of all the others.

The ratio k/k_{LB} has been calculated for each of the plotted points in Figures 2 (a) and (b), where k defines the curve of which the particular point at pressure, p, lies. Thus using this ratio we can compare the overall average DCS incidence for the 1958 and Blackpool Tables projects. Values of this ratio are given in Table 1, using the relevant pressure time intervals. A value close to 1 means the DCS incidence was at the practical minimum.

Table 1. Relative DCS performance of UK projects under the 1958 and Blackpool Tables

Working times		**Values of k/k_{LB} for pressures (psig)**					
(h)		**14-18**	**18-22**	**22-26**	**26-30**	**30-34**	**34-38**
0-2	1958	11.8	2.7	12.4	11.1	12.6	-
	Blpl	3.8	5.4	17.6	4.6	5.5	2.7
2-4	1958	6.7	4.0	7.7	3.5	7.6	-
	Blpl	0.3	4.0	9.0	4.1	4.4	0.6
4-6	1958	2.6	4.5	9.8	11.8	10.0	-
	Blpl	1.8	3.9	11.3	10.2	5.6	3.2
6-8	1958	11.8	6.9	9.6	9.9	9.0	-
	Blpl	4.3	9.2	13.3	15.4	4.1	5.8
8+	1958	15.9	12.2	18.9	19.0	15.7	-
	Blpl	5.2	7.2	11.1	9.7	2.8	1.0

Inspection of Table 1 shows:

1. 60% of the Blackpool values of k/k_{LB} were less than 6, but only 20% of the 1958 Tables project were. 44% of the 1958 values were greater than 11 compared with 16% for the Blackpool values.
2. In 16 of the 25 pressure/time ranges, the Blackpool values are less than or equal to the 1958 ones.
3. In the 18 to 30 psig range many of the Blackpool values are considerably higher than the 1958 ones.

While these results show considerable relative reductions in DCS on the Blackpool Tables projects for many of the exposure times, there is scope for improvement. Using an arbitrary criterion of k/k_{LB} = 6, the pressure/working time ranges where the Blackpool Tables projects were less successful are given in Table 2. In most of these cases, the 1958 projects were relatively better than the Blackpool ones. It is suggested that there is scope for reviewing the decompression procedures for these pressure/time ranges.

Table 2. Working exposures where Blackpool Tables have been less successful

Pressure range (psig)	**Working times (h)**
18 to 22	6-8, 8+
22 to 26	All
26 to 30	4-6, 6-8, 8+

5 Predicting and monitoring DCS incidence

The representation of empirical data from past projects in the way explained in Section 2.1 (in particular by the envelope between the probable maximum and probable minimum bounds) can be used to predict the expected range of DCS. It requires an estimate of the likely pressures to be used and the exposures (working times and numbers of decompressions). These are all matters considered when making up a tender for a project and can be refined in the contract planning stage. Thus, this prediction could form part of the risk assessment that will be required for the health and safety management of the work. Even though the possible range predicted for DCS incidence is wide (about one order of magnitude between probable maximum and minimum values), the method has three advantages over simple bends-rate predictions:

1. It sets a target for what is reasonably practicable.
2. It can be used to allow for the effects of project size (number of exposures) and for acclimatisation.
3. It provides a model against which to monitor the actual occurrence of DCS, either on a shift-by-shift basis or cumulatively during the course of the project.

6 Concluding remarks

1. Comparisons of the incidence of DCS on different projects or under different procedures should be made on a similar basis.
2. The normalised parameter, DCS treatments per standard exposure hour, expressed as $B/tE^{0.5}$, for specific sets of exposures (p, t, E) with resulting DCS treatments, B, is a possible basis of comparison.
3. This approach can be used for estimating the likely amount of DCS that would occur on a future project, e.g. for a risk assessment at the planning stage. It

could also be used as a method of monitoring the DCS incidence during the work. The approach, while based on empiricism, is more rational than estimating from the bends rates observed on past projects.

4. Examination of DCS reported on overseas projects using this approach shows that the incidence of DCS is much the same for workers of different races and for different decompression schedules.
5. Comparison of the UK data suggests that while the Blackpool Tables projects were more successful than those under the 1958 Tables for many exposure sets, there is scope for considerable improvement. In particular the decompression procedures should be reviewed for the 18-30 psig pressure ranges.

References

Anderson, J.M. (1994) An engineer's perspective on legislative changes. In: *Engineering and health in compressed air work* (Jardine, F.M. and McCallum, R.I. eds.), E. & F.N. Spon, London

Campbell, J.G. and Gutteridge, D.R. (1994) Compressed air tunnelling for the Hong Kong Mass Transit Railway. In: *Engineering and health in compressed air work* (Jardine, F.M. and McCallum, R.I. eds.), E. & F.N. Spon, London

Court, D.J. (1994) The use of compressed air for shaft sinking and tunnelling in the Greater Cairo Wastewater Project, Egypt, Contracts 3, 12 and 12/5A. In: *Engineering and health in compressed air work* (Jardine, F.M. and McCallum, R.I. eds.), E. & F.N. Spon, London

Evans, A. (1990) Shift lengths for compressed air work. *Proc. EUBS, 1990.* Jt. Mtg. on Diving and Hyperbaric Medicine, Amsterdam, Aug. (Sterk, W. and Geeraedts, L. eds.). Foundation for Hyperbaric Medicine, Univ. of Amsterdam, Netherlands, pp 299-306

Evans, A. (1992) Further analysis of the Decompression Sickness Registry data. In: *Results of the analysis of compressed air records in the Decompression Sickness Central Registry.* (Jardine, F.M. ed.), Project Report 6, CIRIA, London, pp 9-35

How, J. and Wong, T.N. (1994) Medical aspects of compressed air tunnelling – the Singapore experience. In: *Engineering and health in compressed air work* (Jardine, F.M. and McCallum, R.I. eds.), E. & F.N. Spon, London

Jardine, F.M., Evans, A. and McCaig, R.H. (1991) Studies of the incidence of decompression sickness on UK compressed air projects, *Proc. Tunnelling '91*, Institution of Mining and Metallurgy, London

Mano, Y., Shibayama, M., Yamami, N., Nakayama, T. and Kawashima, M. (1994) Statistical investigation of working pressure and DCS for compressed air workers. In: *Engineering and health in compressed air work* (Jardine, F.M. and McCallum, R.I. eds.), E. & F.N. Spon, London

Nashimoto, I., Gotoh, Y., Kobayashi, K., Nodera, M., Araki, R. and Eda, F. (1994) Analysis of decompression sickness and bubble appearance rates on a caisson project – effects of working shift pattern and post-decompression oxygen breathing. In: *Engineering and health in compressed air work* (Jardine, F.M. and McCallum, R.I. eds.), E. & F.N. Spon, London

Rüegger, M., Buhlmann, A.A. and Völlm, E. (1994) Decompression in tunnel construction using the hydroshield process: experience in Switzerland with decompression tables adapted for construction sites with the ZH-L 16 model. In: *Engineering and health in compressed air work* (Jardine, F.M. and McCallum, R.I. eds.), E. & F.N. Spon, London

Part Nine
Developing Technologies

37 Decompression options for tunnel work at more than 3 bars (gauge) pressure

R.W. Hamilton*, J. Vollbrandt, M. Heineke and G. Luther
**Hamilton Research Ltd., Tarrytown, USA*
GKSS Forschungszentrum Geesthacht, Germany

This paper is dedicated to the memory of the late Albert R Behnke to whom the word's compressed air workers owe a great deal

Abstract
The most cost effective way to prepare for compressed air work at greater than 3 bars (gauge) pressure is to plan the decompression procedures to suit the operational situation rather than to attempt to fit a tunnelling operational or engineering plan to current diving tables. Special decompression procedures are needed when work is required at pressures greater than this. We have compared a variety of decompression approaches to this task using the *DCAP Plus* computer program; most of them take advantage of the benefits of oxygen for improved decompression, but also consider that the control of oxygen exposure is a major planning factor. Work in air with air decompression is inefficient, but is not limited by oxygen exposure. The most effective approach and one on which there is some experience is work in air with oxygen breathing during decompression. Even more effective decompressions are possible with oxygen-enriched air, but this becomes limiting due to oxygen exposure when used for the more stressful exposures, and requires breathing of prepared gas mixtures. For jobs requiring many man-days of work at high pressures the benefits of saturation techniques may offset the cost of the equipment. Saturation-excursion techniques can improve safety and make the final decompression shorter. An important aspect of using innovative decompression techniques is the requirement to validate their reliability; a new set of consensus guidelines makes it feasible to introduce into the workplace new techniques based on documented experience with little or no laboratory validation. It will be less efficient hence more expensive to work at higher pressures, but this can be done with no increase in risk, and for many situations will be the most cost effective approach.
Keywords: Decompression, Decompression tables, Oxygen, Oxygen tolerance, Tunnel work, Compressed air work.

Engineering and Health in Compressed Air Work Edited by F.M. Jardine and R.I. McCallum. Published in 1994 by E & FN Spon. ISBN 0 419 18460 0

1 Introduction

When the opportunity presents itself for advanced engineering and construction work well below the water table, a major consideration in project planning is whether or not workers will have to be exposed to pressure. When the projected pressure exposure is greater than that of about 3 bars (gauge) the concerns include that of how to do the decompressions. Both lore and past experience would suggest that risks of decompression sickness, DCS, become greater as pressure exceeds about 2 bars using currently legislated 'tables', and beyond 3 bars there are few existing compressed air tables or procedures for performing the decompressions, whether prescribed by regulations or not.

Considerable experience from various modes of diving, aerospace, and medical activities, suggest that trouble-free decompressions from pressurised construction operations in the range of 3 bars and beyond are not only possible, they may very well be the most cost-effective way of performing the work. We have performed an analysis of how some techniques familiar in other fields but relatively new to compressed air work might be brought to bear to improve the reliability of tunnel decompression procedures. Some established decompression techniques, many based on creative use of oxygen and oxygen-enriched mixtures, might be adapted for tunnel work. Methods compared include:

- Work in air with air decompression.
- Work in air with air and oxygen decompression.
- Work in air with an intermediate oxygen rich mixture ('enriched air') and oxygen for decompression.
- Work breathing oxygen-enriched air by mask, with or without oxygen during decompression.
- Saturation and saturation-excursion techniques.
- Use of rebreather apparatus.

In order to use new methods several problems may need to be solved. Among these are those of acquiring suitable decompression tables, securing acceptance of the tables in the particular jurisdiction, and managing the use of oxygen. And some provision has to be made to accommodate the legal and ethical aspects involved with introducing new decompression procedures.

Our approach here has been to 'model' these options for comparison. This has been done by calculating a series of comparative profiles with a common set of conditions. We chose an arbitrary 3.6 bars (gauge) as pressure for the worksite, and have selected work periods (called 'bottom times' in diving terms) so that decompression can, where appropriate, be completed in an 8-hour work day. We have tried to deal with both decompression and oxygen exposure.

Many tunnel decompression procedures consider that the duration of the decompression cannot exceed the work time, because of utilisation of the chambers/ locks. For the extra deep jobs different schedules may be necessary to accommodate the need for longer times in decompression, and this analysis is not limited to those situations where decompression time is less than work time. One straightforward method of dealing with this is to provide additional decompression chambers or locks.

Other considerations should not be forgotten, although they are not specifically addressed here. A major issue is fire safety, both in handling oxygen in containers and pipes as a gas, and the possibility of increasing the oxygen level in the compressed air atmosphere. Others are the possible use of mask breathing while working, logistics of mixing, handling, and administering breathing mixtures, the need for special training, and the possible use of 'decanting', a procedure of transferring the worker from the work lock to surface pressure and into another lock for decompression. These are not considered here, but not forgotten either.

2 Methods

2.1 Decompression computations

Many methods are available for computing decompression schedules or tables, but virtually everything that works has an empirical basis and can be traced to laboratory and field experience with decompression. One of the accepted and widely used methods, a variation of the Haldane-Workman-Schreiner computational algorithm (Schreiner and Kelley, 1971), is available to us by means of the *DCAP Plus* computer program ('Decompression Computation and Analysis Program', Hamilton and Kenyon, 1990). Using DCAP as a tool, we have examined a variety of approaches based on the use of oxygen and air-oxygen mixes, specifically to establish the comparative merits of the different approaches. For ascent limits we used an algorithm developed for deep air dives followed by air decompression that has been successful in practice; the 'matrix' of ascent-limiting M-values which stages the profiles is designated MM11F6 (Hamilton, Muren *et al.*, 1988). The procedures calculated with this method are quite conservative in relation to many other computational methods, but the samples presented are not necessarily ready for use. They are intended for comparison only; actual procedures for a particular worksite might vary in either direction.

We considered two types of profile, the maximum time that can be worked in a normal 8- or 9-hour day, and the 'intervention' profile that will be done only occasionally. For the normal work periods we chose the longest time that could be worked and still allow decompression to be completed before the end of the day. For the intervention excursions we arbitrarily chose a 2-hour work period.

When planning decompression from long-duration daily exposures to high pressures of nitrogen, it is necessary to consider the 'multiday' effects. 'Gas loading', a significant but quite hypothetical component of the computation, is found to build up over several days, and this may increase the calculated decompression time. For this reason all the full day calculations reported were done as if carried out over a 5-day work week. Exposure to oxygen was considered on a multiday basis as well.

We 'staged' the decompressions at 0.3 bar intervals, after the pattern used in diving. The results would be essentially the same using linear depressurisation, which would be more in keeping with tunnel practice; operational advantages may result from choosing one method or the other, but there is not much difference in the physiology. The practice of 'decanting', or decompressing to surface pressure followed by repressurisation and execution of a planned decompression profile is not considered here.

2.2 Validation of decompression procedures

An attractive new approach is now offered for dealing with the ethical questions of introducing new decompression tables and procedures without extensive (and expensive!) laboratory testing. A 1987 workshop sponsored by NOAA, U.S. Department of Commerce, brought together many of the leading workers in the field of decompression physiology, medicine, and operations along with administrators, lawyers, the clergy, military leaders, and others, to address the matter of validation of decompression tables (Schreiner and Hamilton, 1988). This workshop came up with guidelines for the validation process, including ways of dealing with such things as 'interpolative' modifications – those within the realm of valid experience – and conservative changes. Many recommendations are what common sense or current validation practice would dictate, and much of what was recommended was recognised as already being current industry practice.

The Validation Workshop protocol calls for documenting a data base for new procedures such that 'new' practices are not extrapolations but rather are interpolations of known experience. Really new ideas call for dedicated laboratory research, done with 'informed' subjects, and under medical control. Interpolative advances are within the job description of the workers, but should be done with special care and training, documentation, feedback, medical backup, and with competent judgement where indicated. A mechanism for making the judgmental decisions (such as when enough tests have been performed, or when to make a change in a procedure) is specified. Since many new tunnel procedures would be conservative compared to the existing ones, it is likely that considerable progress can be made and justified without extensive laboratory testing. Further, in tunnel work the risks of something going wrong leading to a more serious event such as drowning are not much of a factor; most conditions specified by the Workshop are met during any well managed compressed air job. There is now a recognised rationale for making these improvements.

2.3 Oxygen exposure

Virtually all innovations that improve decompression are based on creative use of oxygen. A major aspect of the methods discussed here is tolerance of oxygen exposure and avoidance of oxygen toxicity. This has to be considered as an integral part of any new oxygen-based decompression plan.

Oxygen exposure in these calculations was handled generally according to the Repex algorithm (Hamilton, 1989; Hamilton, Kenyon, and Peterson, 1988). This method uses an empirically determined set of limits based on a the number of days of exposure. Exposures are calculated in Oxygen Tolerance Units; one unit is about the equivalent of an exposure to a partial pressure of oxygen (PO_2) of 1 bar for 1 min (OTU is derived from and equivalent to the earlier UPTD, Unit Pulmonary Toxicity Dose). The breathing gases and other factors were adjusted to keep the exposure within the Repex limits. For 5-day work weeks on a continuous basis Repex allows a maximum of about 420 OTUs per day. This is based on 300 units per day, with 5 days of work and 2 days off each week ((300*7)/5=420). Operational experience in the same configuration has shown this to be acceptable, but for jobs of more than a few weeks of continuous exposure a slightly lower daily dose is recommended. In a long exposure adaptive changes of the same nature as those of acclimatising to life at high altitude will occur, but the acclimatisation will be in response to an excess of oxygen rather than a deficiency. Incidentally, if in a real life situation a continuous daily exposure level is set slightly too high it may be noticed after some time by a few of the workers; it can be adjusted downward at that time without having to make other changes and without any lasting injury or other risk.

2.4 Saturation and saturation-excursion

The term 'saturation' applies to the method highly exploited in diving of maintaining the workers under pressure around the clock for a number of days, then performing a final decompression at the end of the time. This has long been advocated as a possible method for some types of tunnel work, but to our knowledge it has not been implemented. Living full time at work for a couple of weeks is not an uncommon practice, since it is the pattern for ships' crews and offshore platforms, for example. One difference between working at sea and at 3.6 bars, however, is that at the higher pressure with air as the breathing gas the working time is limited by oxygen exposure and only a few hours can be worked each day.

We investigated two saturation options, one of having the living chamber at the pressure of the worksite, here 3.6 bars (gauge) pressure, and the other to have the living chamber at the shallowest pressure from which the workers could make pressure excursions to the worksite without need for decompression stops on return to the living chamber.

For the saturation exposures the time spent breathing air at the hypothetical 3.6-bar worksite is the limiting factor, due to oxygen exposure. For this we assumed the workers would be working every day for at least 10 days and would not take weekends off. The daily exposure should then be about 300 units instead of the 420 used for a 5-day week, and allows about 320 min of work each day; this limit applies no matter what saturation decompression pattern is used.

We chose 2.1 bars as the lowest pressure from which a 320 min excursion to 3.6 bars can be made each day, with no decompression stops required on return to the living chamber. Another set of procedures from CIRIA's Underwater Engineering Group, UK, gives much the same results (Hennessy *et al.*, 1985). For the excursion times and final decompressions we used the Repex procedures for nitrox saturation (Hamilton, Kenyon, and Peterson, 1988).

3 Results

The results of the calculations are given in Table 1.

3.1 Work in air with air decompression

This is the traditional method of performing compressed air work. Using air throughout is the least efficient method as far as working time per day is concerned, but it has the benefit of being familiar, requiring no special equipment or gas handling. At 3.6 bars it requires more than twice as much decompression time as the work time, and even then allows only a little over two hours of work time in a day. If more than one transfer lock (pressure chamber) is available so that alternate shifts can be run then this might become feasible. This option is the only one at 3.6 bars that is not oxygen limited.

3.2 Work in air with air and oxygen decompression

This is the easiest oxygen modification to implement, and gives the most time for the effort. Here the decompression time is about 50% greater than the working time, with a useful working time of greater than 3 hours per day. In practice each worker, in the lock, breathes oxygen by mask for part of the decompression period. For the run listed here oxygen breathing is started at a pressure of 0.9 bar during decompression, and is breathed in continuous cycles of 30 minutes on oxygen followed by 10 minutes off, breathing air from the chamber during the 'off' periods. We also tried oxygen starting at 1.2 bars; it reduced the decompression time but made the oxygen dose even higher. The oxygen breathing and decompression pattern can usually be optimised with respect to decompression time and oxygen exposure.

3.3 Work in air with an intermediate oxygen-rich mixture and oxygen for decompression

This procedure calls for increasing the oxygen earlier in the decompression to reduce gas uptake at the higher pressure stops by breathing an 'enriched air' mixture (sometimes called 'nitrox'). The final oxygen breathing pattern used here was in cycles of 30 min on O_2, 30 min on air. It would require mask breathing earlier in the decompression (since higher levels of oxygen cannot be used in a chamber for fire safety reasons) and would in this case gain very little over just oxygen in the later stages of decompression.

3.4 Work breathing enriched air by mask, with oxygen during decompression

This method offers considerable advantage in decompression time for a short (here 2 h as an example) work period, but is not of much value for daily work because the exposures are limited by oxygen exposure. For short interventions, however, especially where the atmosphere might be contaminated and breathing by mask is required, this method should be considered. The techniques for mixing the enriched air 'EANx' or 'nitrox' mixtures are well established (see for example, Bøe and Hartung, 1983; Mastro and Butler, 1990; Wells, 1989).

Table 1. Calculations for various methods of working at 3.6 bars. All options were started at 0800. DCAP calculations used matrix 11F6 without additional conservatism factors. Work times (min) and gas patterns were adjusted to limit O_2 exposure to about 420 units for 5 runs per week and 300 OTUs for continuous work in saturation, and to make surfacing occur between 16:00 and 17:00. Pairs of values show effect of small changes. Daily runs were taken from the maximum decompression time reached during a 5-day week, calculated continuously. EANx = enriched air, here 34% O_2. Work times for all options except the basic air are oxygen limited.

Type	Work time (min)	Decom. time (min)	Surface (clock) time	OTU
Air, air decompression	120	324	15:24	116
	130	354	16:04	126
	140	399	16:59	136
Air, O_2 decompression	190	309	16:19	402
	200	314	16:34	432
Air, O_2 & EANx decompression	200	299	16:19	416
EANx, O_2 decompression	120	79	11:19	344
Rebreather, 1.4 bar PO_2	120	74	11:14	326
Saturation at 3.6 bar	320	*0	13:20	300
Saturation at 2.1 bar + excursions	320	**0	13:20	300

* decompression at end of saturation = 94 h
** decompression at end of saturation = 46 h

3.5 Use of constant PO_2 rebreather apparatus.
This method considers using devices that supply a constant PO_2 (partial pressure of oxygen) by mask, with a backpack that recycles the gas and maintains the oxygen level. Here this method presents essentially the same decompression picture as breathing enriched air during work; in this specific situation it offers no decompression advantage over the use of enriched air nitrox, but for work at higher pressures it will. Rebreathers may be attractive for interventions. Rebreathers may offer operational advantages because they can reduce the complexity of having to mix gases and supply them by hose, but the complexity of the apparatus may offset this. A rebreather for tunnel work need not have the reliability or redundancy of one used underwater, but it would surely have to be sturdy and simple.

3.6 Saturation at 3.6 bars

Saturation will become more attractive as working depths increase for situations requiring many hours of work each day. The workers can breathe air, and the limitation of daily work duration is oxygen exposure. These decompression techniques are well established and, although time consuming and slow, can be made trouble free. It is not likely to be most effective to have the workers live at the full working pressure, unless the job layout dictates. Having the workers live at working pressure imposes a long final decompression, but no more work time is achieved because the limitation is daily oxygen exposure.

3.7 Saturation-excursion techniques; saturation at 2.1 bars

This is the saturation technique that should be most useful. The 'storage pressure' of the living chamber is chosen so as to require no decompression stops after a work period. Having it at a lower pressure reduces the fire risk, makes it easier to maintain the atmosphere, offers lower risk due to pressure anomalies, and makes the final decompression much shorter. The final decompression does have to take the excursions into account, and both Repex and UEG offer methods for that. The final decompression is too long to consider doing it every weekend.

4 Discussion

In due course the options chosen for further investigation will have to be compared to available tunnel and caisson experience in what does and does not work well (see, for example, Kindwall *et al.*, 1983).

Oxygen breathing has, of course, been used for decompression in diving with great success, and has been used for some compressed air work as well. A notable example of that is the Kiel Supply Tunnel in 1989 (Faesecke, 1990). There the medical authority for the state implemented decompression procedures based on oxygen breathing that met the criteria of the UHMS Validation Workshop, although without that specific intent at the time.

Another possible approach to saturation should be mentioned. Since the daily work time is oxygen-limited and cannot be increased except by providing a low-oxygen breathing mixture, the extra time available each day might be used to decompress back to a living chamber located at a lower pressure than the 2.1 bars used in the example. If the living chamber were at less than 1.4 bars (gauge) then it could be filled with air and the atmosphere controlled by flushing with air instead of controlling oxygen and scrubbing CO_2 as is normally done (1.4 bars gauge = 2.4 bars absolute, which with the volume fraction of oxygen in air of 0.21 gives a PO_2 of 0.5 bar, the threshold for O_2 effects). Special decompression procedures could be generated for this option.

For deeper work than the 3.6 bars used in this example it may be necessary to consider the use of helium mixtures. We have looked at this in detail for work at 80 msw, and satisfactory intervention techniques are available (Luther *et al.*, this conference).

Our purpose here is to point out that manned work and intervention is entirely feasible for tunnel work in the range of 3 bars and beyond. It requires attention to physiology for effectiveness and safety. It will be more expensive to implement than for lower pressures, but it may nevertheless offer the most cost-effective approach. Although this paper focuses primarily on the range of pressures higher than 3 bars, some techniques such as those explored here might also be applied to the higher end of the 'legislated' range of less than 3 bars.

5 References

Bøe, I. and Hartung, K. (1983) Employment of the Polycom 101 gas mixing unit for divers in a major project in Norway. *Dräger Review*, 51, pp26-8

Faesecke, K-P. (1990) Diggin' deeper: compressed air work beyond 30 meters. *Undersea Biomed. Res*, 17(Suppl), 145

Hamilton, R.W. (1989) Tolerating exposure to high oxygen levels: Repex and other methods. *Marine Tech. Soc. J.*, **23**(4), 19-25

Hamilton, R.W. and Kenyon, D.J. (1990) DCAP Plus: New concepts in decompression table research. In: *MTS 90: Science and technology for a new ocean's decade*. Vol. 3. Marine Technology Society, Washington

Hamilton, R.W., Kenyon, D.J. and Peterson, R.E. (1988) *Repex habitat diving procedures: Repetitive vertical excursions, oxygen limits, and surfacing techniques*. Technical Report 88-1B. NOAA office of Undersea Research, Rockville, MD

Hamilton, R.W., Muren, A., Röckert, H. and Örnhagen, H. (1988) Proposed new Swedish air decompression tables. In: *XVth annual meeting of the EUBS: European Undersea Biomedical Society* (ed. T.G. Shields), National Hyperbaric Center, Aberdeen

Hennessy, T.R., Hanson, R.G., Hempleman, H.V. and Shields, T.G. (1985) *Tables for saturation and excursion diving on nitrogen-oxygen mixtures*. UEG UR31. Underwater Engineering Group, CIRIA, London

Kindwall, E.P., Edel, P.O. and Melton, H.E. (1983) *Safe decompression schedules for caisson workers*. Final Report to NIOSH under Grant #5R-01-OH-00947-03. St. Luke's Hospital, Milwaukee

Luther, G., Hamilton, R.W., Nietsch, G.A. and Vollbrandt J. (1994) Procedures for safe work at high pressures in a TBM chamber using special breathing mixtures. In: *Engineering and health in compressed air work* (eds. Jardine, F.M. and McCallum, R.I.), E. & F.N. Spon, London

Mastro, S.J. and Butler G.J. (1990) Air quality requirements for the on-site production of oxygen enriched breathing mixtures. In: *Oceans 90*. Marine Technology Soc., Washington, pp 731-737

Schreiner, H.R. and Hamilton, R.W., eds. (1989) *Validation of decompression tables*. 74(VAL)12-31-87. Undersea Hyperbaric Medical Soc., Bethesda, MD

Schreiner, H.R. and Kelley P.L. (1971) A pragmatic view of decompression. In: *Underwater Physiology IV*, (ed. C.J. Lambertsen), Academic Press, New York

Wells, J.M. (1989) Nitrox diving within NOAA: History, applications and future. In: *Harbor Branch workshop on enriched air nitrox diving*, (eds. R.W. Hamilton, A.W. Hulbert and D.J. Crosson). NOAA Undersea Research Program Technical Report 89-1. NOAA National Undersea Research Program, Rockville, MD

38 Remotely controlled caisson method and its maintenance

T. Hirata, K. Takano, Y. Gotoh*, I. Nashimoto* and W. Sterk**
Study Group for Deep Caisson Work, Tokyo, Japan
**Department of Hygiene, Saitama Medical School, Saitama, Japan*
***DADCODAT, Zuidwolde, The Netherlands*

Abstract
A remotely controlled hyperbaric construction machine system has been developed for caisson work, which avoids the risks of decompression sickness (DCS) and nitrogen narcosis at great depth. The machine can semi-automatically excavate the bottom of a caisson and remove muck. The system consists of an excavator (movable shovel on the caisson ceiling), equipment for loading and dumping muck, TV cameras, and a controller located outside the caisson. The caisson is sunk after the removal of muck from under the cutting edge. Most of the routine work can be done by this newly developed machine, but in the cases of repair and maintenance, technicians must enter the caisson. In such cases and if the working pressure is more than 4 bar, the technicians will breathe a nitrogen-helium-oxygen mixture (trimix). Breathing trimix requires a special or modified decompression procedure, which is now being developed.
Keywords: Caisson Method, Remote Control, Robot, Great Depth, Trimix.

1 Introduction

A pneumatic caisson method is a construction technique by which a caisson block is lowered to a predetermined depth, by excavation of the bottom ground enclosed by a pressurised working room beneath the block, preventing water from entering the work area. Utilising this method, the construction of bridge foundations and shafts can reliably be carried out in a site where underground water is abundant. Therefore, it has often been used in soft ground delta areas in Japan. The pneumatic caisson method, however, carries a potential risk for the workers exposed to the hyperbaric environment. In the last decade construction machinery has been introduced for excavation instead of relying on manpower, to reduce the risk as much as possible. As a result, the number of workers who must enter the caisson has decreased, and even if they have to work under hyperbaric conditions, little manpower is required. As a consequence, the number of victims suffering from decompression sickness has been greatly reduced and serious decompression sickness

Engineering and Health in Compressed Air Work Edited by F.M. Jardine and R.I. McCallum. Published in 1994 by E & FN Spon. ISBN 0 419 18460 0

(Type II DCS) has scarcely been reported.

To eliminate the health hazards related to pressurised caisson work completely and to overcome the increasing lack of skilled workers, a remotely operated caisson construction system which excavates and conveys removed waste without manpower has been developed in the last couple of years. This system has already been tested under operational conditions. Although the remotely controlled caisson method makes caisson workers superfluous, technicians have still to descend for repairs and routine maintenance. If the pressure inside the caisson were in the range up to 4 bar, the conventional compression and decompression procedures that are legally regulated in Japan could be applied. At greater pressures, however, this would be impossible, mainly due to the risk of nitrogen narcosis.

As the depth of the caisson increases, the ambient pressure increases. Theoretically, it is possible to set a caisson block deeper than 50 m when the remotely controlled caisson system machinery is operated properly. In such a case, when technicians have to enter the caisson for repair or maintenance of machinery, various problems related to hyperbaric air exposure such as nitrogen narcosis will hinder work performance.

Recently, a new plan has been introduced, known in the Japanese mass media as the Geofront Plan, to develop below ground to great depth for more efficient land use. This is extremely important in a metropolitan area like Tokyo, where the population has rapidly become very concentrated. Because of such rapid urban growth, the remotely controlled compressed air caisson method has received increasing attention, particularly for the purpose of firm and accurate construction of foundations.

To make the remotely controlled caisson method possible for such great depths, we have developed a trimix breathing gas system to replace the currently used compressed air. Its practicality has been established by testing under operational conditions.

2 The remotely controlled caisson method

A diagram of the remotely controlled caisson method is shown in Figure 1. The main work process inside a compressed caisson block is excavation and removal of waste. To make it an unmanned operation, appropriate technology and a system that remotely operates the process are essential.

In the excavation system of the remotely controlled caisson method, an excavator equipped with a shovel that is hung from the caisson ceiling and movable on a rail is remotely controlled from a central control room located at the surface. An operator in the control room can precisely manipulate the excavator and load excavated waste into a bucket by watching the work site on a monitor display through a TV camera installed on the excavator, and referring to a computer-processed graphic display that depicts location, direction, arm length, rotation and vertical swing angles of the machine. This operation can be performed semi-automatically.

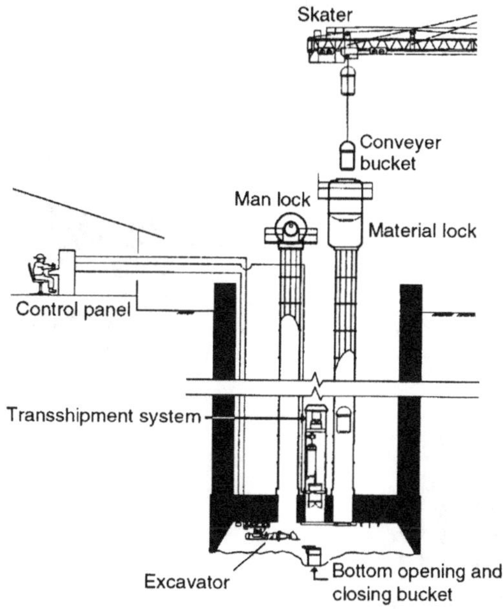

Fig. 1. Diagram of the remotely controlled caisson method

Conveyance of the waste from the work site to an unloading site outside the caisson is automatically carried out by a bucket conveyer system. To minimise the loss of operational efficiency of the excavator during the transportation period of the loaded bucket, a waste transhipment system was developed as shown in Figure 2. (See also the photographs at the end of the paper.)

While the excavator is loading waste into a bottom opening and closing bucket, the conveyer bucket is moving through a material shaft. When the unloaded conveyer bucket moves down, the loaded bottom opening and closing bucket is lifted into the transhipment system. As soon as the conveyer bucket is moved underneath the lifted bucket by a platform car, the lifted bucket opens the bottom and dumps the waste into the empty conveyer bucket. While the platform car carries the loaded conveyer bucket back to the material shaft, the empty bucket closes the bottom and is lowered down to the work site and the excavation and loading process is resumed. With the assistance of the transhipment system, almost twice as much waste can be conveyed outside of the caisson in a given time.

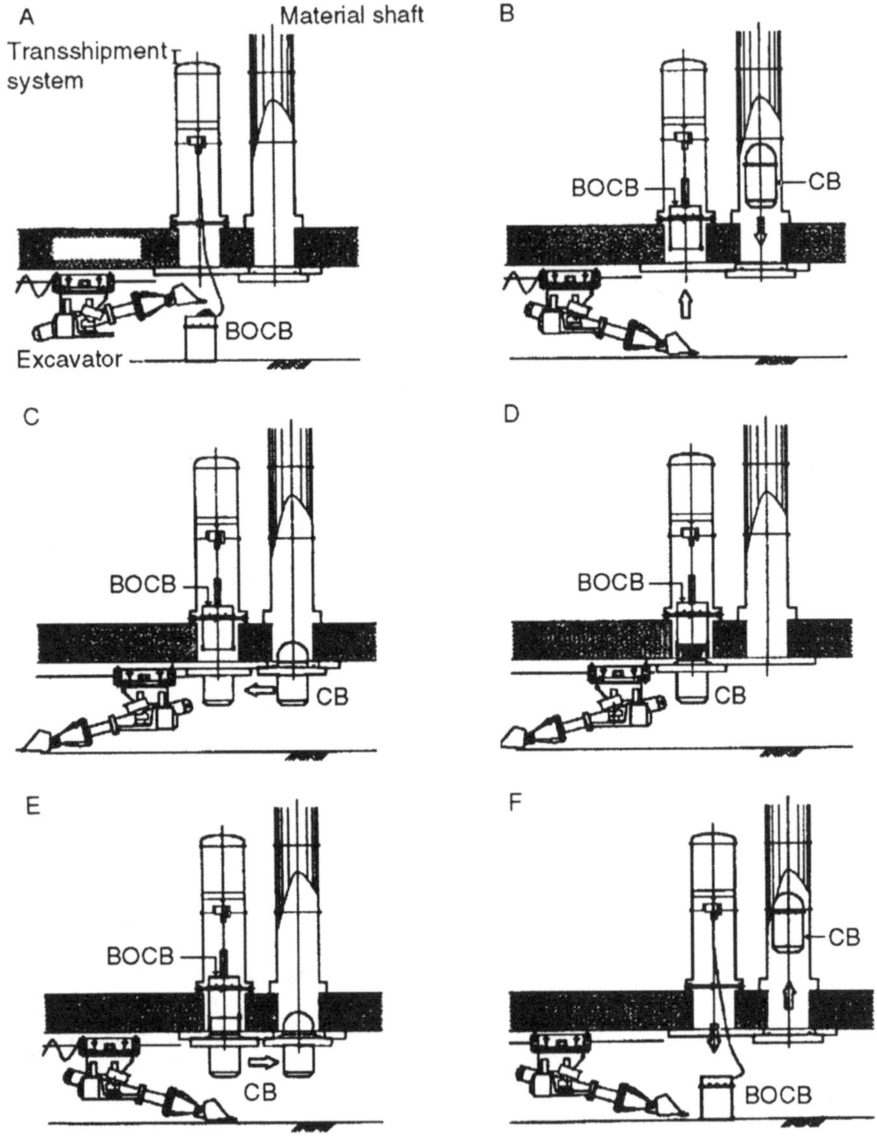

Fig.2. Procedures of waste transhipment system. **A:** Excavator loads waste into bottom opening and closing bucket (BOCB). **B:** BOCB goes up and conveyer bucket (CB) comes down. **C:** Platform car moves CB under BOCB. **D:** Transshipment of waste from BOCB into CB. **E:** Platform car moves CB to material shaft. **F:** CB moves up and BOCB is lowered to be filled again

3 Inspection and maintenance of the caisson system

When the excavator or conveyer system in the caisson has malfunctioned, technicians must enter the caisson to correct the failure. Regular inspection and maintenance are also required to prevent machine failures and make machinery function as smooth as possible. If the pressure inside the caisson is not too high, manned compressed air work can be utilised using a conventional man lock and shaft system, as depicted on the left in Figure 1. However, when the ambient pressure exceeds about 5 bar, almost all workers would suffer from nitrogen narcosis if breathing compressed air. Since workers intoxicated by high nitrogen partial pressure tend to become careless and display poor judgement, they are not only incapable of carrying out their tasks, but they may also be a risk to themselves and the work. To avoid nitrogen narcosis, breathing mixtures which contain a lower percentage of nitrogen should be used instead of compressed air, of which about 80% is nitrogen. To this end, a proper mixture of nitrogen, helium and oxygen (trimix) seems more attractive than a mixture of helium and oxygen (heliox) for various reasons.

4 An entrance and exit method for a caisson with a mixed gas breathing system

Since it is technically difficult and economically unrealistic to fill the inside of a caisson with a mixed gas completely, it is more practical for technicians who enter the caisson to breathe a mixed gas while the caisson is filled with compressed air.

Figure 3 depicts the basic plan we have proposed for an entrance and exit method of a caisson with a mixed breathing gas system. Technicians will be compressed with air while breathing a mixed gas through a demand-type breathing system built in a man lock. When the inside of the man lock is equalised with the worksite ambient pressure, the technicians descend to the work site through a bottom hatch and perform required tasks while breathing the mixed gas through an umbilical hose connected to one of the built-in regulators inside the man lock. After their tasks are completed, they return to the man lock and are decompressed to the surface. In the course of decompression, they switch to ambient air breathing when the pressure is lowered to a certain level. To shorten decompression time, oxygen breathing cycles begin at a given pressure as planned in the decompression schedule. While breathing oxygen through a built-in-breathing-system (BIBS), exhaled gas is exhausted using an overboard dump method to prevent fire hazard.

5 Selection of mixed breathing gas, compression and decompression procedures

For caisson work at greater pressures than approximately 4 to 5 bar, breathing gas mixtures containing less nitrogen than air are obligatory to prevent nitrogen narcosis. In this respect, a mixture of helium and oxygen (heliox) can be used, since helium has no narcotic properties in the practical depth range. However, helium is

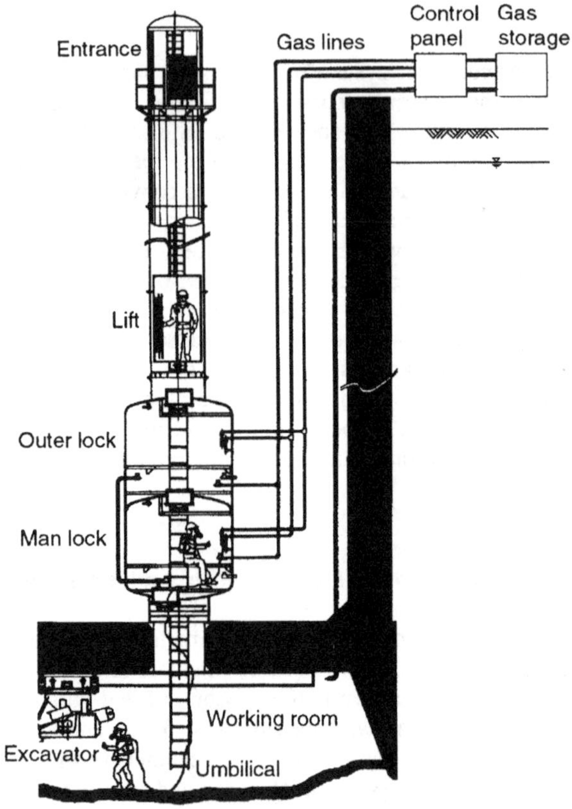

Fig.3. Entrance and exit method for a caisson with a mixed gas breathing system.

expensive, gives considerable voice distortion and has a high heat conductivity. Therefore, a mixture of nitrogen, helium and oxygen (trimix) seems preferable. In this mixture, the nitrogen partial pressure should be lower than 3.2 bar (air at 3 bar gauge pressure, or approximately 4 ATA) to avoid nitrogen narcosis, while the oxygen partial pressure should be lower than 1.6 bar to prevent central nervous system oxygen toxicity. For a given ambient pressure greater than 4.8 bar absolute, the remainder of the mixture should then be helium.

We adopted Sterk's method for decompression procedures in which trimix gas is used. Sterk has developed a neo-Haldanian computational model, which generated for instance the Eastern Scheldt long exposure air tables as well as the Netherlands National Diving Centre (NDC) air and enriched air (nitrox) tables with very few cases of DCS (Sterk, 1991 a and b). Also, trimix wet bell diving tables calculated with this model were successfully tested and used for depths up to 80 m by a diving company, but not published because of 'company secrecy'. So far, however, this model had not been used to produce tables for trimix caisson work. Therefore, we decided to test the capability of the model in this respect.

6 Test of trimix breathing for a 5 kg/cm^2 - 60 minutes hyperbaric exposure

We conducted an experimental chamber dive using trimix breathing to evaluate its practicality for use in hyperbaric caisson work. Since pressure gauges for caisson work in Japan are calibrated in kg/cm^2, we will use this unit of pressure from now on. (1 kg/cm^2=98.0665 kPa=0.980665 bar).

Subjects were six professional divers. Two separate chamber dives were made, each with 3 divers. They were compressed with air to 5 kg/cm^2 gauge pressure and were supposed to remain 60 minutes, while breathing 50% N_2 - 25% He - 25% O_2 trimix through a full-face mask equipped with a demand type regulator. In the second dive, however, one diver had to discontinue trimix gas breathing after 18 minutes because of a failure in his breathing apparatus. Therefore, he was decompressed in the outer lock with air, while the other two divers completed the 60 minutes exposure in the main chamber as planned. The five divers who completed the planned exposure were decompressed according to Sterk's schedule (Figure 4). At 2.1 kg/cm^2 in the course of decompression, they switched to ambient air breathing and began oxygen breathing in 20 minute cycles interrupted by 5 minute air breaks from 1.2 kg/cm^2 up to the surface. The total decompression time was 92 minutes and the accumulated CPTD (Cumulative Pulmonary Toxicity Dose) describing the oxygen (over)dose was 230 for exposure and decompression, which is well below the allowed daily dose (Sterk, 1987). The procedure for calculating the CPTD is outlined by Wright (1972) and can also be found in *The Underwater Handbook* (Shilling *et al.*, 1976).

During the dive, there were no symptoms of nitrogen narcosis or oxygen toxicity. In the course of decompression and in the 2-h observation period thereafter, no signs of decompression sickness, including minor symptoms such as skin rash or niggles, were encountered, nor were there any later reports about such events. Two divers who were involved in a hyperbaric air exposure of 4.9 kg/cm^2 for 30 minutes just two weeks prior to our experiments expressed their opinion that trimix gas breathing was much more comfortable and less tiring than the air breathing in the previous dive.

To further evaluate decompression risk, bubble detection in the blood stream was performed by ultrasound Doppler (Doppler Bubble Detector Model 1032G, made by Institute of Applied Physiology and Medicine, Seattle, Washington, U.S.A.) and echogram methods (Echo Camera SSD-500, made by Aloka, Tokyo, Japan) three times during the 2-h observation after surfacing. Measurements were made over the right heart, the pulmonary artery and the subclavian veins, with the divers at rest and after movement of the upper extremities. No bubbles were detected in any of the five divers who completed the planned schedule.

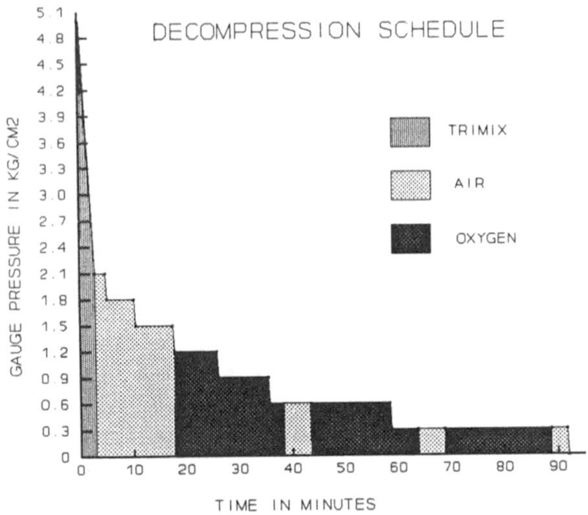

Fig.4. Decompression schedule for 5 kg/cm^2 - 60 minutes exposure to trimix 50% N_2 - 25% He - 25% O_2. Total decompression time is 92 minutes. Total CPTD is 230.

7 Conclusions

From our observations and the divers' comments it is suggested that trimix gas is useful as a breathing medium for deep caisson work, while our approach to the composition of the trimix gas and the decompression procedure seems reliable. Therefore, no obstacles are expected in using the remotely controlled caisson method at greater depths as planned.

8 Acknowledgement

The chamber trimix dive experiments were supported by funds from the Study Group for Deep Caisson Work, Japan.

Photographs of the Ohmoto-gumi remotely controlled caisson method

9 References

Shilling, C.W., Werts M.F. and Schandlemeier N.R., eds (1976) *The Underwater Handbook: A guide to physiology and performance for the Engineer.* Plenum Press, New York, pp. 158-165

Sterk, W. (1987) The use of oxygen in decompression, in: *Decompression in Surface-Based Diving* (eds. I. Nashimoto and E.H. Lanphier), Publ. 73(DEC)6/15/87, Undersea Hyperbaric Medical Soc, Bethesda, MD, pp. 13-19

Sterk, W. (1991a) The Netherlands National Diving Center data base, in: *Operational Dive and Decompression Data: Collection and Analysis* (eds. W. Sterk and R.W. Hamilton), EUBS Report (DATA)17-8-90, Foundation for Hyperbaric Medicine, Amsterdam (ISBN 90-9004500-7), pp. 3-6

Sterk, W. (1991b) Incident analysis: How do we deal with submaximal data?, in: *Operational Dive and Decompression Data: Collection and Analysis* (eds. W. Sterk and R.W. Hamilton), EUBS Report (DATA)17-8-90, Foundation for Hyperbaric Medicine, Amsterdam (ISBN 90-9004500-7), pp. 107-110

Wright, W.B. (1972) *Use of the University of Pennsylvania, Institute for Environmental Medicine procedure for calculation of cumulative pulmonary oxygen toxicity.* Report 2-72, U.S. Navy Experimental Diving Unit, Washington

39 Oxygen decompression in tunnelling

J. Cl. Le Péchon
JCLP Hyperbarie, Paris, France

Abstract
Although safety in compressed air tunnelling is far from being only the management of decompression sickness risks, this is often considered as a critical issue. It was understood very long ago that breathing pure oxygen during decompression results in either shorter or better quality decompressions. This is extensively used in diving as well as in space during Extra Vehicular Activity preparation. A review of the various effects of hyperbaric oxygen breathing in decompression is made: nitrogen washout is accelerated due to low nitrogen level in alveoli; the body dissolved oxygen content is increased, which prevents hypoxia that could eventually be induced by microbubbles; nitrogen content of bubbles is reduced and bubble collapse is enhanced; inflammation due to hypoxia is prevented. Safety aspects are then considered: management of oxygen breathing to prevent oxygen poisoning by proper limitations of PO_2 and duration of exposures; management of fire risk associated with the use of pure oxygen in compression chambers during tunnelling operations; oxygen systems in air-locks. An example of an air/oxygen tunnelling decompression schedule is presented and discussed in view of time saving, safety records and practicability in tunnelling operations.
Keywords: Oxygen, Decompression, Tunnelling, Compressed air work.

1 Introduction

Caisson working and compressed air have been used since the middle of the 19th century and Paul Bert (1878) explained the etiology of caisson diseases, first described by Pol and Wattelle, followed by Boycott, Damant and Haldane in the development of operational decompression procedures.

In the popular understanding, compressed air work is associated with 'the bends' and decompression sickness. However, the health and safety of hyperbaric works carried out in a confined environment include many aspects other than the short-term or long-term effects of decompression disorders (Le Péchon, 1990).

Engineering and Health in Compressed Air Work Edited by F.M. Jardine and R.I. McCallum. Published in 1994 by E & FN Spon. ISBN 0 419 18460 0

The experience and knowledge gained from diving, in particular from the recent advances in the offshore diving industry, and from space Extra Vehicular Activities, allow man to move safely in the field of times and pressures to extreme exposures and durations. In all cases proper gases must be available for breathing, enough time should be given for gas exchanges, and environmental conditions are to be controlled within the right limits.

Application of this experience has still to be introduced into tunnelling technology even in the range of pressure where air remains an acceptable breathing medium (0-5 bars).

2 Main characteristics of decompression sickness

Decompression sickness, which may rather be called decompression illness (Francis and Smith, 1990) when resulting from compressed air work exposure is mainly a painful joint problem, occasionally a neurological syndrome and in some cases delayed symptoms of bone necrosis.

All are caused by microbubbles either located in tendons resulting in pain, or in vessels resulting in a blood disease in reaction to foreign bodies and vascular obstruction. Most of the effects are related to tissue hypoxia caused by vascular gas emboli associated with disseminated microcoagulation (Hallenbeck and Andersen, 1982). Hypoxia prevents the reduction of free radicals, and inflammation processes are triggered.

Decompression sickness is basically an hypoxia stress sickness.

3 Absorption of gases at increased pressure

It is widely accepted among physiologists that a person exposed to increased air pressure will dissolve nitrogen in his tissues according to Henry's and Dalton's laws, to reach a so-called saturation status. When breathing air this process would last about 48 hours and various compartments in the body can be identified according to their rate of nitrogen uptake. The final quantity of nitrogen (which can be expressed in pressure units like PN_2 in the atmosphere) depends only on the pressure of exposure to compressed air (Figure 1). Due to the potential harmful effects of the oxygen also contained in air, such long exposures to compressed air are only possible to pressures lower than 2.5 ATA.

When exposure lasts only a few hours as with compressed air works in tunnelling, the nitrogen uptake into the various compartments of the body is gradual and must be evaluated with a gas-uptake model. Generally, Haldanian models are used with a variable number of compartments from 4 to more than 16 (only 2 are displayed on Figure 1).

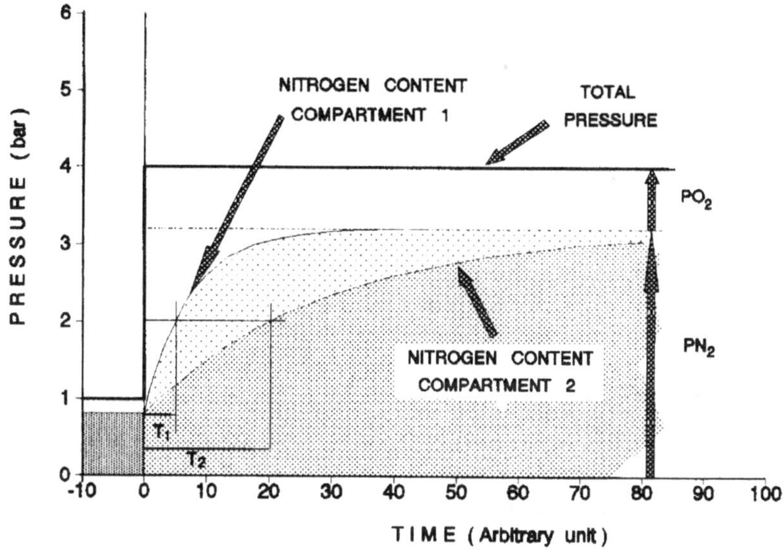

Fig.1. Nitrogen absorption kinetics

4 Gas elimination process

Gas elimination is not symmetrical to the absorption process. When, after a pressured top, the dissolved nitrogen content (expressed in pressure units) in a compartment becomes higher than the ambient pressure, microbubbles may form. Most of these are tolerated in the body until their size and number become unacceptable, then decompression sickness (DCS) is triggered which may develop into symptoms some time later.

Decompression persons from Nitrox saturation exposure is easy and uneventful although many unsuccessful procedures have been tried. This can be done with a slow pressure bleed at decreasing rate, and not on a linear decompression profile or with decompression stops (Le Péchon, 1981).

Decompression from a single short-duration exposure to compressed air is well documented and produces statistically an acceptable or at least accepted number of DCS cases (Shields and Lee, 1986; Imbert and Bontoux, 1989). Most protocols are calculated to control oversaturation (Figure 2) and eventually the bubble growth as a function of the various compartment gas kinetics (Hills, 1977; Van, 1989; Schreiner and Hamilton, 1989).

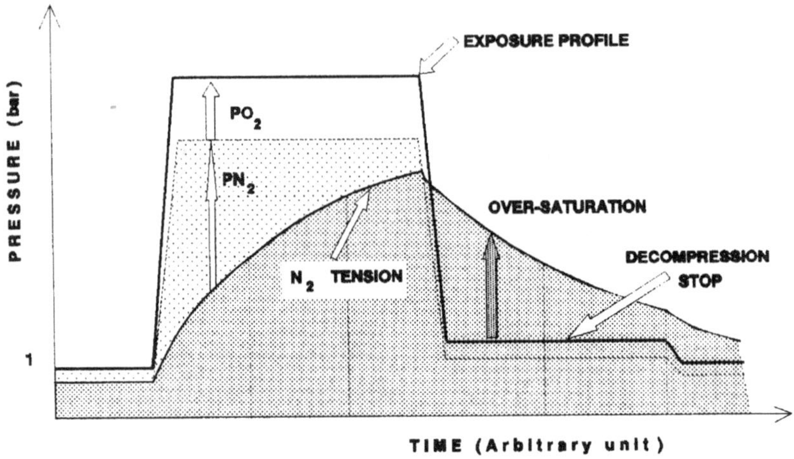

Fig.2. Oversaturation during a stop

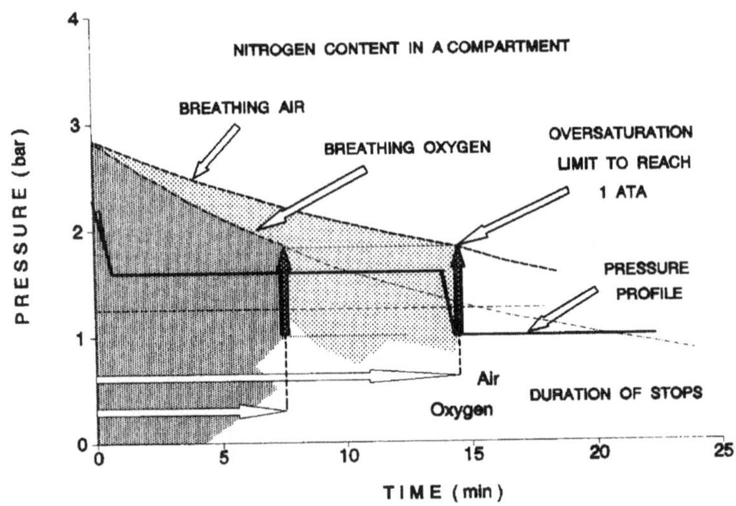

Fig.3. Effects of breathing air or oxygen

Compressed air work and repetitive diving exposures are still a potential health hazard. The durations involved stand in between saturation decompressions and short-exposure decompressions, both familiar and considered safe in the world of commercial diving.

To keep an efficient nitrogen driving force, the nitrogen partial pressure in the lung should be kept minimal; and to prevent bubble growth the external pressure should be kept as high as possible. This can be performed by breathing 'nitrogen-free' gases during decompression and by selecting high pressure stops.

Following both previous conclusions: that decompression sickness is by nature an hypoxic stress and that during decompression one should breathe nitrogen-free gases; the issue becomes very obvious as had already been suggested by Paul Bert in 1878 which is to breathe pure oxygen during decompression.

Unfortunately, breathing pure oxygen under pressure acts as a drug which triggers various physiological reactions in the body among which are:

- Vaso-constriction in many oxygen-sensitive tissues, the central nervous system for example
- Modification of surfactant-producing cells in the pulmonary membrane
- Biochemical induction of seizure.

Oxygen may become toxic when breathed under increased pressure and symptoms may appear after a latency period related to the oxygen partial pressure during exposure (see Figure 4).

The variation and limits of pure oxygen hyperbaric exposure have been evaluated in detail for diving applications (Lambertsen, 1988; Sterk and Hamilton, 1991), medical use in hyperbaric oxygen (HBO) as well as for treatment of decompression sickness (Goodman and Workman, 1965) (see Table 1).

A calculation method of pulmonary oxygen toxicity is available (Unit Pulmonary Toxicity Dose, UPTD) and acute exposures are well documented. In particular, it has been demonstrated that short interruptions of oxygen breathing postpone the occurrence of oxygen intolerance (Clark, 1982).

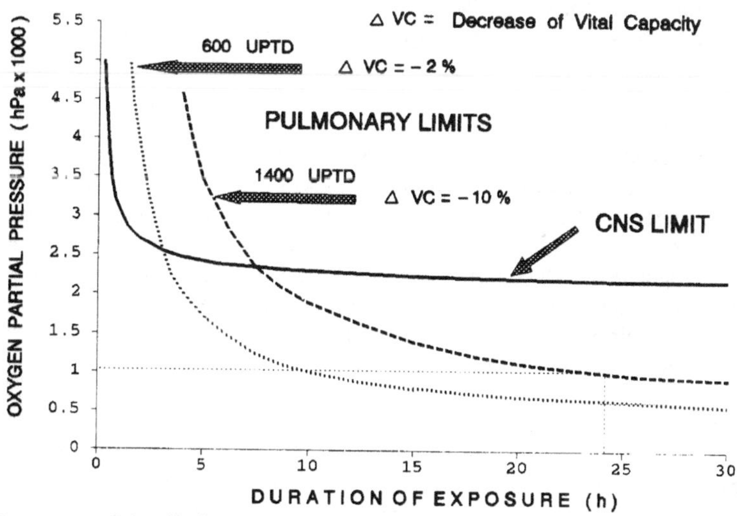

Fig.4. Oxygen toxicity limits

$$UPTD = t\left\{\frac{500}{PO_2 - 500}\right\}^{-0.833} \qquad (1)$$

with: t = Duration of high PO_2 exposure in minutes
PO_2 = Oxygen partial pressure in hectoPascals (hPa)

Table 1. Oxygen exposure limits

Situations	**UPTD Limits**
Single exposure	615
Daily exposures	400
Emergencies and medical	1415

Specific air diving tables with oxygen decompression have been published by several countries or organisations, but very few oxygen tables have been used in tunnelling (Faesecke, 1990; Kindwall, 1989; Nashimoto and Mano, 1974; French regulations, 1990).

It must also be remembered that diving tables are grossly inadequate for caisson work, first because the bottom times are not in the same range, and secondly because the exposed population and the type of work performed under pressure are very different. Validation of diving tables has not been made on a compressed air work population. The correction factor on decompression times used to establish the new French compressed air work tables from air diving tables has been estimated to be about + 25%.

4.1 Acclimatisation

It has been observed (Paton and Walder, 1954) that decompression sickness was more frequent in workers who had not been exposed previously, and that some acclimitisation occurs after several exposures. This is contrary to recreational diving experience which demonstrates that every day diving promotes the incidence of decompression sickness. Decompression tables used in tunnelling operations should be validated for people exposed for the first time. Procedures which show acclimatisation should not be used, their validation being inadequate for this purpose. Then acclimatisation would not appear any more.

4.2 Decanting

Decanting is a technique similar to the so-called surface decompression in diving. Workers are decompressed directly to atmospheric pressure and recompressed in a compression chamber to undergo the necessary stops. In diving, air breathing during the chamber recompression phase is scarcely used, and most of surface decompression procedures include oxygen breathing by masks (Shields and Lee,

1986). Even with oxygen breathing, limitation of exposures had to be introduced due to the high incidence of decompression sickness involving neurological symptoms in air diving in the North Sea (87% carried out with surface decompression and oxygen). In our opinion, decanting should not be carried out on air breathing and should be strictly used only for emergency evacuations from the working chamber.

5 How to use pure oxygen decompression in tunnelling

5.1 Equipment requirements

Oxygen safety. Oxygen constitutes a serious fire hazard and its use must be associated with a number of safety precautions:

- Use oxygen-safe clean plumbing.
- Distribute oxygen at low pressure in solid copper tubing.
- Maintain a maximum of 25% of oxygen by volume in the air lock by ventilation and using an overboard dump system. Monitor oxygen fraction.
- Have an hyperbaric fire extinguisher in air-lock.
- Train caisson workers (this is now compulsory by law in France for any type of hyperbaric work).

Breathing equipment:

- Use ergonomic masks with low-resistance overboard dump systems.
- The masks should be easy to remove from the airlock and easy to clean (most of the diving-type Built-In-Breathing-Systems are not convenient because removing and cleaning are too difficult).
- Use a clean individual mask for each decompression.

Oxygen quality standard. Medical or diving oxygen are acceptable. Any oxygen with analytic control better than 99.9% is acceptable.

5.2 Selection of decompression tables and procedures

When decompression tables with oxygen are not available, the use of oxygen breathing instead of air during standard stops will much improve the quality, people will be less tired, and bone necrosis will not appear.

A reduction of 20% in the stop times carried out on oxygen instead of air is a very conservative short cut if the original tables were safe ones (which remains to be demonstrated for most of them). Start breathing oxygen at the 0.9 bar stop and stay at 0.6 bar for the total time specified at 0.6 and 0.3 bar. The resulting extra oxygen toxicity will remain within acceptable limits.

Long-duration stops on pure oxygen breathing must be interrupted by air breathing for 5 minutes every 25 minutes, this time on air is not to count as stop time in the decompression profile.

Final decompression from 0.6 bar to atmospheric pressure is done during the last 2 minutes of stop time.

5.3 French compressed air decompression tables

The French Ministry of Labour published in 1992 (see Règlementation Française in the list of references), a comprehensive set of decompression tables covering most of the occupational hyperbaric situations and which includes air and oxygen decompressions for compressed air work at pressures up to 4.8 bar.

An example of Compressed Air Work Tables with oxygen breathing is shown in Table 2. All tables have been calculated by Imbert and Bontoux (1989) on the basis resulting from the validation of the French Air Diving Table 1974. They are now the tables to be used in France from June 1992. The mean ratio of decompression times on oxygen tables to the decompression time on air tables for the corresponding exposures is 55%.

Table 2. Times (minutes) for working pressure 3.3 bar – oxygen decompression

Duration of work	Transit time to 1st stop	1.5 bar air	1.2 bar air	0.9 bar oxy	0.6 bar oxy	0.3 bar oxy	Total decomp. time *	Period under pressure
20	10	-	-	-	-	5	15	35
25	9	-	-	-	5	5	19	44
30	8	-	-	5	5	5	23	53
45	8	-	-	10	10	10	43	88
60	7	-	3	15	15	20	65	125
90	6	3	15	25	30	30	124	214

* Total decompression time includes the air breaks between oxygen sessions

The last line is the emergency decompression to be used only if work duration accidentally exceeds the time above.

A formal structure for data collection and analysis is under consideration and effective validation will take place in the next years (Sterk and Hamilton, 1991).

7 Conclusions

Oxygen decompression in tunnelling is an efficient manner to improve the quality of decompressions; according to the French tables it reduces decompression times to about 55%. It requires the enforcement of hygiene precautions in the use of the masks and of stringent safety rules to control the fire risks. It is probably the only safe and practical means for decompressing from working at pressures above 2.5 bar.

No compressed air work is now permitted in France without formal training of personnel.

8 References

Bert, P. (1878) *La Pression Barométrique*, Réédit, 1979, Editions du CNRS, Paris

Boycott, A.E., Damant, G.C.C. and Haldane, J.S. (1908) The prevention of compressed air illness. *J. Hyg.*, Cambridge, **8**, pp 342-443

Clark, J.M. (1982) Oxygen toxicity, In: *The Physiology and Medicine of Diving*, 3rd edition, Bennett, P.B. and Elliott, D.H. eds. Baillière Tindall, London, pp 200-238

Faesecke, K.P. (1990) Diggin' deeper – Compressed air work beyond 30 m. In: *Proceeding XVI° Annual Meeting EUBS*, Amsterdam, pp 295-8

Francis, T.J.R. and Smith, D.J. (1990) Describing Decompression Illness, *42nd UHMS Workshop*, UHMS Publ. 79(DECO)5/15/92. Bethesda MD

Goodman, M.W. and Workman, R.D. (1965) *Minimal recompression, oxygen breathing approach of treatment of decompression sickness in divers and aviators*, BuShips, Project SF011 06 05 Task 11513-2 Research Rept., pp 5-65

Hallenbeck, J.M. and Anderson, J.C. (1982) Pathogenesis of decompression disorders, In: *The Physiology and Medicine of Diving*, 3rd ed., Bennett, P.B. and Elliott, D.H. eds., Baillière Tindall, London, pp 435-60

Hills, B.A. (1977) *Decompression sickness.* Wiley and Sons ed., New York, Volume 1

Imbert, J.P. and Bontoux, M. (1989) A method for introducing new decompression procedures. In: *Validation of decompression tables: 74*, Schreiner, H.R. and Hamilton, R.W. eds., (VAL)161-88, UHMS Bethesda MD; pp 97-105

Kindwall, E.P. (1989) Caisson decompression, In: *Proceedings of UHMS Symposium: The Physiological Basis of Decompression*, Vann, R.D. ed., Rept. 75-6/1/89, UHMS, Bethesda MD; 375-395

Lambertsen, C.J. (1988) Physiologic factors in human organ oxygen tolerance extension. In: *Proceedings XIV EUBS Annual meeting*, Aberdeen

Le Péchon, J.Cl. (1981) Procédures de décompression pour plongées à saturation, principes de calculs, résultats opérationnels. In: *Proceedings 7th EUBS Annual Meeting*, Cambridge, pp 205-17

Le Péchon, J.Cl. (1990) Safety in hyperbaric works. In: *Franchissements souterrains pour l'Europe*, Legrand ed., Balkema, Rotterdam, pp 275-81

Nashimoto, I. and Mano, Y. (1974) Experimental studies on oxygen decompression. *5th International Hyperbaric Congress*, Simon Fraser University, Burnaby, Canada, pp 900-4

Paton, W.D. and Walder, D.N. (1954) *Compressed air illness*, Special Report Medical Research Council No. 281, London.

Pol, B. and Wattelle, T.J.J. (1854) Mémoire sur les effects de la décompression de l'air appliquée au creusement de puits à houille. *Annals d'hygiène publique et de médecine légale* (industrielle et sociale) **1**, pp 241-79

Schreiner, H.R. and Hamilton, R.W. eds. (1989) *Validation of decompression tables*, 74(VAL)161-88, UHMS, Bethesda MD

Shields, T.G. and Lee, W.B. (1986) *The incidence of decompression sickness arising from commercial offshore air-diving operations in the UK sector of the North Sea during 1982/83*, Dept. of Energy Contract TA 93/22/147, Aberdeen, Hyperbaric med. Unit, Robert Gordon Institute of Technology

Sterk, W. (1986) Intermittent hyperoxia in operational diving: what are the safe limits? In: *Proceedings XII EUBS Annual Meeting*, Rotterdam, pp 55-64

Sterk, W. and Hamilton, R.W. Eds. (1991) *Operational Dive and decompression data: collection and analysis*, EUBS Workshop, EUBS Publ. (DATA)17-8-90

Vann, R.D. ed. (1989) The physiological Basis of Decompression, *Proceedings 38th UHMS Workshop*, Duke University, Durham, NCa, June 1989

Walder, D.N. (1982) The compressed air environment. In: *The physiology and medicine of diving*, 3rd ed., Bennett, P.B. and Elliott, D.H. eds., Baillière Tindall, London, 15-30

Reglementation Francaise:

DECRET n° 90-277 du 28 Mars 1990, relatif à la protection des travilleurs intervenant en milieu hyperbare. J.O. du 29 Mars 90, 3826-3830

ARRETE du 28 Janvier 1991, définissant les modalités de formation à la sécurité des personnels intervenant dans les opérations hyperbares. J.O. du 2 Mars 91, 3020-3025

ARRETE du 28 Mars 1991, définissant les recommendations aux médecins du travail chargés de la surveillance médicale des travailleurs intervenant en milieu hyperbare. J.O. du 26 Avril 91, 5640-5642

ARRETE (to be published), Définissant les procédures d'accès, de séjour, de sortie et d'organisation du travail en milieu hyperbare

40 Procedures for safe working at high pressures in a TBM chamber using special breathing gas mixtures

G. Luther, R.W. Hamilton*, M. Heineke, G.A. Nietsch** and J. Vollbrandt
GKSS Forschungszentrum, Geesthacht, Germany
**Hamilton Research Ltd., Tarrytown, USA*
***Dyckerhoff & Widmann AG, Hamburg, Germany*

Abstract
Strait crossing by means of a tunnel boring machine is becoming more frequent and will continue in the future. The normally automatic tunnelling progress needs occasional human intervention in the pressure-exposed part of the shield even at pressures beyond those considered acceptable for breathing air. It is important to establish that with due attention to the physiological and human exposure requirements, this work can be done safely, and it may be less expensive than doing it by other methods. The need is likely to be for short (<2 h) interventions by specialists, but could involve multiple shifts of an entire team. In addition to the usual considerations of working with a tunnel boring machine, the high-pressure work situation involves many other factors like selection and management of breathing gases and breathing equipment, later re-use of pressure chambers, effects of gases on human performance, selection of decompression patterns, fire and oxygen safety, mobilisation time and integration with normal work shifts and training at several levels. Many of the risk factors and redundancies necessary for diving are not applicable here. Decompression procedures are better if designed for the operational situation, not the reverse. Partial saturation profiles not commonly used in diving for economic reasons may work well in the tunnel situation.
Keywords: Tunnel work, Partial saturation, Decompression computation, Heliox, Nitrox, Storebælt-tunnel.

1 Introduction

The Storebælt tunnel will be part of the fixed link between the islands of Funen and Zealand in Denmark. It consists of two tubes with a length of about 8 km each and its deepest point is approx. 80 m below sea level. Only little experience is available about tunnelling beneath straits with comparable depths and water pressures. Although there have been reasons for the management of the tunnelling works to expect largely impervious underground conditions one has to be prepared to encounter water-bearing layers at full pressure corresponding to the depths below sea

Engineering and Health in Compressed Air Work Edited by F.M. Jardine and R.I. McCallum. Published in 1994 by E & FN Spon. ISBN 0 419 18460 0

level. Indeed a fatal water inflow happened during the tunnelling works in 1991. In order to make human intervention at hyperbaric conditions in the work chamber at the shield front possible, ground treatment was foreseen by the designer of the tunnel. However, the contractor wanted to investigate additionally the methods, measurements, and equipment for an access to the work chamber under high hyperbaric conditions. Since both the tunnelling technology and the underwater technology have to contribute their specific knowledge to this task, Dyckerhoff & Widmann AG as a partner of the joint venture for Great Belt Tunnelling and GKSS were asked for a study, which deals with hyperbaric 'dry' intervention in the work chamber of a tunnel boring machine when deeper than 30 m below sea level under consideration of the tunnelling environment.

This paper gives some of the conclusions of that study and points out some of the advantages and problems of the sub-saturation approach. It relates to tunnel work:

(a) in depths deeper than 30 m and/or
(b) with heliox and nitrox as breathing gas mixtures.

2 Basic design aspects using partial saturation

Any work at great depths or pressures must consider the risk of pressure loss. The work chamber behind the Storebælt boring machine has a volume of about 100 m³. The maximum pressure corresponds to 80 m (9 bar). If there should be an opening in the shield or a rupture of a pipe should occur the first pressure drop will be very high. The same effect is true for diving facilities. In Figure 1, the theoretical pressure drop is shown for free openings of 20, 40 and 100 mm diameter. Having an opening of 100 mm, no supply of air for compensation, and an elapsed time of one minute the pressure in the work chamber will drop from 9 bar to about 4 bar. Regarding the tightness of the system this effect should be considered during the design and operation periods.

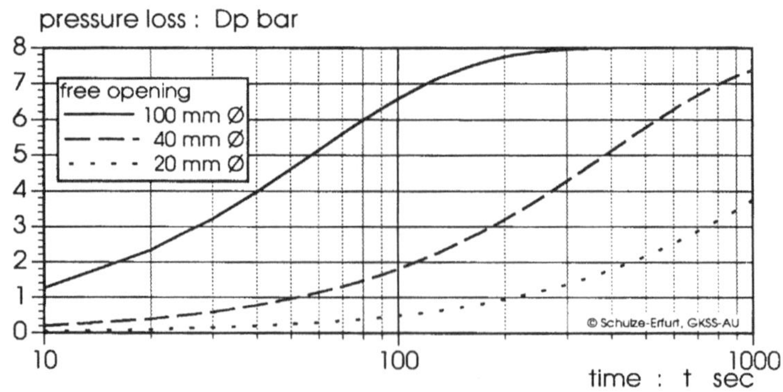

Fig.1. Theoretical pressure loss of a work chamber as function of time (initial air pressure of 8 bar; work chamber volume of 100 m³)

It is possible to use air in all chambers as chamber gas up to a pressure of depth 80 m. The lock chamber, which is directly connected to the shield, should be pressurised starting at 2 bar absolute with helium as backup for a break in gas supply to the built-in-breathing system (BIBS) (first safety arrangement). In case of a dangerous situation like an unforeseen pressure loss in the work chamber the pressure crew has to hurry into the lock chamber. In order that at depths greater than 50 m the crew remains clear thinking, it is recommended that there should be a lower nitrogen concentration in the chamber by adding helium during the preceding compression period (second safety arrangement).

Having air in the work chambers and in the lock chamber means a partial pressure of oxygen of about 1.9 bar. Fire protection has to be integrated into the design. Pressurising the lock chamber with helium is a convenient solution for this chamber. A concept for smooth evacuation of the pressure crew or for transfer to an ill pressure worker has to be established in detail before construction starts.

3 Operational limits of partial saturation

After a stay of 24 to 48 h under pressure the external inert gas pressure is in all tissues of the human body. In this case the human body is saturated. Now it is possible to stay as long as necessary in this depth without any further prolongation of decompression time. This technique can be safely used up to pressures of 60 bar (Luther *et al.*, 1988).

If the time in pressure is shorter than 24 h the tissues of the human body are only partly saturated by the inhaled inert gas (Workman *et al.*, 1975). The pressure in the tissues will increase very quickly in the first hours. That is even true for slow tissues with a theoretical degree of saturation of 99.99% after stay of 24 h in air with 9 bar environmental pressure (see Figure 2). For occasional single interventions which may require one or two hours work it may be much more efficient to use partial saturation, the tunnel equivalent of 'bounce dive'.

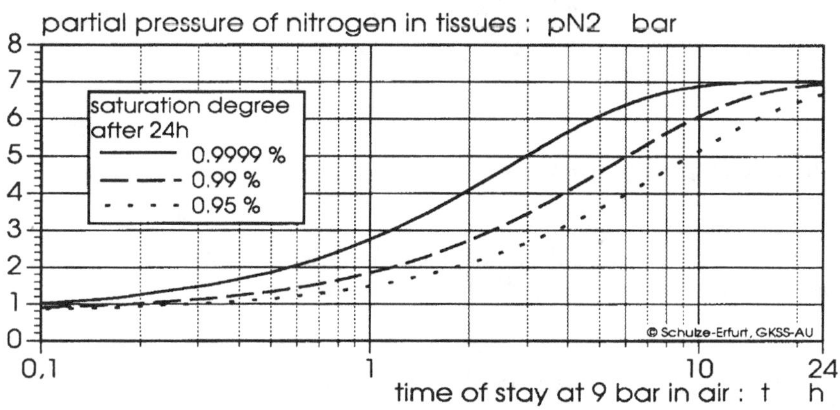

Fig.2. Theoretical saturation curves with nitrogen

If a task will take more than two hours, two or three pressure crews should do the work. If the work requires pressure exposure of more than 6 hours per day over a prolonged period, then saturation techniques may be the solution. Using saturation, decompression profiles over a time of up to 5 days (depending on the pressure behind the shield) should be kept in mind.

4 General demands of compression and decompression profiles

At full saturation the pressure crew lives for a long period in an accommodation/decompression chamber waiting for their missions. Three shifts are necessary to support the locked saturation crew at the surface. Many of the required tasks can be done in a short time (less than 6 hours per day). For the performance of those tasks partial saturation is the right and economic technique, but the decompression procedure has to fit into the work of the tunnel boring process. The partial saturation crew is a part of the tunnel boring crew and has to work under pressure only on request. The following demands are recommended for the integration of the partial saturation technique in the tunnel process. The compression and decompression procedures should:

- be safe regarding time and pressure accuracy
- be simple regarding change of breathing mixtures
- be simple regarding crew transfer between chambers
- offer the use of air as chamber gas wherever possible
- use only one special breathing mixture per compartment
- reduce heliox breathing as far as possible
- regard the qualifications standards and
- regard the experience of tunnel crews.

At partial saturation, safety reasons require a gas storage capacity for full saturation dive at maximum depth to decompress the crew in case of a substantial delay. This applies for the gas storage only, no extra crew or equipment have to be provided.

5 Location of decompression chambers

5.1 General possibilities

Depending on the location where the pressure crew will be decompressed two possibilities can be generally discussed:

1. Decompression within the tunnel near the work site, and
2. Transport and transfer under pressure to a decompression chamber outside the tunnel.

5.2 Decompression in the tube

Decompression of the pressure crew near the work site can be performed in the lock chamber or after transfer in a separate decompression chamber located near the shield.

Decompression in the lock chamber is possible if this chamber and its auxiliaries are equipped for all needed gas mixtures and for saturation decompression, too. Decompression with transfer of the pressure crew to a decompression chamber located near the shield is an excellent system configuration. The transfer between the lock chamber and the separate decompression chamber can be performed by surfacing and recompression, by a trunk system connecting the hatches at the lock chamber and at the decompression chamber or by a 'Shuttle'. The 'Shuttle' is a movable chamber that allows workers to be transferred near the shield from the lock chamber to the well equipped decompression chamber. The transfer in a 'Shuttle' is the best method regarding safety of the pressure workers and efficiency of the tunnel boring requirements.

5.3 Decompression outside the tube

Decompression at the outside can be performed by transport of the pressure crew under pressure to an outside area where the needed logistic support for a safe decompression is located. In this case the workers will be transferred first by the 'Shuttle' to a 'Transfer Chamber' and afterwards by the 'Transfer Chamber' to the outside area. A part of the decompression can be performed during the time the pressure crew is on its way to the tunnel entrance. Figure 3 contains a schematic view of such a system configuration. Two other options for decompressions are to use an existing diving support vessel moored near the tunnel end, or to take workers to an existing hyperbaric complex elsewhere.

5.4 Main disadvantages of single system variants

Decompression of the worker near or at the shield leads to larger restrictions of the tunnel boring process. During decompression periods no boring is possible and if the work behind the shield is not completed, the second or third crew have to wait during the total decompression period.

Surfacing and recompression is an unsafe method especially at higher pressure ranges. A trunk system interconnected to the lock chamber through the arms of the erector, able to be attached to another chamber is a time-consuming procedure and the tightness of the flanges might cause problems. Transfer of an ill worker would be nearly impossible, if a trunk system is used.

The use of a diving vessel or an hyperbaric complex usually involves considerable loss of time, more manpower, dependence on weather conditions, prolonged restrictions or boring work, restricted medical care of and additional logistics for the transport of a pressure crew. This possibility can only be applied in very special cases and is not discussed in this paper as standard procedure.

Decompression on the outside with workers transferred by a special 'Transfer Chamber' leads to a higher establishment cost, but in terms of productivity the overall costs will be lower. The decompression system itself can be used at later

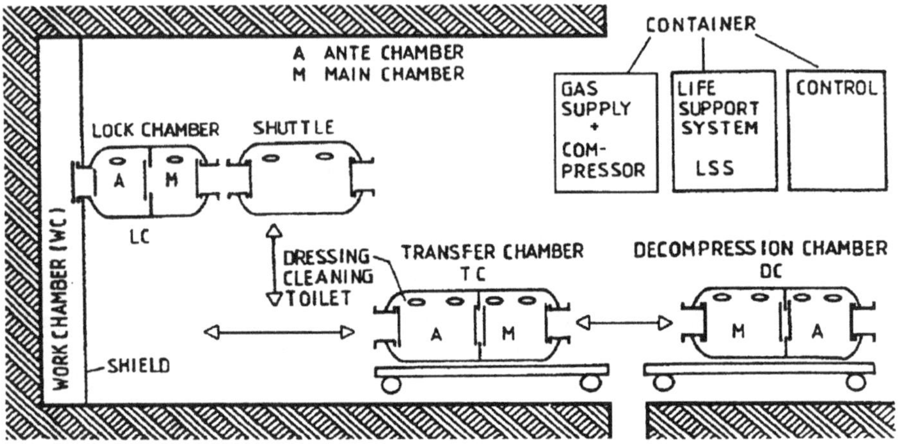

Fig.3. Chamber system for transfer and decompression

tunnel contracts, too. Regarding the health of the worker such a system also provides a higher safety level.

6 DCAP Program as an engineering tool

DCAP stands for decompression computation and analysis program (Hamilton *et al.*, 1987).

6.1 Short introduction into DCAP-Program

For the decompression computation GKSS has a comprehensive computer program called DCAP, which is designed to facilitate the computation of all types of partial decompression tables. DCAP is primarily a computational tool, and although it handles this step of the decompression development process quite effectively, for its output to be useful it has to rely on a basis of experience. More on the background of DCAP as a tool is presented by Hamilton *et al.* in their paper to this conference.

One curious point about the tables being generated is that exposures with air have a shorter decompression than those with heliox. Intuitively one would expect the heliox tables to require less decompression time. Helium is less soluble, therefore less gas has to be unloaded, and experience shows this. For example, decompression from heliox saturation can be 2 to 3 times as fast as from saturation with nitrox. There are several reasons for this apparent discrepancy.

First, the algorithm used has not been optimised for helium. It is more conservative than needed in some areas, but to change it would require that analysis and integration that has not yet been done. Second, because helium moves faster than nitrogen a short exposure results in picking up more helium. At some point as exposures get longer the times will be more advantageous for helium. Mixtures of helium, nitrogen and oxygen might well be found to be optimal.

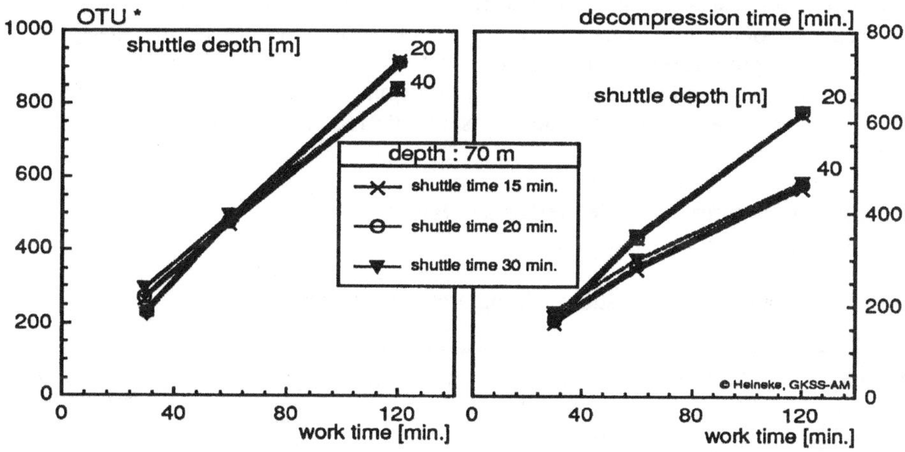

Fig.4. Influence of work time, 'Shuttle' depth and 'Shuttle' time on OTU* and decompression time

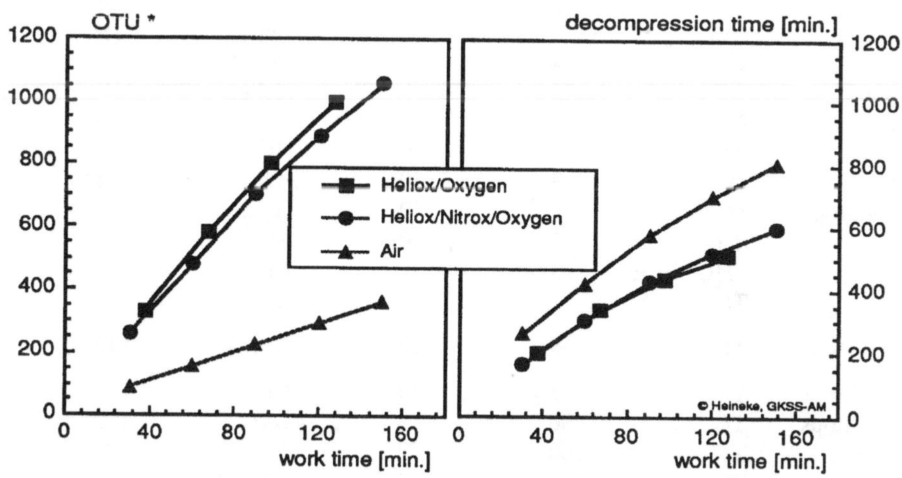

Fig.5. Influence of work time and breathing mixtures on OTU* and decompression time

* OTU: Oxygen Tolerance Units, Hamilton (1989) and Hamilton *et al.* (1991)

6.2 Use of DCAP Program for the tunnel boring process

As pointed out, GKSS is using the DCAP program. This program gives the chance to the engineers to arrange the decompression procedure (stops, oxygen level, decompression time) according to the possibilities of a tunnel boring process. The results of the DCAP calculation laid down in Figures 4 and 5 give a good impression of the optimising capability incorporated in the program. Different shuttle times can be effectively compensated (see Figure 4). This is an important point for managing the operational procedures in front of the shield.

Table 1 contains an overview of a decompression procedure for a depth of 70 m and a real work time of 55 min under pressure. Prerequisites for DCAP calculations were:

Breathing mixtures	Heliox/Nitrox/Oxygen
Shuttle transfer time	15 min

Table 1. Example of an operational decompression procedure

Activity	Chamber	Time demand min	Depth m	gas mix	Time total min
T	MLC	0	0	Air	0
C	MLC	1	10	Air	1
C	MLC	4	50	Heliox	5
T	ALC	3	50	Heliox	8
C	ALC	2	70	Heliox	10
T	WC	4	70	Heliox	14
Work	WC	55	70	Heliox	69
T	ALC	4	70	Heliox	73
D	ALC	2	50	Heliox	75
T	MLC	3	50	Heliox	78
D+Stops	MLC	9	30	Heliox	87
T	SHUTTLE	3	30	Air	90
S	SHUTTLE	15	30	Air	105
T	TC	3	30	Air	108
D+Stops	TC	35	18	Nitrox	143
D+Stops	TC	1	15	Nitrox	144
D+Stops	TC	36	9	Oxycyl.	180
Stops	TC	60	9	Air	240
D+Stops	TC	155	0	Oxycyl.	395

T Transfer of pressure crew, C Compression, D Decompression,
S Shuttle procedures Oxycyl. = Oxygen breathed in cycles
For chambers see Table 2.

7 Operational aspects for partial saturation intervention

7.1 Breathing gases

The chamber gas in all chambers is air. Breathing gases are supplied by BIBS (built-in breathing system) or by rebreather. The gases have to be mixed according to the needs of the decompression profile. In Table 2, types of breathing gas mixtures related to single chambers are listed. The concentration of breathing gas mixtures for work in depths deeper than 30 m depends on the partial saturation profile for compression and decompression. For the higher depths of the Storebælt project the heliox breathing gas might contain 18% oxygen and the intermediate enriched air or 'nitrox' mix might contain 50% of oxygen.

Table 2. Distribution of breathing gas mixtures for chambers used for transfer and decompression

Chamber		Heliox 1	Air	Nitrox	Oxygen
WC	Work chamber	x	(x)		
ALC	Ante lock chamber	x			
MLC	Main lock chamber	x			(x)
	Shuttle		x		(x)
TC	Transfer chamber		x	x	x
	Decompression chamber*		x	x	x

* if available

7.2 Optimal plant configuration and plant operation for tunnel boring work requiring longer work periods

Long interruptions for the tunnel boring process can be avoided if two transfer chambers (TC-1 and TC-2) and one outside decompression chamber (DC) are integrated in the design and operation of such a project. Using this prerequisite 3 shifts can work up to six hours per day behind the shield. The overall procedure corresponds to the following transfer plan.

First pressure crew

First crew is compressed in the lock chamber (LC) at the shield to the pressure level of the work chamber (WC) and works for two hours. First decompression stops are performed in the main compartment of the lock chamber (MLC) to the pressure level of the 'Shuttle'. All inhabitants are transferred to the 'Shuttle'. Afterwards the 'Shuttle' itself is disconnected and transferred to TC-1 without decompression. Further decompression of the pressure crew down to surface pressure is initiated directly after connection and pressure balance between 'Shuttle' and TC-1.

Second pressure crew
About 2 h after the first crew has left MLC the second crew can enter MLC for compression to the pressure behind the shield. The second TC-2 is used for decompression of this crew. Transfer and decompression procedures are the same as for the first crew.

Third pressure crew
About 8 h after the start of the compression of the first pressure crew the third crew can enter MLC for compression. Decompression of this crew is afterwards possible in TC-1. If each crew has to work one hour under pressure behind the shield and needs one hour for tunnel transfers, 11 h after the first crew has entered MLC the tunnel is free for normal operational work, while the third crew is decompressed further at the outside base.

7.3 Gas consumption for 3 shifts

During partial saturation procedures the TCs and one decompression chamber are used for decompression. Maximum consumption of gas is needed at max. depth (80 m) and at max. working time under isobaric conditions.

Table 3. Gas storage for intervention at 80 m in m³ (1.08 bar, 0°C)

Breathing gas	**Demand**			**Total**
	1 Crew	**3 Crews**	**ALC**	
Heliox	200 m³	600 m³	50 m³	650 m³
Nitrox	50 m³	150 m³	-	150 m³
Oxygen	30 m³	90 m³	-	100 m³
Air	yes	yes	yes	

The consumptions in Table 3 are based on two TCs, 3 shifts and a 16 h period under pressure. Gas reserves are included for some delays and 120 min at pressure. ne repressurisation of the lock chamber is included. If Heliox is supplied by a closed circuit breathing system (CCBS) or rebreathers to the pressure worker the consumption of Heliox can be reduced to less than 25% of max. value (150 m³/day instead of 600 m³/day).

7.4 Personnel requirements for partial saturation work

A surface crew of 12 people is needed to perform short intervention in 3 shifts using partial saturation techniques. The shifts can be the normal tunnel boring crew, if the crew members are trained for such a job. For each shift 3 people are needed for pressure work. At least 12 people out of the total workforce of one tunnel should be licensed to work in pressure. Additional training for using partial saturation technique is needed.

The surface crew for the operation and supervision of the pressure work is normally a part of the standard tunnel boring crew. Thus the shift engineer A might be the supervisor A and the senior foreman D the chamber operator D. Table 4 demonstrates how the distribution can be arranged.

Table 4. Distribution of personnel per 3 shifts

	Shift 1 (8h)			**Shift 2 (8h)**			**Shift 3 (8h)**		
	LC	E*	TC	LC	E*	TC	LC	E*	TC
Supervision	A	A	A	B	B	B	C	C	C
Chamber op.	D		D	E		E	F		F
Erector op.		G			H			I	
Transfer technician		G			H			I	
Chamber operation			G			H			I
Life support	K		K	L		L	M		M
Surface personnel		4			4			4	
Pressure crew		3			3			3	

* Erector operation

At full saturation the pressure crew is living under pressure waiting for their missions behind the shield. This possibility is only necessary if a lot of pressure work is assumed for the boring period. Full saturation capability means a surface crew of 11 people additionally needed during stand-by periods for each tunnel side. 15 people are needed to operate the system during pressure work periods.

8 Operational support for a safe decompression

It is essential for the decompressing crew to remain relaxed and limber during decompression. If possible they should get up and move around every half hour or so, moving all limbs and joints and activating the circulation. Cramped conditions are detrimental to decompression. However, most decompression chambers contain bunks or cots, and most experts consider that this gives the crew the best practical conditions for maintaining good circulation. We recommend bunks, with flameproof mattress covers and bedding.

The pressure crew should not be allowed to become cold, but should not be too hot either. Sleeping is acceptable, but should be interrupted by the operator for moving around at least every hour. Intensive or strenuous exercise during decompression is not beneficial and should be avoided.

Dehydration appears to be detrimental to reliable decompression. Individuals decompressing should remain well hydrated, and to do this should make a conscious effort to drink plenty of liquids. Too much coffee is not recommended because it is a diuretic. Juices, soda, sports drink, etc. are good – whatever liquids taste good, will result in a lot being consumed. Milk, if used, should be low fat. For these decompressions prevention of dehydration can be an important factor, and should be identified in the operational plan. Provision should be made for the regular supplying of fluids before starting the operation, and the 'topside' operators should be trained to encourage the crew to drink plenty of appropriate fluids.

9 Conclusion

Partial saturation can efficiently support tunnel boring work. In order to get the maximum benefit of this possibility, partial saturation technique should be integrated in a very early stage of design into the tunnel boring concept, including needed operation procedures. GKSS uses DCAP program as a powerful software program. This program delivers safe prerequisites for an excellent optimising process regarding both the tunnel boring possibilities and decompression needs.

References

Hamilton, R.W. (1989) Tolerating Exposure to High Oxygen Levels: REPEX and other Methods. *MTS Journal*, Vol. 23, pp 19-25

Hamilton, R.W., Vollbrandt, J., Holthaus, J., Heineke, M. and Kühns, M. (1987) GUSI's capability for developing decompression techniques, in: *Proceedings of 2nd International Symposium – Underwater Technology*, June 15-16, (eds. H.G. Schafstall, G.F. Schultheiss and D. Seeliger), GKSS-Forschungszentrum Geesthacht GmbH, pp 20-1 20-13

Hamilton, R.W., Luther, G. and Holthaus, J. (1991) Efficient and safe use of oxygen in decompression, in: *Proceedings of 3rd International Symposium on Underwater Technology*, April 9-10, (eds. H.G. Schafstall, G.F. Schultheiss and D. Seeliger), GKSS-Forschungszentrum Geesthacht GmbH, Geesthacht, pp 13-1-13-8

Luther, G., Bennett, P.B., Elsner, W., Holthaus, J., Schafstall, H-G., Schmidt, K. and Vollbrandt, J. (1988) Efficiency of Working in Depths Down to 600 m, in: *Proceedings of 1988 Offshore Technology Conference*, Houston, May 2-5 (eds. OTC, Inc., Richardson, TX./USA), OTC-report 5706, Vol. 2, pp 227-233

Workman, R.D. and Bornmann, R.C. (1975, second edition) Decompression Theory: American Practice, in: *The Physiology and Medicine of Diving and Compressed Air Work* (eds. Bennett, P.B. and Elliott, D.H.), Baillière Tindall, London, pp 307-330

41 Unmanned excavation systems in pneumatic caissons

S. Shiraishi
Shiraishi Corporation, Tokyo, Japan

Abstract
This paper describes the principal features of a recently developed unmanned excavation system (UES) for pneumatic (compressed air) caissons. Traditional hand-mined caissons became extremely expensive in comparison with alternative construction techniques, so mechanical excavators were developed (caisson shovels) which, although removing the need for hand mining, still required their operators to work in compressed air. The UES is a further development in which the operator is in an atmospheric capsule. It combines the benefit of mechanical excavation but without the time constraints or potential hazards of men having to work in compressed air.
Keywords: Pneumatic caissons, Excavation, Unmanned systems, Atmospheric capsule, Compressed air.

1 Introduction

Pneumatic (compressed air) caissons are seldom built elsewhere in the world other than in Japan. In USA, no pneumatic caisson has reportedly been constructed for about fifty years and, in Europe, there have only been relatively few in recent years.

The reasons for this are the labour intensive nature of the excavation and construction works together with the health hazards for workers. Alternative, less expensive methods were often found for substructure work.

Pneumatic-caisson technology in Japan, however, has been in continuous, steady demand and has developed considerably from the traditional techniques in use in the post-war period.

In recent years, in-caisson excavation works have been mechanised in Japan and pneumatic-caisson technology revolutionised. Pneumatic caissons in Japan are now strongly competitive against such newly developed technological rivals such as cast-in-place concrete piles.

This successful improvement has been accomplished by increasing the efficiency of in-caisson excavation in particular, as well as many other aspects of pneumatic-caisson works. This improved efficiency of in-caisson excavation work has been

Engineering and Health in Compressed Air Work Edited by F.M. Jardine and R.I. McCallum. Published in 1994 by E & FN Spon. ISBN 0 419 18460 0

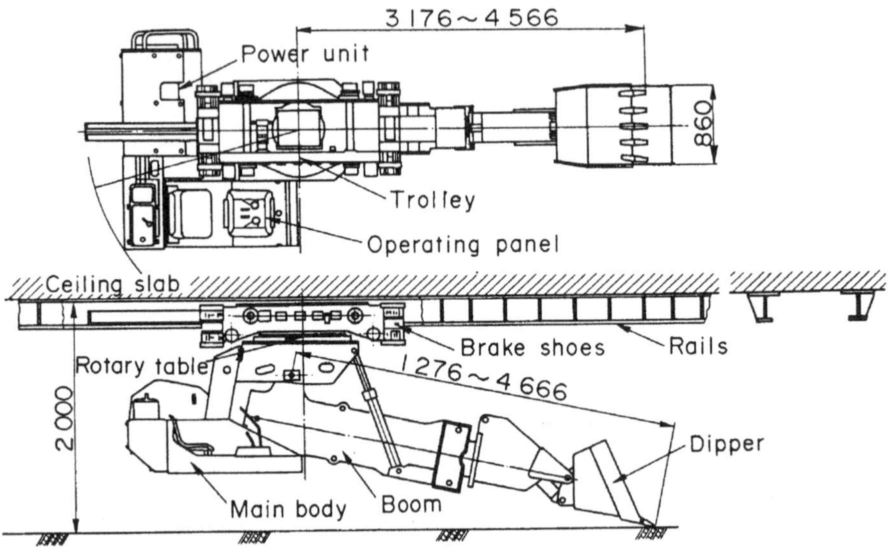

Fig.1. Caisson shovel

achieved by newly developed mechanical excavators called caisson shovels to replace hand-mining. Rapid soil loaders also contribute additional efficiency by speeding up spoil removal from inside caissons which have sufficient working space.

Further improvement has been achieved by relieving men from the potential hazards of work in compressed air, by the application of newly developed unmanned excavation systems (UES) in pneumatic-caisson works. This new system for in-caisson excavation has been designed and developed by Shiraishi Corporation of which the author is the Chairman.

2 Caisson shovel

A caisson shovel is a compact power shovel designed to be operated efficiently in the limited space in a caisson working chamber. As shown in Figure 1, it is suspended from a small trolley running on a pair of I-shaped rails fastened to the ceiling slab of the working chamber, so that it can travel back and forth along the rails like a suspended monorail wagon.

Its suspended body rotates 360 degrees around the vertical axis of its rotary table. An 150-litre dipper is attached to the front end of its telescopic dipper boom which swings 19 degrees upward and 32 degrees downward. Up to 10 tons (98.1 kN) of thrust exerted at the dipper's teeth for cutting into the ground may be actuated by a hydraulic cylinder encased in the hollow-box boom. The reaction for this thrust is obtained by engaging the trolley's brake shoes to the rail flanges.

The maximum working radius of a caisson shovel when its boom is fully extended is 4 m and it covers an 8-m wide strip of ground below the running rails by travelling back and forth. It is powered by a 30-kW electric motor and actuated with hydraulic systems.

A caisson shovel may be disassembled into five pieces, each piece being designed to clear narrow passages in an air-lock when it is brought down into or up from the caisson's working chamber.

Maximum hourly production rate of a caisson shovel in digging and hauling spoil may be as high as 20 cu.m when the digging is easy, hauling distance minimum and its operator skilful.

A caisson shovel, with its powerful shoving capacity and high degree of mobility (whatever the ground surface) can be adapted efficiently to a wide range of soil types varying from soft mud to very hard ground like hardpan or cemented sand. When its dipper is replaced with rotary cutters or a giant rock breaker, soft rock may be broken into pieces and mucked out by another caisson shovel attached with a dipper.

Caisson shovels have been employed extensively since six of them were first put into practical use in 1971 in the then world's largest pneumatic caisson for one of the main-pier foundations of Osaka Harbour Great Bridge. Since then, the accumulated total volume of ground dug out by them in more than 400 caissons exceeds 2 million cu.m.

The use of caisson shovels has brought not only a remarkable saving in man-power, but also appreciable cost savings, because it has replaced costly caisson labour whose wage rate has been steadily increasing.

3 Quick soil loader

The bottleneck of in-caisson excavation lies not in the excavation itself but in the process of removing the excavated soil out of the caisson by earth-bucket handling. As is illustrated in Figure 2(a), the soil scooped up by the caisson-shovel's dipper is carried from the digging point, D, to the point, L, where the empty earth bucket sits below the shaft, S, and the airlock for materials, M. Then, at the point, L, the soil is loaded into the bucket.

For filling up the earth bucket by repeated feeding, a caisson shovel has to shuttle back and forth between the points D and L, taking several minutes per cu.m of soil. The bucket fully loaded with soil is then pulled up through the shaft, S, and is moved further sideways to the point above the soil-storage bin, B, where the bucket is upturned, dumping the soil into the bin.

The empty bucket is then returned to the loading point, L, reversing through the material lock, M, and the shaft, S. In the above cycle, several minutes are lost to productive activity while the bucket is being loaded. For example, if there are 36 cycles in a 6-hr work shift and five minutes lost in each cycle, the total lost production time could be as much as half the shift time.

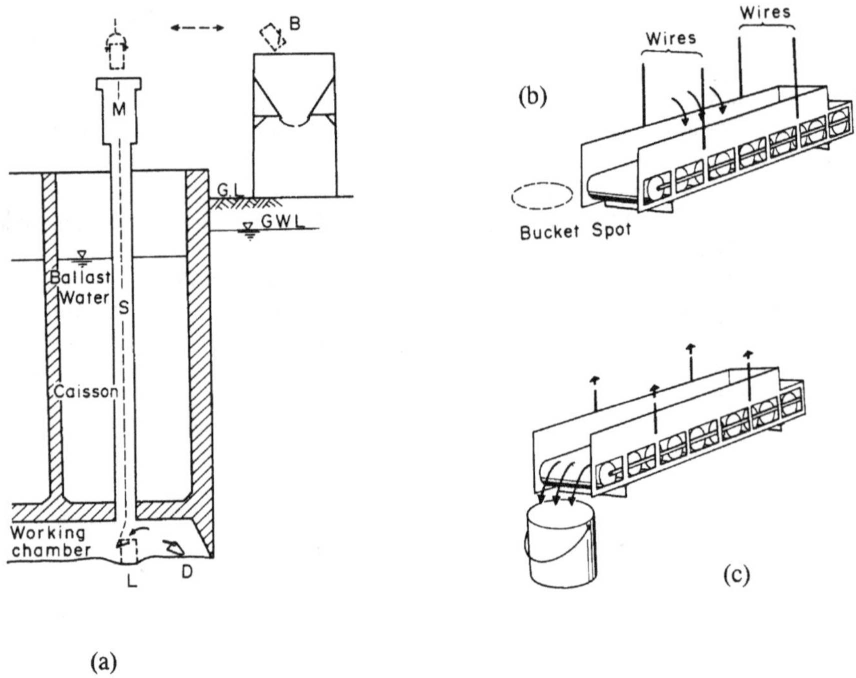

Fig.2. Quick soil loader
(a) Bucket handling cycle, (b) Storing position, (c) Dumping position

A quick soil loader was designed to eliminate this time loss. Instead of feeding soil from the caisson shovel directly to the earth bucket, the soil is loaded into the storage ditch of the quick soil loader during the time that the full bucket is moved from the point L through S and M to B and returned empty to L. The storage ditch with its built-in belt-conveyer bottom is capable of holding 1 cu.m of spoil. When the storage ditch is fully loaded, the quick soil loader may be pulled up and moved over to the empty earth bucket and all the stored soil dumped into the bucket in a matter of seconds by running the belt-conveyer for the ditch-long stroke.

By the use of quick soil loaders, the cycle time of earth-bucket handling may be shortened nearly by half and the efficiency of soil hauling-out activity is therefore almost doubled.

4 Unmanned excavation system

Revolutionary improvement of in-caisson excavation was finally achieved by putting the unmanned excavation systems into practical use.

In an unmanned excavation system (UES), the caisson-shovel operator sits in an atmospheric capsule installed in the working chamber of a caisson and he manoeuvres his caisson shovel with a wireless remote-control kit in a manner similar to the operation of a radio-controlled model car.

In the cylindrical steel capsule the air pressure is kept atmospheric in contrast to the high air pressure in the working chamber. It is designed to withstand external air pressures up to 4 bar in excess of atmospheric pressure with an ample factor of safety.

The capsule can be rotated about its vertical axis by the operator so that its windows of reinforced glass always face towards the caisson shovel and follow its entire movement. Thus he has a good view of the caisson shovel at all times, enabling him to operate the shovel efficiently.

Compared with direct control of the machinery by an operator within the working chamber, remote control from the capsule is more awkward requiring a higher degree of skill from the operator.

As illustrated in Figure 3, the atmospheric capsule is suspended from the piston head sliding in a vertical air cylinder which is firmly anchored to the reinforced concrete ceiling slab of the working chamber of a caisson. Figure 3(a) shows its

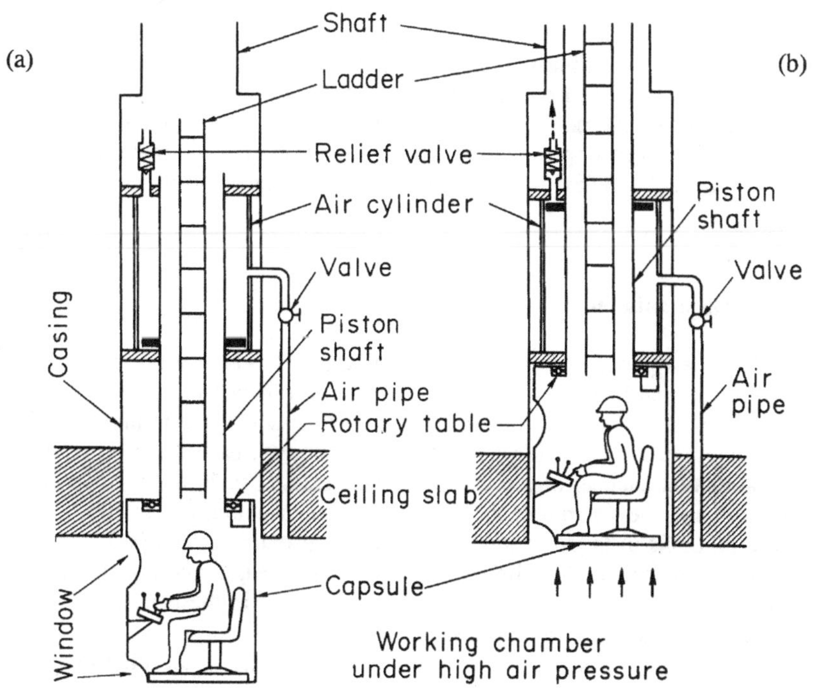

Fig.3. Atmospheric capsule
(a) Operating position, (b) Highest position

lowest operating position, in which compressed air is fed from the working chamber through an air pipe to fill the space in the cylinder above the piston head.

Should the caisson settle in an uncontrolled way and put the capsule at risk of being crushed, the increase in air pressure (as the volume of the working chamber reduces) activates the relief valve of the air cylinder – thus the capsule rises within the casing to be out of harm's way (see Figure 3(b)).

From recent works records of large deep caissons, hourly production rates of an unmanned excavation system vary within a wide range between 3 cu.m and 10 cu.m. This excavation rate depends largely on the skill of operators and other related factors such as soil-haulage distance in the caisson, degree of hardness in digging ground, or the cycle time of earth-bucket handling for discharging soil out of the caisson.

The hourly production rate by average operators in UESs is relatively low compared with the rate of direct operation in compressed air because of the difficulties of remote control of the operations.

However, the operators in the atmospheric capsules of UESs can work their full work-shift while the workable hours allowed for the operators exposed to the compressed air are limited by the need to comply with the exposure limits and decompression times.

For example with an air pressure of 3.5 bar, the operators exposed to the compressed air only work a quarter of their full work-shift hours while the operators in the UESs can work for almost four times longer.

Even if the hourly UES production rate is as little as half that of the direct operation, the shift output per operator is about double. This is in addition to the other benefits of UES of improving the working environment for caisson-shovel operators, in particular relieving them from the risks to health from compressed air work.

5 World's largest caisson

The world's largest pneumatic caisson of 160,000 cu.m in bulk volume was built and sunk down to its final level of 46.5 m below mean sea level from the sea bed in the 12-m deep water of Tokyo Harbour in 1989. This caisson is the foundation of the off-shore anchorage for the 798-m long Tokyo Harbour Great Bridge crossing over the main navigation channel of the harbour. This caisson is 70 m by 45 m in plan and 51 m high and the volume of the ground excavated within it was 109,000 cu.m.

Ten sets of UESs installed in this caisson dug approximately 60,000 cu.m of the ground layers deeper than 27 m below mean sea level. Eight additional manned caisson shovels assisted the UESs by digging the hardpan in the peripheral narrow strips inside and along the cutting edges of the caisson.

Two other sets of UESs were similarly used successfully in the off-shore side main-tower foundation caisson in 1989.

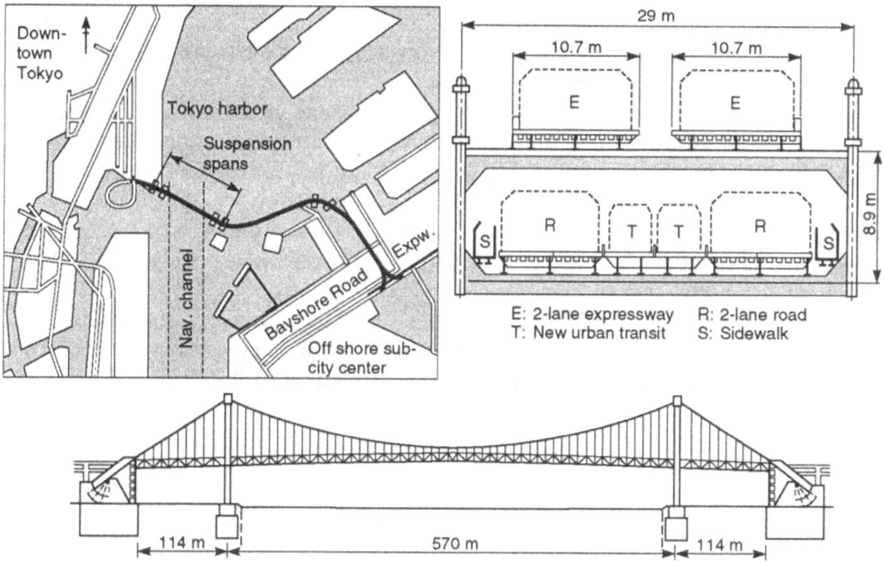

Fig.4. Tokyo Harbour Great Bridge

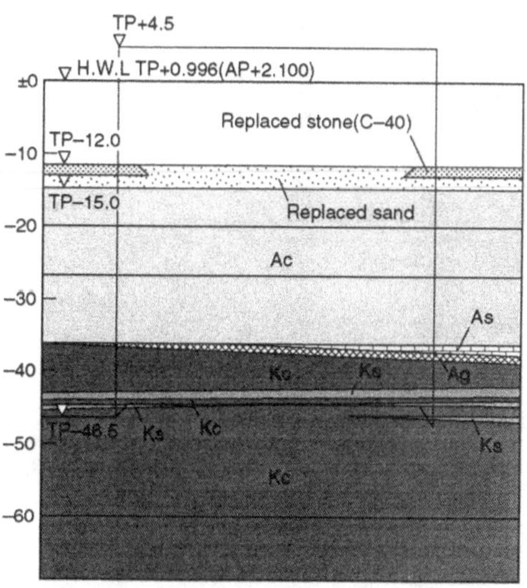

Fig.5. Soil profile

The Tokyo Harbour Great Bridge is being built on Route No. 12 of the Tokyo Expressway linking the city centre of Tokyo to the Marine Sub-city Centre being developed on Reclaimed Land Area No. 13, which lies on the east side of the main navigation channel in the harbour.

This suspension bridge when it is completed will carry the very heavy loads of four lanes of the Expressway, four lanes of freeways and double tracks of new-urban transits as well as sidewalks on its double-decked stiffening truss.

As shown in Figure 4, the centre span of this bridge is 570-m long and its side spans are each 114 m. As shown in Figure 5, the submarine ground at the anchorage site is formed by the very thick layer of very soft clay (Ac) underlain by the combined layers of sand (As) and gravel (Ag) as well as the very hard layers of hardpan (Kc) sandwiching the very dense sand layers (Ks). Groundwater was pumped up from the layers As, Ag and Ks through 6 deep wells installed outside around the caisson in order to reduce the maximum groundwater pressure, and consequently the air pressure in the caisson, to 3.5 bar or less.

Figure 6 shows the anchorage caisson with its working chamber of 2.3-m clear headroom encased at the bottom, and the latticed partition walls inside above the ceiling slab of the working chamber. A 5-m thick top reinforced concrete slab was placed on the top of the partition walls after the entire caisson works had been completed. The critical path activities in the construction of the caisson were not in the in-caisson excavation, but in the building-up works of the massive reinforced-concrete caisson body.

For assembling numerous reinforcement bars, erecting formwork and placing concrete of more than 3,000 cu.m for each 3-m lift of the huge caisson, at least four weeks were required, whereas the 3-m depth of excavation of 9,480 cu.m in ground

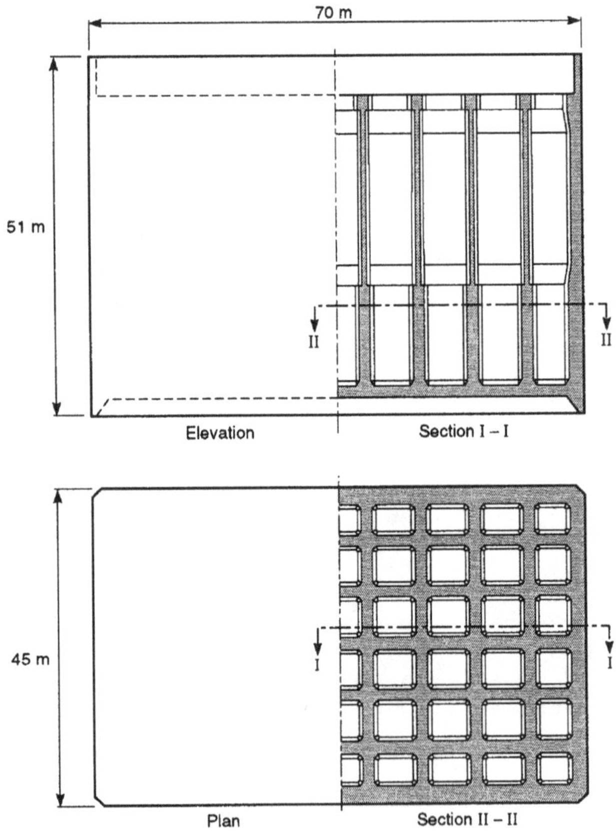

Fig. 6. Anchorage caisson of Tokyo Harbour Great Bridge

volume could be made by 10 sets of UESs in less than two weeks. In other words, the UESs were not occupied full-time, but operated intermittently.

Since 1990 to the present, several other submarine caissons of large size have been constructed for the foundations of the approach-span piers of the Tokyo Harbour Great Bridge and for the foundations of Tsurumi Estuary Great Bridge.

The pneumatic caisson for the West Main-Tower foundation of the Central Nagoya Harbour Great Bridge, whose final depth is designed to be 52.5 m below mean sea level, has been placed on the sea floor under 14.6-m deep water and its in-caisson excavation was started in December, 1991.

Many caisson shovels have been and are being used efficiently in the above mentioned caissons and some of them converted to UESs.

6 Further improvement made to UES

The atmospheric capsules, air locks, shafts, air pipes and valves now in use for in-caisson excavation by UES are designed for air pressures up to 4 bar. By additional reinforcement of the UES, ultra-deep caisson excavation may be made possible in the near future to depths exceeding 80 metres below the groundwater table.

In order to maintain the equipment and instruments installed in the working chamber under such high air pressures in good working condition, occasional inspection or minor repairs are required. Therefore, three or four men would have to enter the working chamber under very high air pressure for a short period, possibly once a week or fortnight for inspection purposes.

Engineering and medical aspects of this compressed air work under the very high pressures for the above purposes are discussed by Oda *et al.* (1994) in their paper to this conference.

Trial runs of a remote-control system of caisson shovel remotely operated from above the ground surface were performed in two 14.6-m dia. circular reinforced concrete caissons, one was 29-m deep and another 26-m deep, which were built and sunk in Futtsu City located about 45 km southeast of Tokyo across Tokyo Bay.

In this new control system, the operator sitting in the control room installed above the ground surface manoeuvres his caisson shovel while he is watching the TV images sent from video cameras installed both on the caisson shovel and inside the working chamber of the caisson.

The operator also observes other graphic displays indicating the position and the attitude of the shovel as are shown in Figure 7(a), together with the undulation of the ground surface being excavated as in Figure 7(b). The ground-surface display is transmitted from ultrasonic sensors attached to the shovel. These sensors receive the ultrasonic wave pulse reflected back from the ground surface and send electric signals to a microcomputer where the pulse-signals are processed to compose the graphic displays.

Operating efficiency of this new remote-control system is considerably less than the unmanned system operated from the capsule installed in the working chamber of a caisson, because it depends on interpreting flat images without the advantage of

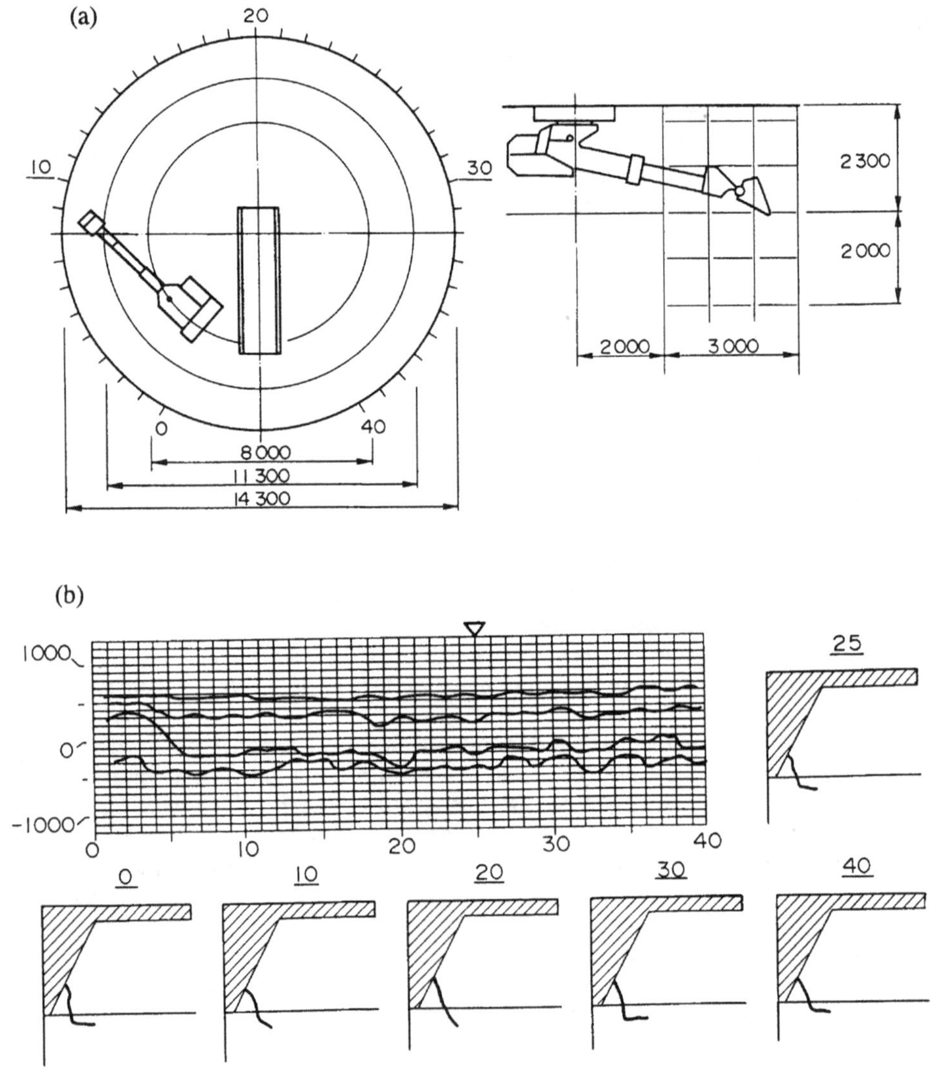

Fig.7. Examples of graphic displays
(a) Position and attitude of the caisson shovel in the working chamber
(b) Excavated ground surface

direct 3-D observation through the capsule-windows.

The new remote control system, however, is very suitable and economical for the use in a relatively small caisson where there is limited space for installing an atmospheric capsule in its working chamber.

7 Contribution and award

The development of unmanned excavation systems used in pneumatic caissons has been beneficial to the Japanese construction industry and its workers. This has been acknowledged by industry and government.

The fundamental policy of the Japanese government is to promote mechanisation and automation of various construction activities. The effort in developing the unmanned excavation systems followed this policy.

Such being the case, Shiraishi Corporation and Kajima Kenestsu Company, Ltd., co-sponsors of the research programme for developing the new remote-control system of in-caisson excavation, were honoured to receive the award of the Presidential Prize of the Japanese Construction Mechanisation Association in 1990.

Furthermore, the author's contributions to promoting modernisation of construction technology were recognised by the honour of receiving the Fifth-Degree Decoration in the award of the Rising-Sun Imperial Medal in November, 1991.

References

Shiraishi, S. (1990) World's largest caisson, *Civil Engineering*, American Society of Civil Engineers, Vol. 60 No. 8, August 1990, pp. 47-49

Shiraishi, S. (1989) Excavation of submarine ground in the world's largest pneumatic caisson in Tokyo Harbour, Symposium on *Underground Excavations in Soils and Rocks*, Sponsored by Asian Institute of Technology, Bangkok, Thailand, November 27-29, 1989, pp. 347-366

Shiraishi, S. (1990) History and recent development of pneumatic caissons, *Conference on Deep Foundation Practice*, CI-Premier Conference Organisation, Singapore, October 30-31, 1990, pp. 221-228

Oda, S. *et al.* (1994) Application of deep-sea diving technology to pneumatic works in very deep caissons, In: *Engineering and Health in Compressed Air Work*, (eds. Jardine, F.M. and McCallum, R.I.), E&FN Spon, London

42 Application of deep-sea diving technology to pneumatic works in very deep caissons

S. Oda, J. Takeuchi, M. Ishii and Y. Mano
Shiraishi Corporation & Tokyo Medical and Dental University, Tokyo, Japan

Abstract
The virtue of unmanned in-caisson excavation systems (Shiraishi, 1993), is that men are relieved from hard digging in the hazardous compressed air environment of caissons. However, occasional man entries under very high air pressure are still required for inspection or minor repairs of the excavating machines and appliances installed in the compressed-air chamber. Because the incidence of such harmful disorders as decompression sickness, nitrogen narcosis, dyspnoea or oxygen toxicity, is higher under air pressures exceeding 4 bar (5 ATA, kg/cm^2g) than under lower air pressures, the limiting pressure had been considered to be 4 bar. In order to overcome this pressure barrier, an experimental research programme was sponsored by Shiraishi Corporation with the cooperation of the Tokyo Medical and Dental University (TM&DU) and the Japan Marine Science and Technology Centre (JAMSTEC) aiming to establish a system which would enable men to enter into compressed air at very high pressures, up to 7 bar, by applying deep-sea diving technology to the compressed-air work. In this system, a mixture of oxygen, helium and nitrogen (Trimix) is used for breathing through masks at the same pressure as the air in the working chamber.
Keywords: Pneumatic caissons, Unmanned systems, Helium, Deep-sea diving, Compressed air, Trimix.

1 Introduction

The entrance of men into compressed air in caissons is strictly controlled by the Rules of Safety and Health for Workmen under High Air Pressure. These rules limit the duration under high pressure, specify the time scheduled for compression and decompression in air locks, and set the maximum limit of air pressure to 4 bar. Those limitations are specified primarily to prevent the occurrence of such harmful ill effects as decompression sickness. The viability of pneumatic caissons as a construction option is affected by the limitations on the allowable working time of men in compressed air, which severely restricts the efficiency of in-caisson

Engineering and Health in Compressed Air Work Edited by F.M. Jardine and R.I. McCallum. Published in 1994 by E & FN Spon. ISBN 0 419 18460 0

excavation and of caisson-sinking work, and on the limitation on air pressure which is equivalent to the attainable depth of caissons below the groundwater table. In order to overcome these handicaps, unmanned excavation systems were developed to relieve men from the hazardous work in compressed air.

Recently, demands for very deep caissons exceeding 40 m below the groundwater table have been increasing in Japan. In order to cope with these demands, the research and development programme of Great-Depth Pneumatic Caissons (GDPC) as illustrated in Figure 1 commenced six years ago.

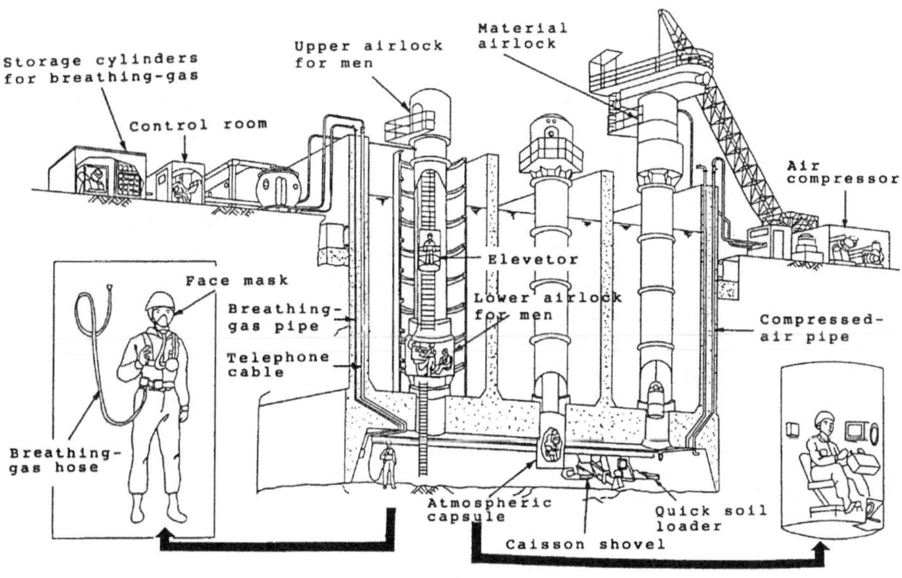

Fig. 1. Great-depth pneumatic caisson

The unmanned excavation systems may be applied efficiently to compressed-air work under very high pressure, exceeding 4 bar, provided that all the pneumatic vessels such as airlocks or shafts are reinforced to bear the air pressure with an ample factor of safety. However, occasional man entries into the very high pressure air for inspections or minor repairs have to be made in order to keep the machines and appliances installed in the compressed-air chamber in good working condition.

For these purposes, three or four men would have to enter into air at pressures up to 7 bar, and to stay there for as long as 90 minutes, possibly once a week or once a fortnight.

The research work for realising the GDPC was commenced in 1986 in both TM&DU and JAMSTEC. Animal tests as well as human experimentation were performed in compressed-air chambers in which the air pressure could be controlled to rise up to 7 bar.

While the above tests were performed, the equipment and instruments required for the prototype GDPC were designed and fabricated.

2 Limitations and hazards of work under compressed air

The possibility of various health hazards to men becomes high when compressed air is breathed by them in a high pressure environment.

2.1 Oxygen (O_2) toxicity

When the partial pressure of oxygen in the air being breathed is too great there can be signs and symptoms of acute toxicity (Paul Bert effect), e.g. epileptiform convulsions, loss of consciousness, or even death. For more prolonged exposures at lower oxygen pressures, the signs and symptoms of chronic oxygen toxicity (Lauren Smith effect), e.g. respiratory problems, can be encountered.

There are individual differences in the permissible limit in partial pressure of oxygen, P_{O2}, or in the duration of exposure to compressed air. It also changes depending on the physical condition of the individual at the time when he is exposed to the compressed air. As shown in Table 1, the safe duration of exposure decreases as the P_{O2} of breathing air rises.

Table 1. Safe limits for oxygen breathing

Partial pressure of oxygen (kg/cm^2)	Equivalent air pressure (kg/cm^2)	Limit on exposure time (min)
1.6	6.6	30
1.5	6.1	40
1.4	5.7	50
1.3	5.2	60
1.2	4.7	80
1.1	4.2	120
1.0	3.8	240

2.2 Nitrogen(N_2) narcosis

When the partial pressure of nitrogen P_{N2} in breathing air is raised sufficiently then conditions similar to intoxication by alcohol may appear. Thinking ability is reduced; repetition of simple actions is impaired.

These kinds of symptoms appear in sensitive men even at the relatively low pressure of 3 bar; and when the pressure rises beyond 6 bar, N_2 narcosis affects everybody, except for very short exposures (Bennett, 1982).

If the pressure exceeds 10 bar, it is liable to induce loss of consciousness. There is no harm in N_2 narcosis itself, but anyone affected by it at work is liable to secondary accidents. Figure 2 shows how calculating abilities decline as the pressure of breathing gas increases. Although breathing N_2-O_2 gas (natural air) severely affects calculating ability, breathing Heliox (He-O_2) gas has little effect on this ability if the pressure is not more than 12 bar.

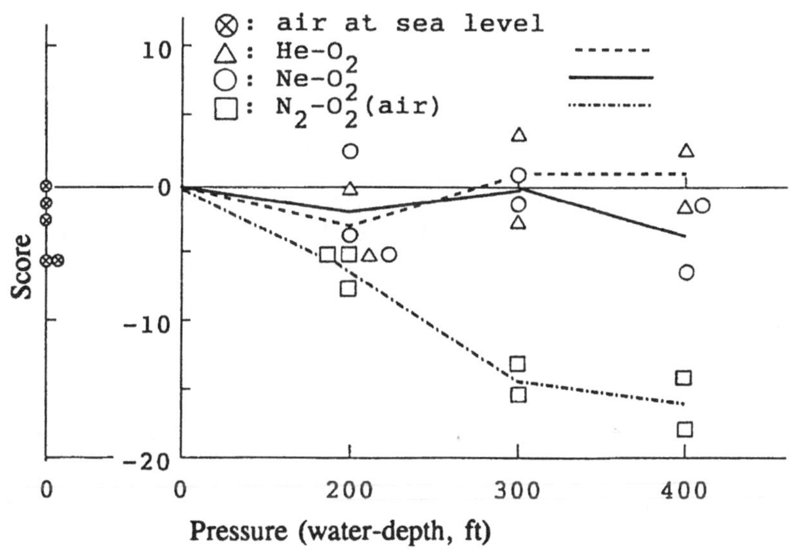

Fig.2. Decrease in calculating ability

2.3 Dyspnoea and carbon-dioxide (CO_2) toxicity

Because respiratory resistance increases in proportion to the square root of the density of breathing gas (Graham's Law), the respiratory resistance doubles in an environmental air pressure of 3 bar (4 ATA) (square root of 4 = 2). Therefore, dyspnoea may occur if the density of breathing gas increases radically. When heavy work is performed while breathing high-density gas, intake of breathing gas may be insufficient and removal of CO_2 in a human body may also become impaired.

2.4 Acute increase in occurrence of decompression sickness

Table 2 shows the incidence of limb bends (symptom of decompression sickness) for various air pressure levels in caissons. Although the recent incidence is showing a decreasing trend, the rate in working air pressures exceeding 3 bar is more than 2% even though the decompression schedules specified in the Rules of Safety and Health

Table 2. Occurrence rate of limb bends in pneumatic caissons

Working air pressure (bar)	Occurrence rate of limb bends (%)		
	before and in 1980	1981-1985	overall average
0 to 1.0	0	0	0.0
1.0 to 2.0	0.27	0.07	0.16
2.0 to 3.0	2.56	0.82	1.37
3.0 to 3.6	4.45	2.21	3.06
1.0 to 3.6	0.83	0.35	0.54

for Workmen under High Air Pressure (High-Pressure Rules) are strictly observed (Mano *et al.*, 1987).

According to the High-Pressure Rules, when the working air pressure is 3.4 bar and the time in compressed air (from the beginning of compression until the beginning of decompression) is 120 minutes, the total decompression time should be 129 minutes as shown in Figure 3. Including the additionally required time of 60 minutes for reducing the gas pressure remaining inside the human body after the decompression is over, the total paid time for the 120-minutes of work under the high air pressure is 309 minutes or nearly 2.6 times the net time under high pressure.

Thus, the efficiency of the pneumatic works declines remarkably as the air pressure rises.

3 Application of deep-sea diving technology

The application of deep-sea saturation diving technology for long-duration work of men under air pressures exceeding 4 bar was studied in the past. However, this scheme was discarded because it would not only have required large-scale equipment similar to those used for saturation diving such as high-pressure vessels for living quarters and elevator-cabins, but also very large amounts of extremely expensive helium which would have been consumed in keeping men for several days under high pressure.

In the Great-depth Pneumatic Caisson projects, it is not necessary to have men working for long hours under very high pressure. Only occasional short-duration work for routine maintenance of digging machines and other appliances installed in the compressed-air chamber is required. Men could do the relatively brief work breathing Trimix through their face masks while they stay under the high pressure. There is neither need for large-size equipment like high-pressure living cabins nor large amounts of helium for long stay under very high pressure. Therefore, very deep caissons can be built and dug economically without consuming too much expensive helium.

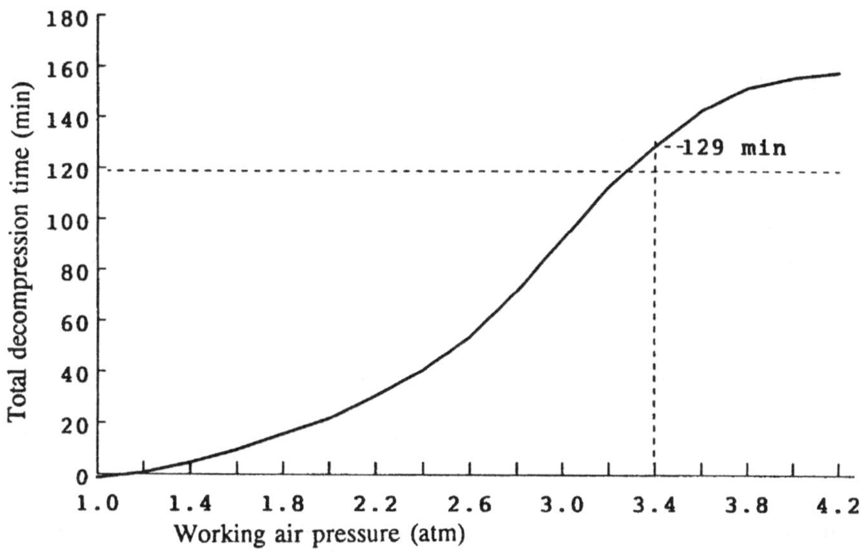

Fig. 3. Total decompression time

4 Deep-sea diving technology

4.1 History and present situation

When submarine F-4 of the US Navy sank to the sea bed at a depth of 91 m off-shore from Honolulu in 1912, effective rescue activity by divers was not possible and they were unable to save the lives of the submarine crew. Since this accident, research efforts were made by the US Navy to improve deep-sea diving technology by developing new breathing gases. In 1924, animal tests and living-body tests were performed using helium mixed gas ($He\text{-}0_2$) for breathing and the great merit of this type of gas was clearly proved.

Further research efforts were made not only in the United States but also in such European countries as France and the United Kingdom until diving down to the great depth of 300 m can be made in routine practice and diving experiments to the record depth of 600 m (60 bar of pressure) were successfully made.

Also in Japan, the JAMSTEC and the Self-Defence Navy have succeeded in making diving experiments down to a depth of 300 m in simulation tanks (pressure chambers) and in natural sea water by using Heliox as the breathing gas. Furthermore, commercial Heliox diving has been improved to the extent that diving to a depth of about 150 m is in normal practice for divers working on the seabed for finishing submarine construction projects (e.g. Honshu-Shikoku, 1978) or in deep-water diving for repair works on dams (Maehara, 1988).

4.2 Merits of helium-mixed gas as breathing gas

(1) Prevention of oxygen toxicity. Helium-mixed gas (HMG) is the artificial mixture of either oxygen and helium (Heliox), or of oxygen, helium and nitrogen (Trimix). When the mixing proportion of those components is adjusted to fit the working pressure properly, oxygen toxicity can be prevented completely.
(2) Nitrogen narcosis is not induced. By breathing HMG that either does not include nitrogen or only has a small amount of it, nitrogen narcosis is not induced. By using the HMG as the breathing gas, no influence of nitrogen narcosis appears and calculating ability does not decline.
(3) Respiratory resistance is reduced. The molecular weight of helium is very low, i.e. 1/7 of nitrogen or 1/8 of oxygen. Therefore, its respiratory resistance is comparatively small. Because dyspnoea or CO_2 toxicity is mainly caused by the increased density of the breathing gas, HMG is the most suitable breathing gas for eliminating these defects. For diving to a depth exceeding 300 m, hydrogen, whose molecular weight is half that of helium, is used instead because, at very great depth, even helium is too dense.
(4) Low occurrence rate of decompression sickness and reduction in decompression time. As shown in Table 2, the incidence of decompression sickness, by breathing natural air when the air pressure exceeds 3 bar, is high at 2%. However, by breathing Heliox, the records of 721 exposures to a pressure of 7.3 bar show an incidence of only 0.83%. Breathing HMG is effective in reducing the rate of occurrence of decompression sickness.

Because helium is less soluble in the fat of the living body than nitrogen, the required decompression time is shorter than is the case when breathing natural air.

Figure 4 illustrates the merit of Heliox in the pressure range exceeding 3.0 bar.

4.3 Disadvantages of breathing helium-mixed gas

(1) Helium voice. The voice issued when breathing gas of very low molecular weight such as helium is high-pitched like that of Donald Duck in Walt Disney's animation pictures. This kind of shrill voice is not only inadequate for conversation between workmen but can also be misleading in an emergency situation.
(2) Loss of bodily heat. Because the specific heat of helium is about 5 times and its heat conductivity about 6 times that of air, the heat loss (from the body surface exposed to the helium mixed gas and in breathing it) acutely increases to make men very sensitive to temperature changes. The optimum temperature of the compressed air in the working chamber of a caisson is usually 18°C-25°C, but if bodily heat is lost, for the above reasons, men shiver as from bitter cold, and heating might be required.
(3) Evaporation of moisture in living body. The relative humidity of the helium-mixed gas in the storage cylinder is nearly 0%. Breathing this very dry gas for a long time may give men a painful thirst. Some humidity adding appliances are required to be incorporated into the breathing-gas feeding lines.

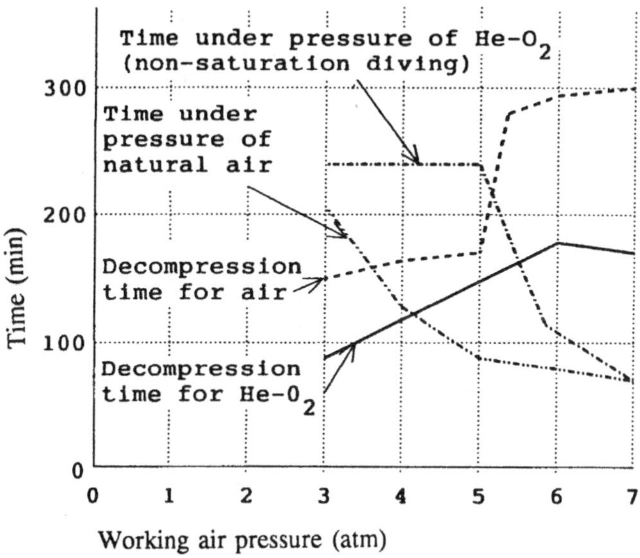

Fig.4. Maximum time under pressure and specified decompression time (Heliox: Diving Manual (US Navy, 1981) Air: High-pressure Rules)

5 Outcome of human experimentation

Four healthy men who had experience in compressed-air work were selected for the living-body tests under very high pressure. The tests were performed in the high-pressure chambers of TM&DU and JAMSTEC over the pressure range of 0-7 bar in an environment of compressed air, either wearing the masks for helium-mixed gas or without any mask and breathing air. During those tests, psychological experiments and kinetic-load tests were performed.

5.1 Physiological functions of tested men

The physiological features and heart-lung functions of the tested men are as shown in Table 3. The maximum oxygen ingestions $V(O_2)_{max}$ (the volume of oxygen ingested or consumed per unit weight (kg) of a tested man in a minute) of all the tested men are higher than the average values according to L.E. Morehouse's classification (39.5-44.5 ml/kg/min for the 24-29 year olds and 37.5-42.5 ml/kg/min for the 30-35 year olds).

As the oxygen ingestion ratio % $V(O_2)_{max}$ is roughly proportional to the heart rate per minute (HR), approximate values of % $V(O_2)_{max}$ are given by the following equation.

$$\% \ V(O_2)_{max} = 0.64 \ (HR) - 28.2 \ (\%)$$

Because the devices for measuring the values of % $V(O_2)_{max}$ are complicated and may not be brought into the high-pressure chambers, the values of % $V(O_2)_{max}$ were derived by the above equation from the measured values of heart rate (HR). The maximum possible amount of activity in a caisson is considered to be the ascent and descent by the vertical ladder between the upper airlock and the elevator or between the lower airlock and the working chamber. As these movements take several minutes, the values of % $V(O_2)_{max}$ were estimated to be 50%; and the 10-minute activity exercise equivalent to 50 % $V(O_2)_{max}$ was performed by the man in the high-pressure chamber (as shown in Figure 5) (Shibayama *et al.*, 1989).

Table 3. Physical functions of tested men

Man	Age (y)	Height (cm)	Weight (kg)	Lung capacity (l)	
A	35	173	74	3.31	
B	27	169	65	4.13	
C	28	167	65	4.24	
D	25	164	58	4.13	
Ave.	29	168	66	3.95	
S.D.	4	4	7	0.43	
Man	**Max. beats (pulse/min)**	**$V(0_2)_{max}$ (ml/kg/min)**	**V_{max} (1/min)**	**Breath rate (/min)**	**V_r (1)**
A	191	44.7	105.9	49.4	2.15
B	192	51.4	148.4	52.6	2.82
C	179	47.4	128.0	83.3	1.54
D	200	46.8	138.0	57.7	2.39
Ave.	190.5	47.6	130.1	60.8	2.22
S.D.	8.7	2.8	18.1	15.4	0.53

Notes: V_{max} and V_r are the maximum breathing capacity rate per minute and the volume of one breath, respectively.

5.2 Appropriateness of decompression chart

For the tests of exposure to high air pressure, the decompression chart developed by the Tokyo Dental Medical University was applied.

After finishing the 30-minute exposure to an air pressure of 7 bar by breathing natural air and after finishing the 60-minute exposure to the 7-bar air pressure by breathing the Trimix, the degree of bubbling induced in the blood was sensed by means of an ultrasonic Doppler air-bubble sensor (Figure 6).

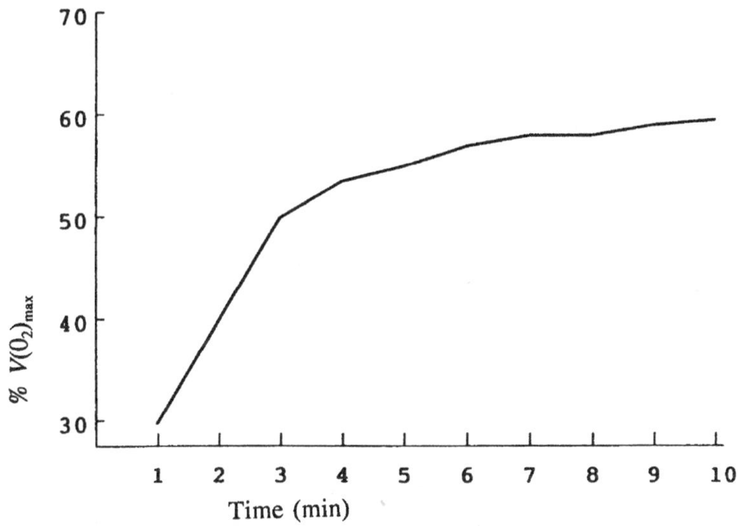

Fig.5. % $V(O_2)_{max}$ in descent and ascent of ladder

Fig.6. Listening to air-bubble noise

This sensor detects the air-bubble noise by the Doppler effect which relates to the degree of air bubbles which appeared in a gelatin specimen simulating human blood. The man exposed to the high air pressure can sense the noise through his headphones.

The result of the air-bubble sensing, made 3 h after the decompression had been finished, was that a slight degree of bubbles corresponding to Stages I or II of Table 4 was observed, but no subjective symptoms of decompression sickness were recognised. No bubble was sensed at 18 h after the decompression had been finished.

It was therefore concluded that the decompression chart used in these tests was appropriate for the safety and health of men under high air pressure.

5.3 Adaptability of living body to high pressure

(1) Mental response (N_2 narcosis and fatigue). The scores of mental tests made in the high-pressure chamber on the men when breathing the Trimix were better than

when breathing air, and symptoms of nitrogen narcosis were not recognised (Table 5, Figure 7).

Table 4. Stages of Doppler air-bubble sensing

Stage	Degree of air bubble appearance
0	No air bubble sensed
I	Slight air-bubble noise at irregular intervals
II	Bubble noise in less than half a pulse-beat period
III	Air-bubble noise in entire period, pulse-beat noise is also heard
IV	Air-bubble noise in entire period, pulse-beat noise is not clearly heard

Table 5. Results of tapping tests

	Breathing air at 0 bar	Breathing air at 7 bar	Breathing Trimix at 7 bar
Score	100	79.6	90.9

Fig.7. Tapping test (count successful taps with a small hammer on tiny pin heads)

Furthermore, from 'subjective-symptom questionnaires', it was found that the degree of fatigue was less in Trimix breathing than in air-breathing, because the Trimix with a lower density than air was easier to breathe (Table 6).

No fatigue was reported to remain after decompression following breathing Trimix.

(2) Loss of bodily heat. During the decompression from a state of Trimix face-mask breathing at an environmental air pressure of 7 bar and at a room temperature of 23.1-31.3°C (average 26.7°C) and with the room temperature being 16.9°C immediately after the decompression was started, body temperatures were measured at the skin on the forehand, breast, handback and in rectum. Because there was no heat loss from the surface of the body, the loss of bodily heat did not seriously affect

Table 6 Scores answered to subjective symptom questionnaires

Environmental pressure	Breathing gas	Group I	Group II	Group III
7 bar (immediately after compression)	Air Trimix	56.7 16.7	43.3 20.0	20.1 13.3
0 bar (immediately after decompression)	Air Trimix	3.3 10.0	0.0 0.0	0.0 6.7

Remarks: Group I; 'sleepiness' and other weariness in general (10 items)
Group II; 'difficulty in concentration' and other mental symptoms (10 items)
Group III; 'mental sensing symptoms'.

the living body. There was no need for heating, wearing warm clothes was sufficient for enduring the coldness.

(3) Helium voice. There was no problem in conversations among the men in the high-pressure chamber and they could speak by telephone to places outside chamber without any confusion. The use of Facsimile was found useful for transmitting complicated messages.

There are expensive kits for correcting helium voice tuned to pressures higher than 10 bar, but it seems not necessary to provide such an expensive kit for a pressure of 7 bar.

(4) Evaporation of water content from the body. The feeling of thirst is not sensed when sitting quietly for one hour or so and breathing Trimix through a face mask. However, some feeling of thirst is sensed by moderate activity (equivalent to table tennis play) for 10 minutes. Therefore, moistening kits as shown in Figure 8 are deemed necessary for mask-breathing of two hours or longer (including the decompression time). The trial use of the moistening kits was successful.

5.4 Heart-lung function and breathing mask type

Movement-load tests by running an ergometer (work-load meter by rotating a bicycle wheel) were made by each tested man at pressures of 0, 1, 3, 5 and 7 bar at 40-50% of the % $V(O_2)_{max}$ of each man. In those tests, comparisons were made between the two types of masks, one covering the entire face (face mask; Figure 9(a)) and another covering only the mouth and nose (oronasal mask; Figure 9(b)). It was confirmed that both of those two types are suitable for their purpose. Although the face mask is inconvenient for the man who wears eye glasses, it is nevertheless better because of its smaller respiratory resistance (Figure 10) and greater durability. The face mask, which belongs to the pressure-demand type (positive pressure type), has a further advantage because the pressure inside the mask is slightly higher than the environmental pressure and no environmental gas can intrude into the mask. Therefore, the face mask is preferred for use in practice.

5.5 Amount of breathing-gas consumption

The relation between the amount of movement or the value of % $V(O_2)_{max}$ and the amount of breathing-gas consumption is as shown in Table 7 (the values given were measured at the corresponding environmental pressure).

Table 7. Relation of % $V(O_2)_{max}$ to breathing gas consumption

Amount of movement	Amount of breathing-gas consumption (1/min)	Standard deviation (1/min)
State of rest	13.9	4.4
30% $V(0_2)_{max}$	34.0	10.7
50% $V(0_2)_{max}$	45.3	11.7

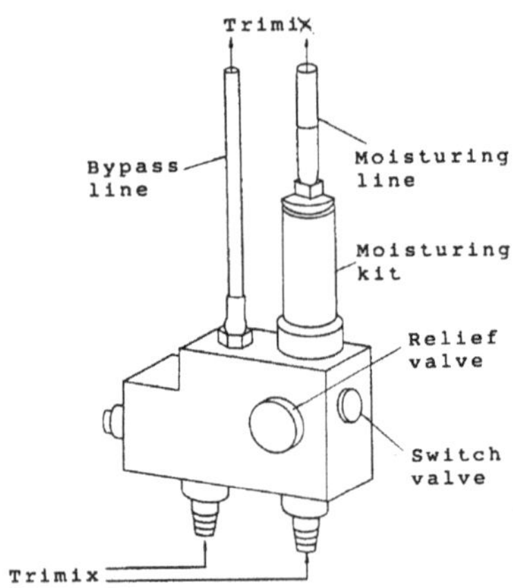

Fig.8. Moistening kit

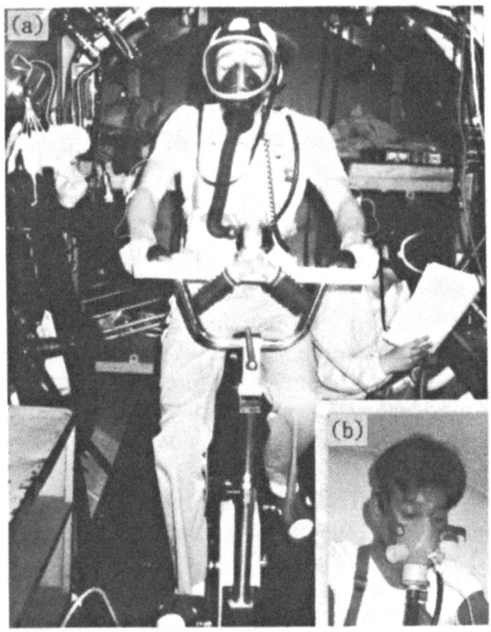

Fig.9. (a) Ergometer test at 7 bar with face mask. (b) With orinasal mask

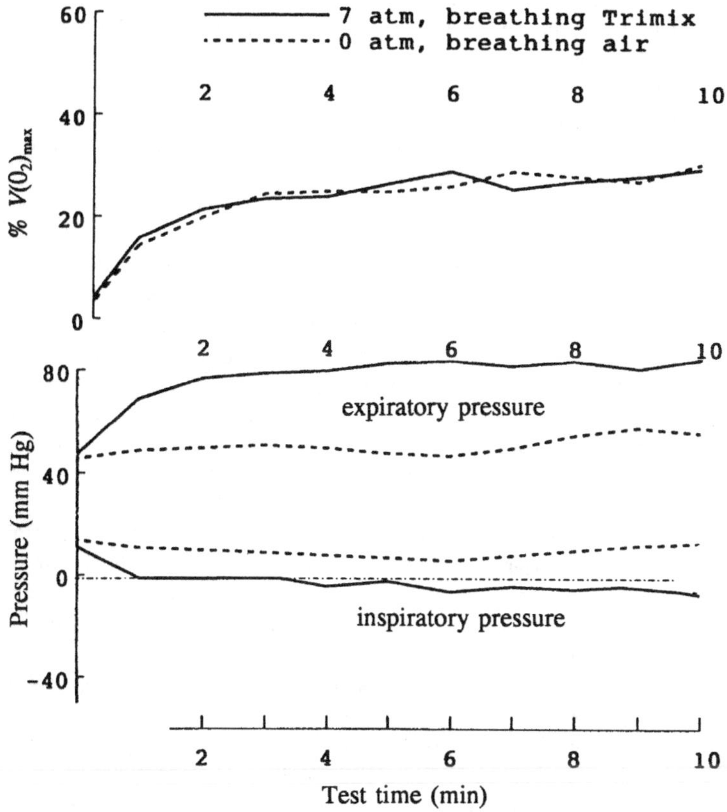

Fig.10. Respiratory resistance during ergometer rotating tests

6 Conclusions

The results of human testing are summarised as follows.

(1) Because, even after exposure to an air pressure of 7 bar for 90 minutes, the decompression could be completed without any occurrence of such a disease as decompression sickness or any other abnormal symptom, it may be stated that the application of deep-sea diving technology was successfully made, and that the decompression chart applied to the living-body tests is appropriate for practical use.

(2) Such disorders as oxygen toxicity or nitrogen narcosis which might have occurred when air was breathed at high pressure could be prevented by breathing the helium mixed gas (Trimix in this case).

(3) The problems of bodily heat loss and helium voice are not as serious as had been anticipated. There is no problem in the pressure range up to 7 bar.

(4) The loss of moisture in the living body by evaporation due to the dry breathing gas supplied from the storage cylinders may be replenished by incorporating moistening kits into the breathing-gas line.

(5) A type of face mask suitable for moderate activity work of 40-50% $V(0_2)_{max}$ under a pressure of 7 bar has been developed and evaluated. It proved successful because of its low respiratory resistance and durability.

(6) The standard amount of breathing gas required for a workman in a caisson may be set at 45 l/min based on the amount of his movement within it.

There is a definite difference between diving work in water and work in compressed air. The environment of the former is water whereas the latter's environment is air. This difference in environment gives radically different consequences. In deep-sea diving, the rupture of a breathing line means death of the diver whereas, in compressed air, even if the breathing line is cut, the man can transfer to the air lock, make his decompression in there and safely come back to atmospheric pressure. There was an example where a man, after breathing air at a pressure of 7 bar for 30 minutes, came out safely through decompression in the airlock.

Because the coefficient of heat-conductivity of water is very high and 25 times that of air, one of the most important requirements in diving work is to maintain a comfortable temperature of the diver's body, whereas, in the environment of compressed air, there is not much loss of heat from the surface of living bodies and there is little problem in keeping them warm.

Therefore, it may be stated that there is not much difficulty in the application of deep-sea diving technology to work in compressed air.

'Great Depth Pneumatic Caissons', when ready for wide-spread practical use, will counteract the prejudice against caissons as bad working environments or an inhumane use of labour. The extension of the attainable depth limit will widen the scope and opportunities for pneumatic-caisson builders and for the construction industry.

References

Bennett, P.B. (1982) Inert gas narcosis, in: *The Physiology and Medicine of Diving*, Baillière Tindall, 3rd Ed., pp 239-261

Committee on Civil Engineering Constructions, Japan Society of Civil Engineers (1985) *Construction technology of marine works - Substructures of Kojima Sakaide Route of Honshu-Shikoku Connecting Bridges*, September, 1985 (in Japanese)

Honshu-Shikoku Connecting Bridge Public Corporation, (1978) *Study on excavating methods in deep caissons*, Final Report, June 1978 (in Japanese)

Ishii, K. *et al.* (1979) (Trans.) *Movement physiology*, Kyorin Shoin, December, 1979, 227-244 (in Japanese)

Maehara, M. *et al.* (1988) Replacing work of intake-tower's screen by means of saturation diving, *Civil Works in Electric Power*, No. 216, September, 1988, 27-36 (in Japanese)

Mano, Y. *et al.* (1987) Occurrence rate of decompression sickness in compressed air works in recent five years, *Industrial Medicine*, Vol. 29, 271-278 (in Japanese)

Powell, M.R. *et al.*, (1982) Ultrasonic surveillance of decompression, in *The Physiology and Medicine of Diving*, (Bennett, P.B. and Elliott, D.H. eds.) Ballière Tindall, 3rd Ed., 404-434

Shibayama, M. *et al.* (1989) Work intensity based on the functions of breathing masks used in great-depth pneumatic caissons, *Physiology and Human*, Vol. 8, No. 3, July 1989, 155-164 (in Japanese)

Shibayama, M. *et al.* (1990) Work efficiency in compressed air at the pressure of 8 ATA by breathing air and Trimix, *Physiology and Human*, Vol. 9, No. 2, April 1990, 100-108 (in Japanese)

Shiraishi, S. (1994) Unmanned excavation systems in pneumatic caissons, In: *Engineering and Health in Compressed Air Work*, (Jardine, F.M. and McCallum, R.I. eds.), E&FN Spon, London

US Navy (1981) *Mixed-gas diving*, US Navy diving manual, Vol. 2

Workman, R.D. (1969) American decompression theory and practice, in: *The Physiology and Medicine of Diving*, (Bennett, P.B. and Elliott, D.H. eds.) Baillière Tindall, 3rd Ed., 271

Discussion on Session 9: Developing technologies

The chairman of the session, **Mr R P Lewis**, introduced those presenting the papers of the session. **Dr R W Hamilton** reviewed the options for tunnel work at pressures higher than 3 bar (gauge). **Mr J C Le Péchon** examined the use of oxygen decompression in tunnelling. **Mr G Luther** explained how partial saturation techniques when integrated into the tunnelling system could provide a solution for man intervention in the face of a TBM at very high pressure. **Mr S Oda** described the development of hyperbaric procedures and techniques for use when man entry is needed for maintenance of a caisson excavating system. **Mr T Hirata** explained the method of working of a remotely controlled caisson excavation system and the corresponding hyperbaric procedures when, for inspection and maintenance, men have to enter the excavation. **Dr Y Mano** introduced the paper by Dr S Shiraishi, who could not be present, which describes the development and use of the unmanned excavation system for very deep caissons.

The report below is the discussion that followed these presentations.

Dr W Sterk asked Mr Oda to give details of the composition of the trimix and of the decompression schedule they used: these were not given in either the paper or the presentation.

Dr Y Mano replied that in the experimental work the partial pressures of oxygen and nitrogen were 0.6 and 2.4 atm. respectively, although these will be changed when in operational use.

Mr Lewis asked if trimix has been used in Europe.

Mr Le Péchon answered that trimixes gases have been used in France since the early 1980s in deep diving. The French Navy always use trimix (tables developed by Dr L Barthelemy and Mr J Parc). C G Doris, under my supervision, has been using similar trimix gases (O_2 = 1.4 bar, N_2 = 3.2 bars + helium as required) for all bounce dives since 1973 and until 1986.

In tunnelling, for a feasibility study I carried out for the Channel Tunnel project (Robbins T1), compressed gases operations were to be carried at 10 bars. For non-saturation interventions, trimixes with breathing masks and full body protection were considered, air being used to dry up the working chamber. Heliox was considered for works in saturation. Due to the costs involved, ground treatment was selected by the contractor and used when needed throughout the project. There has been no hyperbaric works at any time during this project.

Dr Sterk said that trimix breathing mixtures are used by Dutch diving companies to depths of about 80m.

Dr I Nashimoto, referring to the paper by Oda *et al.* asked why they chose a pO_2 of 0.6 bar for their breathing (trimix) mixture during experiments at 8 bar absolute? The bottom time of the chamber dives was 60 to 90 minutes. They were not saturation dives. Dr Nashimoto thought that the pO_2 of 0.6 bar is too low. At atmospheric pressure this means that the O_2 concentration was only 7.5%. If the pressure inside the working chamber or the man-lock reduces rapidly in case of trouble, it could easily induce severe hypoxia in the workers: it would be very risky. A further question for Dr Mano is why are they changing the pO_2 value in their trimix from 0.6 bar to 1.6 bar in actual caisson work? In the paper, they present in Table 1 the safe limits for oxygen breathing, as apparently taken from the US Navy for normal exposures. Within these limits, an exposure to a pO_2 of 1.6 bar is only allowed for 30 minutes. However, as reported by Sterk (1987) (The use of oxygen in decompression. In: Nashimoto, I., Lanphier, E.H., eds., *Decompression in surface-based diving*. Report 73(DEC)6/15/87, Bethesda, MD: Undersea Hyperbaric Medical Soc; 13-19), these limits are considered too conservative in commercial diving and usually the US Navy limits for exceptional exposure are taken. In these limits, Sterk discovered some inconsistencies and advised setting the maximum allowable oxygen dose by calculating CPTD to prevent whole-body oxygen toxicity effects, while for actual diving the maximum pO_2 should be 1.5 bar to prevent CNS toxicity. What are the reasons for choosing a pO_2 of 1.6 bar for actual caisson work?

Dr Mano's answered that the design of their decompression table for trimixed gas was chosen for a pO_2 of 0.6 bar, but the pO_2 to be used in actual operations would be 1.6 bar. The reason for the changed pO_2 level is that although there are tables more than sufficiently safe for decompression after saturation, their decompression table still has to gain credit by use with short bottom-time of trimix gas exposures, i.e. it needs 'banking' by experience. In operation, the total decompression time will become a little bit longer than Dr Sterk's table, but Dr Mano prefers to use a safer table. The reason for choosing a pO_2 of 1.6 bar rather than Dr Sterk's recommendation of 1.5 bar is that the work they plan for is inspection or maintenance of an unmanned excavation system and not heavy labouring work as in hand-mining. The tasks are light so that 1.6 bar will be safe, although, of course, the operatives are not immersed in water like divers.

Professor D N Walder asked if Mr Le Péchon could give an indication of DCS rates with oxygen decompression.

Mr Le Péchon said that as yet it was unknown; for example, over 2 months last year with its use with a slurry machine, there were very few interventions.

Dr J D King asked Mr Luther about the possibility of their proposed system being used at Storebaelt.

Mr Luther answered that he had been told that ground treatment was likely to be the chosen method.

Mr D R Parks of Mott MacDonald and who is working on the project confirmed it was not the contractor's intention to use the transfer pod for working at compressed air pressures above 3 bar on the Storebaelt Project. Current developments to achieve this were by dewatering in the tills and grouting ahead of the TBMs in the marls, where higher potential pressures were expected. In this way it was hoped that compressed air pressures less than 3 bar could be achieved for the deeper sections of tunnels.

Mr D R Lamont. In diving a reserve supply of breathing mixture is required by divers using gas-line-fed breathing apparatus. This reserve is provided by means of a self-contained cylinder supply. Would such a reserve supply be required for work inside a tunnelling machine head where an air-line-fed breathing supply was being used and when the breathing mixture was nitrox, heliox or trimix? Would a self-contained supply inhibit mobility?

Mr Luther answered that air might be the right gas for the working chamber, but this has to be reviewed.

Mr Le Péchon commented that an integrated set-up would be needed, probably involving a re-breather system.

Dr Sterk. A test for experience would be to do without trimix for 5 min, assuming this is enough time to get away.

Mr Lewis asked if oxygen will be used regularly, even solely, for decompression.

Mr Le Péchon gave his opinion that contractors would choose because the incentive to use oxygen is very strong, e.g. where the limit is 6h/day including decompression times and oxygen saves 50% of the decompression time. When there is a major repair it would be necessary to go to oxygen, although for short working periods it would not be needed.

Mr A P Smith. In view of the comments in CIRIA 44 about the use of compressed oxygen not being recommended on compressed air sites, have those responsible for the document and its revision, and also the HSE, heard anything from the many proponents of the system to make them change their minds and allow its use in UK tunnelling?

Dr R W Hamilton responded that the **only** accepted therapy for DCI today is with hyperbaric oxygen and condemned as nonsense any idea that oxygen therapy is not worthwhile.

Professor Walder, accepting that CIRIA Report 44 needs updating, said that he remains uneasy about the use of oxygen at high, frequent dosages. Nobody has come up with a DCS rate that is less than the Blackpool tables – the results in the papers in this conference, for example, show worse rates.

Dr King commented that he also had been involved in the Compressed Air Working Group's proposed revision of CIRIA Report 44 and so recognised that situations have changed. HBO therapy may be appropriate with the right people and the right equipment, provided that the workforce are also persuaded. He noted that in treating over 2,000 cases of decompression sickness, they all got better – without oxygen therapy. Nevertheless the projects like the Dartford Tunnel, with many men labouring, will be rare. The use of machines will not only reduce the workforce, but also require technicians rather than labourers. For this entirely different group and the different circumstances, there is a case for the use of oxygen in routine decompression.

Dr I J Ribeiro. In responding to Dr Walder I would like to emphasise the very low incidence of decompression sickness in our experiment with oxygen decompression in Brazil. We had just one case of DCS in 902 man decompressions from 1.5 bar and working shifts of 7 and 8.5 h. Such an incidence of 0.11% was four times lower than that obtained when using air decompression from the same working pressure and shift duration. Additionally we were able to save time by using oxygen decompression, for it was reduced to one half of that using air decompression tables. Therefore, I believe that, from both the medical and engineering points of view,

oxygen decompression is the best method of decompression for compressed air workers, particularly when dealing with working pressures above 2 bar absolute.
Dr Hamilton. The methods for the use of oxygen are not necessarily linked with DCS incidence. When you change a procedure you plan on some decompression illness. Any significant exposure to pressure entails a certain risk of DCI. It often occurs following exposures that are well under limits considered acceptable. DCI is not an accident; there appears to be a certain irreducible limit. But we can more or less dial up the incidence level considered acceptable by adjusting the procedures.
Dr H V Hempleman commented on the use of oxygen for speeding decompression. Some compressed air contracts last for several months, and sometimes as long as two years. If a worker breathes high pressures of oxygen every day for many weeks or months, we have no knowledge of how this may affect his health (someone later remarked that there was evidence for damage to retinal blood vessels). Patients receiving hyperbaric oxygen therapies, or professional divers, are not comparable.
Dr Sterk pointed out that on the Ooster Schelde works, some 10 to 12 divers were daily working at depths of 25 m (and more) on the seabed for 4 to 5 h over two years. In all there were about 1,500 exposures with no other ill effects than two cases of limb bends, and here the oxygen decompression was daily.
Mr Marks asked if there were cases of aseptic bone necrosis without decompression sickness or at pressures below 1 bar, but the session which had already overrun its time had to be concluded before any answers were forthcoming.

(Reporters: J Buchanan and F M Jardine)

Part Ten
Challenges for the Future

43 Discussion session: Challenges for the future

The session was chaired by Mr A J Howard (Mott MacDonald Group), and the discussion was led by the following panel of delegates:

Mr D Court (Edmund Nuttall Ltd, UK)
Dr I Nashimoto (Saitama Medical School, Japan)
Dr G C Voigt (Gustav C Voigt & Associates, USA)
Dr R W Hamilton (Hamilton Research Ltd, USA).

Mr Howard asked each of the panellists to speak for a few minutes, in a provocative way if they wished, on how, from their own perspective, they expected compressed air work to develop in the future. An open discussion would then follow.

Mr Court. As a civil engineer whose task is to construct the work which has been contracted for, within budget for the client, Mr Court sees compressed air as one of the tools available to him, which will be used for many years yet. This use will particularly be for low-cost, typically small-diameter tunnels, but also for access to TBMs. The days of the large compressed air contracts, such as Dartford, are over.

He agreed with the HSE about the importance of risk assessment and proper control of the operations under a responsible person. While there are costs associated with improved safety and health-control measures, these result in savings overall, as with an effective quality system. As such, the three key requirements are a clear objective, experience, and authority – to set up and run an effective system.

Compressed air is a safe method, provided civil engineers control the risks. With more information about the health effects, for which we look to the medical experts, and with their advice on prevention measures, this should continue to improve. Perhaps there will always be a residual element of decompression illness, but there is still more that can be done to minimise this, both in respect of equipment and people.

Finally, Mr Court referred to decanting, which was used to a considerable extent on the Cairo contracts with which he was involved. Here, the discipline of the local labour force was good and this probably contributed to the relatively low DCS rates. In his view there is still a place for decanting.

Dr Nashimoto said that he had been treating DCS cases since 1953; he has carried out his work on sites, entering caissons and had experienced the 'bends'. In Japan there has been great progress in the mechanisation of caisson excavations. The incidence rates of DCS were very high thirty years ago, and Dr Nashimoto used to see all the symptoms frequently. One reason for the move to mechanisation is that young people disliked caisson work, so now those doing it are the older, long experienced men, but relatively few. Thus mechanised methods are being used increasingly, which will be further developed for very deep caissons. While their purpose will be to eliminate the need for men under pressure in the caisson excavation process, the equipment will need maintenance. Hence the current efforts to develop the technology for these entries.

Engineering and Health in Compressed Air Work Edited by F.M. Jardine and R.I. McCallum. Published in 1994 by E & FN Spon. ISBN 0 419 18460 0

In that there will be relatively few exposures, there will be very much less decompression illness, but the challenge is to design safe procedures for these higher pressure exposures.

In two comments on current practice, Dr Nashimoto first supported the criticism of the split-shift that had been expressed by Dr Kindwall and, secondly, pointed out that the actual DCS incidence rates on some sites were probably higher than had been reported to the Japanese Ministry of Labour.

Dr Voigt explained that he had only recently become involved with compressed air work. He had been brought in to advise a tunnelling contractor who had unexpectedly encountered ground saturated with old gasoline, so that the tunnel air was dangerously contaminated. After cleaning the air of hydrocarbons, they put the tunnel under compressed air. Dr Voigt thus came to realise not only the scope of tunnelling activities, but also about the capability of the compressed air method and how it has been used world wide. He quoted a saying of Will Rogers that it is 'not what we don't know that hurts us, but what we know that just ain't so'. It is clear that we do not have answers to all the problems associated with this work. In particular we need to know more about the changes that take place in the brain and CNS following compressed air work. We have not yet defined the clinical correlates. Any attrition of cells over that of normal aging has to be looked at carefully and steps taken to reduce it.

Dr Voigt remarked on one of the questions posed by Mr Belman in the Session 8 discussion, that about women working in tunnels, and mentioned the hiring of the first women safety adviser for tunnels, and the reaction against this. Interestingly the patron saint of tunnels is a women, St Barbara, whose shrines are quite often seen at tunnel portals. The point, however, is whether there is a hazard from hyperbaric exposure to the female reproductive system. A recent diving symposium noted that there was not yet sound epidemiological evidence, but some studies suggested that there is an increase in birth defects following hyperbaric exposure.

Dr Hamilton chose to speculate on what the future will bring. First he was pleased to see the emphasis on controlling risk – rather than controlling the method by too rigid regulation. (The following is Dr Hamilton's note of his contribution).

The most important issue presented at this conference, and the one with the most far-reaching impact on both the health of workers and the business of working in compressed air, is that concerning the use of oxygen for decompression. First, it appears clear from experience in diving, hyperbaric medicine, and some compressed air work that decompression can benefit from the use of oxygen. This has more meaning at pressures above 2 or 2.5 bars gauge, but it applies for all decompressions. Extended use of oxygen, because it improves decompression in general, should have an important beneficial effect on reducing the prevalence of dysbaric osteonecrosis (DON); if we can reduce DCI to a minimum level, elimination of DON will follow. Several presentations at the conference included the use of oxygen, but some of these were done some time in the past, and with all due respect they may not necessarily have used the most effective decompression procedures. Experience from other places where it has been used, including both stressful decompressions and treatment of the decompression illnesses, shows clearly the benefit of oxygen breathing. Treatment of DCI is universally accepted as being better with oxygen than with air alone. In fact, a doctor electing to treat DCI using air alone in the U.S. today would be subject to litigation if there were any residual injury.

I commend Dr Nashimoto's efforts to achieve the benefits of oxygen in a restrictive legislative environment by having workers breathe it after returning to surface pressure. I believe the future will show that this procedure is quite beneficial.

Along with its benefits, the use of oxygen requires special care. If oxygen is to be used, it has to be done with proper precautions. Handling oxygen is hazardous if not done correctly, but an extensive and well established technology exists on proper use of oxygen. Oxygen breathing in a chamber or lock requires certain special procedures, including training of both workers and lock operators, and scavenging of expired breathing gas ('overboard dump'), and an adequate fire suppressing system; the technology for these conditions has been well worked out and is practised extensively in the medical and commercial diving communities with good results. Daily breathing of high doses of oxygen can cause lung inflammation and other effects; doses have to be controlled.

One point to be emphasised in reference to the rather strong case made for oxygen during the conference is the caution not to be too glib about it. Using oxygen is beneficial, but it requires care and understanding of its hazards, and appropriate decompression procedures.

A second new trend in CA work is the concept of risk assessment. This says, essentially, that each project or situation should be assessed individually for the risk to workers. This is in happy contrast to previous and other systems where rigid rules are known to legislate unsafe conditions (see, for example, the papers by Kindwall and Holthaus). Risk assessment could allow a refreshing injection of flexibility into the CA business. As mentioned in my talk this morning, there are now responsible procedures for the ethical introduction of new decompression procedures ('tables') in any jurisdiction. I propose that the overall Compressed Air Operation be planned for any major project. This would enable the decompression methods used to be matched to the needs of the project, and thus to incorporate the most effective and appropriate of current procedures. It makes much better sense to tailor the breathing gas and decompression to the project, rather than to try to fit the construction operation into a rigid format dictated by outdated legislation. This option will not be immediately available everywhere, but it can be done; a strong case for a variance can be supported with existing evidence.

An example of what can be done – I am truly sorry that we did not have a report of it at this meeting – is the Kiel Supply Tunnel, completed in 1990. Dr Karl-Peter Faesecke, as the occupational safety minister for Schleswig-Holstein, was able to introduce – and modify as necessary – oxygen breathing procedures for work to as high as 3.9 bars gauge, and the project was successfully completed (1990). (Faesecke, K-P. (1990). Diggin' deeper - compressed air work beyond 30 meters. In: Sterk, W., Geeraedts, L., eds. *Proceedings XVIth Meeting of the European Undersea Biomedical Society*. Amsterdam; Foundation for Hyperbaric Medicine).

Regarding safety, another new trend that will have a seriously beneficial impact is the concept of requiring a Manager of Compressed Air Operations. This allows a system approach, as just mentioned, and even with conventional air work but especially with creative decompression techniques, which should improve both safety and effectiveness by integrating worker exposure – safety and effectiveness – into the overall project. This concept should be followed whether required by law or not. Having a competent and well trained person in charge of the compressed air operation will go a long way to eliminate problems caused by improper actions, and it will pay off for the contractor.

The new French regulations (reported by Le Péchon) call for adequate training of all workers to be exposed to pressure in the course of their job, and of those managing decompression as well. The proposed amount of training will be expensive initially but will pay off in the end in improved safety. Experience in North Sea diving operations showed clearly that human factors were the cause of most accidents; requiring competence in both divers and supervisors resulted in drastic reductions in the accident rate.

The issue of reporting has been mentioned (D H Elliott; Marks). There are two aspects to this which could be improved. First is to make every effort to 'decriminalise' the act of admitting to having symptoms of DCI, to offer impunity to the worker who reports. This requires efforts by management, the doctor, unions, and the authorities. Dr John King says the doctor can establish confidence enough to get some honest reports without penalty to the worker.

Another developing aspect of reporting is the use of electronic monitoring and recording, to ensure that accurate pressure-gas-time exposure profiles for all workers are acquired, and to facilitate their analysis. Monitoring and recording are critical elements of the ongoing assessment that, according to others at this Conference, is now considered essential in all such occupational exposures. The necessary equipment is not expensive and is certainly cost-effective.

Regarding the use of helium in breathing mixtures, some experience is relevant. Divers routinely work in air to pressures of 6 bars (although there is a 5 bar – 50 msw – limit in the North Sea); many have exceeded that, but it is generally conceded that the ability to deal with complex situations and emergencies can be seriously compromised. Immediate life-safety is not much of an issue in tunnel work, but some jobs will require the necessary sharpness, others will not. Extensive experience (> 1,000 exposures) has built up in the last 4 or 5 years on the use of helium-nitrogen-oxygen 'trimix' for diving in the pressure range 6 to 8 bars gauge. Performance is good, and decompressions are reliable. A big problem with breathing helium-containing mixtures at work is the need to breathe by mask. Rebreather backpacks will be used when this is needed, but in my opinion most work can be done with air; it is the decompression that requires the creativity.

One last comment about decanting. As has been said earlier, it is acceptable if done properly. Although we are limited in our understanding of the inherent physiology, one thing generally agreed on by decompression researchers is that some account needs to be taken for the surface interval. Oxygen should, of course, be used for the chamber phase.

The discussion was then opened to the floor.

Mr R P Lewis referred to Mr Court's observation about having a good management system in place for the compressed air operations. In contrast to large projects with their specific organisation, there is a difficulty on the smaller contracts with fewer and junior staff, and where there is likely to be centralised and so more remote management of systems.
Mr Court agreed, there often being a relatively junior person on the site supervised by visits from a contracts manager. As with quality systems the responsibility lies at the higher level. One difficulty is that civil engineers tend not to be 'nuts and bolts' men, and so need help on the mechanical systems of the compressed air set-ups. For these smaller contracts, a company-wide policy is needed for the management of the compressed air work.
Mr Howard asked for views on whether there should be better definition of the health and safety requirements for equipment set-ups.
Mr J C Le Péchon said that while CEN standards set out general guidelines, so that the basic principles will be met, individual set-ups of equipment will be project specific.
Dr J D King pointed out that the CEN standards apply to new equipment; there is still serviceable 1930s equipment in use.
Mr Court said that in Cairo in 1985, seven Gowring-type locks from the 1930s were in use, but on stripping down the equipment for cleaning, it was downgraded after pressure testing.
Mr D R Lamont explained that while the Provision and Use of Work Equipment Regulations are to come into force on 1 January 1993 and apply initially to new equipment, by 1 January 1997 all equipment will be covered.
Mr Lewis noted that it is the supplier's responsibility to undertake correct, safe design and manufacture of equipment, but how it is used is outside the supplier's control. Whose responsibility is training and operation of the equipment?
Mr Howard asked about responsibility for first use of the equipment.
Mr Lamont answered that for its first use the supplier has responsibility, but subsequently this falls upon the user of the equipment.
Mr Court. There is a worry because compressed air equipment, which is complex, can be hired to users inexperienced in its operation. They need to receive training.
Mr G Luther remarked that the logic of the risk assessment approach should include these questions, e.g. the equipment should be tested and certificated and the total procedure examined.
Dr R D Vann asked if manufacturers should build in oxygen equipment, and **Mr Lewis** distinguished between building-in equipment and making provision so that it could, if needed, be added later.
Mr Le Péchon thought yes; like the provision for heating and cooling it does not have to be used, but is there when needed.
Mr Howard asked if compressed air work was being 'outlawed'.
Mr Le Péchon remarked that the situation was rather like what had happened in diving during the early days of deep water oil exploration. At first the divers were

considered to be needed only to assist the drilling – that later they could be dispensed with – but in practice they continued to be needed. We do not know how to do without compressed air, but its practices will improve, again like diving, as there is increasing need to go deeper.

Dr Voigt thought that contractors are often reluctant to use compressed air, but do not have a better alternative. In a specific tunnel, where there was hydrocarbon contamination to the ground, the contractor first tried cartridge respirators to clean the air (which was at about 10% LEL). The filters had to be changed every 30-40 minutes, but then adopted compressed air, in effect to exclude the contamination. There will be many more situations of ground contamination affecting tunnels in urban areas and where there is methane from wastes.

Dr King remarked that compressed air remains a useful engineering tool and, in his view, will still be a viable option for the next 30 years or so. He accepted that its use may decrease, but this is no reason to relax efforts to minimise its effects on health.

Mr Court expressed concern about the approach of 'tacking on' a lock to a slurry tunnel boring machine. It smacked of being 'just in case' and could be more of a risk if there was not adequate planning and preparation beforehand. Again this means a proper risk assessment initially so that, if the locks are to be put on, all the arrangements for their use will be clear.

Mr J Buchanan drew attention to EN29000 on quality systems – which he thought to be heeded better in UK than elsewhere in Europe – but which gives the basis for firms to set up properly assessed and certificated quality systems.

Mr L Lofthouse noted that London Underground Limited has introduced the ISRS scheme for its own safety management and requires contractors to follow it as well.

Mr Howard in closing this part of the discussion, commented that it was better to run a risk of decompression illness than of drowning. He then moved the discussion on to medical implications.

Dr A C Palmer. Decompression illness results from marauding, disseminated bubbles, but it is a multi-system disease affecting many organs. There is a need to look not only at the clinical manifestations, but also at the pathology by systematic study of what organs are affected. This research has yet to be properly financed.

Mr L Marks. As a legal practitioner in this field for over 25 years (and acting exclusively for plaintiff injured tunnel workers), I have never experienced any particular difficulty in establishing claims, in appropriate cases, where the injury involved the disease of bone necrosis ('bone rot'), on which subject, there is a considerable amount of published material.

However, I wish to draw to the attention of this conference (and in the presence of so many respected medical experts here today), that I have, on occasion, experienced unusual difficulty in linking other physical and organic injuries (e.g. neurological injuries, etc.) with work in compressed air (whether or not any breaches of regulations or good practice were involved), because there is so little researched and/or published material on these other aspects of injury, and I therefore call upon the medical experts in this field not to simply concentrate on manifestations of bone necrosis but to furnish us all with much more information and evidence, on these other aspects, in order, *inter alia*, to assist the predicament in which, in my respectful view, a considerable number of seriously ill ex-tunnel workers in compressed air now find themselves, without the effective right to compensation from either their employers' insurers or the DSS Industrial Compensation scheme.

The number of such uncompensated cases that have come to me for my consideration have led me to believe that there is a serious unresearched and hitherto unidentified problem here, which, as a matter of public policy, needs to be highlighted and urgently alleviated.

Mr F M Jardine asked the medical practitioners present for their views on the new approach to classification of decompression illness proposed by Surgeon Commander Sykes. Speaking as a geotechnical engineer concerned to understand the many variations in the ground, the basis of engineering classification has to be accurate, systematic description, carefully recorded. The simple distinctions of soil or rock is not much help; how hard is the rock? is the soil sand or clay? The analogy with decompression illness can be extended, doctors like engineers are in the business of finding solutions, and improved description and classification are the key to identifying the situation for the solution is needed.

Dr Voigt agreed with Dr Palmer that there is disturbance in all organs, which is characterised by physical symptoms, just as measles shows as a rash but affects other parts of the body. Changes in the micro-circulation, e.g. micro-thrombi, emboli, endothelial alterations, may be reversible but they must be considered very seriously.

Professor R B Philp said that he had had an opportunity to examine enzyme levels in divers with DCS and found elevated SGOT and SGPT within 72 h and liver and muscle damage not clinically visible.

Dr Voigt remarked that this is the same as in any other disseminated intravascular coagulopathy.

Professor Philp referred to a pair of divers, one of whom was treated for DCS, but the other (who did not have DCS) was found to have liver damage shown by enzyme level changes.

Dr Voigt. The number of ear injuries is greater than the number of decompression sickness cases and result in more lost time, especially on low pressure contracts. But training can prevent these ear injuries. Another area for improvement is in the large number of physical injuries, e.g. from trips and falls.

Mr Lofthouse emphasised the role of medical lock-keepers in training.

Dr King, referring to Dr Palmer's saying that pathological studies would be needed, pointed out that, thankfully, there have been very few deaths in compressed work.

Dr H V Hempleman reported that, in UK, there had been two compressed air deaths since the war.

Dr King remarked that post-mortem pathology would have to be where deaths were from other causes, such as when ear changes had been found in a diver who had committed suicide.

Dr R L Marcus commenting on the descriptive approach to classification and accepting that decompression illness is a multi-system syndrome, still thought the Type I/Type II classification is useful in terms of risk assessment.

Dr P Bryson, speaking from his viewpoint of diving medicine, made a plea for greater collaboration across hyperbaric disciplines particularly in reporting. The proposed classification should be adopted for two reasons; it is likely to be used internationally, and it helps the person treating the disease. An important question is whether or not there should be anonymous reporting. We do not know what the real incidence of decompression illness is.

Mr Howard recognised there may be a case for anonymous reporting, but care will be needed to set up a satisfactory procedure.

Mr Buchanan felt that anonymous reporting could be suspect.

Dr Hamilton repeated his previous plea that decompression illness should be 'decriminalised', that there should be no penalty for receiving treatment.

Mr Le Péchon noted the system to be used in France for recording hyperbaric exposures and that there is a hidden link to payments. It is intended that log-sheets are collected by OPBTP (French 'equivalent' to CIRIA) for tunnelling and INPP (National Commercial Diving Institute) for divers. Pressure records are kept by companies for 20 years.

In using data collected by divers or workers, the payment schedule is a significant factor inducing bias in the pressure and time quoted when bonuses are paid from logged information. Then there can appear in the scientific literature detailed statistics which represent payment thresholds rather than actual dives and which might be used to validate decompression tables; thus, we have to be careful about this.

Mr Lamont commented that as workers treated for Type II DCS are declared unfit for further work in compressed air, there is a disincentive against reporting.

Dr King considered that confidential reporting is the basis from which to work. If the doctor is seen to be impartial and not linked either to the employer or the trade union, men will talk about episodes when they did not seek treatment. At one site, Dr King noted that he had been told of untreated cases which were about one fifth of the number who received treatment.

Mr Howard then asked how the risk of decompression illness compared with that of other industrial health hazards.

Professor Nashimoto pointed out that Type II DCS had reduced. If ten years ago there were typically two or three cases, for the last three years there were none.

Professor Walder commented that a number of industrial diseases had been recognised in the last twenty to thirty years. For example, relatively recently, the condition of vibration white finger is now notifiable and there has been a spate of compensation claims. Others are pneumoconiosis, asbestosis, etc.

Dr Marcus speaking as an occupational health physician said that cases of decompression sickness are a very small proportional of industrial illnesses in comparison with those from exposure to solvents or heavy metals and to repetitive strain injuries.

Dr Hempleman remarked that even a simple injury such as stubbing a toe results in various biochemical changes in the body. But the body has a built-in resilience to cope with these minor injuries. Hence it is necessary to be careful when interpreting reports of other signs of minimal injuries, especially if they have not reached a level

with which the body cannot cope.
Dr Voigt agreed that it is important to take a very moderate approach to any perceived potential occupational hazard. Repetitive strain injury is now the single most common claim, sometimes even after as little as three weeks employment, with many people operated on, but we may not be in any better a position to prevent it. There are two stages in the approach to a potential hazard: (1) a scientific interest in defining the problem and (2) in relation to occupational health, seeing how to complete the project work with minimal ill results. There has to be a balance between what is significant to the individual, to the project and to medical research. Clearly, though, there are not enough resources to deal with all occupational health issues, and decompression illnesses are relatively low down the list.
Mr Le Péchon noted that frequency indexes recording industrial accidents when in terms of working hours lost or the injured person being excluded from work might not be a suitable basis for comparison of injuries, e.g. treatment for decompression sickness might not be included in such an index if the sufferer is back at work the next day. Again, rather than duration under pressure, the important matter is the decompression. Mr Le Péchon referred to a gravity index used for comparison in diving which took the number of excursions and the working time into account.
Dr Palmer asked how many scuba divers there were in the USA who could be at risk, and was told about 2 million.
Dr Voigt, following someone's remark about the high risk of football injuries, noted that as with sports diving, there is a wide difference between our acceptance of voluntary risks and our attitude to imposed risks.
Mr Howard brought the discussion on to the question of regulations and how these should be improved.
Mr J Hill. In the current UK context there are probably no more than six companies with the in-house capability to undertake compressed air work; two companies provide equipment for hire. In perhaps nine out of ten instances, the compressed air work is not at the risk pressures, i.e. it is less than 1 bar. And in the other cases the work is usually under the supervision of the same physician. Is it realistic therefore to seek to reduce even more what is already a small risk?
Mr V J Belman addressed the question of risk and acceptability. In the next 20 to 30 years we will only have one or two projects per year at above 1 bar and they will be of short duration, perhaps no more than 2,000 exposures per year. As such we might predict one or two cases of Type II DCS or bone necrosis per year. Would not the community consider it reasonable to consider this to be acceptable as it would cost so much to eliminate these entirely or to make the regulations even more stringent?
Mr Lewis reminded the audience that much hinges on what is 'reasonably practicable'.
Dr Bryson pointed out that it has recently been shown that regular treatment by oxygen of divers going below 90 m can result in an increase in myopia which, while reversible, can take three months to do so. Treatments should be kept to less than forty.

Dr Hamilton noted that in the USA there have been cases where this condition did not reverse. He also mentioned that the US OSHA, when developing standards for divers, realised that even a very low incidence was in essence an acceptance of some DCI. As the Standard turned out it did not specify an acceptable limit, but rather said that the employer had to investigate all cases of DCI and take appropriate steps to prevent further occurrences.

Mr Howard asked about current methods of training and awareness.

Mr Damsgaard referred to the Storebaelt project, where training is given about symptoms, whom to contact, and how to behave regarding alcohol, sleep, etc. and, in effect, the ground rules for what to do and what not to do before going into compressed air. In addition, training is given about first aid.

Mr Howard wondered if this deterred the workforce from going into 'air'.

Mr Damsgaard thought rather that it reduced their anxiety. The training lasts five days and includes a lock test, first with a supervisor, and then on their own, before going into the TBM. Very few out of some 200-300 declined. Nevertheless they became much more careful after there had been some cases of the bends.

Mr Le Péchon noting that a French company is involved in this training and would like to have the course recognised and the training made a requirement. The training is about what people should know and be able to do under pressure and includes a demonstration of evacuation procedures.

Professor D N Walder said that while training courses used to be held at Newcastle for medical lock attendants and managers, there is so little and irregular demand that it is difficult to organise courses of training.

Dr Voigt commented that in these circumstances the training would have to be for the permanent safety personnel of company so that they could train others.

Mr Buchanan felt that 'hands on' experience is lost in this way. In the situation of today's compressed air project, the training – as with the project – has to be on an *ad hoc* basis, people being trained specially for it.

Mr Lewis added that this is an extra difficulty for small projects.

Mr Lofthouse noted that Dr King specifically trains the lock keepers on jobs he supervises and that he (Mr Lofthouse) and a colleague could also provide a training service for medical lock keepers and men.

Mr Howard commented on the turnover in workmen, suggesting that at the Tyne, of ten men taken on, five went into air, but only one remained at the end of the first week. This has a bearing on the cost of training.

Mr Hill noted that the situation had now changed because of the small size of the jobs. Perhaps in UK some 30 managers would need training and this would not be a problem. Typically there would only be 5 or 6 men per shift on most jobs, so numbers are relatively small.

Mr M H Olsen, referring to the Storebaelt project, explained that men for work on the TBM head, were selected from the workforce but were given additional medical checks and training.

Dr Bryson commenting that at the Diving Diseases Research Centre they exclude from the pressure chamber women who are or may later wish to be pregnant. In his view, they should also avoid work in compressed air.

Dr Voigt noted that the International Women Divers Association has recently stated that pregnant women should not exceed 2 atm abs. There are some data which could suggest a risk to foetal development. Dr Voigt also remarked on the risk from chemical exposure.

At this point the discussion closed.

The following written contribution on matters discussed was received from Mr Le Péchon after the conference:

1. Women and compressed air. In France women have not been allowed to work in compressed air since the end of last century. The reason was morality and not physiology. The 1990 *décret* has cancelled this rule. Women are informed during mandatory training and on the occasion of medical fitness checks that they should not go under pressure during pregnancy and consider themselves as 'temporarily unfit'.

2. The future in compressed air in TBMs. A great similarity exists between diving from drill barges or drillship in the 1970s, which made the fortune of early diving companies, and compressed air work in shield boring machines. Improved drilling technology has taken all these divers to the beach. Improved TBM technology will considerably reduce the need for compressed air works in tunnels; it is only a matter of technological improvement, investment in research and time.

3. Fire risks and oxygen breathing. It must be remembered that:

   (a) Using oxygen requires proper plumbing, cleaning and operational procedures. The risk involved is outside the locks.
   (b) Breathing oxygen requires a dump system, ventilation and continuous oxygen percent control outside.
   (c) Flammability is not proportional to PO_2 (in nitrogen) but increases sharply with the amount of oxygen available at the fire site (barbecue effect) which is directly related to partial pressure. Therefore at 1.6 bar (a) breathing oxygen, within an atmosphere slightly contaminated with oxygen (23% for example - atmospheric PO_2 = 0.37 bar), is much safer than in air at 3 bars absolute (PO_2 = 0.63 bar), where the flammability and the fire risk is much greater although this appears to be considered (wrongly) by most of the attendees as a safe environment in relation to fire.

(Reporters: J Buchanan, R H McCaig, M W Vowells, F M Jardine)

Conference Chairman's closing remarks

I.H. Elliott
Mott MacDonald Group

It has been an exceptional privilege to have been able to spend these past three days at this conference in good company discussing a fascinating and important subject – and not having to deal with my in-tray. We have covered a lot of ground in this period. While appreciating that it has been an intense three days and that many of you have long journeys home ahead, I think it worth while to reflect briefly on what I believe has been a most successful conference.

After Viscount Ullswater and the Master of St Catherine's College, Dr Brian Smith, started us off so well, John Anderson gave the first keynote lecture with an excellent overview of the past, present and future safety legislation and how it relates to the use of compressed air.

The first two sessions provided an interesting mix of construction projects which illustrates the present range of compressed air applications. These included the work in Hong Kong, perhaps one of the last examples of a major, all compressed air project, involving up to 700 men and 16 working faces. I do not recall any mention of the NATM (New Austrian Tunnelling Method) which if it is to be used in unstable water-bearing ground may continue to need compressed air. We also heard about the work in Cairo where slurry machines were used, which raises the subject of intermittent intervention into the machine head. And at Lowestoft we saw a good example of what might be described as a 'small' all compressed air project, working at relatively low pressures. Reference was made to the importance of quick transfer into decant locks (5 minute transfer) and the use of T-locks which can reduce the need for decanting.

The second keynote lecture by Professor Dennis Walder (who set up the Decompression Sickness Centre Registry in 1966) then took us into the fundamental medical consideration of dysbaric osteonecrosis. Professor Walder explained the mechanism of bubbles forming in the sinusoids and the probable relationship between the development of bends and bone necrosis. He suggested that it was unlikely that bone rot would develop unless a person had previously suffered from bends (although it might not have been reported), and that it mainly occurred as a result of working at pressures over 2 bar for periods above 4 hours. Conversely, working at not more than 1 bar appeared in this regard to be relatively risk free.

Engineering and Health in Compressed Air Work Edited by F.M. Jardine and R.I. McCallum. Published in 1994 by E & FN Spon. ISBN 0 419 18460 0

There followed a series of papers on various aspects of bone necrosis. We were warned about use of statistics and that we should consider carefully how they are presented and whether the data truly reflect what has happened, particularly bearing in mind the possibility of under reporting. We heard not only about observations of intramedullary pressure from work carried out with sheep and the hypothesis of the 'compartment syndrome' in the bone marrow contributing to bone necrosis, but also of UBOs (unidentified bright objects) in brain scans. There appeared to be a broad consensus that introducing oxygen breathing into the decompression procedure was medically beneficial, although it would require careful consideration of the practicalities and benefits for each project and its use at higher pressures.

On Tuesday, Session 4 dealt with a number of aspects of compressed air design; an example was given of where compressed air was used in the hope that this would help to control ground movements. This had not been entirely successful perhaps due to the decision to keep pressures generally well below the balancing pressure and to use steel ribs and timber laggings for the temporary lining. When considering what alternatives are available, we recognise that these may sometimes be more dangerous than the proper use of compressed air. Another paper gave a comprehensive evaluation of the many factors which contribute to air losses, demonstrating the difficulty of estimating air requirements with the lack of published data. We also heard about the well known phenomenon of compressed air being 'trapped' by low permeability strata above the tunnel. There was discussion about the desirability of installing air locks on tunnelling machines – perhaps this is where a decant system would be the best option? The session finished with a very useful description of the modern portable container plant now available which makes it practical for air to be used for short periods and on smaller projects.

The fifth session provided some good examples of the practical application of medical procedures for compressed air working. Some decompression sickness is perhaps inevitable on most projects, but the level of incidence can be limited by establishing high standards of management to make sure that proper working practices are adopted. Also there must be sufficient resources (plant and staff) to deal with accidents and emergencies. There is an increasing awareness of the importance of the personal environment on work performance and health. We need to recognise that working in compressed air greatly decreases the body's ability to shed heat and remember that heat acclimatisation is an important factor. We heard of the apparently successful use of oxygen during the final stages of decompression which reduces both the decompression time and the incidence of bends albeit at a relatively low 1.5 bar pressure. Cramped conditions in small air locks associated with tunnelling machines may well increase the likelihood of bends.

The third keynote lecture was given by Professor David Elliott who described the complexities and interrelationship of the wide range of conditions which may be generally considered as decompression illness. Dr Elliott also provided a summary of the development and role of the Decompression Sickness Panel formed in the UK in 1957. Pleas were made for more research, a return to descriptive reporting, and the need to establish some centralised data bank particularly since the UK Registry is no longer active.

The next two sessions (6 and 7) then provided a marvellous opportunity, particularly for the engineers in the audience, to hear about the medical experience, treatment and implications of decompression illness. Compressed air is being adopted over a wider range of circumstances requiring working and medical practices to develop. Even given that recognised procedures are not always fully observed no evidence could be found to show that decanting is any more harmful than direct decompression. There was some evidence that memory capability is impaired for workers with a history of decompression illness but personality remains stable. A suggestion that this may be caused by alcohol intake was refuted by a view that there are more alcoholic doctors than divers! Changes in blood components have been observed during decompression contributing to the formation of small thrombi which could be a factor in the development of bone necrosis. The possibility of using aspirin to reduce this effect was noted.

On Wednesday morning we came back to the all important codes and regulations. We heard of a new, very much prescriptive, specification for air locks being developed for the EC. It was suggested that these are formulated by majority voting, but hopefully someone takes overall responsibility? In Europe in future it will be necessary to carry out a safety risk assessment to identify appropriate protective measures. Examples were given of countries where the regulations are badly out of date in that they have not kept up with medical knowledge or the new methods of tunnel construction. This is often due to legislative delay.

The last session in which papers were presented provided a number of examples of the development of new technologies in the application of compressed air. The use of pure oxygen during decompression can increase the proportion of working time in a given period. This and other methods can be considered for working over 3 bar including saturation and partial saturation techniques. Another development is the use of 'unmanned' caissons working at over 4 bar. For repair and maintenance work men used a nitrogen-helium-oxygen mixture (trimix) based on deep-sea diving technology.

We live in a rapidly changing world and our regulations and procedures should be formulated in a flexible way so that we can apply the best knowledge available at the time. Major projects with extensive use of compressed air may be a thing of the past, but the compressed air technique will continue to be needed for the foreseeable future. Good management is probably the most important factor in achieving high safety standards.

We have a lot of people to thank for being able to hold this conference. Dr Hempleman and the MRC Decompression Sickness Panel who encouraged the idea, our co-sponsors BTS and HSE, the Advisory Committee, all of the CIRIA staff involved including Louise Harrison, Fin Jardine and Jim Buchanan, the College staff dealing with our slides and providing us with refreshments at regular intervals, the Authors, the Referees, and Key Note Lecturers, and of course all of you – the Delegates.

Appendices

Appendix A: Conference glossary of engineering, diving and medical terms, abbreviations and units of pressure

Compiled by F.M. Jardine
CIRIA, London, UK

Introductory Note. All the terms in the following list are used in the papers to this conference. The deliberately simplified definitions or explanations are presented for quick reference in alphabetical order of term in a single list, i.e. medical and engineering terms are not separated. The intention is to help those attending the conference or reading the papers to a quick, albeit necessarily cursory, appreciation of an unfamiliar subject. Most of the listed terms are medical, perhaps reflecting the needs of the compiler, though a number of engineering expressions have a physiological connotation too, e.g. rib, crown, blister, eyes, piles and skin. The compiler asks for the forbearance of specialists dissatisfied with the inadequacy of definitions in their subjects but also for their corrections. The sources of definitions are referenced at the end of the glossary.

absolute (pressure) — the sum of all pressures acting on an object; the pressure above zero pressure: i.e. the sum of atmospheric and gauge (or hydrostatic) pressure.

ad hominem — (legal) of an argument or appeal founded on the preferences or principles of a particular person rather abstract ideas.

adipocytes — a fat-containing cell of adipose tissue, i.e. where fat is stored and cells are distended by fat.

adit — horizontal passage or opening, usually off the main drive of a tunnel.

adjuvant therapy — helpful or auxiliary treatment of a medical condition.

aetiology — the science of the cause of a disease or the causal process. (In U.S. spelling, etiology).

Engineering and Health in Compressed Air Work Edited by F.M. Jardine and R.I. McCallum. Published in 1994 by E & FN Spon. ISBN 0 419 18460 0

air	by implication, the colloquial expression 'in air' means in compressed air. The term, free air, is used when working at normal atmospheric pressure.
air consumption	the volume rate of air supply to compressed air workings, usually expressed as cubic feet or cubic metres of air volume at atmospheric pressure per minute, i.e. cfm or m^3/min.
air deck	the structural bulkhead within or at the top of a shaft, below which work is carried out in compressed air and through which access is gained for men and materials *via* locks fitted on to the air deck.
air lock	a chamber for transfer between different pressure environments in which air pressures are raised or lowered to equalise with that in the direction of transfer.
air loss	the amount or volume rate of air lost through the ground from compressed air workings.
algorithm	process or rule for calculation.
alluvium	earth materials deposited by a river.
ambient	pertaining to the surrounding environment.
aneurysm	local dilation (expansion) of an artery.
annular	of grouting, in tunnels and shafts the gap between their linings and the surrounding ground. (med) a ligament which surrounds another structure or provides an attachment.
angiogenesis	the development of a system of blood vessels.
angiography	radiological study of the blood vessels.
anterior	at or towards the front.
appellant	(legal) one who appeals to a higher court about the decision of a lower court.

appositional	being in a correct position, e.g. as in setting a fractured bone.
aqua tower	cooling-water system for a compressor set-up.
arch, arch rib	a frame, usually of steel joists, placed at intervals to support the sides and roof of a tunnel.
arteriole	a small artery.
artery	a tube of muscle and elastic fibres which distributes blood from the heart to the capillaries.
arthralgia	pains in a joint or joints, e.g. compression arthralgia.
aseptic	free from septic matter or disease-producing bacteria.
aseptic necrosis of bone	death or decay of bone in the absence of septic matter; also called dysbaric osteonecrosis when presumed cause is pressure change; old term is caisson disease of bone.
at rest earth pressure	the horizontal pressure exerted within a soil mass in a state of equilibrium.
attractant	a substance which attracts.
audiovestibular	of the entrance cavity to the cochlea of the ear.
aural	of the ear.
aural barotrauma	damage to a part or parts of either or both ears resulting from changes in ambient pressure.
autochthonous	indigenous, inherited; or of behaviour, activated by its own drive; or originated in that part of the body where found.
axilla	armpit.
axoplasm	the protoplasm of an axon (which is a long and single nerve-cell process that usually conducts impulses away from the cell body).
backhoe	mechanical excavator whose loading bucket is drawn down and back towards the machine.

Bade shield — A shield provided with 'breasting panels', hydraulically operated to provide physical support to the face. In one case these were associated with a fully mechanised cutting wheel. In another, there were three work platforms to which the panels were fixed and from which excavations were by shovel loaders.

ballast, ballasting — the addition of weights to alter the buoyancy of a caisson, often in the form of concrete or metal blocks called kentledge.

baromedicine — study and practice of medicine in relation to the effects of changes of ambient pressure, *cf* hyperbaric medicine.

barotrauma — injury, damage or insult to tissues caused by unequal pressures.

batholith — an enormous body of granitic rock intruded into the earth's crust.

bends — term applied to certain types of decompression illness, e.g. limb bends; used colloquially for any instance of decompression sickness though mainly for Type I.

bentonite — a clay, largely composed of the mineral montmorillonite, used in water slurries for ground support and overcoming friction when sinking caissons, and in grouts for sealing pore spaces in the ground.

blast, blasting — the use of explosives in the excavation of rock.

blind shield — developed from Greathead shield. The shield advances by pushing aside the surrounding ground. To reduce the pressure needed for this, doors in the bulkhead can be opened to allow ground to enter the tunnel.

blister lock — a small man-lock attached to main chamber of a vertical lock, it accommodates only one or two people standing; and is used for access, in the decanting process, and for very short duration decompressions.

blow-out — a sudden, uncontrolled loss of compressed air from a caisson, shaft or tunnel, which can increase rapidly with potentially disastrous consequences.

blow-up — in diving, a sudden, uncontrolled ascent or drop in pressure.

boiler lock (boiler lock section) — a steel cylinder flanged at each end. Several, bolted together, are used to form air locks or access shafts to air decks in a compressed air tunnel or shaft.

bounce dive — a short, non-saturation dive beginning and ending at the surface; alternative term is intervention dive.

breach — (legal) the breaking of a legal obligation.

breakthrough (break-out) — the operations (or the place) involved where a tunnel is driven out or into a walled or lined shaft.

bulla (pl. bullae) — a large blister or vesicle (a hollow, thin-walled, rounded, long protuberance); in the lung, usually a large thin-walled air space.

bulkhead — the structural panel separating parts of a tunnel at different pressures, containing pressure-tight doors and service ducts.

bus mains — the pipe arrangement for delivery of air at the required pressure.

buoyancy — the vertical upward thrust (on a caisson) from its displacement of water in the ground.

caisson — a cylindrical or rectangular ring-wall for keeping water or soft ground from flowing into an excavation and which forms part of a foundation or a shaft to a tunnel; pneumatic caissons are those in which the working chamber is full of compressed air.

calcified — with lime deposits.

cancellous — of tissue, the honeycomb arrangement of bone cells beneath the compact layer especially at the ends of long bones and which contains red bone marrow.

cardio-vascular — concerning the heart and blood vessels.

cartilage — a tough connective tissue, gristle, usually covering the articular surfaces of bone.

castellation	the shape of battlements on a castle, notched; of a retaining nut, with rectangular upstands.
cathead	a grooved pulley wheel, high up on a pile frame.
caulking	blocking of a joint, e.g. between tunnel lining rings, to make it water-tight by insertion of sealing material.
centrifuge modelling	the physical testing of a small-scale model under carefully chosen conditions and, in particular, increased gravitational forces.
cerebral	of the cerebrum, the largest part of the brain.
chiller/dryer	the equipment in the ventilation system whereby air is exhausted, cooled and dehumidified.
cholesterol	a steroid alcohol present in animal cells and body fluids important in physiological processes.
choroidal	of a vascular membrane containing large, branched pigment cells that lies between the retina and the sclera (the white) of the eye; also the choroidal artery which supplies blood to part of the brain.
claustrophobia	fear of confined spaces.
cochlea	the spiral canal of the inner ear labyrinth containing the organ of hearing.
cognitive	of conscious intellectual activity.
cognitive dysfunction	impaired or abnormal functioning of intellectual activity.
cohesive	of a soil with sufficient clay in its constituents to bind its particles.
colluvium	weathered debris at the foot of a slope which has originated from erosion of the slope.
Common Law	the unwritten law of England, embodied in past judgements, administered by courts of the crown.

Compartment syndrome	set of symptoms of compartmentation - intracellular partitioning of cellular substances and metabolic activities by membranes.
compensation grouting	a responsive method of injecting grout into the ground to offset ground movements such as settlement caused by tunnelling to prevent or minimise the effect on nearby structures.
complementation (complement)	substance in blood which completes the destruction of bacteria invading the body - also the process.
compressor	the machine for compressing air to the required pressure; may be reciprocating, centrifugal or free-piston in method.
conjunctivitis	inflammation of the mucous membrane covering the eyeball and lining the eyelids.
consistency	the strength of a clay or cohesive soil.
consolidation	the process whereby moisture is squeezed out of soils by applied stress.
consolidation test	on soils, the measurement of compressibility and rate of pore-water drainage under load.
consumption (of air)	the volume rate of compressed air supply.
contiguous (bored) pile wall	the formation of a concrete wall in the ground by constructing bored piles which touch (or intersect if secant piling).
corbel	a support bearing on a wall or column but extending out from it.
corestone	relatively unweathered portion of parent rock within a mass of usually highly or completely decomposed rock.
corpora amylacea	starchy bodies.
cortex	the external layer of an organ, e.g. brain or bone.

Corti, organ of — long, ribbon-like end-organ within the cochlea, within which the nervous impulses responsible for hearing originate when stimulated by vibrations.

cranial — of the bony cavity which contains the brain.

crossover — a track which joins two parallel railway tracks, hence, in tunnelling, the S-shaped tunnel or enlargement to link two parallel tunnels.

cross-passage — the excavation to provide access between two parallel tunnels.

crown — the highest part of the roof of a tunnel.

cutaneous — pertaining to, of, the skin.

cutting edge — the leading edge of a tunnelling shield (or a soil sampler) driven into the ground ahead.

decanting — the procedure of rapid decompression of men in a lock from a compressed air working chamber to atmospheric pressure for transfer to another chamber (decant lock) for prompt recompression and staged decompression.

decant lock — a two-compartment chamber, separate from the working chamber, used after decanting for recompression and subsequent staged decompression.

decomposed (rock) — of a rock which has undergone a degree of *in-situ* weathering and alteration, even to a soil-like condition.

deltaic deposits — soils deposited in a river delta, i.e. a complex interweaved layering of silts, sands and clays.

defendant — (legal) a person sued in a court of law (also defender).

denaturation — modification of molecular structure so as to destroy or diminish some of the original properties and, especially, the biological activity.

denatured protein — protein with its molecular structure modified so that its properties and biological activity are diminished or destroyed.

diabetes mellitus	disease, due to deficiency of insulin, with excessive excretion of glucose in urine.
diaphysis	the shaft of a long bone, e.g. a femur.
differential settlement	unequal downward movement of different parts of a structure or of the ground surface as a result of compression or movement of the ground.
diorite	coarse grained igneous rock.
diplopia	double vision.
Disse, Space of	a space between the chains of cells which make up the lobules or functional units of the liver.
dissipation	the reduction of excess pressures in soil pore water by drainage.
distal	further (away) i.e. from the point considered to be the centre of a physiological system, *cf* proximal.
distention	enlargement.
dolerite	a medium grained igneous rock usually found in relatively thin or sheet-like intrusions into the earth's crust.
dorsal	of the back.
drawdown	the lowering of a groundwater table by pumping or drainage into an excavation or tunnel.
dump system	a system built into a chamber which transfers exhaled gas out of the chamber to prevent oxygen build-up; also overboard dump system.
dump valve	a one-way valve for discharging expelled air.
dysbaric	of or relating to a change of ambient pressure.
dysbaric osteonecrosis	aseptic necrosis of bone.
dysbarism	any illness or injury pertaining to change of ambient pressure.

dysfunction	impaired or abnormal functioning (as of an organ in the body).
dyspnoea (dyspnea)	difficult or laboured breathing.
earth pressure balance (tunnelling) machine	a full-face tunnel-boring machine for water-bearing soft ground, in which a slurry of soil and water in the chamber which forms the face and holds the cutters is maintained at a pressure to equal that exerted by the ground.
eductor wells	a system of ground dewatering in fine-grained soils using a vacuum created by the Venturi effect rather than by pumping from conventional deep wells or well points.
effective stress	the stress within a soil mass carried by the soil grains.
electro-encephalogram (EEG)	a record of the electrical activity of the brain.
embolism	obstruction of a blood vessel by a travelling blood clot or particle of matter.
embolus (pl. emboli)	substance carried by the blood stream until it causes obstruction by blocking a blood vessel.
embryological	of features and phenomena exhibited in the formation and development of an embryo.
emphysematous bullae	large blisters containing air or other gas in tissues or cavities of the body; can be present on the lung surface in vesicular emphysema (a chronic disease of the lungs).
encephalomyelopathy	any disease that affects the brain and spinal cord.
endocrine	internally secreting, of certain glands in mammals which pour their secretion into the blood, and so can affect distant parts of the body.
endolymph	the fluid inside the membranous labyrinth of the ear.
endosteum	the lining membrane of bone cavities.

endothelial	of the membranes lining serous, synovial and other internal surfaces.
engorgement	congestion of a tissue or organ with blood.
enzyme	a protein which acts as a catalyst, i.e. to bring about chemical change without itself changing.
ependymal	of the membrane lining the cerebral ventricles and the central canal of the spinal cord.
epidemiological	of the study of the origins and occurrence of a disease.
ergometer	apparatus for measuring the work performed by a group of muscles.
erythema	superficial redness of the skin.
erythematous	relating to or marked by erythema.
estuarine	of soils, deposited in a river estuary.
etiology	the science of the cause of a disease, or the causal process (also spelt, aetiology).
euphemistic	using a less distasteful or more favourable term for something harsher or more offensive.
excursion	movement of diver from working place or habitat at the saturation depth.
extroversion	turning inside out.
eyes (of tunnel)	the holes in the walls of a shaft from which tunnels are driven.
face	of a tunnel, the ground being excavated, i.e. the current end of an advancing tunnel drive; of a tunnel-boring machine, its front, which holds the excavating machinery.
facie	see *prima facie*.
fat emboli	globules of fat released into the blood supply, e.g. from a fractured bone.

femoral	of the femur, i.e. thigh bone.
femur	the thigh bone.
fibrin	a protein formed from fibrinogen in blood plasma in the process of clotting: it is separated from the serum and entangles the corpuscles.
fibrinogen	soluble albuminous substance present in the blood which is converted into fibrin by the action of the enzyme, thrombin, when the blood clots.
fibrinolysis	the usual breakdown of fibrin by enzymes.
fill	earth materials placed or dumped by man.
fluvial	of a river.
foramina	opening or hole in bone.
formation	the prepared surface on which a structure or its foundation is placed.
Formation	a geological system or suite of rocks which is the basic unit of classification of bodies of rock.
fovea	small pit or depression.
Franki pile	a proprietary type of pile, but term used to describe a driven, cast-in-place pile which has an enlarged base formed by hammering the fresh concrete out from the base of the pile.
freezing (of ground)	the technique of freezing weak or water-bearing ground to stabilise it prior to excavation.
gamma emissions	(spontaneous) emissions of photons.
gasket	a seal between bearing surfaces in a pipe joint or between segments of a tunnel lining.
gauge (pressure)	the difference between absolute pressure and atmospheric pressure; gauge pressure is the excess or overpressure.

glial	of or relating to neuroglia, which is supporting tissue intermingled with the essential elements of nervous tissue.
gliosis	excessive development of neuroglia, the supporting tissues within nervous tissue.
granite	a coarse grained acid igneous rock.
granulation tissue	the new tissue grown when wounds heal whose edges are not in apposition, i.e. not correctly fitting together.
grano-diorite	a coarse grained acid igneous rock.
Greathead shield	The basic design from which all tunnelling shields have been derived, named after the designer, and first used in 1879. The first shield to move all in one piece, it consisted of a steel cylinder stiffened in front with a cast-iron ring with a rounded cutting-edge. Between the cutting edge and the cylinder was a steel bulkhead with a doorway which could be closed by dropping planks across it (held by channels). Cast iron lining was built within the cylinder at the rear and hand-operated ratchets jacked the shield forward off the lining.
ground freezing	stabilisation of weak or water-bearing ground by the use of refrigerants, e.g. liquid nitrogen.
ground loss	the volume of ground settlement around an excavation or a tunnel which results from the movement of the surrounding ground towards the excavation.
grout	a fluid, suspension or solution, which is injected into the ground in order to render it less permeable or to strengthen it.
ground treatment	a general term to cover the artificial improvement of ground *in situ*, particularly by some form of grouting and often with chemical grouts which set and strengthen or make impermeable the ground.
grout injection (grouting)	the controlled intrusion of grout into the ground by pressure.

haemo-concentration	increased concentration of cells in the blood usually resulting from loss of fluid to the tissues.
haemocrit	(1) an instrument for determining (usually by centrifuge methods) the relative amounts of plasma and corpuscles in blood, (2) the ratio of the volume of packed red blood cells to the volume of whole blood.
haemostasis	stoppage or sluggishness of blood.
haematological	of the study of the blood.
half-time	the time required for half the amount of a substance introduced into a living system to be eliminated or disintegrated by natural processes; in diving the required time for a tissue to absorb or eliminate 50% of the equilibrium amount of an inert gas.
heave	upward displacement of the ground.
heliox	a breathing mixture of helium and oxygen.
hemiplegia	paralysis of one half of the body.
histopathological	of the tissue changes that affect a part or accompany a disease.
histology	the study of tissues in the body.
hominen	see *ad hominen*.
hyaline	(1) commonest type of cartilage, white and glassy, (2) nitrogenous substance that forms the main constituent of the walls of certain cysts and which yields a sugar on decomposition; (3) any of several translucent substances like hyaline that collect around cells.
hyalinisation	the process of becoming hyaline or undergoing hyaline degeneration or the resulting state.
hydrophilic (compound)	having an affiniuty for water; of a gasket material used to seal joints in tunnel linings, which, as it comes in contact with water, expands to create a better seal.

hydrostatic	of the pressure at any point in a liquid at rest (i.e. the depth of the liquid multiplied by its unit weight); of groundwater pressures, when the pressure in the pore water is equal to that exerted by the height of the column of groundwater.
hyperaemic (hyperemic)	excess of blood in any part.
hyperaesthesia	excessively acute feeling in any part (of the body).
hyperbaric	of pressures greater than one atmosphere.
hyperbaric medicine	study and practice of medicine in relation to the effects of changes of ambient pressure; more usual term for baromedicine.
hypercapnia	an increased amount of carbon dioxide in the blood causing over-stimulation of the respiratory centre.
hyperoxic	of a bodily condition characterised by a greater oxygen content of the tissues and organs than normally exists at sea level.
hypertension	abnormally high tension, i.e. pressure, particularly of the blood in its vessels.
hyperventilation	over-breathing, it lowers the dissolved carbon dioxide in blood plasma.
hypoxia	deficiency of oxygen reaching the tissues of the body.
igneous (rock)	rock that has crystallised from molten silicate liquid or magma.
iliac	of the ilium, or hip bone.
immersed tube tunnel	a tunnel formed from prefabricated tunnel sections and sunk to a prepared position on a river (or sea) bed.
in-bye	a coal-mining term, the entry or supply-route side to the face being worked.

indictable	(legal) liable to be accused of a crime, or of the offence which creates such a liability, and which can only be decided in a crown court.
infarct	the area of necrosis in an organ produced by the blocking of a blood vessel, usually due to emboli.
infarction	the formation of an infarct, or the infarct itself.
infusion	introduction of fluid into the body, i.e. subcutaneous or intravenous injection.
inguinal	relating to the groin.
inguinal lymphatics	vessels that contain or convey lymph in the groin area.
inter-atrial septal defect	a defect of the partition (septum) between the two upper chambers (atria) of the heart.
interceptor (sewer)	a sewer which receives sewage from a number of transverse sewers or outlets.
intraosseous	into a bone.
in vitro	in glass, hence test tube, or by artificial experiment.
in vivo	in a live body, hence by experiment on living animals.
ischaemia (ischemia)	decrease in blood flow from obstruction leading to poor or absent oxygenation of tissues; also, localised anaemia, i.e. reduced quantity or quality of red corpuscles in blood.
jet grouting	a technique for forming roughly cylindrical columns of strengthened soil in the ground by injection of, usually, cement slurry under pressure from a rotating bit.
juxta-articular	next to the bone surface at a joint.
lacuna (pl. lacunae)	a small cavity or depression in any part of the body, e.g. one of the minute cavities in bone or cartilage occupied by osteocytces (cells characteristic of adult bone).
lacunar infarcts	area of necrosis in bone or cartilage cavities.

lagging temporary protection, often timber boards, of excavated faces between main supports.

laminar (of fluid flow) a state of flow when streamlines are parallel to flow direction.

latency the period during which a condition is temporarily concealed or is not manifest.

launch chamber the working space at or from the base of an access shaft for the erection of a tunnelling machine and from which tunnel-driving starts.

lesion a local diseased condition.

limitation (legal) the statutory specification of a period within which an action must be brought.

lining a concreted, bricked or sprayed concrete surfacing of a shaft or tunnel.

lipid fats and other substances, e.g. waxes, oils, that with proteins and carbohydrates constitute the principal structural components of living cells.

lipofuscin a dark brown naturally occurring pigment found in tissues of the aged.

lock i.e. air lock, a chamber for equalising pressures.

lumen the space inside a tube, i.e. the bore of a blood vessel.

lymph the pale coagulable fluid from the blood which has transduced through capillary walls to supply nutrient to cells.

lymphatic a vessel that contains or conveys lymph: if small, has no distinct walls or walls composed only of endothelial cells; if large, resembles a vein.

macro-permeability for a large mass of the ground, the ease with which a fluid (usually water) flows through it acting under a pressure difference.

macrophage	a large cell (e.g. in liver, spleen, bone marrow and lymph glands) which has the power to ingest cell debris and bacteria.
macular (rash)	discoloured area of skin, or spot, not raised above the skin surface.
man-lock	an airlock solely for the entry and exit of personnel.
marine deposit	a fine grained soil (silt or clay) laid down on the sea bed.
marrow	the substance contained in the middle of long bones and in the cancellous (honeycomb) bone tissue; yellow marrow is a fatty substance; red marrow is found in all cancellous tissue and blood cells are made in it.
mastic	a bituminous sealing compound.
mediator	an intervening agency, e.g. in transmitting or carrying as an intermediate mechanism.
membrane	a thin elastic tissue covering the surface of certain organs and lining body cavities.
meningeal	relating to the meninges, the membranes covering the brain and spinal cord.
micro-angiography	minutely detailed radiological examination of blood vessels.
microthrombus(i)	a very small thrombus or stationary blood clot.
monoarticular	affecting only one joint of the body.
moraine	earth materials brought together by the action of a former glacier or ice sheet.
morbidity	the presence of a diseased condition, or its prevalence or incidence.
morphological	of the study of the structure of organisms.
muck lock	an air lock for the removal of excavated soil and for bringing in materials.

mucous	pertaining to, or secreting, mucus, a slippery protective material.
myalgia	pain in the muscles.
myelin	the fatty covering of medullated (i.e. sheathed) nerve fibres.
myelopathy	any disease affecting the spinal cord.
narcosis	state of unconsciousness or unnatural sleep, produced by a narcotic drug, or effect of gases breathed at pressure.
necropsy	a post-mortem examination of the body.
negligence	(legal) inattention to duty or carelessness.
neoprene	a type of artificial rubber, used to form gaskets, seals, membranes and wet suits.
neovascularisation	the new development of vessels.
neuralgic	with a sharp stabbing pain along the course of a nerve caused by inflammation or functional disturbance of the nerve.
neurotic(ism)	characterised by exaggerated anxiety and tension often without any external cause.
nitrox	a breathing mixture containing nitrogen and oxygen in various proportions.
nullah	(in Far East) an open storm water drain or natural gully.
nystagmus	involuntary rapid movement of the eyeballs.
obiter	(legal) an expression of opinion by a judge in court on a matter of the law, but not binding authority and not an essential part of the reasons determining a decision.
occlusion	obstructive closure of the lumen or bore of a vessel, usually a blood vessel.

oedema	swelling due to effusion of fluid into the tissues.
Organ of Corti	see Corti, organ of; a complex structure in the ear.
osteocyte	the characteristic cell of living bone found in isolation in the lacunae (cavities) of the bone substance.
osteotomy	the cutting of bone, usually to correct a deformity.
otitis externa	inflammation of the outer ear.
otitis media	inflammation of the middle ear.
overboard dump system	a system to discharge exhaled gases out of a chamber to prevent oxygen build-up.
over-pressure	pressure in excess of normal atmospheric pressure.
oxygen therapy	the use of oxygen breathing in decompression (also hyperbaric oxygen therapy).
oxygen toxicity	the deleterious or poisonous quality (or effect) of oxygen at high partial pressures.
papular (rash)	a rash with pimples, i.e. raised above the skin.
paraparesis	an incomplete paralysis, especially of lower limbs.
partial pressure	the pressure exerted by a (specified) component in a mixture of gases.
patency	openness.
pathogenesis	the causation of a disease.
pea gravel	single-sized, fine gravel used as backfilling in small spaces because of its ease of pouring.
perched water table	the body of groundwater in a near-surface permeable stratum lying above an impermeable one.
perfusion	the pumping of a fluid (e.g. blood) through an organ or tissue.

periosteum	the fibrous membrane covering the surface of bones. It consists of two layers, the inner forms new cells by which bone grows in girth and, in close contact, a fibrous layer supplied with many blood vessels.
perivascular	of, occurring in, the tissues surrounding a blood vessel.
permeability	the ease with which fluids can flow through a porous medium under a pressure gradient.
permeability, coefficient of	for ground materials and, usually unstated, groundwater flow, the coefficient of proportionality between flow velocity and pressure gradient.
pharmocokinetics	the characteristic interactions of a drug and the body in terms of its absorption, distribution, metabolism etc.
pharyngitis	inflammation of the pharynx, the muscular pouch lined with mucous membrane at the back of the mouth and nose.
piezometer	an instrument installed in the ground which measures the pressure of the water in the pore spaces.
pile	a timber, steel or concrete column installed to depth in the ground to form part of a foundation of a structure.
pilot tunnel	a small diameter tunnel driven as a heading in advance of the full-bore tunnel; often used to determine the need for and to carry out ground treatment.
pitch circle	the common circle on which lie the centres of, for example, the holes for bolts connecting two rings of tunnelling or piles forming the walls of a circular shaft.
plaintiff	(legal) the party who brings a case (lawsuit) into court, i.e. against the defendent; alternative word is pursuer.
plasma	the clear straw-coloured fluid of the blood, after all the cells have been removed.
plasmin	an enzyme which dissolves the fibrin of blood clots.
plasminogen	the precursor of plasmin that is found in blood plasma and serum.

platelet	(of blood) thrombocytes, i.e. cells in blood concerning with clotting.
plea in law	(legal) an action at law or the presentation of an action in court.
pleading	to state the case or claim of a party to a lawsuit.
plenum chamber	in a tunnelling machine, the compartment filled with fluid pressurised to counter balance earth and groundwater pressures.
pneumatic	acting by pressure of air or gas, e.g. a pneumatic caisson.
pneumoconiosis	a disease of the lungs from prolonged exposure to dust-ridden atmospheres.
polyarticular	affecting many joints in the body.
pore (water) pressure	the pressure of water in a saturated soil.
prima facie	at first sight, based or founded on the first impression; thus, (legal) there appearing to be a case to answer.
procoagulant	a substance which promotes the coagulation (clotting) of blood.
proof before answer	(legal) in Scottish law, evidence given before a judge upon an issue.
proteinaceous	of, resembling or derived from protein.
proximal	nearer (to the point considered to be the centre of a system), *cf* distal.
psychometric	of the use of quantitative devices for assessing psychological trends.
psychosomatic	a condition relating to the mind and the body.
psychotic(ism)	of a severe mental illness affecting the whole personality.

pulmonary	pertaining to or affecting the lungs.
pursuer	the party who brings a case lawsuit into court; see plaintiff.
pyrimidine	an organic base of penetrating odour, or a derivative which is a constituent of DNA.
quadriplegia	paralysis in which all four limbs are affected.
quartz	a very common mineral form of silica (silicon dioxide), familiar as grains of many sands.
radionuclide	a radioactive nuclide, a species of atom characterised by the constitution of its nucleus.
receiver	an air container, in effect a reservoir, to store compressed air for the short but frequent periods when air demand exceeds supply.
recharge wells	boreholes specifically constructed for the introduction of water into the ground to re-establish or maintain the original groundwater level or pressure which, for example, may otherwise be lowered by nearby dewatering.
retinal	of the innermost coat of the eyeball, the retina, formed of nerve cells and fibres and from which the optic nerve leaves the eyeball.
rhinitis	inflammation of thc mucous membrane of the nose.
ribs	frames usually of steel to support a tunnel roof.
running ground	very wet (or occasionally very dry) sand or silt which flows like a liquid.
running tunnel	a tunnel constructed for an underground railway line.
sand jack	temporary support made from steel boxes filled with sand, which may be released by opening a plug in the side to empty the sand from the box.

saturation — (1) the condition in which the partial pressure of a gas dissolved in a fluid equals its maximum possible partial pressure under the existing temperature and pressure conditions. (2) state of a porous medium, such as soil, when all its pores are filled with fluid, e.g. water.

scantlings — ribs or similar form of stiffening which increase the rigidity of a structure or component.

seepage — entry of water into a tunnel or excavation or through a dam.

segment — one of the pre-fabricated units which together form the permanent, usually circular, lining of a tunnel.

segmental lining — the permanent support and lining of a tunnel, assembled of pre-cast concrete units or cast iron.

sequelae — morbid conditions following a disease and resulting from it.

serous — of, or like, serum, the clear fluid residue of blood.

serum — plasma (the clear fluid residue of blood) after removal of fibrin and other clotting agents.

settlement — downward movement of a structure or of the ground surface due to compression or movement of the ground.

shield — a steel hood or frame which protects men driving a tunnel; it may have an open or closed face.

signet(ed) — (legal) in Scottish law, to stamp or seal a document and, in effect to commence the process that may lead to a court case.

sinter — compressed particles of metal or ceramic to a coherent solid body.

sinus — a cavity in a bone.

sinusitis — inflammation of any of the air-containing cavities of the skull which communicate with the nose.

sinusoid	like a sinus; used to describe the irregular channels by which blood vessels connect.
skin friction	the resistance along the outer surfaces of a body being installed in the ground.
sludging	aggregation and settling out of the formed elements of blood leading to the formation of a semi-solid or solid precipitated mass as a result of a general or localised reduced rate of flow.
slurry	a fluid mixture of solids and water, particularly one which contains bentonite in colloidal suspension.
slurry tunnelling machine	a tunnel boring machine which uses a slurry to balance earth and water pressures and, with recirculation, to remove the excavated soil from the face.
spheroidal graphite iron	a type of cast iron used for tunnel segments.
spirometry	measurement of the air exhaled and inhaled during respiration (breathing).
split-shift	working for two periods in compressed air with a rest period between at atmospheric pressure.
splitting (hydraulic)	use of fluid pressure internally to break up boulders.
snorer	a colloquial tunnelling term, the drainage pipe from a sump in the invert for discharge of water to free air outside the tunnel; a one-way valve is fitted at the sump; when the valve is open, water is pumped out by the air pressure in the tunnel.
stability	the ability of an excavation to be self supporting.
stability number	an index, using geometric and soil strength parameters, of the comparative stability of an unsupported length of tunnel.
Starling (resistor)	in physiology, Starling's hypothesis is that the flow of fluids across capillary walls depends on the balance between the force of blood pressure on the walls, which tends to force fluids out, and the osmotic pressure across the walls, which tends to force them in under the

	greater concentration of dissolved substances in the blood, so that the declining gradient in blood pressure from the arterial to the venous end of the capillary results in an outflow of fluids at its arterial end with an increasing inflow towards the venous end.
step-plate junction	an arrangement of tunnel lining rings in groups of increasing diameters to allow for bifurcation from one tunnel to two tunnels. A headwall built within the largest diameter then surrounds two tunnels of the same size as the initial tunnel at the other end of the junction.
stop	designated pressure (or depth in diving) and time to remain at that pressure in a formal decompression procedure, called a stage in compressed air working.
stratigraphy	the branch of geology concerned with the description and classification of bodies of rock and their correlation with one another.
subaqueous	of a tunnel, under a river, estuary or sea.
subchondral	immediately below the articular cartilage.
subclavian (vein)	vein situated under the clavicle (shoulder blade).
sump	a low point in an excavation for collection and pumping out of seepage water.
surfactant	a material which has the property of altering the interfacial tension of liquids or solids, specifically in the lungs to reduce surface tension in alveoli (small cavities of the lungs) to keep airways open.
syndrome	a group of symptoms typical of a distinctive disease, and of that disease only.
tail	rear of a tunnelling shield within which tunnel lining is erected.
tail seal	the fittings on the rear of a tunnelling shield machine between it and the erected lining to prevent water or slurry from the face or grout interfering with work behind the shield.

tail skin — another term for the tail of a tunnelling shield, i.e. the rear and within which the tunnel lining is erected.

temporary works — operations and structures involved in order to make possible or expedite the construction, e.g. dewatering before excavation, or scaffolding; usually the choice, design and responsibility of the contractor.

therapeutic — of the science and art of healing and treatment of a disease.

therapy — treatment of a disease.

thrombogenic — capable of forming a blood clot.

thrombus (pl. thrombi) — stationary blood clot.

till — poorly sorted debris produced by the erosion of rocks by moving ice (usually called glacial till and also referred to less correctly as boulder clay).

time barred — of an intended legal action ruled to have been brought too late.

tort — (legal) the breach of duty imposed by law whereby some person acquires a right of action for damages in the civil courts.

total stress — the pressure on a plane in a mass of soil borne by the soil grains and pore fluids in combination, *cf* effective stress.

trabecula (pl. trabeculae) — band or plate of fibrous tissue forming part of the supporting framework; e.g. of bones, the thin interconnected spikes of bone in the medullary cavities of all bones which provide the framework for the bone marrow.

transcutaneous — across the skin.

treatment (of ground) — artificial improvement usually by injection of fluids which set in the ground and strengthen it or make it impermeable.

triage	the action of sorting according to quality or need, e.g. as with emergency medical uses.
trimix	a breathing gas mixture of three gases, usually oxygen, helium and nitrogen.
trunk sewer	a sewer which receives many tributary branches or discharges from other large sewers and serves a large area.
trunking	cylindrical duct or pipe, usually refers to ventilation ducting.
trunking (structural)	term used to describe the circular tubes fixed to the air deck, which formed the muck-lock and shaft to the man lock, containing the air pressures and providing structural support.
tympanic membrane	the ear drum, i.e. resonating membrane of the inner ear.
tympanum	the ear drum.
ultrasound detector	device for detecting reflected high frequency sound waves, used in non-destructive testing and body imaging. In hyperbaric work used to detect bubbles in blood and tissues.
ultravascular	beyond, in space, the range of the vessels which convey fluids.
underpinning	providing new, deeper support under a wall or column; in shafts, the installation of steel lining rings one below the other as excavation proceeds.
vasculature	the disposition or arrangement of blood vessels in an organ.
vascular	of vessels which convey fluids or provide for the circulation of fluids.
vascular stasis	slowing or stoppage of normal flow.
vasoconstriction	constriction of blood vessels.

vasodilation	dilation of blood vessels.
vasomotor (rhinitis)	rhinitis (inflammation of the mucous membranes of the nose) with constriction or expansion of arteries.
vasospasm	sharp, and often persistent, contraction of a blood vessel.
venous	of the veins.
venous sinusoids	irregular channels by which branches of two blood vessels join.
ventral	of the belly, but of the aspect of an animal which is normally turned towards the ground.
ventricle	(1) a thick-walled chamber of the heart (2) one of the fluid-filled cavities of the brain.
vestibular	at an entrance, e.g. the cavity at the entrance to the cochlea of the ear.
vortex drop shaft	a shaft in a sewerage system where sewage flows enter tangentially and mix; it serves to regulate flow to the next part of the system.
weathering (of rock)	the degradation of rocks on the earth's surface by environmental agencies since their formation.
yoke	a beam or collar which spreads load or a mechanism for tying components together.

Sources

Webster's Medical Dictionary, Merriam Webster Inc., published in Europe by Chapman and Hall, London, 1986

Ballière's Nurse's Dictionary, (Barbara F Cape), Baillière, Tindall and Cassell, London, 16th Edn. 1964

Penguin Dictionary of Civil Engineering, (J.S. Scott), Penguin Books, Harmondsworth, UK, 3rd Edn., 1980

A Dictionary of Earth Sciences (ed. Stella E. Steigler) Pan Books, London, 1978, in association with The Macmillan Press

Histology, by A.W. Ham and T.S. Leeson, Pitman Medical Publishing Co., London, 4th Edn., 1961

Biomedical and physiological terms used in diving, Norsk Hydro (undated)

Glossary of Diving and Hyperbaric Terms, Undersea Medical Society Inc., Bethesda, MD. 1978

Glossary of terms used in water pollution control, Institute of Water Pollution Control, Maidstone, UK, 1975

Chambers Directory of Science and Technology, (eds. T.C. Collocott and A.B. Dobson), Chambers, Edinburgh, revd. edn., 1974

The Compact Oxford English Dictionary, Oxford University Press

List of abbreviations

AGE	Arterial gas embolism
ATA	atmosphere absolute
BIBS	built-in breathing system(s)
bp	blood pressure
BS	British Standard
BSI	British Standards Institution
BTS	British Tunnelling Society
CA	compressed air
CAW	compressed air worker
CAWG	Compressed Air Working Group (of the Medical Research Council's Decompression Sickness Panel)
CIRIA	Construction Industry Research and Information Association
CNS	central nervous system
CPR	cardio-pulmonary resuscitation
CPTD	cumulative pulmonary toxicity dose
DCS	decompression sickness
DCI	decompression illness
DHMC	Diving and Hyperbaric Medical Centre (Singapore)
DO, DON	dysbaric osteonecrosis
EC	European Commission
EEG	electro-encephalogram
EPB (EPBM)	earth-pressure balance (...tunnelling machine)
EUBS	European Undersea Biomedical Society
FO_2 (FN_2) etc.	proportion of oxygen (nitrogen) in a gas mixture
GDPC	Great-Depth Pneumatic Caissons
HBO	hyperbaric oxygen (therapy)
HMG	helium mixed gas
HPNS	High-pressure neurological syndrome
HSE	Health and Safety Executive
ICE	Institution of Civil Engineers
ISO	International Standards Organisation
JAMSTER	Japan Marine Science and Technology Centre
LEL	Lower explosive limit (of a gas in an atmosphere: for methane in air it is 5% by volume)
MO	Medical Officer
MRC	Medical Research Council
MRI	magnetic resonance imaging
MRT	Mass Rapid Transit (of Singapore)
MRTC	Mass Rapid Transit Corporation (Singapore)
MTR	Mass Transit Railway (of Hong Kong)
NMI	Nuclear magnetic investigation
NUMAC	Northern Universities Multiple Access Computing (facility)
OTU	oxygen tolerance units

PDOB	post-decompression oxygen breathing
PO_2, pO_2	partial pressure of oxygen in a gas mixture
ppm	parts per millsion
RN	Royal Navy
RSN	Royal Singapore Navy
SFS	Singapore Fire Service
SMC	site medical centre
TBM	tunnel boring machine
TM&DU	Tokyo Medical and Dental University
TWA	
UBO	unidentified bright object
UHMS	Undersea Hyperbaric Medical Society
UPTD	unit pulmonary toxicity dose
URTI	upper respiratory tract infection
USN	United States Navy
VMR	vasomotor rhinitis

Pressure units and conversions

Various units of pressure are used in different countries and in the disciplines of diving and engineering. In UK the preferred pressure unit for compressed air is bar (the SI pressure unit of kiloPascals being too fine), but was previously pounds (force) per square inch. Elsewhere, metric units such as kilograms (force) per square centimetre are or have been used. In diving, the natural unit is metres of sea water, but standard physical atmospheres are also used, as is necessary when dealing with altitude. For strict comparison, all pressures should be expressed in absolute terms. In practice, however, it is usual to ignore the absolute value of atmospheric pressure, assuming it is sensibly constant, and to express the pressure of a compressed air environment as that in excess of the current atmospheric pressure, when it is reported as gauge pressure.

	bar	**kPa**	**atm**	**psi**	**kgf/cm²**	**msw (sp. gr. = 1.02)**
1 bar	= 1	100	0.987	14.5	1.02	10
1 kPa	= 0.01	1	0.0099	0.145	0.0102	0.10
1 atm	= 1.013	101.3	1	14.7	1.03	10.11
1 psi	= 0.069	6.895	0.068	1	0.070	0.69
1 kgf/cm²	= 0.98	97.98	0.967	14.21	1	9.80
1 msw	= 0.096	9.62	0.095	1.30	0.098	1

Simple conversions are:

1 bar $\simeq$ 1 atm $\simeq$ 1 kgf/cm^2 $\simeq$ 10 m water $\simeq$ 15 psi

Appendix B: List of delegates

NAME	ORGANISATION
Ahrenkiel, Dr J	Naval Technical School, Denmark
Anderson, Mr J M	Health & Safety Executive, UK
Bandmann, Mr M	Tiefbau-Berufsgenossenschaft, Germany
Belman, Mr V J	Fairclough Civil Engineering Ltd, UK
Brown, Mr R I	J Murphy & Sons Ltd, UK
Bryson, Dr P	DDRC, UK
Buchanan, Mr J	CIRIA, UK
Campbell, Mr J G	Charles Haswell & Partners Ltd, UK
Catton, Dr M J	Appointed Doctor Services, UK
Chandler, Mr D	Naval Medical Research Institute, USA
Cole, Mr B	North West Water Engineering, UK
Cole, Mr J M	Delta Civil Engineering Ltd, UK
Cose, Mr B	South West Water
Court, Mr D J	Edmund Nuttall Ltd, UK
Cupit, Mr J C	Sir Robert McAlpine & Sons Ltd, UK
Dabill, Mr D W	HSE Research & Laboratory Services, UK
Dales, Mr M J	Cairo Wastewater Consortium, UK
Damsgaard, Mr B	A/S EM.Z Svitzer, Denmark
Davies, Mr H	London Underground Ltd, UK
Denman, Mr D	Italstrade (UK) Ltd, UK
Elba, Dr S K	Lilley Construction Ltd, Egypt
Elliot, Professor D H	MRC Decompression Sickness Panel, Robens Institute, UK
Elliott, Mr I H	Mott MacDonald, UK
Ewold, Mr J	Odense University Hospital
Evans, Dr A	University of Newcastle upon Tyne, UK
Field, Mr C	London Underground Ltd, UK
Fisher, Mr P	Costain Civil Engineering Ltd, UK
Forrest, Mr I N	Delta Civil Engineering Co Ltd, UK
Foster, Mr C	Costain Civil Engineering, UK
Fernando, Dr J S	Department of Transport, UK
Gaba, Mr A R	Soil Mechanics Associates, UK
Gittoes, Mr G	Sir William Halcrow & Partners Ltd, UK
Gotoh, Dr Y	Saitama Medical School, Japan
Greenbaum, Mr L G	Undersea Hyperbaric Medical Society
Hamilton, Mr R W	Hamilton Research Ltd, USA
Harrison, Miss L	CIRIA, UK
Hazel, Ms K	Delta Civil Engineering Ltd
Hempleman, Dr H V	Compressed Air Working Group, UK
Hill, Mr J	Delta Civil Engineering Ltd
Hoad, Mr R S	Sir William Halcrow & Partners, UK

Holthaus, Professor J	GKSS Forschungszentrum Geesthacht, Germany
Hooper, Mr P	Higgs & Hill, UK
How Yew Chen, Col (Dr) J	Royal Singapore Navy, Singapore
Howard, Mr A J	Mott MacDonald, UK
Jaques, Mr P A	J Murphy & Sons Ltd, UK
Jardine, Mr F M	CIRIA, UK
Jasper, Mr A	Higgs & Hill plc, UK
Jennion, Mr R F	Kilroe Civil Engineering Ltd, UK
Kindwall, Mr E P	Medical College of Wisconsin, USA
King, Dr J D	London Hyperbaric Medical Services, UK
Lamont, Mr D R	Health & Safety Executive, UK
Leach, Dr J	Lancaster University
Le Péchon, Mr J-C	JCLP Hyperbarie, France
Lehner, Dr C E	University of Wisconsin-Madison, USA
Lewis, Mr R P	Markham & Company Ltd, UK
Lofthouse, Mr L	London Underground Jubilee Line Extension Project, UK
Luther, Mr G	GKSS Forschungszentrum GmbH, Germany
Mannche, Dr K H	Arzt f Chirurgie, Bremen, Germany
Mano, Dr Y	Tokyo Medical & Dental University, Japan
Mantle, Mr D	Deacon Engineering, UK
Marcus, Dr R L	Consultant Occupational Physician, UK
Marks, Mr L	Finers Solicitors
May, Mr M	Thames Water Utilities Ltd, UK
McCaig, Dr R H	Health & Safety Executive, UK
Mercer, Ms S	Decon Engineering, UK
Miller, Mr A	Thames Water Utilities Ltd, UK
Nashimoto, Dr I	Saitama Medical School, Japan
O'Connell, Mr J R	Costain Civil Engineering, UK
Oda, Mr S	Shiraishi Corporation, Japan
Olsen, Mr M H	M T Group, Denmark
O'Shea, Mr B	Thames Water Utilities Ltd, UK
Palmer, Dr A C	University of Cambridge, UK
Parks, Mr D R	Mott MacDonald, UK
Parsons, Dr K C	University of Technology, Loughborough, UK
Philp, Professor R	University of Western Ontario, Canada
Pooley, Mr J	University of Newcastle upon Tyne, UK
Reed, Mr R J	Lilley Construction, UK
Remnant, Mr P	Specialist Plant Associates Ltd, UK
Ribeiro, Dr I J	Centro Medico Hiperbarico de Sao Paulo (CEMHIPER), Brazil
Ruegger, Mr M	Schweizerische Unfallversicherunsanstallt (SUVA), Switzerland
Ripley, Mr K J	Delta Civil Engineering Co Ltd, UK
Sawa, Mr Y	Ohomoto Gumi Company Ltd, Japan

Sawin, Dr A	Howard Humphreys, UK
Schmidt, Mr B R	Tiefbau-Berufsgenossenschaft, Germany
Shutter, Mr G B	Mott MacDonald, UK
Simpson, Mr P T	Charles Haswell & Partners, UK
Slocombe, Mr R	Specialist Plant Associates Ltd, UK
Smith, Mr A P	Costain Civil Engineering, UK
Smith, Dr D E B	St Catherine's College, Oxford, UK
Sterk, Dr W	Dadcodat, Netherlands
Sykes, Surg. Cmdr. J J W	Institute of Naval Medicine, UK
Tagg, Mr A P	Costain Civil Engineering, UK
Thalmann, Captain E D	Naval Medical Research Institute, USA
Thogersen, Mr K	Cowiconsult, Denmark
Umney, Mr A R	G Maunsell & Partners, UK
Vann, Mr R D	Duke University Medical Center, USA
Vincett, Mr J L	Tony Gee & Partners, UK
Voigt, Dr G C	Gustav C Voigt & Associates, USA
Vollbrandt, Mr J	GKSS Forschungszentrum Geesthacht, Germany
Vowells, Mr M W	Health & Safety Executive, UK
Walder, Professor D N	University of Newcastle upon Tyne, UK
Wanscher, Dr M	Odense University Hospital, Denmark
Watts, Mr R	Balfour Beatty Civil Engineering Ltd, UK
Widriksen, Mr B	Great Belt A.S., Denmark
Wright, Mr F B	University of Salford, UK

Author index

Subject index

FSC
www.fsc.org
MIX
Papier | Fördert
gute Waldnutzung
FSC® C083411

Kolibri 360 GmbH
Ferdinand-Jühlke-Straße 7
99095 Erfurt, Deutschland
produktsicherheit@kolibri360.de